Modern Linear and Nonlinear Econometrics

Dynamic Modeling and Econometrics in Economics and Finance

VOLUME 9

Series Editors

Aims and Scope

The series will place particular focus on monographs, surveys, edited volumes, conference proceedings and handbooks on:

- Nonlinear dynamic phenomena in economics and finance, including equilibrium, disequilibrium, optimizing and adaptive evolutionary points of view; nonlinear and complex dynamics in microeconomics, finance, macroeconomics and applied fields of economics.

- Econometric and statistical methods for analysis of nonlinear processes in economics and finance, including computational methods, numerical tools and software to study nonlinear dependence, asymmetries, persistence of fluctuations, multiple equilibria, chaotic and bifurcation phenomena.

- Applications linking theory and empirical analysis in areas such as macrodynamics, microdynamics, asset pricing, financial analysis and portfolio analysis, international economics, resource dynamics and environment, industrial organization and dynamics of technical change, labor economics, demographics, population dynamics, and game theory.

The target audience of this series includes researchers at universities and research and policy institutions, students at graduate institutions, and practitioners in economics, finance and international economics in private or government institutions.

Modern Linear and Nonlinear Econometrics

by

Joseph Plasmans

University of Antwerp, Belgium,
and Tilburg University, The Netherlands

Springer

A C.I.P. Catalogue record for this book is available from the Library of Congress.

ISBN-10 0-387-25760-8 (HB)
ISBN-13 978-0-387-25760-0 (HB)
ISBN-10 0-387-25761-6 (e-book)
ISBN-13 978-0-387-25761-7 (e-book)

Published by Springer,
P.O. Box 17, 3300 AA Dordrecht, The Netherlands.

www.springer.com

Printed on acid-free paper

Printed in the Netherlands.

To my wife Sim and also my children Roel, Tom and Inge, and Bart

Preface

The basic characteristic of this new book in econometrics is that it presents a unified approach of modern linear and nonlinear econometrics. It covers a wide range of contemporary topics in applied econometrics in a concise and intuitive way, while focusing on four major parts of modern econometrics, i.e. linear and nonlinear econometric estimation and testing, linear and nonlinear time series analysis, models with categorical and limited dependent variables, and, finally, a thorough analysis on linear and nonlinear panel data modeling.

Some distinctive features for this book are:

- a unified approach of both linear and nonlinear theoretical aspects and applications in modern econometrics;
- emphasis on sound theoretical and empirical relevance and intuition, paying attention to the links between alternative approaches;
- focus on econometric and statistical methods for the analysis of linear and nonlinear processes in economics and finance, including computational methods and numerical tools;
- completely worked out empirical illustrations are provided throughout, the macroeconomic and microeconomic (household and firm level) data sets of which are available from the internet; these empirical illustrations are taken from finance (e.g. CAPM), international economics (e.g. exchange rates), innovation economics (e.g. patenting), labor and educational economics (e.g. demand for teachers according to gender), business cycle analysis, monetary economics, housing economics, and many others;
- exercises are added to the chapters to gain experience and intuition and several of these exercises involve the use of actual data that are typical for current empirical work and that are available on the internet.

The book is (almost) completely self-contained and written in a clear way for final year bachelor and master students, but also for practitioners in business, industry, and public institutions who want to learn about recent developments in the field of econometrics. A basic introductory text in Econometrics or Statistics is a valuable asset in reading and using this book. The present volume originates from various lecture notes used for courses in applied econometrics, nonlinear econometrics, and time series analysis, both at the University of Antwerp and at Tilburg University.

'Learning by doing' is an important paradigm in econometrics. Therefore, readers are strongly encouraged to solve theoretical exercises, to (re-)estimate a number of models and to carry out a number of specification tests. With modern software the actual computation of even the more complicated estimators and test statistics is often very simple, sometimes even too simple to be safe. But this book will help the reader to remain on the right track. The contents and form of this book are designed in a way that enables scholars to use it as a textbook in the area of econometrics.

Acknowledgements

I am largely indebted to several people and institutions.

In the first instance, to the many colleagues, friends and students who advised me in the course of the recent years, both at the University of Antwerp and at Tilburg University. I am especially indebted to my good colleague and friend Bertrand Melenberg and to research assistant A. Daniela Cristina who gave me very precious advice and tips when I threatened to loose the right track. Also other research assistants, principally Ruslan Lukach, but also Bruno Merlevede, Tomasz Michalak, Sigrid Suetens, Tom Van Ourti, and Lucas Ferraz have assisted me on various occasions. Students over recent generations have inspired me with their constant attention and often intelligent questions, and constantly gave me feedback on the contents of the preliminary versions of this book. Of course I am not committing any of these persons and I retain sole responsibility for any remaining shortcomings.

I also express my acknowledgement to the Department of Economics of the University of Antwerp which provided a good research and educational environment and also financial means to accomplish this project. Also, the excellent and stimulating research environment present in the Department of Econometrics and Operations Research of Tilburg University was greatly appreciated. Finally, Herma Drees, Cathelijne van Herwaarden and Elseliene van de Klooster from Springer are very much appreciated for editorial assistance.

Whilst planning and writing this book I have not always been present where I ought to have been, namely with my family, and even in those moments that I was present, I suffered from absentmindedness. My wife Sim and also my children Roel, Tom and Inge, and Bart were always ready to give the serene support I needed to finish the project. Therefore, I dedicate this book to them.

Antwerp, March 2005

Contents

List of Tables

List of Figures

List of symbols and abbreviations

Vectors and matrices are always shown in boldface, usually in lower case and upper case, respectively.

$\leq$ ($\geq$)	less (greater) than for a vector, some elements being equal
$\leqq$ ($\geqq$)	less (greater) than or equal to of a scalar
$\simeq$	approximation up to first order
$\cong$	good approximation (at least up to second order)
$:=$	defined as
$\overset{as}{\sim}$	asymptotically distributed as
$\overset{as}{\approx}$	approximately asymptotically
$\sim$	distributed as
$\neq$	different from
$\Rightarrow$	leads to
$\Longleftrightarrow$	if and only if
$\|x\|$	modulus of x, absolute value
$\overline{x}$	average of x
$\|\mathbf{X}\|$	determinant of matrix $\mathbf{X}$
$\mathbf{X}'$	transpose of matrix $\mathbf{X}$
$\mathbf{X}^*$	adjoint of matrix $\mathbf{X}$
$\forall$	all
$\exists$	there exists at least one
$\varepsilon_t \overset{iid}{\sim} (0, \sigma_\varepsilon^2)$	all ε_ts are *iid* with mean 0 and variance σ_ε^2
$\boldsymbol{\varepsilon}_t \overset{iid}{\sim} (\mathbf{0}, \mathbf{\Sigma})$	vectors $\boldsymbol{\varepsilon}_t$s are *iid* with mean $\mathbf{0}$ and var.-cov. matrix $\mathbf{\Sigma}$
$acov(x, z)$	asymptotic covariance between x and z
$\mathbf{acov}(\mathbf{x}, \mathbf{z})$	asymptotic covariance matrix between the vectors $\mathbf{x}$ and $\mathbf{z}$
ACF	autocorrelation function
AIC	Akaike information criterion
AD(1,0)	autoregressive (of order 1) distributed lag (of order 0)
ADF	augmented Dickey-Fuller
AO	additive outlier
AR	autoregressive
ARCH	autoregressive conditional heteroskedasticity
arg max (min)	argument that maximizes (minimizes)
ARIMA	autoregressive integrated moving average
ARMA	autoregressive moving average
$avar(x)$	asymptotic variance of x
$\mathbf{avar}(\mathbf{x})$	asymptotic variance-covariance matrix of vector $\mathbf{x}$
BEKK	Bollerslev-Engle-Kroft-Kroner
BHHH	Berndt-Hall-Hall-Hausman
BIC, SC, SBC	Bayesian, Schwartz information criterion
BJ	Box-Jenkins
BLUE	best linear unbiased estimator
$\mathbb{C}$	the set of complex numbers

CACF	cross autocorrelation function
CAPM	capital asset pricing model
CCC	constant conditional correlation
CH	Canova-Hansen
$\backsim \chi^2_k$	is Chi-Squared distributed with k degrees of freedom
CML	conditional maximum likelihood
$x_m \xrightarrow{d} x$	convergence in distribution of x_m to x
$cov(x, y)$	covariance between x and y
CUSUM	cumulated sum
DCC	dynamic conditional correlation
d.f.	degrees of freedom
DF	Dickey-Fuller
DFP	Davidon-Fletcher-Powell
DGP	data generating process
DHF	Dickey-Hasza-Fuller
DW	Durbin-Watson
Δ^d	d^{th} difference
Δ_S	seasonal difference
Δ_d	d-period difference
$\mathbf{g} := \boldsymbol{\nabla}$	gradient (vector)
$\frac{\partial L}{\partial \beta}\|_{\beta=\hat{\beta}}$	partial derivative of L with respect to β evaluated at $\hat{\beta}$
$\frac{\partial L}{\partial \beta}\|_{NLML}$	partial derivative of L evaluated at the NLML estimator
$E(\cdot)$	unconditional expected value
$E_{t-1}(\cdot)$	expected value conditional on $t-1$ information set
$\in$	set inclusion
$\theta^{(k)}$	value of θ at the kth iteration
ECM	error correction mechanism
EGARCH	exponential *GARCH*
e.g.	for example
EGARCH-M	exponential *GARCH*-in-mean
EGLS	estimated GLS
EML	exact maximum likelihood
EMU	European Monetary Union
$\sim F_{k_1,k_2}$	is F-distributed with d.f. k_1 and k_2
FE	fixed effect
FGLS	feasible GLS
FOC	first order condition
GARCH	generalized *ARCH*
GARCH-M	*GARCH* in Mean
GDP	gross domestic product

GED	generalized exponential distribution
GES	generalized exponential smoothing
GLS	generalized least squares
GMM	generalized method of moments
GSD	generalized Student t-distribution
H_0	null hypothesis
H_A	alternative hypothesis
HEGY	Hylleberg-Engle-Granger-Yoo
$\sim I(d)$	follows an integrated process of order d
$\mathbf{X}^{-1}$	inverse matrix $\mathbf{X}$
iid	independently and identically distributed
$\mathbf{I}_p$	identity matrix of order p
$\Im_t$	information set at time t
$IGARCH$	integrated $GARCH$
i.e.	that is
IRF	impulse response function
IVs	instrumental variables
JB	Jarque-Bera
K	kurtosis
KPSS	Kwiatkowski-Phillips-Schmidt-Shin
LLS	linear least squares
ℓ	joint likelihood function
L	loglikelihood function (usually equal to $\ln \ell$)
LS	least squares
LSDV	least squares dummy variables
$LSTAR$	logistic smooth transition autoregressive
MA	moving average
$\max_x(\min_x)$	maximize (minimize) as a function of x
ML	maximum likelihood
MoM	method of moments
MSE	mean squared error
MSFE	mean squared forecast error
MOSUM	moving sum
$\mathbb{N} := \{0, 1, 2, 3, ...\}$	the set of natural numbers
NLLS	nonlinear least squares
$N(\mu, \sigma^2)$	normal distribution with mean μ and variance σ^2
NLLS	nonlinear least squares
$\sim N(\mu, \sigma^2)$	is Gaussian with mean μ and variance σ^2
OLS	ordinary least squares
PA	partial adjustment
PACF	partial autocorrelation function
PP	Phillips-Perron
pdf	probability density function

$\underset{\beta\to\alpha}{plim}$	probability limit for $\beta \to \alpha$
PPP	purchasing power parity
QML	quasi-ML
$\mathbb{R}$	the set of real numbers
$\mathbb{R}^n$	the set of vectors with n entries
RE	random effect
R^2	coefficient of determination
SAR	seasonal AR
$SARIMA$	seasonal $ARIMA$
SES	simple exponential smoothing
$SETAR$	self-exciting threshold autoregressive
SGSD	skewed generalized Student t-distribution
SI	seasonal index
SK	Skewness
SMA	seasonal MA
$SOMI$	single output multiple input
$SOSI$	single output single input
$STAR$	smooth transition autoregressive model
SUR	seemingly unrelated regression
TAR	threshold autoregressive
$TARCH$	threshold $ARCH$
2SLS	two stage least squares
U(0,1)	uniform distribution on the interval (0,1)
UIP	uncovered interest rate parity
US(A)	United States (of America)
UK	United Kingdom
$var(x)$	variance of x
$\mathbf{var}(\mathbf{x})$	variance-covariance matrix of vector $\mathbf{x}$
$var_{t-1}(x)$	variance of x conditional on $\Im_{t-1}$
$\mathbf{var}_{t-1}(\mathbf{x})$	variance-covariance matrix of $\mathbf{x}$ conditional on $\Im_{t-1}$
VaR	value at risk
VAR	vector autoregression
vec	vectorized
WN	white noise
$\mathbb{Z} := \{0, \pm 1, \pm 2, \pm 3, ...\}$	set of equidistant time points $0, \pm 1, \pm 2, \pm 3, ...$

Part I

Linear and Nonlinear Econometric Inference: Estimation and Testing

This part consists of three chapters. Chapter 1 provides an introduction to the basic algorithms of linear and nonlinear estimation with a focus on the econometric specification of the models. More technical nonlinear optimization procedures are delegated to the Appendix. Least squares and maximum likelihood estimators are at the core of this chapter. Chapter 2 digs further in modern econometric estimation methods, as the method of moments and the generalized method of moments. These estimation methods encompass more common estimation methods as (generalized) least squares, instrumental variables, maximum likelihood, simultaneous equations estimation methods, etc. Also several applications on real-world economic cases are discussed, involving, among others, rational expectations models. Since there cannot be econometric inference without testing, Chapter 3 focuses on testing linear and nonlinear hypotheses with a basic split between nested and nonnested model tests.

Chapter 1

Estimation in Linear and Nonlinear Models

1.1 Introduction to linear and nonlinear optimization

Linear modeling is known to most economists and social scientists, but in practice there are numerous examples of nonlinear models, such as nonlinear autoregressive models, nonlinear moving average models, bilinear models, time-varying models, limited dependent variables, production models, exchange rate models.

In general, two types of linear/nonlinear estimation methods will be analyzed in this introductory chapter. On the one hand, a (generalized) least squares (LS) type of estimation method yielding estimators[1] corresponding to the shortest (weighted) distance between observations and the unknown linear/nonlinear regression line and, on the other hand, a maximum likelihood (ML) type of estimation method aiming for a modal estimator that maximizes the joint probabilty density function (pdf) of the observations and that possesses elegant statistical properties under the hypothesis of, for example, a normal pdf.

Consider the general nonlinear regression model for $i = 1, 2, \ldots, n$:

$$y_i = f(\mathbf{x}_i, \boldsymbol{\beta}) + \varepsilon_i,$$

where the error terms are assumed to be distributed according to a (not necessarily normal, e.g. Pareto, Cauchy) distribution function characterized by two types of parameters (expectation and variances-covariances), or the $n \times 1$ vector

[1] An *estimator* must be computable from the data only and must not be formulated in terms of unknowns. In general, the estimator is itself a random variable and an *estimate* is a specific numerical value for the estimator, usually arising from a sample of realized values on (a) stochastic variable(s). Hence, an estimate is a realization of an estimator.

$\boldsymbol{\varepsilon}$ is assumed to be distributed according to a certain distribution with mean zero and (variance-)covariance matrix $\boldsymbol{\Sigma}$, or $\boldsymbol{\varepsilon} \backsim (0, \boldsymbol{\Sigma})$, $\mathbf{x}_i$ an m dimensional vector of 'explanatory variables' at observation i, and $\boldsymbol{\beta}$ a p dimensional vector of unknown parameters, or $\boldsymbol{\beta} \in \mathbb{R}^p$.

Stacking vectors as $\mathbf{y} := (\, y_1, y_2, \ldots, y_n)'$, $\boldsymbol{\varepsilon} := (\varepsilon_1, \varepsilon_2, \ldots, \varepsilon_n)'$, and $\mathbf{f}(\mathbf{X}, \boldsymbol{\beta}) := [f(\mathbf{x}_1, \boldsymbol{\beta}), f(\mathbf{x}_2, \boldsymbol{\beta}), \ldots, f(\mathbf{x}_n, \boldsymbol{\beta})]'$, consider for linear(arized) estimation a first order Taylor approximation around a parameter vector $\boldsymbol{\beta}^0$:

$$\mathbf{y} = \mathbf{f}(\mathbf{X}, \boldsymbol{\beta}) + \boldsymbol{\varepsilon} = \mathbf{f}(\mathbf{X}, \boldsymbol{\beta}^0) + \left(\left. \frac{\partial \mathbf{f}}{\partial \boldsymbol{\beta}} \right|_{\boldsymbol{\beta}=\boldsymbol{\beta}^0} \right) (\boldsymbol{\beta} - \boldsymbol{\beta}^0) + \ldots + \boldsymbol{\varepsilon} \tag{1.1}$$

with $\left. \frac{\partial \mathbf{f}}{\partial \boldsymbol{\beta}} \right|_{\boldsymbol{\beta}=\boldsymbol{\beta}^0}$ an $n \times p$ Jacobian matrix $\mathbf{Z}(\boldsymbol{\beta}^0)$, or

$$\bar{\mathbf{y}}(\boldsymbol{\beta}^0) \simeq \mathbf{Z}(\boldsymbol{\beta}^0)\boldsymbol{\beta} + \boldsymbol{\varepsilon} \tag{1.2}$$

with $\bar{\mathbf{y}}\,(\boldsymbol{\beta}^0) := \mathbf{y} - \mathbf{f}(\mathbf{X}, \boldsymbol{\beta}^0) + \mathbf{Z}(\boldsymbol{\beta}^0)\boldsymbol{\beta}^0$, which is called the *pseudo-linear model* (or *quasi-linear* or *linearized* regression model).

The corresponding *generalized least squares* (GLS) estimation aims to solve:

$$\begin{aligned} \min_{\boldsymbol{\beta}} \boldsymbol{\varepsilon}' \boldsymbol{\Sigma}^{-1} \boldsymbol{\varepsilon} &= \min_{\boldsymbol{\beta}} \left\{ (\mathbf{y} - \mathbf{f}(\mathbf{X}, \boldsymbol{\beta}))' \, \boldsymbol{\Sigma}^{-1} \, (\mathbf{y} - \mathbf{f}(\mathbf{X}, \boldsymbol{\beta})) \right\} \\ &\simeq \min_{\boldsymbol{\beta}} \left\{ \left(\bar{\mathbf{y}}(\boldsymbol{\beta}^0) - \mathbf{Z}(\boldsymbol{\beta}^0)\boldsymbol{\beta} \right)' \boldsymbol{\Sigma}^{-1} \left(\bar{\mathbf{y}}(\boldsymbol{\beta}^0) - \mathbf{Z}(\boldsymbol{\beta}^0)\boldsymbol{\beta} \right) \right\} \end{aligned} \tag{1.3}$$

which yields the GLS estimator for a known error covariance matrix $\boldsymbol{\Sigma}$:

$$\hat{\boldsymbol{\beta}}_{GLS} = ((\mathbf{Z}(\boldsymbol{\beta}^0))'\boldsymbol{\Sigma}^{-1}\mathbf{Z}(\boldsymbol{\beta}^0))^{-1}(\mathbf{Z}(\boldsymbol{\beta}^0))'\boldsymbol{\Sigma}^{-1}\bar{\mathbf{y}}(\boldsymbol{\beta}^0) \tag{1.4}$$

and the *feasible GLS* (FGLS) estimator if some initial estimate $\hat{\boldsymbol{\Sigma}}$ of $\boldsymbol{\Sigma}$ is available:

$$\hat{\boldsymbol{\beta}}_{FGLS} = ((\mathbf{Z}(\boldsymbol{\beta}^0))'\hat{\boldsymbol{\Sigma}}^{-1}\mathbf{Z}(\boldsymbol{\beta}^0))^{-1}(\mathbf{Z}(\boldsymbol{\beta}^0))'\hat{\boldsymbol{\Sigma}}^{-1}\bar{\mathbf{y}}(\boldsymbol{\beta}^0). \tag{1.5}$$

If the error covariance matrix is a scalar covariance matrix, or $\boldsymbol{\Sigma} := \sigma_\varepsilon^2 \mathbf{I}_n$, the appropriate estimator (1.4) and (1.5) is the ordinary least squares (OLS) estimator $\hat{\boldsymbol{\beta}}_{OLS} = ((\mathbf{Z}\,(\boldsymbol{\beta}^0))'\mathbf{Z}(\boldsymbol{\beta}_0))^{-1}(\mathbf{Z}(\boldsymbol{\beta}^0))'\bar{\mathbf{y}}(\boldsymbol{\beta}^0)$.

1.1.1 Nonlinear least squares

In this section the nonlinear regression model $\mathbf{y} = \mathbf{f}(\mathbf{X}, \boldsymbol{\beta}) + \boldsymbol{\varepsilon}$ with $\boldsymbol{\varepsilon} \sim (\mathbf{0}, \sigma_\varepsilon^2 \mathbf{I}_n)$ is studied, from which it is clear that the precise pdf of $\boldsymbol{\varepsilon}$ is not fully specified. Then the nonlinear least squares (NLLS) estimator is derived from:

$$\min_{\boldsymbol{\beta}} \boldsymbol{\varepsilon}'\boldsymbol{\varepsilon} = \min_{\boldsymbol{\beta}} (\mathbf{y} - \mathbf{f}(\mathbf{X}, \boldsymbol{\beta}))'(\mathbf{y} - \mathbf{f}\,(\mathbf{X}, \boldsymbol{\beta})) = \min_{\boldsymbol{\beta}} S\,(\boldsymbol{\beta})$$

and the NLLS estimators of $\boldsymbol{\beta}$, $\hat{\boldsymbol{\beta}}_{NLLS}$, and σ_{ε}^2, $\hat{\sigma}^2_{\boldsymbol{\varepsilon},NLLS}$, should directly be derived from the resulting normal equations:

$$\left(\mathbf{y} - f(\mathbf{X}, \hat{\boldsymbol{\beta}}_{NLLS})\right)' \mathbf{Z}\left(\hat{\boldsymbol{\beta}}_{NLLS}\right) = \mathbf{0} \tag{1.6}$$

and $\hat{\sigma}^2_{\boldsymbol{\varepsilon},NLLS} = \frac{S(\hat{\boldsymbol{\beta}}_{NLLS})}{n-p}$ with estimated variance-covariance matrix

$$\hat{\boldsymbol{\Sigma}}_{\hat{\boldsymbol{\theta}},NLLS} = \hat{\sigma}^2_{\boldsymbol{\varepsilon},NLLS}.((\mathbf{Z}(\hat{\boldsymbol{\beta}}_{NLLS}))'\mathbf{Z}(\hat{\boldsymbol{\beta}}_{NLLS}))^{-1}.$$

1.1.2 Nonlinear maximum likelihood

Assume now normality of the stochastic error terms, or $\boldsymbol{\varepsilon} \sim N(\mathbf{0}, \boldsymbol{\Sigma})$, which leads to the following nonlinear maximum likelihood (NLML) problem:

$$\max_{\boldsymbol{\beta},\boldsymbol{\Sigma}} \ell(\boldsymbol{\beta},\boldsymbol{\Sigma}) = \max_{\boldsymbol{\beta},\boldsymbol{\Sigma}} \frac{1}{(2\pi)^{n/2} |\boldsymbol{\Sigma}|^{1/2}} \exp\left\{-\frac{1}{2}(\mathbf{y}-\mathbf{f}(\mathbf{X},\boldsymbol{\beta}))' \boldsymbol{\Sigma}^{-1} (\mathbf{y}-\mathbf{f}(\mathbf{X},\boldsymbol{\beta}))\right\}.$$

To maximize the joint likelihood function, take logarithms:

$$L(\boldsymbol{\beta}, \boldsymbol{\Sigma}) := \ln \ell(\boldsymbol{\beta}, \boldsymbol{\Sigma}) = -\frac{n}{2} \ln 2\pi - \frac{1}{2} \ln |\boldsymbol{\Sigma}| - \frac{1}{2}(\mathbf{y}\text{-}\mathbf{f}(\mathbf{X},\boldsymbol{\beta}))'\boldsymbol{\Sigma}^{-1}(\mathbf{y}\text{-}\mathbf{f}(\mathbf{X},\boldsymbol{\beta})). \tag{1.7}$$

In the sequel, both the ML estimator from the maximization of (1.7) and the statistical properties of this ML estimator will be analyzed.

Nonlinear ML estimator

Considering in first instance a scalar covariance matrix $\boldsymbol{\Sigma} := \sigma_{\varepsilon}^2 \mathbf{I}_n$, the necessary conditions for a maximum of (1.7) are:

$$\left.\frac{\partial L}{\partial \boldsymbol{\beta}}\right|_{NLML} = \frac{1}{\hat{\sigma}^2_{\varepsilon,NLML}} \left.\frac{\partial \mathbf{f}}{\partial \boldsymbol{\beta}}\right|'_{\boldsymbol{\beta}=\hat{\boldsymbol{\beta}}_{NLML}} (\mathbf{y} - \mathbf{f}(\mathbf{X}, \hat{\boldsymbol{\beta}}_{NLML})) = \mathbf{0} \tag{1.8}$$

and

$$\left.\frac{\partial L}{\partial \sigma_{\varepsilon}^2}\right|_{NLML} = -\frac{n}{2\hat{\sigma}^2_{\varepsilon,NLML}} + \frac{1}{2\hat{\sigma}^4_{\varepsilon,NLML}} S\left(\hat{\boldsymbol{\beta}}_{NLML}\right) = 0, \tag{1.9}$$

so that the estimated residual variance satisfies $\hat{\sigma}^2_{\boldsymbol{\varepsilon},NLML} = S\left(\hat{\boldsymbol{\beta}}_{NLML}\right)/n$. [2]

Replacing this expression for the variance $\hat{\sigma}^2_{\boldsymbol{\varepsilon},NLML}$ in (1.8):

$$\frac{n}{S\left(\hat{\boldsymbol{\beta}}_{NLML}\right)} \left.\frac{\partial \mathbf{f}}{\partial \boldsymbol{\beta}}\right|'_{NLML} (\mathbf{y} - \mathbf{f}(\mathbf{X}, \hat{\boldsymbol{\beta}}_{NLML})) = \mathbf{0}, \tag{1.10}$$

so that by solving this system (1.10) of nonlinear equations $\hat{\boldsymbol{\beta}}_{NLML}$ is found.

[2] Mind the difference between $\hat{\sigma}^2_{\boldsymbol{\varepsilon},NLLS} = \frac{S(\hat{\boldsymbol{\beta}}_{NLLS})}{n-p}$ and $\hat{\sigma}^2_{\boldsymbol{\varepsilon},NLML} = \frac{S(\hat{\boldsymbol{\beta}}_{NLML})}{n}$.

Statistical properties of the ML estimator

Sample statistics follow probability distributions depending on the sample size n, where finite and large samples can be analyzed. *Large sample (asymptotic) theory* is studied when analyzing what happens as $n \to \infty$. Of course, n is never infinitely large, but asymptotic results often provide good approximations when samples are large. Denoting a sequence of random variables $x_1, x_2, x_3, \ldots$ with a corresponding sequence of distribution functions $F_1(u), F_2(u), F_3(u), \ldots$, where $F_n(u) := P(x_n \leqq u)$, *converges in distribution* to a random variable x as $x_n \xrightarrow{d} x$ and we call the distribution of x the *limiting* (or *asymptotic*) *distribution* of x_n.[3] If $\lim_{n\to\infty} P(|x_n - c| < \epsilon) = 1$ for all $\epsilon > 0$, or equivalently $\lim_{n\to\infty} P(x_n = c) = 1$, we say that x_n *converges in probability* to c and we write that $x_n \xrightarrow{p} c$ or $\operatorname{plim}_{n\to\infty} x_n = c$, where the number c is called the *probability limit* of x_n; if x_n would be an estimator of c we would say then that it is a *consistent* estimator of c.[4] The *central limit theorem* establishes the limiting distribution (function) of $x_n \xrightarrow{d} x$ implying e.g. $x_n \overset{as}{\sim} N(\mu, \sigma^2)$, where the symbol $\overset{as}{\sim}$ means *asymptotically distributed.* Mind that a common limiting distribution does not imply that the stochastic variates are (asymptotically) the same: the sequence of normalized and standardized sample means $z_n := \frac{\overline{x}_n - \mu}{\sigma/\sqrt{n}}$ is according to the central limit theorem asymptotically normally distributed with zero mean and unit variance, or $z_n \overset{as}{\sim} N(0,1)$, but also $-z_n \overset{as}{\sim} N(0,1)$ and clearly z_n and $-z_n$ are not the same!

While the general linear LS (LLS) estimator $\widehat{\boldsymbol{\theta}}_{LLS}$ of the $p \times 1$ vector $\boldsymbol{\theta}$ belonging to the parameter space $\boldsymbol{\Theta}$ is *unbiased*, or $\text{bias}\left(\widehat{\boldsymbol{\theta}}_{LLS}\right) := E\left(\widehat{\boldsymbol{\theta}}_{LLS}\right) - \boldsymbol{\theta} = \mathbf{0}$ for all $\boldsymbol{\theta} \in \boldsymbol{\Theta} \subseteq \mathbb{R}^p$ and *efficient*, i.e. it has lowest variance of all unbiased estimators,[5] it has also certain desirable *large sample properties* as the NLLS and NLML estimators have. More specifically, under certain regularity conditions, the estimator $\hat{\boldsymbol{\theta}}_{NLML} := (\hat{\boldsymbol{\beta}}'_{NLML}, \hat{\sigma}^2_{\varepsilon,NLML})'$, derived from (1.8) and (1.9), is *consistent* or $\operatorname{plim}_{n\to\infty} \hat{\boldsymbol{\theta}}_{NLML} = \boldsymbol{\theta}$,[6] *asymptotically unbiased* or $\lim_{n\to\infty} E(\hat{\boldsymbol{\theta}}_{NLML}) =$

[3] A simple example is obtained by looking at Student's $t(n)$-pdf, where the standard normal pdf is obtained by letting $n \to \infty$. For more information on statistical properties of the ML estimators, see e.g. Berenson *et al.* (2003).

[4] A stronger mode of stochastic convergence corresponds to uniform convergence in the deterministic case: if $P\left(\lim_{n\to\infty} x_n = c\right) = 1$ we say that x_n *converges almost surely* (or *with probability 1*) to c for which we write $x_n \to c$. The two types of stochastic convergence, i.e. convergence in probability and almost-sure convergence, have two corresponding important implications. Consider therefore a random sample $\{x_n\}$ from some unspecified distribution with mean μ and variance σ^2, both central moments assumed to exist, and let $\overline{x}_n := \frac{1}{n}\sum_{i=1}^n x_i$ denote the sample mean based on n observations. The *strong law of large numbers* states that $\overline{x}_n \to \mu$, i.e. the sample mean converges almost surely to the population mean as $n \to \infty$, while the *weak law of large numbers* states that $\overline{x}_n \xrightarrow{p} \mu$.

[5] Since the LLS estimator $\widehat{\boldsymbol{\theta}}_{LLS} = \widehat{\boldsymbol{\theta}}$ is unbiased and efficient, it has also lowest *mean squared error* (MSE) because $\text{MSE}\left(\widehat{\boldsymbol{\theta}}\right) = E\left(\widehat{\boldsymbol{\theta}}\text{-}\boldsymbol{\theta}\right)\left(\widehat{\boldsymbol{\theta}}\text{-}\boldsymbol{\theta}\right)' = \text{var}\left(\widehat{\boldsymbol{\theta}}\right) + \text{bias}\left(\widehat{\boldsymbol{\theta}}\right)\text{bias}\left(\widehat{\boldsymbol{\theta}}\right)'$. It is known as the best linear unbiased estimator (BLUE) then.

[6] One could also consider *strong consistency* by requiring almost-sure convergence instead

$\boldsymbol{\theta}$, *asymptotically efficient* with covariance matrix being the inverse information matrix $(\mathbf{J}(\boldsymbol{\theta}))^{-1}$, where the information matrix is given by $\mathbf{J}(\boldsymbol{\theta}) := -E\left(\frac{\partial^2 L}{\partial \boldsymbol{\theta} \partial \boldsymbol{\theta}'}\right)$, involving the *Cramèr-Rao lower bound*, and from the central limit theorem *asymptotically normally distributed* or $\sqrt{n}(\hat{\boldsymbol{\theta}} - \boldsymbol{\theta}) \overset{as}{\sim} N(0, \boldsymbol{\Sigma}_{\hat{\boldsymbol{\theta}}_{NLML}})$. To evaluate this information matrix, we take partial derivatives of the theoretical values of (1.8) and (1.9):

$$\begin{aligned}
\frac{\partial^2 L}{\partial \beta_i \partial \beta_j} &= \frac{1}{\sigma_\varepsilon^2} \sum_{h=1}^{n} \left[\left(y_h - f\left(\mathbf{x}_h, \boldsymbol{\beta}\right)\right) \frac{\partial^2 f\left(\mathbf{x}_h, \boldsymbol{\beta}\right)}{\partial \beta_i \partial \beta_j} - \frac{\partial f\left(\mathbf{x}_h, \boldsymbol{\beta}\right)}{\partial \beta_i} \frac{\partial f\left(\mathbf{x}_h, \boldsymbol{\beta}\right)}{\partial \beta_j} \right] \\
\frac{\partial^2 L}{\partial \left(\sigma_\varepsilon^2\right)^2} &= \frac{n}{2\sigma_\varepsilon^4} - \frac{1}{\sigma_\varepsilon^6} \sum_{h=1}^{n} \left(y_h - f\left(\mathbf{x}_h, \boldsymbol{\beta}\right)\right)^2 \\
\frac{\partial^2 L}{\partial \beta_i \partial \left(\sigma_\varepsilon^2\right)} &= -\frac{1}{\sigma_\varepsilon^4} \sum_{h=1}^{n} \left(y_h - f\left(\mathbf{x}_h, \boldsymbol{\beta}\right)\right) \frac{\partial f\left(\mathbf{x}_h, \boldsymbol{\beta}\right)}{\partial \beta_i}.
\end{aligned}$$

Since the asymptotic mean and variance of a random variable are usually the mean and variance of the asymptotic distribution, the asymptotic variance-covariance matrix of the ML estimator can be expressed as:

$$\mathbf{avar}(\hat{\boldsymbol{\theta}}_{NLML}) = \tfrac{1}{n} \lim_{n\to\infty} E\{n(\hat{\boldsymbol{\theta}}_{NLML} - \lim_{n\to\infty} E(\hat{\boldsymbol{\theta}}_{NLML}))(\hat{\boldsymbol{\theta}}_{NLML} - \lim_{n\to\infty} E(\hat{\boldsymbol{\theta}}_{NLML}))'\} =: \boldsymbol{\Sigma}_{\hat{\boldsymbol{\theta}}_{NLML}}.$$

Hence, the asymptotic variance-covariance matrix of the ML estimator satisfies the Cramèr-Rao lower bound:

$$\boldsymbol{\Sigma}_{\hat{\boldsymbol{\theta}}_{NLML}} = (\mathbf{J}(\boldsymbol{\theta}))^{-1} = \begin{bmatrix} \left\{ \frac{1}{\sigma_\varepsilon^2} \sum_{h=1}^{n} \frac{\partial f(\mathbf{x}_h, \boldsymbol{\beta})}{\partial \beta i} \frac{\partial f(\mathbf{x}_h, \boldsymbol{\beta})}{\partial \beta_j} \right\} & \mathbf{0} \\ \mathbf{0}' & \frac{n}{2\sigma_\varepsilon^4} \end{bmatrix}^{-1}. \tag{1.11}$$

In practice, $\boldsymbol{\Sigma}_{\hat{\boldsymbol{\theta}}_{NLML}}$ is not observable and can be estimated by ML as:

$$\hat{\boldsymbol{\Sigma}}_{\hat{\boldsymbol{\theta}}_{NLML}} = \begin{bmatrix} \hat{\sigma}^2_{\varepsilon,NLML}((\mathbf{Z}(\hat{\beta}_{NLML}))'\mathbf{Z}(\hat{\beta}_{NLML}))^{-1} & \mathbf{0} \\ \mathbf{0}' & \frac{2\hat{\sigma}^4_{\varepsilon,NLML}}{n} \end{bmatrix}, \tag{1.12}$$

such that, under normality assumption, you can observe that no second order derivatives are needed to evaluate the covariance matrix of the estimators.

Alternatively, $\boldsymbol{\Sigma}_{\hat{\boldsymbol{\theta}}_{NLML}}$ can, under certain regularity conditions, also consistently (but less asymptotically efficiently) be estimated by the inverse negative Hessian matrix: $\left(-\frac{\partial^2 L}{\partial \boldsymbol{\theta} \partial \boldsymbol{\theta}'}\right)^{-1}\Big|_{\boldsymbol{\theta}=\hat{\boldsymbol{\theta}}_{NLML}}$. In this case, second order derivatives are needed.

Under reg condits,
a less efficient estimator
J(θ)

of convergence in probability.

1.2 Econometric methods of nonlinear estimation

Most of the methods or algorithms of nonlinear estimation that derive the NLLS estimator from (1.6) or the NLML estimator from (1.10) are based on the *gradient principle*, analyzed in Section A.2 of Appendix A and saying that if you want to minimize the objective function, you have to choose a *downhill direction.* According to Lemma A.1 of that Section A.2 the best downhill direction in which the objective function decreases most is the direction opposite to that of the gradient vector $\mathbf{g}$ (being equal to $\nabla S(\boldsymbol{\beta}) := \frac{\partial S}{\partial \boldsymbol{\beta}}$ for NLLS and to $\nabla L(\boldsymbol{\beta}) := \frac{\partial L}{\partial \boldsymbol{\beta}}$ for NLML).

In Appendix A several examples of nonlinear unconstrained optimization methods based on the gradient principle are discussed and mutually compared: steepest descent, Newton(-Raphson), rank one correction, and Quasi-Newton with, among others, the Davidon-Fletcher-Powell algorithm. But the convergence properties of the regression-based methods of unconstrained optimization that are essential for nonlinear econometric estimation as the Gauss, Quasi-Gauss, Marquardt, and the quadratic hill-climbing methods and the method of scoring are discussed in detail in this section. Algorithms for constrained nonlinear estimation based on the transformation of parameters, the transformation of the objective function (*barrier* and *penalty function* methods), and the transformation of the optimization procedure itself (extended (or augmented) Lagrange function method) are presented in Section A.3 of Appendix A.

1.2.1 Gauss method

Consider the nonlinear regression problem for the model $\mathbf{y} = f(\mathbf{X}, \boldsymbol{\beta}) + \boldsymbol{\varepsilon}$ with $\boldsymbol{\varepsilon} \sim (\mathbf{0}, \boldsymbol{\Sigma})$:

$$\min_{\boldsymbol{\beta},\boldsymbol{\Sigma}} \varphi(\boldsymbol{\beta}) = \min_{\boldsymbol{\beta},\boldsymbol{\Sigma}} (\mathbf{y} - \mathbf{f}(\mathbf{X}, \boldsymbol{\beta}))' \boldsymbol{\Sigma}^{-1} (\mathbf{y} - \mathbf{f}(\mathbf{X}, \boldsymbol{\beta}))$$

$$\nabla\varphi(\boldsymbol{\beta}) := \frac{\partial \varphi}{\partial \boldsymbol{\beta}} = -2\left(\frac{\partial \mathbf{f}}{\partial \boldsymbol{\beta}}\right)' \boldsymbol{\Sigma}^{-1} (\mathbf{y}\text{-}\mathbf{f}(\mathbf{X}, \boldsymbol{\beta})) = -2(\mathbf{Z}(\boldsymbol{\beta}))' \boldsymbol{\Sigma}^{-1} (\mathbf{y}\text{-}\mathbf{f}(\mathbf{X}, \boldsymbol{\beta})) = \mathbf{0},$$

so that the Hessian matrix satisfies:

$$\begin{aligned} \mathbf{H}(\boldsymbol{\beta}) \quad &:= \frac{\partial^2 \varphi}{\partial \boldsymbol{\beta} \partial \boldsymbol{\beta}'} = 2(\mathbf{Z}(\boldsymbol{\beta}))' \boldsymbol{\Sigma}^{-1} \mathbf{Z}(\boldsymbol{\beta}) - 2\frac{\partial^2 \varphi}{\partial \boldsymbol{\beta} \partial \boldsymbol{\beta}'} \boldsymbol{\Sigma}^{-1} (\mathbf{y}\text{-}\mathbf{f}(\mathbf{X}, \boldsymbol{\beta})) \\ &= 2(\mathbf{Z}(\boldsymbol{\beta}))' \boldsymbol{\Sigma}^{-1} \mathbf{Z}(\boldsymbol{\beta}) \text{-} 2\sum_{i=1}^{n}\sum_{j=1}^{n} \sigma^{ij} \underbrace{(y_i\text{-}f(\mathbf{x}_i, \boldsymbol{\beta}))}_{\varepsilon_i \approx 0} \frac{\partial^2 f(\mathbf{x}_i, \boldsymbol{\beta})}{\partial \boldsymbol{\beta} \partial \boldsymbol{\beta}'}. \quad (1.13) \end{aligned}$$

The *Gauss estimation method* is a *gradient method* (see Section A.2 of Appendix A) and is characterized by the parameter sequence $\boldsymbol{\theta}^{(k+1)} := \boldsymbol{\theta}^{(k)} + \boldsymbol{\zeta}^{(k)}$ with $\boldsymbol{\zeta}^{(k)} := -\mathbf{P}^{(k)} \mathbf{g}^{(k)} = -\left(\mathbf{H}\left(\boldsymbol{\beta}^{(k)}\right)\right)^{-1} \mathbf{g}^{(k)}$, or the direction matrix $\mathbf{P}^{(k)}$ satisfies from (1.13): $\mathbf{P}^{(k)} = \left(\left(\mathbf{Z}\left(\boldsymbol{\beta}^{(k)}\right)\right)' \boldsymbol{\Sigma}^{-1} \mathbf{Z}\left(\boldsymbol{\beta}^{(k)}\right)\right)^{-1}$.

If the error terms are asumed to be white noise (no correlations among error terms belonging to different observations), or $\mathbf{\Sigma} := \sigma_\varepsilon^2 \mathbf{I}_n$, then:

$$\begin{aligned}\boldsymbol{\beta}^{(k+1)} &= \boldsymbol{\beta}^{(k)} + \frac{1}{2}\left((\mathbf{Z}(\boldsymbol{\beta}^{(k)}))'\mathbf{Z}(\boldsymbol{\beta}^{(k)})\right)^{-1} 2(\mathbf{Z}(\boldsymbol{\beta}^{(k)}))' \left(\mathbf{y} - \mathbf{f}(\mathbf{X}, \boldsymbol{\beta}^{(k)})\right) \\ &= ((\mathbf{Z}(\boldsymbol{\beta}^{(k)}))'\mathbf{Z}(\boldsymbol{\beta}^{(k)}))^{-1}(\mathbf{Z}(\boldsymbol{\beta}^{(k)}))'(\mathbf{y}\text{-}\mathbf{f}(\mathbf{X}, \boldsymbol{\beta}^{(k)})) + \mathbf{Z}(\boldsymbol{\beta}^{(k)})\boldsymbol{\beta}^{(k)}),\end{aligned}$$

which is the OLS estimator of the pseudo-linear model (1.2) at iteration k:

$$\mathbf{y} - \mathbf{f}(\mathbf{X}, \boldsymbol{\beta}^{(k)}) + \mathbf{Z}(\boldsymbol{\beta}^{(k)})\boldsymbol{\beta}^{(k)} =: \bar{\mathbf{y}}(\boldsymbol{\beta}^{(k)}) \simeq \mathbf{Z}\left(\boldsymbol{\beta}^{(k)}\right)\boldsymbol{\beta} + \varepsilon, \tag{1.14}$$

or the Gauss algorithm of the nonlinear regression problem yields a row of linear regression estimators, where the linear regressions are linear approximations to the original nonlinear regression problem.

Example 1.1 *As a theoretical example, consider the (two-factor) production function* $y_i = f(\mathbf{x}_i, \boldsymbol{\beta}) + \varepsilon_i = \beta_1 + \beta_2 x_{i2} + \beta_2^2 x_{i3} + \varepsilon_i$ *with the error terms independently and identically distributed (iid) with zero mean and constant variance, or* $\varepsilon_i \overset{iid}{\sim} (0, \sigma_\varepsilon^2)$.

From the resulting definition of the $n \times 2$ Jacobian matrix $\mathbf{Z}(\boldsymbol{\beta}) := \begin{bmatrix} 1 & x_{12}+2\beta_2 x_{13} \\ \vdots & \vdots \\ 1 & x_{n2}+2\beta_2 x_{n3}\end{bmatrix}$,

you get the moment matrix:

$$(\mathbf{Z}(\boldsymbol{\beta}))'\mathbf{Z}(\boldsymbol{\beta}) = \begin{bmatrix} n & \sum_{i=1}^n (x_{i2} + 2\beta_2 x_{i3}) \\ \sum_{i=1}^n (x_{i2} + 2\beta_2 x_{i3}) & \sum_{i=1}^n (x_{i2} + 2\beta_2 x_{i3})^2 \end{bmatrix}$$

and the expression $\bar{\mathbf{y}}(\boldsymbol{\beta})$ for the pseudo-linear model (see (1.2) and (1.14)):

$$\bar{\mathbf{y}}(\boldsymbol{\beta}) = \begin{bmatrix} y_1 - (\beta_1+\beta_2 x_{12}+\beta_2^2 x_{13}) + (\beta_1+\beta_2 x_{12}+2\beta_2^2 x_{13}) \\ y_2 - (\beta_1+\beta_2 x_{22}+\beta_2^2 x_{23}) + (\beta_1+\beta_2 x_{22}+2\beta_2^2 x_{23}) \\ \vdots \\ y_n - (\beta_1+\beta_2 x_{n2}+\beta_2^2 x_{n3}) + (\beta_1+\beta_2 x_{n2}+2\beta_2^2 x_{n3}) \end{bmatrix} = \begin{bmatrix} y_1+\beta_2^2 x_{13} \\ y_2+\beta_2^2 x_{23} \\ \vdots \\ y_n+\beta_2^2 x_{n3}\end{bmatrix},$$

so that the Gauss estimator of this pseudo-linear model is obtained after k iterations for which $\boldsymbol{\beta}^{(k+1)} = \boldsymbol{\beta}^{(k)} = \hat{\boldsymbol{\beta}}^{(k)}$, yielding the OLS estimator $\hat{\boldsymbol{\beta}} = ((\mathbf{Z}(\boldsymbol{\beta}))'\mathbf{Z}(\boldsymbol{\beta}))^{-1}(\mathbf{Z}(\boldsymbol{\beta}))'\bar{\mathbf{y}}(\boldsymbol{\beta})$, where the iteration index k is omitted, or:

$$\hat{\boldsymbol{\beta}} = \frac{1}{|Z'(\beta)Z(\beta)|}\begin{bmatrix} \sum_{i=1}^n (x_{i2}+2\beta_2 x_{i3})^2 \sum_{i=1}^n (y_i+\beta_2^2 x_{i3}) - \sum_{i=1}^n (x_{i2}+2\beta_2 x_{i3}) \sum_{i=1}^n (y_i+\beta_2^2 x_{i3})(x_{i2}+2\beta_2 x_{i3}) \\ n\sum_{i=1}^n (y_i+\beta_2^2 x_{i3})(x_{i2}+2\beta_2 x_{i3}) - \sum_{i=1}^n (y_i+\beta_2^2 x_{i3}) \sum_{i=1}^n (x_{i2}+2\beta_2 x_{i3}) \end{bmatrix}.$$

Example 1.2 *Consider the nonlinear model* $y_i = \theta_1 + \theta_2 x_{i2} + \theta_2^2 x_{i3} + \varepsilon_i$ $(i = 1, ..., 20)$ *with data given in Figure A.3 of Appendix A.*

The Gauss iterations are obtained in Figure 1.1. Notice the sensitivity of the nonlinear optimization procedure to the choice of the starting values ($\boldsymbol{\theta}^0 = (3,2)'$ and $\boldsymbol{\theta}^0 = (1.5, 0.5)'$ lead to the local minimum 16,1 and not $\boldsymbol{\theta}^0 = (3,-1)'$).

Figure 1.1: Iterations of the Gauss algorithm (Example 1.2)

k	$\theta_1^{(k)}$	$\theta_2^{(k)}$	$F(\theta^{(k)})$
1	3.000000	2.000000	264.3918
2	0.723481	1.404965	16.6635
3	0.837007	1.259230	16.0880
4	0.861002	1.238408	16.0818
5	0.864359	1.236040	16.0817
6	0.864740	1.235780	16.0817
7	0.864782	1.235752	16.0817
8	0.864787	1.235749	16.0817
9	0.864787	1.235749	16.0817
1	3.000000	-1.000000	25.5156
2	2.498561	-0.989894	20.4856
3	2.498566	-0.985678	20.4824
4	2.498571	-0.983903	20.4823
5	2.498574	0-.983154	20.4823
6	2.498575	-0.982837	20.4823
7	2.498576	-0.982703	20.4823
8	2.498576	-0.982646	20.4823
9	2.498576	-0.982623	20.4823
10	2.498576	-0.982612	20.4823
11	2.498576	-0.982607	20.4823
12	2.498576	-0.982605	20.4823
13	2.498576	-0.982605	20.4823
1	1.500000	0.500000	20.2951
2	1.067414	1.213585	16.6646
3	0.868351	1.233424	16.0818
4	0.865161	1.235496	16.0817
5	0.864828	1.235721	16.0817
6	0.864792	1.235746	16.0817
7	0.864788	1.235748	16.0817
8	0.864787	1.235748	16.0817
9	0.864787	1.235748	16.0817

1.2.2 Method of scoring

The *method of scoring* is an estimation method that is based on the NLML procedure, with first order partial derivatives given by (1.8) and (1.9), which together form the score vector, and the Cramèr-Rao lower bound as estimated variance-covariance matrix (1.12). Hence, the appropiate direction matrix for the method of scoring is defined from Section 1.1.2 as $\mathbf{P}^{(k)} := \left[-E\left(\frac{\partial^2 L}{\partial \boldsymbol{\theta} \partial \boldsymbol{\theta}'} \right) \right]^{-1} \Big|_{\boldsymbol{\theta}=\boldsymbol{\theta}^{(k)}}$.

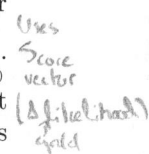

The advantages of the method of scoring are that it is a simple and efficient algorithm which has good local convergence properties. However, in many cases $\mathbf{P}^{(k)}$ is not positive definite or does not even exist because the underlying matrix is singular. Moreover, if the linear approximation model is bad, large errors and slow convergence emerge.

1.2.3 Quasi-Gauss method

The *Quasi-Gauss method* is used when there is singularity of the underlying matrix or nonpositive definiteness of the direction matrix $\mathbf{P}^{(k)}$. The algorithm is based on a regular and positive definite approximation of the Hessian matrix of the regression function.

For $\mathbf{G}_i^{(k)}$ an approximation to the Hessian $\left(\frac{\partial^2 f(\mathbf{x}_i, \boldsymbol{\beta})}{\partial \boldsymbol{\beta} \partial \boldsymbol{\beta}'} \Big|_{\boldsymbol{\beta}=\boldsymbol{\beta}^{(k)}} \right)$ of the regression function at the kth iteration and a corresponding correction matrix $\mathbf{M}_i^{(k)}$ in $\mathbf{G}_i^{(k+1)} := \mathbf{G}_i^{(k)} + \mathbf{M}_i^{(k)}$, we find from a linearization of the gradient of the regression function *at the new point* (see also the Quasi-Newton procedure in Section A.2.3 of Appendix A):

$$\begin{aligned} \nabla f\left(\mathbf{x}_i, \boldsymbol{\beta}^{(k)}\right) \quad : &= \frac{\partial f(\mathbf{x}_i, \boldsymbol{\beta})}{\partial \boldsymbol{\beta}} \Big|_{\boldsymbol{\beta}=\boldsymbol{\beta}^{(k)}} \\ &\simeq \frac{\partial f(\mathbf{x}_i, \boldsymbol{\beta})}{\partial \boldsymbol{\beta}} \Big|_{\boldsymbol{\beta}=\boldsymbol{\beta}^{(k+1)}} + \frac{\partial^2 f(\mathbf{x}_i, \boldsymbol{\beta})}{\partial \boldsymbol{\beta} \partial \boldsymbol{\beta}'} \Big|_{\boldsymbol{\beta}=\boldsymbol{\beta}^{(k+1)}} \left(\boldsymbol{\beta}^{(k)} - \boldsymbol{\beta}^{(k+1)}\right), \end{aligned}$$

so that the correction matrix $\mathbf{M}_i^{(k)}$ should satisfy:

$$\nabla f\left(\mathbf{x}_i, \boldsymbol{\beta}^{(k+1)}\right) - \nabla f\left(\mathbf{x}_i, \boldsymbol{\beta}^{(k)}\right) \simeq \frac{\partial^2 f(\mathbf{x}_i, \boldsymbol{\beta})}{\partial \boldsymbol{\beta} \partial \boldsymbol{\beta}'} \Big|_{\boldsymbol{\beta}=\boldsymbol{\beta}^{(k+1)}} \left(\boldsymbol{\beta}^{(k+1)} - \boldsymbol{\beta}^{(k)}\right)$$

or $\mathbf{M}_i^{(k)} \left(\boldsymbol{\beta}^{(k+1)} \text{-} \boldsymbol{\beta}^{(k)}\right) \simeq \nabla f\left(\mathbf{x}_i, \boldsymbol{\beta}^{(k+1)}\right) \text{-} \nabla f\left(\mathbf{x}_i, \boldsymbol{\beta}^{(k)}\right) \text{-} \mathbf{G}_i^{(k)} \left(\boldsymbol{\beta}^{(k+1)} \text{-} \boldsymbol{\beta}^{(k)}\right)$,

i.e. $\mathbf{M}_i^{(k)} \left(\boldsymbol{\beta}^{(k+1)} - \boldsymbol{\beta}^{(k)}\right) \left(\boldsymbol{\beta}^{(k+1)} - \boldsymbol{\beta}^{(k)}\right)' \simeq \boldsymbol{\mu}_i^{(k)} \left(\boldsymbol{\beta}^{(k+1)} - \boldsymbol{\beta}^{(k)}\right)'$

with $\boldsymbol{\mu}_i^{(k)} := \nabla f\left(\mathbf{x}_i, \boldsymbol{\beta}^{(k+1)}\right) \text{-} \nabla f\left(\mathbf{x}_i, \boldsymbol{\beta}^{(k)}\right) \text{-} \mathbf{G}_i^{(k)} \left(\boldsymbol{\beta}^{(k+1)} \text{-} \boldsymbol{\beta}^{(k)}\right)$, so that:

$$\mathbf{M}_i^{(k)} \simeq \frac{\boldsymbol{\mu}_i^{(k)} \left(\boldsymbol{\beta}^{(k+1)} - \boldsymbol{\beta}^{(k)}\right)'}{\left(\boldsymbol{\beta}^{(k+1)} - \boldsymbol{\beta}^{(k)}\right)' \left(\boldsymbol{\beta}^{(k+1)} - \boldsymbol{\beta}^{(k)}\right)},$$

which is similar to the rank one correction procedure in Section A.2.3 of Appendix A for quasi-Newton.

The advantage of this method is a reasonable convergence, while a small disadvantage is that it requires a pretty large computer memory.

1.2.4 Marquardt method

This method is a combination between the Gauss and steepest descent algorithms (see also Section A.2.1 of Appendix A). The direction matrix belonging to the *Marquardt method* satisfies $\mathbf{P}^{(k)} = \left((\mathbf{Z}(\beta^{(k)}))'\mathbf{\Sigma}^{-1}\mathbf{Z}(\boldsymbol{\beta}^{(k)}) + \alpha^{(k)}\mathbf{C}^{(k)}\right)^{-1}$ with $\alpha^{(k)}$ a sufficiently large positive number.

If $\alpha^{(k)}$ tends to zero, the Marquardt algorithm converges to the Gauss algorithm, and if $\alpha^{(k)}$ tends to infinity, Marquardt converges to the steepest descent algorithm.

Example 1.3 *Comparison between the Gauss and Marquandt algorithms (based on Judge et al. (1980), pp. 737-739).*

Consider the CES production function $Q_i = A[\delta L_i^{-\rho} + (1-\delta)K_i^{-\rho}]^{-\frac{\nu}{\rho}} \exp(\varepsilon_i)$ or $\ln Q_i = \beta_0 + \beta_1 \ln\left[\beta_2 L_i^{\beta_3} + (1-\beta_2)\, K_i^{\beta_3}\right] + \varepsilon_i$ $(i = 1, 2, \ldots, n)$, with data given in Figure A.1 in Appendix A and a starting vector $\boldsymbol{\beta}^0 := (1, -1, 0.5, -1)'$. Comparing Figure 1.2 as iterations of the Gauss algorithm with those of the Marquardt algorithm in Figure 1.3, you can observe that both algorithms converged after 15 iterations, but that the iteration process proceeded much quicker in the Marquandt algorithm; the last Marquardt steps were practically equal to those of the Gauss iterations.

1.2.5 Quadratic hill-climbing method

If the Hessian of the objective function to be minimized is transformed as $\mathbf{P}^{(k)} := \left(\mathbf{H}^{(k)} + a^{(k)}\mathbf{I}_p\right)^{-1}$, we get the *quadratic hill-climbing* direction matrix. Note that this method is not invariant to transformations of the parameter space.

Hence, the Marquardt method is generally superior for nonlinear regressions. In general, the Marquardt method converges very well, even when the initial parameters are rather far from the optimal values.

1.2.6 Concluding remarks

In this chapter the regression-based methods that are essential for nonlinear econometric estimation: Gauss, Quasi-Gauss, the method of scoring, quadratic hill-climbing, and Marquardt have been mutually compared. Other unconstrained nonlinear optimization procedures are discussed in Appendix A: steepest descent, Newton(-Raphson), rank one correction, and Quasi-Newton with, among others, the Davidon-Fletcher-Powell algorithm. The unconstrained nonlinear optimization algorithms differ in the choice of the direction matrix $\mathbf{P}^{(k)}$

Figure 1.2: Iterations of the Gauss algorithm

True Parameter Values	β_0^*=0.0	β_1^*=-0.5	β_2^*=0.3	β_3^*=-2.0	
Iterations	β_0	β_1	β_2	β_3	$S(\beta)$
0	1.00000000	-1.00000000	0.50000000	-1.00000000	37.09647696
1	0.53348846	-0.48109035	0.45060133	-1.49993613	35.48664241
2	0.32051558	-0.30765599	0.38316020	-2.30968355	22.69061393
3	0.12479022	-0.28742856	0.30140828	-3.41818117	1.84546513
4	0.12404406	-0.30792127	0.31714961	-3.20435160	1.83336003
5	0.12293317	-0.35563222	0.34972967	-2.80035238	1.82033702
6	0.12508517	-0.32429484	0.33021412	-3.08911305	1.77400355
7	0.12401054	-0.34250499	0.34052971	-2.95160409	1.76210788
8	0.12471260	-0.33275379	0.33459590	-3.03898265	1.76117653
9	0.12434641	-0.33824362	0.33784883	-2.99373468	1.76105698
10	0.12456252	-0.33519741	0.33602421	-3.02017060	1.76104269
11	0.12444618	-0.33689004	0.33703533	-3.00586988	1.76104014
12	0.12451237	-0.33594673	0.33647149	-3.01396582	1.76103953
13	0.12447601	-0.33647084	0.33678488	-3.00950550	1.76103936
14	0.12449640	-0.33617890	0.33661040	-3.01200194	1.76103931
15	0.12448510	-0.33634124	0.33670746	-3.01061741	1.76103929

Figure 1.3: Iterations of the Marquardt algorithm

True Parameter Values	β_0^*=0.0	β_1^*=-0.5	β_2^*=0.3	β_3^*=-2.0	
Iteration	β_0	β_1	β_2	β_3	$S(\beta)$
0	1.00000000	-1.00000000	0.50000000	-1.00000000	37.09647696
1	0.09415663	-0.56438540	0.39608434	-1.47811571	8.21029114
2	0.10527940	-0.47751281	0.35597976	-2.05526646	2.14540896
3	0.11438473	-0.43433472	0.35332137	-2.33452360	1.79265263
4	0.11778508	-0.40169905	0.34893435	-2.52753974	1.77154836
5	0.11985791	-0.38073528	0.34546303	-2.66794650	1.76445604
6	0.12126599	-0.36685455	0.34296606	-2.76791870	1.76233250
7	0.12223435	-0.35747645	0.34116087	-2.83892982	1.76159005
8	0.12290661	-0.35105837	0.33986459	-2.88928169	1.76129103
9	0.12337607	-0.34662508	0.33893837	-2.92494346	1.76115900
10	0.12421153	-0.33763945	0.33607238	-2.99759855	1.76111747
11	0.12437263	-0.33727961	0.33686984	-3.00261134	1.76104041
12	0.12448186	-0.33621046	0.33652837	-3.01170161	1.76103945
13	0.12447476	-0.33644528	0.33674359	-3.00972804	1.76103932
14	0.12449246	-0.33622429	0.33663096	-3.01161499	1.76103930
15	0.12448612	-0.33632386	0.33669540	-3.01076606	1.76103929

and in the corresponding properties. An overview of these direction matrices for iteration k is given by the following list.

If $\mathbf{P}^{(k)} = \mathbf{I}_p$, the *steepest descent* algorithm with a *constant metric* is obtained; all the other gradient methods show a *variable metric*, such as:

$\mathbf{P}^{(k)}=\left(\frac{\partial^2\varphi}{\partial\boldsymbol{\theta}\partial\boldsymbol{\theta}'}\Big|_{\boldsymbol{\theta}=\boldsymbol{\theta}^{(k)}}\right)^{-1} = \left(\mathbf{H}^{(k)}\right)^{-1}$ for the *Newton(-Raphson)* algorithm with variable metric $\left(\mathbf{H}^{(k)}\right)^{-1}$;.

$\mathbf{P}^{(k)}=\mathbf{P}^{(k-1)} + \frac{\boldsymbol{\eta}^{(k-1)}\boldsymbol{\eta}'^{(k-1)}}{\boldsymbol{\eta}'^{(k-1)}(\mathbf{g}^{(k)}-\mathbf{g}^{(k-1)})}$ for the *rank one correction* algorithm;

$\mathbf{P}^{(k)}=\mathbf{P}^{(k-1)}+\frac{\boldsymbol{\xi}^{(k-1)}\boldsymbol{\xi}'^{(k-1)}}{\boldsymbol{\xi}'^{(k-1)}\left(\mathbf{g}^{(k)}-\mathbf{g}^{(k-1)}\right)} - \frac{\mathbf{P}^{(k-1)}(\mathbf{g}^{(k)}-\mathbf{g}^{(k-1)})(\mathbf{g}^{(k)}-\mathbf{g}^{(k-1)})'\mathbf{P}^{(k-1)}}{(\mathbf{g}^{(k)}-\mathbf{g}^{(k-1)})'\mathbf{P}^{(k-1)}(\mathbf{g}^{(k)}-\mathbf{g}^{(k-1)})}$ for the *Davidon-Fletcher-Powell (DFP)* algorithm;

$\mathbf{P}^{(k)}=\lim_{n\to\infty}\left\{\left[-E\left(\frac{\partial^2 L}{\partial\boldsymbol{\theta}\partial\boldsymbol{\theta}'}\right)\right]^{-1}\Big|_{\boldsymbol{\theta}=\boldsymbol{\theta}^{(k)}}\right\}$ for the *method of scoring*;

$\mathbf{P}^{(k)}=\left[(\mathbf{Z}(\beta^{(k)}))'\boldsymbol{\Sigma}^{-1}\mathbf{Z}(\beta^{(k)})\right]^{-1}$ for the *Gauss* algorithm;

$\mathbf{P}^{(k)}=\left[(\mathbf{Z}(\beta^{(k)}))'\boldsymbol{\Sigma}^{-1}\mathbf{Z}(\beta^{(k)})+\lambda^{(k)}\mathbf{C}^{(k)}\right]^{-1}$ for the *Marquardt* algorithm;

$\mathbf{P}^{(k)}=\left(\mathbf{H}^{(k)}+a^{(k)}\mathbf{I}_p\right)^{-1}$ for the quadratic hill-climbing method.

Various contrained optimization methods are also discussed in Appendix A. The most efficient regression-based procedure for unconstrained nonlinear estimation appears to be the Marquardt procedure, while for constrained nonlinear estimation a barrier or penalty function method and the extended (augmented) Lagrange function method are to be preferred.

Chapter 2

Generalized Method of Moments

As seen in Subsection 1.1.2 the linear OLS (LLS) estimator is best linear unbiased, consistent and asymptotically normally distributed and, under certain regularity conditions, the NLLS and NLML estimators are consistent, asymptotically efficient and asymptotically normally distributed, provided that the regressors are not correlated with the disturbances. However, failure to attain this property of noncorrelation, which is so in many real world cases, provides a no longer (asymptotically) efficient estimator. The generalized method of moments (GMM) estimators discussed in this chapter take account of this often observed correlation.

Consequently, GMM is an increasingly popular estimation method in economics and econometrics. It has desirable asymptotic or large sample properties and, one may note two main reasons for its current popularity. The first one is that GMM nests most of the commonly known estimators (such as OLS, instrumental variables (IVs) or ML) and provides a useful framework for their comparison and evaluation. The second one is that GMM provides a 'simple' alternative to other estimation methods, especially when it is difficult to write down the ML estimator.

We begin in Section 2.1 with some examples. Section 2.2 sets up a classical starting point for the GMM: the Method of Moments (MoM) with applications on OLS, IVs, GLS and ML. The GMM estimator is defined in Section 2.3 with various examples and Section 2.4 concludes.

2.1 Endogeneity bias: two examples

Working's example A regressor that is not orthogonal to the error is said to be endogenous, and this causes OLS to break down. Endogeneity naturally arises in a simultaneous equations model of market equilibrium (when modeling e.g. markets of agricultural products). Working's (1927) model specifies:[1]

$$\begin{aligned} q_i^d &= \alpha_0 + \alpha_1 p_i + u_i, && \text{(demand)} \\ q_i^s &= \beta_0 + \beta_1 p_i + v_i, && \text{(supply)} \\ q_i^d &= q_i^s =: q_i && \text{(equilibrium)}, \end{aligned} \tag{2.1}$$

where p_i is the price for observation i (can be period, individual, country, etc.); u_i includes other unobserved determinants of demand; and v_i represents other factors affecting supply with $E(u_i) = E(v_i) = 0$ and variances σ_u^2 and σ_v^2 with (mostly) $cov(u_i, v_i) = 0$. Solving system (2.1) of simultaneous equations for (q_i, p_i) yields $p_i = \frac{\beta_0 - \alpha_0}{\alpha_1 - \beta_1} + \frac{v_i - u_i}{\alpha_1 - \beta_1}$ and $q_i = \frac{\alpha_1 \beta_0 - \alpha_0 \beta_1}{\alpha_1 - \beta_1} + \frac{\alpha_1 v_i - \beta_1 u_i}{\alpha_1 - \beta_1}$.

The covariances between the price p_i and the demand and supply error terms, u_i and v_i, are:

$$E(p_i u_i) = cov(p_i, u_i) = -\frac{\sigma_u^2}{\alpha_1 - \beta_1} \neq 0 \text{ and } E(p_i v_i) = cov(p_i, v_i) = \frac{\sigma_v^2}{\alpha_1 - \beta_1} \neq 0,$$

so that the price is correlated positively with the demand shifter u_i if the demand curve is downward sloping ($\alpha_1 < 0$) and negatively with the supply shifter v_i if the supply curve is upward sloping ($\beta_1 > 0$).

A regressor is said to be *endogenous* if it is not predetermined, i.e. if it is not orthogonal to the error term. Given that both $cov(p_i, u_i)$ and $cov(p_i, v_i)$ are different from zero, the endogeneity of p_i is a fact. Note that the data for the demand equation and the data for the supply equation are identical, and so must be the OLS estimates of these equations. The slope estimate converges in probability to:

$$\frac{cov(p_i, q_i)}{var(p_i)} = \begin{cases} \alpha_1 + \frac{cov(p_i, u_i)}{var(p_i)} \\ \beta_1 + \frac{cov(p_i, v_i)}{var(p_i)} \end{cases}$$

and is neither consistent for α_1 nor for β_1. This is an example of *endogeneity bias* (or *simultaneity bias*). It is a direct consequence of the correlation between (the) explanatory variables and the error term.

How can this be solved? The answer is that without extra data, there is no hope. Suppose, however, that the supply shifter, v_i, can be partly observed. Specifically, $v_i = \beta_1 z_i + \zeta_i$ with z_i observable, uncorrelated with ζ_i and uncorrelated with u_i. The variations in z_i cause price variations and these, in turn, help us to identify the demand curve in Figure 2.1. A variable which has the properties of being predetermined or uncorrelated relative to the error and correlated with the endogenous variable, is an *instrumental variable* or IV. Hence, z_i is an instrument for the demand equation because $cov(z_i, u_i) = 0$ and $cov(z_i, p_i) = \frac{\beta_2}{\alpha_1 - \beta_1} var(z_i) \neq 0$. It follows directly from the demand equation that $cov(z_i, q_i) = \alpha_1 cov(z_i, p_i)$ and this suggests the estimator $\hat{\alpha}_{1,IV} = \frac{\widehat{cov}(z_i, q_i)}{\widehat{cov}(z_i, p_i)}$, where $\widehat{cov}(\cdot)$ denotes the sample covariance. This estimator is an IV estimator, with z_i as IV for p_i.

[1] Hayashi (2000) considers this case at length.

Figure 2.1: Supply and demand shifts

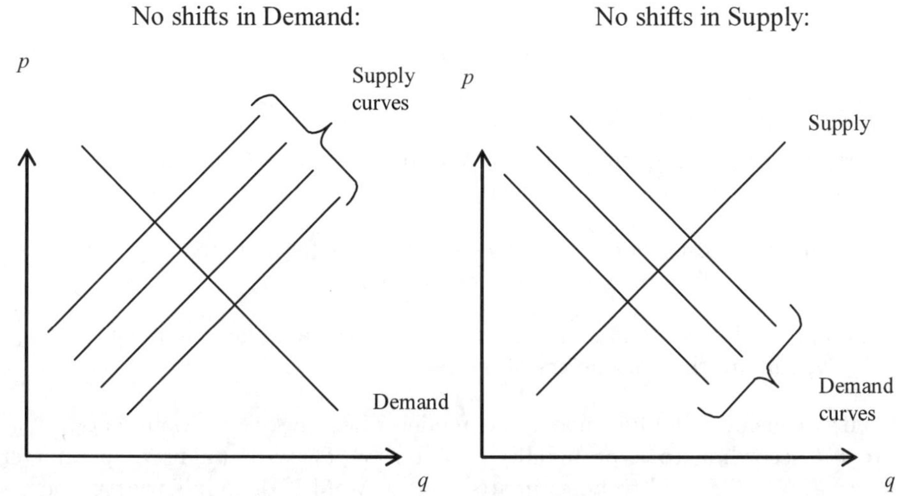

Since endogeneity bias arises when a regressor is correlated with the error, the presence of an instrument, uncorrelated with the error and correlated with the regressor, solves the problem.

A simple macroeconomic example Endogeneity arises in many contexts, e.g.in macroeconometrics and in errors-in-variables situations. Haavelmo (1943) illustrates the endogeneity problem by the following simple consumption output model:

$$C_i = \alpha_0 + \alpha_1 Y_i + u_i, \quad 0 < \alpha_1 < 1$$
$$Y_i = C_i + I_i, \quad \text{where } I_i \text{ is investment}$$

Solving for income Y_i gives $Y_i = \frac{\alpha_0}{1\text{-}\alpha_1} + \frac{I_i}{1\text{-}\alpha_1} + \frac{u_i}{1\text{-}\alpha_1}$. Hence, the covariance between income and the error term u_i is:

$$cov(Y_i, u_i) = \frac{\sigma_u^2}{1\text{-}\alpha_1} > 0$$

showing that Y_i is endogenous, and $cov(Y_i, I_i) = \frac{var(I_i)}{1\text{-}\alpha_1} > 0$. Hence I_i is a valid instrument provided that $cov(I_i, u_i) = 0$. The IV estimator is $\hat{\alpha}_{1,IV} = \frac{\widehat{cov}(I_i, C_i)}{\widehat{cov}(I_i, Y_i)}$.

2.2 The method of moments

In general the MoM is based on the estimation of the moments of a pdf. Defining the general nonlinear regression model in (1.1), the theoretical (first) *moment restrictions* are given by:

$$E_{\boldsymbol{\theta}^0}\left[\mathbf{f}(\mathbf{X}, \boldsymbol{\theta}^0)\right] = \mathbf{0}, \tag{2.2}$$

with $\boldsymbol{\theta}^0$ the true value of the parameter $\boldsymbol{\theta}$, which contains all the unknown parameters.[2] In general the population moment restrictions can be estimated when taking a sample of size n:

$$\frac{1}{n}\sum_{i=1}^{n}\mathbf{f}(\mathbf{x}_i,\hat{\boldsymbol{\theta}})=\mathbf{0}, \tag{2.3}$$

which are known as the *sample or estimated moment conditions.*

However, it is frequently difficult to directly estimate the parameters of interest. An alternative method consists in:

- finding functions of the parameters under the form of auxiliary parameters that can easily be consistently estimated;
- appropriately combining the estimates of the auxiliary parameters to approximate the parameters of interest.

If we consider sampling models, i.e. models where the observations $y_1, y_2, \ldots, y_n$ are *iid* according to a probability distribution $F_{\boldsymbol{\theta}}$ with unknown parameter $\boldsymbol{\theta} \in \Theta \subseteq \mathbb{R}^p$, $\boldsymbol{\theta}$ is said to be estimated by the MoM if the auxiliary parameters are chosen as the first moments of the observations (probability distribution). Specifically, if there exist unique relationships between the first k moments and the parameter vector $\boldsymbol{\theta}$:

$$E_{\boldsymbol{\theta}}(y)=b_1(\theta), E_{\boldsymbol{\theta}}(y^2)=b_2(\boldsymbol{\theta}),\ldots, E_{\boldsymbol{\theta}}(y^k)=b_k(\boldsymbol{\theta}),$$

then, *in a first step*, you can estimate the auxiliary parameters $b_1, b_2, \ldots, b_k$ by the sample moments:

$$\hat{b}_1=\tfrac{1}{n}\sum_{i=1}^{n}y_i, \hat{b}_2=\tfrac{1}{n}\sum_{i=1}^{n}y_i^2,\ldots,\hat{b}_k=\tfrac{1}{n}\sum_{i=1}^{n}y_i^k,$$

which can be assembled in the vector $\hat{\mathbf{b}} := (\hat{b}_1, \hat{b}_2, \ldots, \hat{b}_k)'$; *in a second step*, the MoM obtains a consistent estimator for $\boldsymbol{\theta}$ by $\hat{\boldsymbol{\theta}} = b^{-1}(\hat{\mathbf{b}})$.

2.2.1 Some examples

Example 2.1 *Consider the logistic pdf* $e^{(y-\theta)}/(1+e^{-(y-\theta)})^2$.

Verify that the ML estimator of θ is difficult to compute. The MoM estimator of θ is based on the first moment $E_\theta(y) = b_1 = \theta : \hat{\theta} = \hat{b}_1 = \bar{y}$. This logistic pdf proved itself suited to study households' (skew) income and durable goods consumption behavior exceeding a certain threshold θ.

Example 2.2 *Consider the uniform pdf on the closed interval* $[\theta_1, \theta_2]$.

[2] Notice that, in general, $E_{\boldsymbol{\theta}^0}\left[\mathbf{f}(\mathbf{X},\boldsymbol{\theta})\right] \neq \mathbf{0}$ for $\boldsymbol{\theta} \neq \boldsymbol{\theta}^0$.

If a sample of n observations would be available, appropiate estimators for θ_1 and θ_2 would be $\widehat{\theta}_1 = \min\{y_1, y_2, \ldots, y_n\}$ and $\widehat{\theta}_2 = \max\{y_1, y_2, \ldots, y_n\}$, respectively. Regarding the MoM estimators, the two first noncentral moments are $E_{\boldsymbol{\theta}}(y) = b_1 = \frac{\boldsymbol{\theta}_1 + \boldsymbol{\theta}_2}{2}$ and $E_{\boldsymbol{\theta}}(y^2) = b_2 = \frac{\theta_1^2 + \theta_2^2 + \theta_1\theta_2}{3}$, so that the MoM estimators can be found by solving the following two equations in two unknowns:

$$\hat{b}_1 = \bar{y}_n = \frac{\hat{\theta}_1 + \hat{\theta}_2}{2},\ \hat{b}_2 = \bar{y}_n^2 = \frac{\hat{\theta}_1^2 + \hat{\theta}_2^2 + \hat{\theta}_1\hat{\theta}_2}{3};\ \hat{\theta}_1 < \hat{\theta}_2$$
$$\hat{\theta}_1 = \hat{b}_1 - \sqrt{3(\hat{b}_2 - \hat{b}_1^2)}, \hat{\theta}_2 = \hat{b}_1 + \sqrt{3(\hat{b}_2 - \hat{b}_1^2)}.$$

2.2.2 MoM encompasses many estimation methods

Most known estimators can be constructed as (special cases of) MoM estimators.

Linear least squares

This linear least squares (LLS or OLS) estimator is an MoM estimator since the estimation of the conventional linear model $y_i = \mathbf{x}_i'\boldsymbol{\beta} + \varepsilon_i$ $(i = 1, \ldots, n)$ is based on the assumption that $E(\mathbf{x}_i\varepsilon_i) = \mathbf{0}$, which is precisely the moment or orthogonality condition. From this we can derive that

$$E(\mathbf{X}'\boldsymbol{\varepsilon}) = E(\mathbf{X}'(\mathbf{y} - \mathbf{X}\boldsymbol{\beta})) = E(\mathbf{X}'\mathbf{y}) - E(\mathbf{X}'\mathbf{X})\boldsymbol{\beta} = \mathbf{0},$$

or $\boldsymbol{\beta} = E(\mathbf{X}'\mathbf{X})^{-1}E(\mathbf{X}'\mathbf{y})$, which is estimated as $\hat{\boldsymbol{\beta}}_{OLS} = (\mathbf{X}'\mathbf{X})^{-1}\mathbf{X}'\mathbf{y} = \hat{\boldsymbol{\beta}}_{MoM}$.

Linear instrumental variables

Consider the familiar linear model $\mathbf{y} = \mathbf{X}\boldsymbol{\beta} + \boldsymbol{\varepsilon}$ with the white noise error term $\boldsymbol{\varepsilon}$ with zero mean and covariance matrix $\sigma_\varepsilon^2\mathbf{I}_n$, $\mathbf{I}_n$ being the $n \times n$ unitary matrix with n the number of observations, and $E(\mathbf{X}'\boldsymbol{\varepsilon}) \neq 0$, so that OLS is inconsistent. As put forth by the examples in Section 2.1, the basic issue is the solution of the endogeneity bias by appropiately choosing a matrix $\mathbf{Z}$ of IVs, that are (strongly) correlated with the right-hand side endogenous variable(s) but uncorrelated with the error term $\boldsymbol{\varepsilon}$. The latter property leads to the IV orthogonality condition $E(\mathbf{Z}'\boldsymbol{\varepsilon}) = 0$.

Johnston and DiNardo (1997), pp. 331-332, provide a nice economic example of this situation. Assume we have to determine the logarithmic firm employment as a linear function of the logarithmic entrepreneurial real contract wages. Hence, there is (besides the constant term) only one explanatory variable in the above linear model so that $\mathbf{X}$ may be written as $\mathbf{X} := [\boldsymbol{\iota}, \mathbf{x}]$, where $\boldsymbol{\iota}$ is the unit vector and $\mathbf{x}$ is an n vector of observations on the firm's real contract wages. But since the firm employment results from a rationing scheme on the labor market (e.g. as the minimum of the entrepreneurial labor demand and supply depending on the size of the firm's real wage rate), both employment and wages are the result of both supply and demand changes so that $E(\mathbf{x}'\boldsymbol{\varepsilon}) \neq 0$. Since contract wages are negotiated in advance, a possible IV is unexpected inflation, which shifts the real wage rate. If, e.g., inflation was unexpectedly high, this would lower the real wage rate and employers would move down their labor demand curve, and, hence, mostly their effective employment itself.

Assume now that we have two IVs, the unexpected inflation and the (expected) short term interest rate which is simply taken as the current short term interest rate (the money market is assumed to be efficient). We can put these variables into a matrix form $\mathbf{Z} := [\boldsymbol{\iota}, \mathbf{z}_1, \mathbf{z}_2]$.

The three orthogonality conditions for this problem can be written as:

$$E(\mathbf{Z}'\boldsymbol{\varepsilon}) = E(\mathbf{Z}'(\mathbf{y} - \mathbf{X}\boldsymbol{\beta})) = \mathbf{0}, \tag{2.4}$$

which are estimated as $\frac{1}{n}(\mathbf{Z}'(\mathbf{y} - \mathbf{X}\hat{\boldsymbol{\beta}})) = \mathbf{0}$. Remember that the MoM estimator should satisfy these moment restrictions. Since, in general, the number of IVs is larger than the number of explanatory variables the matrix $(\mathbf{Z}'\mathbf{X})$ is not square because we have three moment restrictions and two parameters in this example.

Given this overidentification, one has several options. First, one could drop one of the equations which boils down in deleting one of the IVs. Second, the squared deviations from the satisfaction to each moment condition could be minimized without any weights in a simple LS framework and, third, the deviations from the moment equations could be weighted according to how precisely (measured by the variance) each of these equations is estimated. Proceeding with this (more) efficient third procedure, an appropriate MoM estimator could be found by solving the weighted LS problem:

$$\min_{\boldsymbol{\beta}} \left\{ \frac{1}{n}[\mathbf{Z}'(\mathbf{y} - \mathbf{X}\boldsymbol{\beta})]'\mathbf{V}^{-1}\frac{1}{n}[\mathbf{Z}'(\mathbf{y} - \mathbf{X}\boldsymbol{\beta})]' \right\} \tag{2.5}$$

with $\mathbf{V}^{-1}$ the inverse variance-covariance matrix, $\mathbf{V}^{-1} = [\mathbf{var}[\frac{1}{n}(\mathbf{Z}'\boldsymbol{\varepsilon})]]^{-1}$, where it is assumed that restrictions which will be estimated less precisely (so, showing higher variances) are given less weight than those that are estimated more precisely, provided that a consistent estimator $\hat{\mathbf{V}}^{-1}$ of this inverse variance-covariance matrix is available ($\hat{\mathbf{V}} := \frac{\hat{\sigma_\varepsilon}^2}{n^2}(\mathbf{Z}'\mathbf{Z})$). Solving the minimization problem (2.5), we find the desired estimator as:

$$\tilde{\boldsymbol{\beta}} = [(\mathbf{X}'\mathbf{Z})(\mathbf{Z}'\mathbf{Z})^{-1}(\mathbf{Z}'\mathbf{X})]^{-1} \; (\mathbf{X}'\mathbf{Z})(\mathbf{Z}'\mathbf{Z})^{-1}(\mathbf{Z}'\mathbf{y}),$$

which is precisely the two stage least squares (2SLS) estimator of $\boldsymbol{\beta}$ (see e.g. Greene (2003), p. 399).

Linear generalized least squares

Since the linear generalized least squares (GLS) estimator is derived from the normal equations $E(\mathbf{X}'\boldsymbol{\Sigma}^{-1}\boldsymbol{\varepsilon}) = \mathbf{0}$ with estimate $\frac{1}{n}\mathbf{X}'\boldsymbol{\Sigma}^{-1}\hat{\boldsymbol{\varepsilon}} = 0$ (for known $\boldsymbol{\Sigma}$, otherwise with a consistent estimate $\hat{\boldsymbol{\Sigma}}$), this is exactly equal to the estimated moment equations. Hence the linear GLS estimator is an MoM estimator.

Maximum Likelihood

The (non)linear ML estimator is derived from (1.8) and (1.9), which can be interpreted as the estimated moment equations, or in general $\frac{\partial L}{\partial \boldsymbol{\theta}}\Big|_{\boldsymbol{\theta}=\hat{\boldsymbol{\theta}}_{ML}} = \mathbf{0}$ so that the ML estimator is an MoM estimator.

Example 2.3 *Compare the ML and MoM estimators of the parameters in the general gamma pdf.*

Consider the loglikelihood function for this general gamma pdf:

$$L := \ln \ell = n[\gamma ln\lambda - ln\Gamma(\gamma)] - \lambda \sum_{i=1}^{n} y_i + (\gamma - 1) \sum_{i=1}^{n} \ln y_i,$$

where the gamma function is $\Gamma(\gamma) := \int_0^\infty t^{\gamma-1} e^{-t} dt$. Since an exponential (parametric) family of distributions is one whose loglikelihood is of the form $\ln \ell(\boldsymbol{\theta}|\,\mathbf{X}) = a(\mathbf{X}) + b(\boldsymbol{\theta}) + \sum_{i=1}^{K} c_i(\mathbf{X}) s_i(\boldsymbol{\theta})$, where $a(\cdot)$, $b(\cdot)$, $c_i(\cdot)$, and $s_i(\cdot)$ are differentiable functions, this gamma pdf belongs to an exponential family with $a(\mathbf{X}) := 0, b(\boldsymbol{\theta}) := n[\gamma ln\lambda - ln\Gamma(\gamma)]$ and two sufficient statistics, $\sum_{i=1}^{n} y_i$ and $\sum_{i=1}^{n} \ln y_i$. The ML estimators of γ and λ and corresponding standard errors, found by maximizing the above gamma loglikelihood function for the income data in Table 2.1 are $\hat{\gamma} = 2.4106$ (0.87683) and $\hat{\lambda} = 0.07707$ (0.0271).

Table 2.1: Relationship between income and education

Observation	*Income*	*Education*	*Observation*	*Income*	*Education*
1	20.5	12	11	55.8	16
2	31.5	16	12	25.2	20
3	47.7	18	13	29.0	12
4	26.2	16	14	85.5	16
5	44.0	12	15	15.1	10
6	8.28	12	16	28.5	18
7	30.8	16	17	21.4	16
8	17.2	12	18	17.7	20
9	19.9	10	19	6.42	12
10	9.96	12	20	84.9	16

Source: This example is considered in Greene (2003), p. 953.

Since the ML estimator is an MoM estimator, the LS criterion for deriving an initial set of MoM estimates can be used. Consider therefore the following relevant *moment conditions* for the gamma pdf:

$$E_{\boldsymbol{\theta}}\left\{y_i - \tfrac{\gamma}{\lambda}\right\} = 0, \qquad E_{\boldsymbol{\theta}}\left\{y_i^2 - \tfrac{\gamma(\gamma+1)}{\lambda^2}\right\} = 0,$$
$$E_{\boldsymbol{\theta}}\left\{\ln y_i - \Phi(\gamma) + \ln \lambda\right\} = 0, \qquad E_{\boldsymbol{\theta}}\left\{\tfrac{1}{y_i} - \tfrac{\lambda}{\gamma-1}\right\} = 0,$$

where the unknown parameter vector $\boldsymbol{\theta}$ is defined as $\boldsymbol{\theta} := (\gamma, \lambda)'$ and use has been made of the property that

$$E_{\boldsymbol{\theta}}\{\ln y_i\} = \tfrac{d \ln \Gamma(\gamma)}{d\gamma} - \ln \lambda = \Phi(\gamma) - \ln \lambda.$$

The estimates of the four moment equations are the estimated (sample) moment equations. Let $y_1 := y$, $y_2 := y^2$, $y_3 := \ln y$, $y_4 := 1/y$, then the first sample moment equation can be written as:

$$\overline{m}_1(\gamma, \lambda) := \tfrac{1}{n}\sum_{i=1}^{n}\left(y_{i1} - \tfrac{\gamma}{\lambda}\right) = \tfrac{1}{n}\sum_{i=1}^{n}(y_{i1} - b_1) = \overline{y} - b_1 = \hat{b}_1 - b_1,$$

and the other sample moment equations $\overline{m}_2(\gamma, \lambda)$, $\overline{m}_3(\gamma, \lambda)$, $\overline{m}_4(\gamma, \lambda)$ can be written down in a similar way.

LS (and, hence, also MoM) estimators of $(\gamma, \lambda)'$ can be obtained by solving the following optimization problem:

$$\min_{\gamma,\lambda}\left\{\sum_{i=1}^{4}\left[\overline{m}_i(\gamma, \lambda)\right]^2 = \sum_{i=1}^{4}\left[\hat{b}_i - b_i\right]^2 = \left[\overline{\mathbf{m}}(\gamma, \lambda)\right]'\left[\overline{\mathbf{m}}(\gamma, \lambda)\right]\right\}. \tag{2.6}$$

This is a minimum distance estimator with a unitary matrix $\mathbf{I}_4$ as a weighting matrix. Following Greene (2003), pp. 538-540, this nonlinear optimization problem can conveniently be solved by multiplying the sample moment equations by λ, λ^2, 1, $(\gamma - 1)$, respectively (equal denominators). The following sample moment equations directly follow from the numerators:

$$\lambda\hat{b}_1 - \gamma = 0, \lambda^2\hat{b}_2 - \gamma(\gamma + 1) = 0, \hat{b}_3 - \Phi(\gamma) + \ln \lambda = 0, (\gamma - 1)\hat{b}_4 - \lambda = 0.$$

This is a system of four nonlinear equations in two unknowns; any pair could be used to estimate the two unknown parameters, but notice that the six pairs in general produce six different LS estimates of $\boldsymbol{\theta} := (\gamma, \lambda)'$. This can be illustrated as follows. Consider the sample of twenty observations on income in Table 2.1. The four relevant sample moments restrictions of the income series serve as the above mentioned auxiliary sample parameters:

$$\tfrac{1}{20}\sum_{i=1}^{20}\left[y_i; y_i^2; \ln y_i; \tfrac{1}{y_i}\right] = [31.278; 1453.96; 3.2214; 0.050014] =: \hat{b}_1, \hat{b}_2, \hat{b}_3, \hat{b}_4,$$

so that the following empirical MoM estimators for $\boldsymbol{\theta} := (\gamma, \lambda)'$ result:[3]

$$\begin{array}{ll}
\hat{\boldsymbol{\theta}}_{(1)}(\hat{b}_1, \hat{b}_2) = (2.0568; 0.06575)' & \hat{\boldsymbol{\theta}}_{(2)}(\hat{b}_1, \hat{b}_3) = (2.4106; 0.07707)' \\
\hat{\boldsymbol{\theta}}_{(3)}(\hat{b}_1, \hat{b}_4) = (2.77198; 0.0886)' & \hat{\boldsymbol{\theta}}_{(4)}(\hat{b}_2, \hat{b}_3) = (2.4806; 0.079835)' \\
\hat{\boldsymbol{\theta}}_{(5)}(\hat{b}_2, \hat{b}_4) = (2.60904; 0.0807)' & \hat{\boldsymbol{\theta}}_{(6)}(\hat{b}_3, \hat{b}_4) = (3.0358; 0.101811)'.
\end{array}$$

Remark that the second estimate $\hat{\boldsymbol{\theta}}_{(2)}$ equals the ML estimate.

Consideration of the first p moments may lead to a substantial loss of information so that the estimators of $\boldsymbol{\theta}$ become asymptotically inefficient. Thus

[3] Since there are four moment restrictions and two parameters of interest $\binom{k}{p} = \binom{4}{2} = 6$ MoM estimators are obtained.

it may be preferable to retain a number k of moments (i.e. k is the number of moment restrictions) that is larger than the number p of parameters of interest. Then the main difficulty is to find a suitable method for solving approximately the system of (moment) equations relating the parameters of interest to the estimates of the auxiliary parameters since this system has more equations (k) than unknowns (p) then. An appropriate estimation method for such overidentified system is the generalized MoM or GMM estimator (see also (2.5)).

2.3 Generalized method of moments

Recall the orthogonality or moment conditions (2.4) for observation i ($i = 1, \ldots, n$) as:

$$E(\mathbf{g}_i) = 0, \tag{2.7}$$

where $\mathbf{g}_i := \mathbf{z}_i(y_i - \mathbf{x}_i'\boldsymbol{\beta})$, $\mathbf{z}_i$ being the k dimensional vector of IVs and $y_i = \mathbf{x}_i'\boldsymbol{\beta}+\varepsilon_i$ ($i = 1, 2, \ldots n$) with $\mathbf{x}_i$ an m dimensional vector of explanatory variables.

The idea of GMM is to estimate $\boldsymbol{\beta} \in \mathbb{R}^p$ by the solution of the sample analogue to the moment condition (2.7), i.e. $\bar{\mathbf{g}} = \mathbf{0}$, where $\bar{\mathbf{g}} = \frac{1}{n}\sum_{i=1}^n \mathbf{g}_i$.[4] The value of $\boldsymbol{\theta}$ that makes this sample mean as close to zero as possible is called the GMM estimator of $\boldsymbol{\theta}$.

- When $k = m(= p)$, an exact solution is possible by IVs,

$$\hat{\boldsymbol{\beta}}_{IV} = \left(\frac{1}{n}\sum_{i=1}^n \mathbf{z}_i\mathbf{x}_i'\right)^{-1} \frac{1}{n}\sum_{i=1}^n \mathbf{z}_i' y_i = \mathbf{S}_{zx}^{-1}\mathbf{s}_{zy},$$

  where $\hat{\boldsymbol{\beta}}_{IV}$ is called the *IV estimator*. It can also be written as $\hat{\boldsymbol{\beta}}_{IV} = (\mathbf{Z}'\mathbf{X})^{-1}\mathbf{Z}'\mathbf{y}$, where $\mathbf{X}$ is the $n \times m$ regressor matrix and $\mathbf{Z}$ the $n \times k$ instrument matrix.[5]

- When $k > m$, an exact solution of $\bar{\mathbf{g}} = \mathbf{0}$ is in general not possible. The idea is to estimate $\boldsymbol{\beta}$ by the solution, $\hat{\boldsymbol{\beta}}$, obtained from minimizing the following objective where we take account of the definition for g_i:

$$n\bar{\mathbf{g}}'\mathbf{W}_n\bar{\mathbf{g}} = n\ (\mathbf{s}_{zy} - \mathbf{S}_{zx}\boldsymbol{\beta})'\,\mathbf{W}_n\,(\mathbf{s}_{zy} - \mathbf{S}_{zx}\boldsymbol{\beta}) \tag{2.8}$$

  with $\mathbf{W}_n$ a $k \times k$ weighting matrix ($\mathbf{W}_n$ depends on the data but has to be symmetric, positive definite, and should converge in probability to a symmetric, positive definite matrix $\mathbf{W}$). The resulting first order condition for a minimum of this weighted sum is:

$$\mathbf{S}_{zx}'\mathbf{W}_n(\mathbf{s}_{zy} - \mathbf{S}_{zx}\hat{\boldsymbol{\beta}}) = \mathbf{0},$$

so that solving for $\hat{\boldsymbol{\beta}}$ yields the GMM estimator (for given $\mathbf{W}_n$),

[4] Observe that $\bar{\mathbf{g}} = \mathbf{0}$ is a system of k equations in $m(= p)$ unknowns.
[5] Note that OLS is IV with $\mathbf{z}_i := \mathbf{x}_i$.

$$\hat{\boldsymbol{\beta}} = \hat{\boldsymbol{\beta}}(\mathbf{W}_n) = (\mathbf{S}'_{zx}\mathbf{W}_n\mathbf{S}_{zx})^{-1}\mathbf{S}'_{zx}\mathbf{W}_n\mathbf{s}_{zy}.$$

Hence, the GMM estimator solves, for a given $\mathbf{W}_n$, either exactly (if $k = m$) or approximately (if $k > m$), the sample orthogonality condition:

$$k = m: \hat{\boldsymbol{\beta}} = \hat{\boldsymbol{\beta}}_{IV} = \mathbf{S}_{zx}^{-1}\mathbf{s}_{zy} = (\mathbf{Z}'\mathbf{X})^{-1}\mathbf{Z}'\mathbf{y}. \tag{2.9}$$
$$k > m: \hat{\boldsymbol{\beta}}(\mathbf{W}_n) = (\mathbf{S}'_{zx}\mathbf{W}_n\mathbf{S}_{zx})^{-1}\mathbf{S}'_{zx}\mathbf{W}_n\mathbf{s}_{zy} = (\mathbf{X}'\mathbf{Z}\ \mathbf{W}_n\mathbf{Z}'\mathbf{X})^{-1}\mathbf{X}'\mathbf{Z}\mathbf{W}_n\mathbf{Z}'\mathbf{y}.$$

Naturally, we wish to choose from the GMM estimators indexed by $\mathbf{W}_n$, one that has the least asymptotic variance. This *efficient* or *optimal* GMM estimator is found by taking the optimal weighting matrix as the Cramèr-Rao lower bound (see 1.11) for the asymptotic variance-covariance matrix for the GMM estimators. The optimal weighting matrix in this lower bound is given by:

$$\hat{\mathbf{W}} = (\mathbf{S}'_{zx}\mathbf{S}^{-1}\mathbf{S}_{zx})^{-1} = (\mathbf{X}'\mathbf{Z}\mathbf{S}^{-1}\mathbf{Z}'\mathbf{X})^{-1}, \tag{2.10}$$

where the covariance matrix $\mathbf{S}$ is estimated by its sample analogue $\hat{\mathbf{S}} = \frac{1}{n}\sum \hat{\varepsilon}_i^2 \mathbf{z}_i\mathbf{z}'_i$.

A more mathematical formulation of the GMM estimators is given in Appendix B.

2.3.1 Example: GMM estimators based on Euler equations

Euler equations are the first order conditions of dynamic optimization problems (often with future rational expectations). GMM treats these first order conditions as moment conditions. In response to the Lucas policy critique, models are now often formulated as rational expectations models with taste and technology parameters,[6] which are believed to be constant over time (see e.g. Hall (1993)). Suppose e.g., following the original example in Hansen and Singleton (1982), that the representative consumer has an expected utility function over current and future consumption and tries to solve the following problem:

$$\max_{c_{t+\tau}} E_t \left\{ \sum_{\tau=0}^{T} (1+\delta)^{-\tau} U(c_{t+\tau}) \right\}$$

subject to the intertemporal budget constraints:

$$c_{t+\tau} + q_{t+\tau} = w_{t+\tau} + (1 + r_{t+\tau})q_{t+\tau-1},$$

where E_t is the mathematical expectation given the information available at time t, δ the (constant) rate of subjective time preference, $c_{t+\tau}$ the per capita (real) consumption in period $t+\tau$, $q_{t+\tau}$ the financial wealth at the end of period $t+\tau$, $w_{t+\tau}$ the wage earnings at period $t+\tau$, and $r_{t+\tau}$ the return on financial wealth (invested in a portfolio of assets during $t+\tau$). Hence, the budget constraint says that individual labor income plus individual asset income should

[6] A conditional expectation is said to be *rational* if its difference with respect to the observation is a white noise error term, i.e. $E_{t-1}(y_t) - y_t = \varepsilon_t$, where $\varepsilon_t \overset{iid}{\sim} (0, \sigma_\varepsilon^2)$.

be spent on consumption or saved in q_{t+T}. Note that this maximization problem is hard to be solved analytically. Nevertheless, it is still possible to estimate the unknown parameters through the first order conditions. It is straightforward to show that solving the above expected utility maximization problem leads to the following *Euler equation* or first order condition:

$$(1+\delta)^{-1}E_t\left\{(1+r_{t+1})\,U'(c_{t+1})\right\} = U'(c_t)$$

where $U'(\cdot)$ is the marginal utility of consumption, or, in other words:

$$\tfrac{1}{1+\delta}E_t\left\{\tfrac{(1+r_{t+1})U'(c_{t+1})}{U'(c_t)} - 1\right\} = 0.$$

To see the intuition behind this claim, suppose that this condition failed to hold. Say, e.g., that the conditional expectation where the sequential marginal utilities are involved were larger than one. Suppose that the representative stockholder saves one more currency unit (dollars, euros, pounds,...) at date t and piles that unit in assets, using the return to boost period $t+1$ consumption. Following this strategy would cause consumption at date t to fall by one currency unit (which reduces the objective function by an amount given by the numerator), while consumption at date $t+1$ would rise by $(1+r_{t+1})$ currency units (increasing the objective function by an amount given in the numerator). If, as assumed, the marginal utility were larger than the future discounted value, then the representative stockholder's objective would be improved under this change. Only when the above restriction is exactly satisfied, the representative stockholder is as well off as possible.

The above condition is essentially a 'conditional' moment restriction, which can be exploited to estimate the unknown parameters if we make some (regularity) assumptions about the utility function U. Since for any function $g(\cdot)$ we have that $E\left\{E\left[x_1|\,x_2\right]g(x_2)\right\} = E\left\{x_1 g(x_2)\right\}$, $E\left\{x_1|\,x_2\right\} = 0$ implies $E\left\{x_1 g(x_2)\right\} = 0$, so that the above 'conditional' moment condition can be transformed into a set of 'unconditional' moment conditions. Suppose, therefore, that $\mathbf{z}_t$ is included in the information set at period t, we can write that:

$$\tfrac{1}{1+\delta}E\left\{\left[\tfrac{(1+r_{t+1})U'(c_{t+1})}{U'(c_t)} - 1\right]\mathbf{z}_t\right\} = 0,$$

so that $\mathbf{z}_t$ can be interpreted as a vector of IVs. Parameterizing the per capita utility function in a risk averse way as in:

$$\begin{aligned} U(c_t) &:= (1-\gamma)^{-1}c_t^{1-\gamma} && \text{for } \gamma > 0 \text{ and } \gamma \neq 1 \\ &= \log\, c_t && \text{for } \gamma = 1, \end{aligned}$$

where γ is the constant relative risk aversion parameter. Then, the above set of unconditional moment restrictions can be rewritten as:

$$\tfrac{1}{1+\delta}E\left\{\left[(1+r_{t+1})\left(\tfrac{c_{t+1}}{c_t}\right)^{-\gamma} - 1\right]\mathbf{z}_t\right\} = 0.$$

This set of unconditional moment restrictions is derived, which can identify the unknown ('taste') parameters δ and γ, given sample observations on $\mathbf{x}_{t+1} := \left(\frac{c_{t+1}}{c_t}, r_{t+1}\right)'$ and $\mathbf{z}_t$. Notice, however, that these unconditional moment restrictions contain conditional expectations, which may be rational expectations. Hence, GMM can be applied using IVs.

2.3.2 Three empirical examples

Example 2.4 *An empirical example based on the gamma pdf*

From the unweighted objective (2.6) in Example 2.3, the objective function for the GMM estimator satisfies:

$$\left\{[\overline{\mathbf{m}}(\gamma, \lambda)]' \, [\mathbf{J}'\mathbf{J}] \, [\overline{\mathbf{m}}(\gamma, \lambda)]\right\},$$

with $\mathbf{J}$ a diagonal matrix with diagonal elements (weights) λ, $\lambda^2, 1, (\gamma - 1)$ respectively. Using the data in Table 2.1 initial weighted LS estimates are found as $\hat{\gamma}_{LS}= 2.0583$ and $\hat{\lambda}_{LS}= 0.065799$ (compare to the ML estimates $\hat{\gamma}_{ML}= 2.4106$ and $\hat{\lambda}_{ML}= 0.07707$). These initial estimates are consistent but not (asymptotically) efficient. Using these estimates the best GMM estimator with (2.10) as optimal weighting matrix can now be derived by solving:

$$\min_{\gamma,\lambda} \; [\overline{\mathbf{m}}(\gamma, \lambda)]' \left[\mathbf{J}'[\mathbf{J}\hat{\mathbf{W}}\mathbf{J}']^{-1}\mathbf{J}\right] [\overline{\mathbf{m}}(\gamma, \lambda)] \, .$$

This minimization yields as 'best' GMM estimators $\tilde{\gamma}_{GMM} = 3.3589\,(0.44967)$ and $\tilde{\lambda}_{GMM} = 0.12449$ (0.0291) with the (asymptotically efficient) standard errors between brackets.

Example 2.5 *Intertemporal asset pricing models*

Virtually all known asset pricing models (including the capital asset pricing model (CAPM)) can be written as:

$$E_t\left\{m_{t+1}\left(1 + r_{j,t+1}\right)\right\} = 1 \quad (j = 1, 2, \ldots, K),$$

with j an index for the assets (stocks, bonds,...) and where m_{t+1} is a stochastic discount factor. Replacing the return of the risky asset in Example 2.3.1 by the return on a portfolio of assets (with optimally chosen weights), the first order Euler conditions lead to the conditional and unconditional moment restrictions in that example.

Hansen and Singleton (1982) estimated these unconditional moment equations using real aggregate US consumption divided by the US population as a measure of per capita consumption c_t. For $r_{1,t}$, they used the inflation-adjusted return that an investor would earn if one currency unit was invested in each stock listed at the New York Stock Exchange, while $r_{2,t}$ was a value-weighted inflation-adjusted return corresponding to the return an investor would earn if the investor owned the entire stock of each company listed. They estimated as IVs a constant term, lagged consumption growth rates, and lagged rates of return and for the explanatory variables:

$$\mathbf{x}_t := \left(1, \frac{c_t}{c_{t-1}}, \frac{c_{t-1}}{c_{t-2}}, \ldots, \frac{c_{t-l+1}}{c_{t-l}}, r_{1,t}, r_{1,t-1}, \ldots, r_{1,t-l+1}, r_{2,t}, r_{2,t-1}, \ldots, r_{2,t-l+1}\right)'$$

and they estimated the two-asset pricing model in a linear way, using l lags of variables.

Cochrane (1996) has investigated a range of asset pricing models using GMM as estimation procedure. He uses 10 portfolios of the New York Stock Exchange sorted by market value and the return on a riskless asset, approximated by the 3-month Treasury Bill return.

Example 2.6 *An example for the impact of financial asset returns on the macroeconomy.*

A slightly alternative formulation of first order Euler conditions with the related GMM estimator for Japanese and US data is provided in Valckx and Plasmans (1998):

$$E_t\left[m_t(\boldsymbol{\theta})R_t\right] = 1 \text{ or } E_t\left[m_t(\boldsymbol{\theta})R_t - 1\right] = 0$$

with $m_t(\cdot)$ denoting the intertemporal marginal rate of substitution and R the gross asset returns (as endogenous variable) at time t. The parameter vector $\boldsymbol{\theta}$, containing among others, the discount rate and the relative risk aversion parameter (see Examples 2.3.1 and 2.5) is to be estimated. As before, this expression is called a *moment condition* since it shows that the statistic $m_t(\boldsymbol{\theta})R_t$ will converge in probability to some constant, here one.

GMM exploits the idea that the disturbances in $E_t\left[\varepsilon_t(\boldsymbol{\theta})\right] = 0$ are derived from a rational expectations model, and are uncorrelated with any information available at time t. Denoting $\mathbf{z}_t$ as a vector of IVs that are orthogonal to the disturbances ε_t at time t (see (2.7)),[7] lagged values of m and R are taken as IVs. The number of lags is usually chosen relatively small, since an increase of it leads to an increase in bias (see Tauchen (1986)).

2.3.3 Exercises

Consider the data on schooling available at http://www.ua.ac.be/joseph.plasmans, containing observations from the 2003 household survey of Encuesta Permanente de Hogares for Argentina. Estimate the returns to education: $log(w_i) =$

[7] Given initial parameter estimates using an arbitrary initial weighting matrix, e.g. $\mathbf{I}_T$, and given a corresponding initial GMM estimator $\hat{\boldsymbol{\theta}}$, another weighting matrix is obtained, either as a robust White estimator of the asymptotic variance-covariance matrix (see White (1984)) or as a Newey-West heteroskedasticity and autocorrelation consistent covariance estimator (see Newey and West (1987) and Section B.1 in Appendix B):

$$\hat{S}_T^{White} = \frac{1}{T}\left[\frac{1}{T}\sum_{t=1}^{T} \mathbf{z}_t\mathbf{z}_t'\hat{\varepsilon}_t^2\right]$$

$$\text{or } \hat{S}_T^{Newey-West} = \frac{1}{T}\left\{\frac{1}{T}\sum_{t=1}^{T} \mathbf{z}_t\mathbf{z}_t'\hat{\varepsilon}_t^2 + \frac{1}{T}\sum_{l=1}^{L}\omega(l)\sum_{t=1}^{T}\hat{\varepsilon}_t\hat{\varepsilon}_{t-l}\left[\mathbf{z}_t\mathbf{z}_{t-l}' + \mathbf{z}_{t-l}\mathbf{z}_t'\right]\right\},$$

where $\omega(l) := 1 - l/(L+1)$ and L the maximum lag length of the autocorrelations in $\hat{\varepsilon}_t^2$.

$\beta_1 + \beta_2 S_i + \beta_3 E_i + \beta_4 E_i^2 + \gamma' \mathbf{x}_i + \varepsilon_i$, where $log(w_i)$ is the log of individual earnings, S_i the years of schooling, E_i the years of (working) experience measured as $age - S_i - 6$, $\mathbf{x}_i$ contains the control dummy variables: gender, household_head, smsa (living a metropolitan area).

i) Make descriptive statistics of the wage variables. Create the variable E_i and make a scatterplot of $log(w_i)$ versus E_i.

ii) Estimate the earnings function and interpret.

iii) Why is there an endogeneity problem with S_i? As an instrumental variable use the dummy variable '*nearc*' indicating whether the person lives near a college or not. Is this a suitable instrument?

iv) Estimate the earnings equation using the IV estimator. Do the coefficients have the expected sign? Which variables are significant?

v) Compare the standard errors of the estimates of β_2 with OLS and with IVs and comment on the goodness of fit.

2.4 Some concluding remarks

Hall (1993) points out two advantages of GMM. First, it is computationally convenient, and second, it avoids potential bias due to misspecification of the distribution of the sample observations (e.g. of m and R in the empirical example 2.6).

Moreover, the GMM estimator encompasses (many/almost all) other estimators. Furthermore, it can allow for heteroskedasticity and autocorrelation and it can estimate parameters even if ML estimators do not exist.

It has to be stressed that the validity of the instruments in $\{\mathbf{z}_t\}$, $t = 1, 2, \ldots, T$ is beyond doubt if the model leads to a conditional moment restriction as put forward in Subsection 2.3.1 and if they are in the conditioning set. Hence, if at period t a representative agent maximizes expected utility given all publicly available information, then any variable that is observed to the agent at period t provides a valid instrument. An explicit test on the validity of IV (GMM) moment restrictions is given in the next chapter.

Chapter 3

Testing in Linear and Nonlinear Models

In general we can distinguish *nested* and *nonnested* (testing) models. Tests are called nonnested if there are separate or non-encompassing families of hypotheses. In other words, tests are nested when the null hypothesis can be written as a special case of a general hypothesis and are nonnested when the null cannot be encompassed by a general hypothesis. Hence, nonnested models may arise if two alternative economic theories lead to different models for the same phenomenon. Examples of nested model tests are the *Lagrange Multiplier* test (if the null hypothesis is simpler than the alternative hypothesis so that the null is used), the *Wald* test (if the reverse is true) and the *Likelihood Ratio* test (where both hypotheses are utilized).

An example of a nonnested model test is the *Cox* test which was first formulated by Pesaran (1974) for single equation linear regression models, both with autocorrelated and nonautocorrelated disturbances. Pesaran and Deaton (1978) extended this analysis to nonlinear regression models.

Before applying nested and nonnested tests, consider the loglikelihood function:

$$L(\boldsymbol{\theta}; \mathbf{y}) := \ln \ell(\boldsymbol{\theta}; \mathbf{y}) = \ln f(\mathbf{y}, \boldsymbol{\theta}).$$

A necessary condition for the maximization of this loglikelihood function is $\left.\frac{\partial L(\boldsymbol{\theta};\mathbf{y})}{\partial \boldsymbol{\theta}}\right|_{\boldsymbol{\theta}=\hat{\boldsymbol{\theta}}} =: \mathbf{s}\left(\hat{\boldsymbol{\theta}}; \mathbf{y}\right) = \mathbf{0}$, where $\mathbf{s}\left(\hat{\boldsymbol{\theta}}; \mathbf{y}\right)$ is known as the *score vector*. An estimate of the (asymptotic) variance-covariance matrix of the corresponding maximum likelihood estimator is given by the estimated Cramèr-Rao lower bound for $(\mathbf{J}(\boldsymbol{\theta}))^{-1}$ where the estimated information matrix $\mathbf{J}(\hat{\boldsymbol{\theta}})$ is obtained from the evaluation of the partial derivatives in the covariance matrix (1.11) at $\boldsymbol{\theta} = \hat{\boldsymbol{\theta}}$. Remark that this covariance matrix only involves first order partial derivatives. Since for complicated nonlinear functions this inverse information matrix is not easily computed, an alternative for the information matrix could be considered. As indicated before, such an alternative is the inverse Hessian (matrix) of the loglikelihood function, or $\hat{\mathbf{J}}(\hat{\boldsymbol{\theta}}) = -\left.\frac{\partial^2 L(\boldsymbol{\theta};\mathbf{y})}{\partial\boldsymbol{\theta}\partial\boldsymbol{\theta}}\right|_{\boldsymbol{\theta}=\hat{\boldsymbol{\theta}}}$. Another alternative is the *BHHH* estimator proposed by Berndt, Hall, Hall, and Hausman (1974), which is based on the result that the matrix of the expected negative second order partial derivatives is the variance-covariance matrix of the vector of the first order partial derivatives, or from Section 1.1 the following estimate may be considered when the expectations are difficult to compute $\hat{\hat{\mathbf{J}}}(\hat{\boldsymbol{\theta}}) = ((\mathbf{Z}(\hat{\boldsymbol{\theta}}))'\mathbf{Z}(\hat{\boldsymbol{\theta}}))$ with the $n \times p$ Jacobian matrix $\mathbf{Z}(\hat{\boldsymbol{\theta}}) := \left.\frac{\partial \mathbf{f}}{\partial \boldsymbol{\theta}}\right|_{\boldsymbol{\theta}=\hat{\boldsymbol{\theta}}}$, so that only the regression parameters are involved here ($\boldsymbol{\theta} := \boldsymbol{\beta}$).This estimator is extremely convenient in most cases, because it only needs first order partial derivatives, which are also needed for deriving ML estimators, and it is always nonnegative definite.

3.1 Nested model tests

First consider the simple case of a null hypothesis that the parameters are equal, or $H_0 : \boldsymbol{\theta} = \boldsymbol{\theta}^0$, where $\boldsymbol{\theta}^0$ is the value of the unknown $\boldsymbol{\theta}$ under the null. This hypothesis will be generalized later to the nonlinear hypothesis testing $H_0 : \mathbf{q}(\boldsymbol{\theta}) = \mathbf{q}(\boldsymbol{\theta}^0) = \mathbf{0}$, where $\mathbf{q}(\cdot)$ is generally a nonlinear and differentiable function of $\boldsymbol{\theta}$ ($\mathbb{R}^p \overset{\mathrm{q}}{\Rightarrow} \mathbb{R}^k$).

3.1.1 Lagrange multiplier test

The *Lagrange Multiplier* (*LM*) *test* for $H_0 : \boldsymbol{\theta} = \boldsymbol{\theta}^0$ is derived from a constrained maximization problem. Consider therefore the Lagrange function:

$$\mathcal{L} := L(\boldsymbol{\theta};\mathbf{y}) - \boldsymbol{\lambda}'\left(\boldsymbol{\theta} - \boldsymbol{\theta}^0\right),$$

with $\boldsymbol{\lambda}$ the Lagrange parameter measuring the shadow price of the constraint (if this price is low, the constraint must be rejected as being inconsistent with the data). Then the first order condition is:

$$\tfrac{\partial L}{\partial \boldsymbol{\theta}} = \boldsymbol{\lambda}^0 \quad \text{and } \boldsymbol{\theta} = \boldsymbol{\theta}^0 \quad \text{or} \quad \boldsymbol{\lambda}^0 = \mathbf{s}\left(\boldsymbol{\theta}^0;\mathbf{y}\right),$$

i.e. the Lagrange parameter is equal to the score vector at the null, so that the LM test is (sometimes) also called the *score test.* This test tests the null of the restricted score vector being equal to zero. More specifically, using the central limit theorem, the score vector is asymptotically normally distributed if the null is true, or $\mathbf{s}\left(\boldsymbol{\theta}^0;\mathbf{y}\right) \overset{as}{\sim} N(\mathbf{0}, \mathbf{avar}\left(\mathbf{s}(\boldsymbol{\theta}^0;\mathbf{y})\right))$, where $\mathbf{avar}\left(\mathbf{s}(\boldsymbol{\theta}^0;\mathbf{y})\right)$ is the asymptotic variance-covariance matrix of the score vector. Then the normalized score vector at the null is asymptotically standard normally distributed, or $\left(\widehat{\mathbf{avar}}\left(\mathbf{s}(\boldsymbol{\theta}^0;\mathbf{y})\right)\right)^{-1/2}\mathbf{s}\left(\boldsymbol{\theta}^0;\mathbf{y}\right) = (\mathbf{J}(\boldsymbol{\theta}^0))^{-1/2}\mathbf{s}\left(\boldsymbol{\theta}^0;\mathbf{y}\right) \overset{as}{\sim} N(0, \mathbf{I}_p)$, and since the sum of mutually independent squared standard normally distributed random variables is χ^2-distributed, the following LM statistic satisfies:

$$\xi_{LM} = \left(\mathbf{s}\left(\boldsymbol{\theta}^0;\mathbf{y}\right)\right)'\left(\mathbf{J}\left(\boldsymbol{\theta}^0\right)\right)^{-1}\mathbf{s}\left(\boldsymbol{\theta}^0;\mathbf{y}\right) \overset{as}{\sim} \chi^2_p, \tag{3.1}$$

so that the LM test statistic is asymptotically χ^2-distributed with p degrees of freedom (being the number of the components in $\boldsymbol{\theta}^0$).

3.1.2 Wald test

Consider the unconstrained ML estimator $\hat{\theta}$, which is according to the central limit theorem and under certain regularity conditions under the null asymptotically normally distributed as $\hat{\boldsymbol{\theta}} \overset{as}{\sim} N(\boldsymbol{\theta}, \boldsymbol{\Sigma}_{\hat{\boldsymbol{\theta}}})$ so that it can be normalized as $(\boldsymbol{\Sigma}_{\hat{\boldsymbol{\theta}}})^{-1/2}(\hat{\boldsymbol{\theta}} - \boldsymbol{\theta}) \overset{as}{\sim} N(0, \mathbf{I}_p)$. Testing under the null that $\boldsymbol{\theta} = \boldsymbol{\theta}^0$, the *Wald test statistic* results then as:

$$\xi_W = \left(\hat{\boldsymbol{\theta}} - \boldsymbol{\theta}^0\right)' \boldsymbol{\Sigma}_{\hat{\boldsymbol{\theta}}}^{-1} \left(\hat{\boldsymbol{\theta}} - \boldsymbol{\theta}^0\right) = \left(\hat{\boldsymbol{\theta}} - \boldsymbol{\theta}^0\right)' \mathbf{J}\left(\hat{\boldsymbol{\theta}}\right)\left(\hat{\boldsymbol{\theta}} - \boldsymbol{\theta}^0\right) \overset{as}{\sim} \chi_p^2, \tag{3.2}$$

or with $\hat{\mathbf{J}}(\hat{\boldsymbol{\theta}})$ or $\hat{\hat{\mathbf{J}}}(\hat{\boldsymbol{\theta}})$ instead of $\mathbf{J}(\hat{\boldsymbol{\theta}})$.

3.1.3 Likelihood ratio test

The *Likelihood Ratio (LR) test statistic* is based on the difference between the maximum of the loglikelihood function under H_0 and the maximum of the loglikelihood function under H_A. The idea behind this test is to test whether the difference between the restricted and the unrestricted maxima of the loglikelihood function is small. Hence, the LR test statistic is:

$$\xi_{LR} := -2\ln\lambda = -2\ln\frac{\ell(\boldsymbol{\theta}^0;\mathbf{y})}{\ell\left(\hat{\boldsymbol{\theta}};\mathbf{y}\right)} = -2\left(L(\boldsymbol{\theta}^0;\mathbf{y}) - L\left(\hat{\boldsymbol{\theta}};\mathbf{y}\right)\right) \overset{as}{\sim} \chi_p^2. \tag{3.3}$$

Theorem 3.1 *The Likelihood Ratio test of the linear regression problem; consider the LR* $\lambda := \frac{\ell(\tilde{\boldsymbol{\beta}},\tilde{\sigma}_\varepsilon^2;\mathbf{y},\mathbf{X})}{\ell(\hat{\boldsymbol{\beta}},\hat{\sigma}_\varepsilon^2;\mathbf{y},\mathbf{X})}$, *where* $0 \leqq \lambda \leqq 1$ *and* $\tilde{\boldsymbol{\beta}}$ *and* $\tilde{\sigma}_\varepsilon^2$ *are the restricted ML estimators (under the null), then:* $-2\ln\lambda \overset{as}{\sim} \chi_p^2$.

Proof For known σ_ε^2 and a linear model:

$$\begin{aligned}
-2\ln\frac{\ell\left(\widetilde{\boldsymbol{\beta}};\mathbf{y},\mathbf{X}\right)}{\ell\left(\widehat{\boldsymbol{\beta}};\mathbf{y},\mathbf{X}\right)} &= 2\ln\ell\left(\widehat{\boldsymbol{\beta}};\mathbf{y},\mathbf{X}\right) - 2\ln\ell\left(\widetilde{\boldsymbol{\beta}};\mathbf{y},\mathbf{X}\right) \\
&= 2\left[-\frac{n}{2}\ln 2\pi - \frac{n}{2}\ln\sigma_\varepsilon^2 - \frac{1}{2\sigma_\varepsilon^2}(\mathbf{y}-\mathbf{X}\widehat{\boldsymbol{\beta}})'(\mathbf{y}-\mathbf{X}\widehat{\boldsymbol{\beta}})\right] \\
&\quad -2\left[-\frac{n}{2}\ln 2\pi - \frac{n}{2}\ln\sigma_\varepsilon^2 - \frac{1}{2\sigma_\varepsilon^2}(y-\mathbf{X}\widetilde{\boldsymbol{\beta}})'(\mathbf{y}-\mathbf{X}\widetilde{\boldsymbol{\beta}})\right] \\
&= \frac{1}{\sigma_\varepsilon^2}[S\left(\widetilde{\boldsymbol{\beta}}\right) - S(\widehat{\boldsymbol{\beta}})]
\end{aligned}$$

with $S\left(\hat{\boldsymbol{\beta}}\right) := \left(\mathbf{y} - \mathbf{X}\hat{\boldsymbol{\beta}}\right)'\left(\mathbf{y} - \mathbf{X}\hat{\boldsymbol{\beta}}\right)$.

Then the LR test statistic satisfies (using $\operatorname*{plim}_{n\to\infty} \mathbf{X}'\boldsymbol{\varepsilon} = \mathbf{0}$):

$$-2\ln\lambda = \tfrac{1}{\sigma_\varepsilon^2}\left[\left(\mathbf{y}-\mathbf{X}\widetilde{\boldsymbol{\beta}}\right)'\left(\mathbf{y}-\mathbf{X}\widetilde{\boldsymbol{\beta}}\right) - \left(\mathbf{y}-\mathbf{X}\hat{\boldsymbol{\beta}}\right)'\left(\mathbf{y}-\mathbf{X}\hat{\boldsymbol{\beta}}\right)\right]$$

$$
\begin{aligned}
&= \tfrac{1}{\sigma_\varepsilon^2}\left(\mathbf{y}'\mathbf{y} - 2\widetilde{\boldsymbol{\beta}}'\mathbf{X}'\mathbf{y} + \widetilde{\boldsymbol{\beta}}'\mathbf{X}'\mathbf{X}\widetilde{\boldsymbol{\beta}} - \mathbf{y}'\mathbf{y} + 2\hat{\boldsymbol{\beta}}'\mathbf{X}'\mathbf{y} - \hat{\boldsymbol{\beta}}'\mathbf{X}'\mathbf{X}\hat{\boldsymbol{\beta}}\right) \\
&\overset{as}{\approx} \tfrac{1}{\sigma_\varepsilon^2}\left(-2\widetilde{\boldsymbol{\beta}}'\mathbf{X}'\mathbf{X}\boldsymbol{\beta} + 2\hat{\boldsymbol{\beta}}'\mathbf{X}'\mathbf{X}\boldsymbol{\beta} + \widetilde{\boldsymbol{\beta}}'\mathbf{X}'\mathbf{X}\widetilde{\boldsymbol{\beta}} - \hat{\boldsymbol{\beta}}'\mathbf{X}'\mathbf{X}\hat{\boldsymbol{\beta}}\right) \\
&\overset{as}{\approx} \tfrac{1}{\sigma_\varepsilon^2}\left(-2\widetilde{\boldsymbol{\beta}}'\mathbf{X}'\mathbf{X}\hat{\boldsymbol{\beta}} + \hat{\boldsymbol{\beta}}'\mathbf{X}'\mathbf{X}\hat{\boldsymbol{\beta}} + \widetilde{\boldsymbol{\beta}}'\mathbf{X}'\mathbf{X}\widetilde{\boldsymbol{\beta}}\right) = \tfrac{1}{\sigma_\varepsilon^2}\left(\hat{\boldsymbol{\beta}} - \widetilde{\boldsymbol{\beta}}\right)'\mathbf{X}'\mathbf{X}\left(\hat{\boldsymbol{\beta}} - \widetilde{\boldsymbol{\beta}}\right) \\
&\overset{as}{\approx} \left(\hat{\boldsymbol{\beta}} - \widetilde{\boldsymbol{\beta}}\right)'\left(\sigma_\varepsilon^2\left(\mathbf{X}'\mathbf{X}\right)^{-1}\right)^{-1}\left(\hat{\boldsymbol{\beta}} - \widetilde{\boldsymbol{\beta}}\right),
\end{aligned}
$$

which is the Wald test statistic since under the null $\widetilde{\boldsymbol{\beta}} \sim N(0, \sigma_\varepsilon^2 (\mathbf{X}'\mathbf{X})^{-1})$, or $-2\ln\lambda \overset{as}{\sim} \chi_p^2$.■

All three test statistics measure the distance between H_0 and H_A: the Wald test statistic in terms of $\left(\boldsymbol{\theta}^0 - \hat{\boldsymbol{\theta}}\right)$, the LR test statistic in terms of $L\left(\boldsymbol{\theta}^0\right) - L\left(\hat{\boldsymbol{\theta}}\right)$ and the LM test statistic in terms of $\mathbf{s}(\boldsymbol{\theta}^0; \mathbf{y})$.

Theorem 3.2 *If L is given by a quadratic function, i.e. if the underlying model is linear in the parameters, then the three nested test statistics are equivalent.*

Proof (for known variance and simple H_0):

Rewriting the loglikelihood function L for the linear model $\mathbf{y} = \mathbf{f}(\mathbf{X}; \boldsymbol{\beta}) + \boldsymbol{\varepsilon} = \mathbf{X}\boldsymbol{\beta} + \boldsymbol{\varepsilon}$ as a quadratic function L:

$$
\begin{aligned}
L\left(\boldsymbol{\beta}^0; \mathbf{y}\right) &= -\frac{n}{2}\ln 2\pi - \frac{n}{2}\ln\sigma_\varepsilon^2 - \frac{1}{2\sigma_\varepsilon^2}\textstyle\sum_i \varepsilon_i^2 \\
&= b - \frac{1}{2\sigma_\varepsilon^2}\left(\boldsymbol{\beta}^0 - \widehat{\boldsymbol{\beta}}\right)'\mathbf{A}\left(\boldsymbol{\beta}^0 - \widehat{\boldsymbol{\beta}}\right),
\end{aligned}
$$

then the three nested test statistics satisfy:

$$
\begin{aligned}
\xi_W = (\widehat{\boldsymbol{\beta}}^0 - \widehat{\boldsymbol{\beta}})'\mathbf{A}(\widehat{\boldsymbol{\beta}}^0 - \widehat{\boldsymbol{\beta}}) = \xi_{LR} \\
\xi_{LM} = (\mathbf{s}(\widehat{\boldsymbol{\beta}}^0; \mathbf{y}))'\mathbf{A}^{-1}\mathbf{s}(\widehat{\boldsymbol{\beta}}^0; \mathbf{y}) = (\widehat{\boldsymbol{\beta}}^0 - \widehat{\boldsymbol{\beta}})'\mathbf{A}(\widehat{\boldsymbol{\beta}}^0 - \widehat{\boldsymbol{\beta}}),
\end{aligned}
$$

so that the three nested test statistics are equal to each other.■

If the nonlinear model $\mathbf{y} = \mathbf{f}(\mathbf{X}; \boldsymbol{\beta}) + \boldsymbol{\varepsilon}$ can be approximated well by a linear function, this equivalence result will largely stay valid but if the nonlinear regression function $\mathbf{f}(\mathbf{X}; \boldsymbol{\beta})$ cannot conveniently be linearized the equivalence property will break down and the three nested test statistics will differ from each other.

3.1.4 Confidence intervals and hypotheses tests

Consider the nonlinear regression model $y_i = f(\mathbf{x}_i, \boldsymbol{\beta}) + \varepsilon_i$ with the error terms following an *iid* normal distribution $\varepsilon_i \overset{iid}{\sim} N\left(0, \sigma_\varepsilon^2\right)$ or, applying the central limit theorem, under certain regularity conditions the NLLS and NLML estimators are asymptotically normally distributed.

In the sequel, equality of individual coefficients, parameter vectors and general nonlinear restrictions are tested.

Test of individual coefficients

If we test under the null $\beta_i = \beta_i^0$, the familiar test statistic is $z_i := \frac{\hat{\beta}_i - \beta_i^0}{\sqrt{\hat{\sigma}_\varepsilon^2 z^{ii}}} \overset{as}{\sim} N(0,1)$ and the corresponding $100(1-\alpha)$ per cent (asymptotic) confidence interval for β_i^0 satisfies from Section 1.1:

$$\left(\hat{\beta}_i - z_{\alpha/2}\sqrt{\hat{\sigma}^2_{\varepsilon,\substack{NLLS\\NLML}} z^{ii}}, \hat{\beta}_i + z_{\alpha/2}\sqrt{\hat{\sigma}^2_{\varepsilon,\substack{NLLS\\NLML}} z^{ii}} \right),$$

with z^{ii} the ith diagonal element of $[(\mathbf{Z}(\boldsymbol{\beta}))'(\mathbf{Z}(\boldsymbol{\beta})]^{-1}$ and $\hat{\sigma}^2_{\varepsilon,NLLS}$ the NLLS residual variance and $\hat{\sigma}^2_{\varepsilon,NLML}$ being the NLML residual variance.

Test of parameter vector

a. In a linear model testing under the null whether $\boldsymbol{\beta} = \boldsymbol{\beta}^0 = \mathbf{0}$, leads to the familiar test statistic:

$$\frac{(S(\boldsymbol{\beta}) - S(\hat{\boldsymbol{\beta}}))/p}{S(\hat{\boldsymbol{\beta}})/n-p} \overset{(as)}{\sim} F_{p,T-p}(\alpha),$$

with $\hat{\boldsymbol{\beta}}$ being the LLS (or ML) estimator.

b. In a nonlinear model there are three alternatives:

i. $\frac{\left(S(\boldsymbol{\beta}) - S\left(\hat{\boldsymbol{\beta}}_{\substack{NLLS\\NLML}}\right)\right)/p}{S\left(\hat{\boldsymbol{\beta}}_{\substack{NLLS\\NLML}}\right)/n-p} = c \overset{as}{\sim} F_{p,T-p}(\alpha)$, where $\hat{\boldsymbol{\beta}}_{NLLS}$ is the NLLS estimator of $\boldsymbol{\beta}$ and $\hat{\boldsymbol{\beta}}_{NLML}$ is the NLML estimator of $\boldsymbol{\beta}$;

ii. a second order Taylor approximation of $S(\boldsymbol{\beta})$ around $\boldsymbol{\beta} := \hat{\boldsymbol{\beta}}_{\substack{NLLS\\NLML}}$ such that:

$$\frac{\left(\hat{\boldsymbol{\beta}}_{\substack{NLLS\\NLML}} - \boldsymbol{\beta}\right)' \left.\frac{\partial^2 S(\boldsymbol{\beta})}{\partial\boldsymbol{\beta}\partial\boldsymbol{\beta}'}\right|_{\boldsymbol{\beta}=\hat{\boldsymbol{\beta}}_{\substack{NLLS\\NLML}}} \left(\hat{\boldsymbol{\beta}}_{\substack{NLLS\\NLML}} - \boldsymbol{\beta}\right)/2p}{S\left(\hat{\boldsymbol{\beta}}_{\substack{NLLS\\NLML}}\right)/n-p} \overset{as}{\sim} F_{p,n-p}(\alpha)$$

iii. based on the pseudo-linear model (1.2):

$$\frac{\left(\hat{\boldsymbol{\beta}}_{\substack{NLLS\\NLML}} - \boldsymbol{\beta}\right)' (\mathbf{Z}(\boldsymbol{\beta}^0))'\mathbf{Z}(\boldsymbol{\beta}^0) \left(\hat{\boldsymbol{\beta}}_{\substack{NLLS\\NLML}} - \boldsymbol{\beta}\right)/p}{\left(\overline{\mathbf{y}}(\boldsymbol{\beta}^0)\text{-}\mathbf{Z}(\boldsymbol{\beta}^0)\hat{\boldsymbol{\beta}}_{\substack{NLLS\\NLML}}\right)' \left(\overline{\mathbf{y}}(\boldsymbol{\beta}^0)\text{-}\mathbf{Z}(\boldsymbol{\beta}^0)\hat{\boldsymbol{\beta}}_{\substack{NLLS\\NLML}}\right)/n-p} \overset{as}{\sim} F_{p,n-p}(\alpha).$$

where $\mathbf{Z}(\boldsymbol{\beta}^0)$ is the $n \times p$ Jacobian matrix evaluated at the null, $\left.\frac{\partial \mathbf{f}}{\partial \boldsymbol{\beta}}\right|_{\boldsymbol{\beta}=\boldsymbol{\beta}^0}$.

Examples of nested hypothesis tests (conjugate confidence intervals)

1. $H_0 : \boldsymbol{\beta}^0 = \boldsymbol{\alpha}$ and $H_1 : \boldsymbol{\beta}^0 \neq \boldsymbol{\alpha}$. are tested with a conventional F-test statistic:

$$\frac{(S(\boldsymbol{\alpha}) - S(\hat{\boldsymbol{\beta}}_{ML}))/p}{S(\hat{\boldsymbol{\beta}}_{ML})/n-p} = c \overset{as}{\sim} F_{p,n-p}(\boldsymbol{\alpha})$$

2. $H_0 : \mathbf{q}(\boldsymbol{\beta}^0) = \mathbf{0}$ and $H_1 : \mathbf{q}(\boldsymbol{\beta}^0) \neq \mathbf{0}$.

For example, we can introduce in the CES production function $\ln Q_i = \beta_0 + \beta_1 \ln \left[\beta_2 L_i^{\beta_3} + (1-\beta_2) K_i^{\beta_3}\right] + \varepsilon_i$ of Example 1.3 the null hypothesis that there are constant returns to scale, or

$$H_0 : \beta_3^0 = \tfrac{1}{\beta_1^0} \text{ or } q(\boldsymbol{\beta}^0) = \beta_1^0 - \tfrac{1}{\beta_3^0} = 0.$$

Note that in general $\mathbf{q}(\boldsymbol{\beta})$ is not arbitrary, e.g. for $y_i = \beta_0 + \beta_1 x_i^{\beta_2} + e_i$ the $H_0 : \beta_2 = 0$ leads to singularity (not defined).

3.1.5 Recapitulating and extending nested hypotheses

First, assume linear constraints so that under the null $H_0 : \mathbf{R}\boldsymbol{\beta} = \mathbf{b}$ and $H_1 : \mathbf{R}\boldsymbol{\beta} \neq \mathbf{b}$. This leads to a constrained ML estimator for both linear and nonlinear models. However, in general we have a set of k nonlinear and differentiable restrictions $\mathbf{q}(\boldsymbol{\beta}) = \mathbf{0}$.

To begin with, assume that you have to determine the variance-covariance matrix of $\hat{\boldsymbol{\alpha}}$ in $\hat{\mathbf{q}}\left(\hat{\boldsymbol{\beta}}\right) = \hat{\boldsymbol{\alpha}}$, where $\hat{\boldsymbol{\beta}}$ is the underlying LLS estimator of $\boldsymbol{\beta}$ and its variance-covariance matrix is known from this linear LLS estimation.

Lemma 3.1 *If $\hat{\boldsymbol{\beta}}$ and $\hat{\boldsymbol{\Sigma}}_{\hat{\boldsymbol{\beta}}}$ are known, the variance-covariance matrix of a (non)linear transformation of $\hat{\boldsymbol{\beta}}$ is given by $\boldsymbol{\Pi}(\hat{\boldsymbol{\beta}})\hat{\boldsymbol{\Sigma}}_{\hat{\boldsymbol{\beta}}}(\boldsymbol{\Pi}(\hat{\boldsymbol{\beta}}))'$, where $\boldsymbol{\Pi}(\hat{\boldsymbol{\beta}})$ is the $k \times p$ Jacobian matrix of the restrictions.*

Proof Asymptotically, under the null that $\mathbf{q}(\boldsymbol{\beta}^0) = \boldsymbol{\alpha}^0 = \mathbf{0}$, i.e. by applying the central limit theorem under the null $\mathbf{q}\left(\hat{\boldsymbol{\beta}}\right) \overset{as}{\sim} N(0, \Sigma_{\mathbf{q}(\hat{\boldsymbol{\beta}})})$, where $\Sigma_{\mathbf{q}(\hat{\boldsymbol{\beta}})}$ can be estimated through a linearization of $\mathbf{q}\left(\hat{\boldsymbol{\beta}}\right)$ at the null:

$$\mathbf{q}\left(\hat{\boldsymbol{\beta}}\right) = \hat{\boldsymbol{\alpha}} = \boldsymbol{\alpha}^0 + \left(\left.\frac{\partial \mathbf{q}(\hat{\boldsymbol{\beta}})}{\partial \hat{\boldsymbol{\beta}}}\right|_{\hat{\boldsymbol{\beta}}=\boldsymbol{\beta}^0} \right) \left(\hat{\boldsymbol{\beta}} - \boldsymbol{\beta}^0\right) + ...$$

so that $\hat{\boldsymbol{\alpha}} - \boldsymbol{\alpha}^0 \simeq \boldsymbol{\Pi}(\boldsymbol{\beta}^0)\left(\hat{\boldsymbol{\beta}} - \boldsymbol{\beta}^0\right)$, where $\boldsymbol{\Pi}$ is the $k \times p$ Jacobian matrix of $\mathbf{q}$ or $\boldsymbol{\Pi}(\boldsymbol{\beta}^0) := \left(\left.\frac{\partial \mathbf{q}(\hat{\boldsymbol{\beta}})}{\partial \hat{\boldsymbol{\beta}}}\right|_{\hat{\boldsymbol{\beta}}=\boldsymbol{\beta}^0} \right)$, so that, taking account of the asymptotic properties of the ML estimator, a consistent estimate of the asymptotic variance-covariance matrix of $\hat{\boldsymbol{\alpha}} = \mathbf{q}\left(\hat{\boldsymbol{\beta}}\right)$ is given by $\hat{\boldsymbol{\Sigma}}_{\hat{\boldsymbol{\alpha}}} \simeq \Pi(\hat{\boldsymbol{\beta}})\hat{\boldsymbol{\Sigma}}_{\hat{\boldsymbol{\beta}}}(\Pi(\hat{\boldsymbol{\beta}}))'$.■

The three nested test statistics (3.1), (3.2), and (3.3) can, taking account of Lemma 3.1, be extended to the case of general restrictions $\mathbf{q}(\boldsymbol{\theta}) = \mathbf{0}$, with $\mathbf{q}(\cdot)$ a continuously differentiable function of $\boldsymbol{\theta}$:

a. an LM test statistic (see (3.1)), where the Lagrange function becomes now $\mathcal{L} = L(\boldsymbol{\theta}; \mathbf{y}) + \boldsymbol{\lambda}'\mathbf{q}(\boldsymbol{\theta})$ so that the necessary maximization condition under the null satisfies $\frac{\partial \mathcal{L}}{\partial \boldsymbol{\theta}}\Big|_{\boldsymbol{\theta}=\tilde{\boldsymbol{\theta}}} = \frac{\partial L}{\partial \boldsymbol{\theta}}\Big|_{\boldsymbol{\theta}=\tilde{\boldsymbol{\theta}}} + \left(\frac{\partial \mathbf{q}}{\partial \boldsymbol{\theta}}\right)' \tilde{\boldsymbol{\lambda}} = \mathbf{0}$ and the restricted score vector results as $\mathbf{s}(\tilde{\boldsymbol{\theta}};\mathbf{y}) = \frac{\partial L}{\partial \boldsymbol{\theta}}\Big|_{\boldsymbol{\theta}=\tilde{\boldsymbol{\theta}}} = -\left.\frac{\partial \mathbf{q}}{\partial \boldsymbol{\theta}}\right|'_{\boldsymbol{\theta}=\tilde{\boldsymbol{\theta}}} \tilde{\boldsymbol{\lambda}}$, where $\tilde{\boldsymbol{\theta}}$ is the restricted estimator of the unknown parameter (under the null); hence, the LM (also called RAO) test statistic is now given under the null from (3.1) by:

$$\xi_{LM} = \tilde{\lambda}' \left.\frac{\partial \mathbf{q}}{\partial \boldsymbol{\theta}}\right|_{\boldsymbol{\theta}=\tilde{\boldsymbol{\theta}}} \tilde{\boldsymbol{\Sigma}}_{\tilde{\boldsymbol{\theta}}} \left.\frac{\partial \mathbf{q}}{\partial \boldsymbol{\theta}}\right|'_{\boldsymbol{\theta}=\tilde{\boldsymbol{\theta}}} \tilde{\boldsymbol{\lambda}} \overset{as}{\sim} \chi^2_k, \tag{3.4}$$

where the restricted estimated covariance matrix $\tilde{\boldsymbol{\Sigma}}_{\tilde{\boldsymbol{\theta}}}$ satisfies $\tilde{\boldsymbol{\Sigma}}_{\tilde{\boldsymbol{\theta}}} = (\mathbf{J}\left(\tilde{\boldsymbol{\theta}}\right))^{-1}$;

b. testing the null $\mathbf{q}(\boldsymbol{\theta}) = 0$, the Wald test statistic satisfies from (3.2) and Lemma 3.1:

$$\begin{aligned} \xi_W &= \left(\mathbf{q}\left(\hat{\boldsymbol{\theta}}\right)\right)' \left(\hat{\boldsymbol{\Sigma}}_{q(\hat{\boldsymbol{\theta}})}\right)^{-1} \mathbf{q}\left(\hat{\boldsymbol{\theta}}\right) \\ &\simeq \left(\mathbf{q}\left(\hat{\boldsymbol{\theta}}\right)\right)' \left(\left.\frac{\partial \mathbf{q}}{\partial \boldsymbol{\theta}}\right|_{\boldsymbol{\theta}=\hat{\boldsymbol{\theta}}} \hat{\boldsymbol{\Sigma}}_{\hat{\boldsymbol{\theta}}} \left(\left.\frac{\partial \mathbf{q}}{\partial \boldsymbol{\theta}}\right|_{\boldsymbol{\theta}=\hat{\boldsymbol{\theta}}}\right)'\right)^{-1} \mathbf{q}\left(\hat{\boldsymbol{\theta}}\right) \overset{as}{\sim} \chi^2_k; \end{aligned} \tag{3.5}$$

c. the LR test statistic (see (3.3)):

$$\xi_{LR} = -2\left(\ln \ell\left(\tilde{\beta}, \tilde{\boldsymbol{\Sigma}}; \mathbf{y}, \mathbf{x}\right)\right) - \ln \ell\left(\hat{\beta}, \hat{\boldsymbol{\Sigma}}; \mathbf{y}, \mathbf{x}\right) \overset{as}{\sim} \chi^2_k. \tag{3.6}$$

where $\tilde{\boldsymbol{\beta}}, \tilde{\boldsymbol{\Sigma}}$ and $\hat{\boldsymbol{\beta}}, \hat{\boldsymbol{\Sigma}}$ are the restricted and unrestricted estimators of $\boldsymbol{\beta}$ and $\boldsymbol{\Sigma}$, respectively.

Note that for testing the null $H_0 : \mathbf{q}(\boldsymbol{\theta}^0) := \boldsymbol{\theta} - \boldsymbol{\theta}^0 = 0$, the test statistics (3.1), (3.2), and (3.3) can directly be derived from (3.4), (3.5), and (3.6), respectively. This is also the case for testing individual coefficients.

For example, the Wald test statistic (3.5) implies for a single restriction and a univariate parameter:
$\xi_W = (\widehat{\theta} - \theta^0)(avar(\widehat{\theta} - \theta^0))^{-1}(\widehat{\theta} - \theta^0) = \frac{(\widehat{\theta}-\theta^0)^2}{avar(\widehat{\theta})} = z^2$, which is asymptotically χ^2-distributed with one degree of freedom and which is also the asymptotic distribution of the square of the standard normal test statistic $z = \frac{|\widehat{\theta}-\theta^0|}{\sqrt{\widehat{avar}(\widehat{\theta})}}$.

3.1.6 Examples of nested model tests

In this section, several examples of nested model tests will be presented.

Example 1. LM, Wald, and LR tests based on a gamma pdf. The gamma pdf used here is slightly different from that in Example 2.3

$$f\left(x_i, y_i; \beta, \rho\right) = \frac{(\beta+x_i)^{-\rho}}{\Gamma(\rho)} y_i^{\rho-1} e^{\frac{-y_i}{\beta+x_i}},$$

with x_i education and y_i income as in Table 2.1 of Example 2.3 with $\Gamma(\rho) := \int_0^\infty t^{\rho-1}e^{-t}dt = (\rho-1)!, \rho \in \mathbb{N}$. Test whether the null hypothesis H_0: $\rho = 1$ is valid.

1. Calculate ML estimates of the parameters.

The joint loglikelihood function for n *iid* observations is:

$\ln \ell(\beta,\rho) = -\rho \sum_i \ln(\beta + x_i) - n \ln \Gamma(\rho) - \sum_i \frac{y_i}{\beta+x_i} + (\rho-1)\sum_i \ln y_i$ or

$\frac{\partial \ln \ell}{\partial \beta} = -\rho \sum_i \frac{1}{\beta+x_i} + \sum_i \frac{y_i}{(\beta+x_i)^2} = 0$ and

$\frac{\partial \ln \ell}{\partial \rho} = -\sum_i \ln(\beta+x_i) - \frac{n \ln \Gamma'(\rho)}{\Gamma(\rho)} + \sum_i \ln y_i = 0$. Given the data, the unrestricted ML estimates are $\hat{\beta} = -4.719$ and $\hat{\rho} = 3.151$, where we need the derivative of the restriction $q(\theta) = \rho - 1 = 0$ for the Wald test, or $\frac{\partial q}{\partial \rho} = 1$. The restricted estimators satisfy $\tilde{\rho} = 1$ and $\tilde{\beta} = 15.603$.

2. For the Wald test we need the information matrix $\mathbf{J}(\boldsymbol{\theta})$ for $\boldsymbol{\theta} := (\beta, \rho)'$

Direct evaluation of the second order partial derivatives of the loglikelihood as in Section 1.1.2 yields:

$\frac{\partial^2 L}{\partial \beta^2} = \rho \sum_i \frac{1}{(\beta+x_i)^2} - 2\sum_i \frac{y_i}{(\beta+x_i)^3}$

$\frac{\partial^2 L}{\partial \rho^2} = -n \frac{(\Gamma(\rho)\,\Gamma''(\rho) - \Gamma'(\rho)\,\Gamma'(\rho))}{(\Gamma(\rho))^2}$

$\frac{\partial^2 L}{\partial \beta \partial \rho} = -\sum_i \frac{1}{\beta+x_i} = \frac{\partial^2 \ln L}{\partial \rho\, \partial \beta}$

3. The test statistics

The LM test statistic from (3.1) is $\xi_{LM} = \left(\frac{\partial L(\tilde{\boldsymbol{\theta}})}{\partial \tilde{\boldsymbol{\theta}}}\right)' (\mathbf{J}(\tilde{\boldsymbol{\theta}}))^{-1} \left(\frac{\partial L(\tilde{\boldsymbol{\theta}})}{\partial \tilde{\boldsymbol{\theta}}}\right)$, or

$$\xi_{LM} = \begin{bmatrix} 0.000 & 7.914 \end{bmatrix} \begin{bmatrix} 0.02166 & 0.6689 \\ 0.6689 & 32.834 \end{bmatrix}^{-1} \begin{bmatrix} 0.000 \\ 7.914 \end{bmatrix} = 5.120 > \chi_1^2(0.05) = 3.842.$$

The Wald test statistic from (3.5) is:

$$\xi_W = \begin{bmatrix} 0 \\ 3.151 - 1 \end{bmatrix} \left[\begin{bmatrix} 0 \\ 1 \end{bmatrix}' \begin{bmatrix} 5.495 & -1.652 \\ -1.652 & 0.6309 \end{bmatrix} \begin{bmatrix} 0 & 1 \end{bmatrix}' \right]^{-1} \begin{bmatrix} 0 & 3.151 - 1 \end{bmatrix}$$

$$= (3.151 - 1)\,[1 \cdot 0.6309 \cdot 1]^{-1}\,(3.151 - 1) = 7.3384 > \chi_1^2 = 3.842$$

The LR test statistic from (3.6) is:

$$\xi_{LR} = -2\left(\ln L\left(\tilde{\boldsymbol{\theta}}\right) - \ln L(\boldsymbol{\theta})\right) = -2\,(-88.436 - (-82.916)) = 11.04 > \chi_1^2(0.05) = 3.842$$

Hence, all three statistics reject the null at the 5 per cent level, but in a different intensity!

Example 2. Test on a subset of parameters According to Breusch and Pagan (1980), we assume a partitioning of the parameter vector $\boldsymbol{\theta}$, where the possible restriction is only analyzed for the first subvector $\boldsymbol{\theta}^{(1)}$, or $\boldsymbol{\theta} := ((\boldsymbol{\theta}^{(1)})', (\boldsymbol{\theta}^{(2)})')'$ with $H_0 : \boldsymbol{\theta}^{(1)} = \boldsymbol{\theta}_{10}$, $H_1 : \boldsymbol{\theta}^{(1)} \neq \boldsymbol{\theta}_{10}$, and $\hat{\boldsymbol{\theta}}$ and $\tilde{\boldsymbol{\theta}}$ are the unrestricted and restricted estimators, respectively. In other words, $\begin{pmatrix} \mathbf{I}_{p_1} & \vdots & \mathbf{0} \end{pmatrix} \binom{\boldsymbol{\theta}^{(1)}}{\boldsymbol{\theta}^{(2)}} = \binom{\boldsymbol{\theta}_{10}}{\mathbf{0}}$ and partitioning the information matrix and the score vector accordingly:

$$\mathbf{J}_{\boldsymbol{\theta}} := \begin{pmatrix} \mathbf{J}_{11} & \mathbf{J}_{12} \\ \mathbf{J}_{21} & \mathbf{J}_{22} \end{pmatrix}, \quad \frac{\partial L}{\partial \boldsymbol{\theta}}\Big|_{\boldsymbol{\theta}=\tilde{\boldsymbol{\theta}}} = \begin{pmatrix} \frac{\partial L}{\partial \boldsymbol{\theta}^{(1)}}\Big|_{\boldsymbol{\theta}^{(1)}=\tilde{\boldsymbol{\theta}}^{(1)}} \\ \mathbf{0} \end{pmatrix},$$

the LM or Rao score test satisfies from (3.1) :

$$\xi_{LM} = \left.\frac{\partial L}{\partial \boldsymbol{\theta}^{(1)}}\right|'_{\boldsymbol{\theta}^{(1)}=\widetilde{\boldsymbol{\theta}}^{(1)}} \left(\widetilde{\mathbf{J}}_{11} - \widetilde{\mathbf{J}}_{12}\widetilde{\mathbf{J}}_{22}^{-1}\widetilde{\mathbf{J}}_{21}\right)^{-1} \left.\frac{\partial L}{\partial \boldsymbol{\theta}^{(1)}}\right|_{\boldsymbol{\theta}^{(1)}=\widetilde{\boldsymbol{\theta}}^{(1)}} . \tag{3.7}$$

Example 3. Test on an arbitrary degree of autocorrelation Assume the linear autocorrelation model: $\mathbf{y} = \mathbf{X}\boldsymbol{\beta} + \mathbf{u}$, $\mathbf{u} = \rho_j \mathbf{u}_{-j} + \boldsymbol{\varepsilon}$ with $\boldsymbol{\varepsilon} \sim N_T\left(\mathbf{0}, \sigma_\varepsilon^2 \mathbf{I}_T\right)$, so that by substituting we get a nonlinear model:

$$\mathbf{y} = \mathbf{X}\boldsymbol{\beta} + \rho_j \mathbf{u}_{-j} + \boldsymbol{\varepsilon} = \rho_j \mathbf{y}_j + \left(\mathbf{X} - \rho_j \mathbf{X}_{-j}\right)\boldsymbol{\beta} + \boldsymbol{\varepsilon} = \mathbf{f}\left(\mathbf{X}, \boldsymbol{\theta}\right) + \boldsymbol{\varepsilon} ,$$

with $\boldsymbol{\theta} := \left(\boldsymbol{\beta}', \rho_j\right)'$.

Test now $H_0 : \rho_j = 0$ using the LM test. Since $\boldsymbol{\varepsilon} \sim N_T\left(0, \sigma_\varepsilon^2 \mathbf{I}_T\right)$:

$$L\left(\boldsymbol{\theta}\right) = -\tfrac{T}{2}\ln 2\pi - \tfrac{T}{2}\ln \sigma_\varepsilon^2 - \tfrac{1}{2\sigma_\varepsilon^2}\left(\mathbf{y} - \mathbf{f}\left(\mathbf{X}, \boldsymbol{\theta}\right)\right)'\left(\mathbf{y} - \mathbf{f}\left(\mathbf{X}, \boldsymbol{\theta}\right)\right),$$

with necessary maximization conditions under H_0:

$$\left.\tfrac{\partial L}{\partial \boldsymbol{\theta}}\right|_{\boldsymbol{\theta}=\widetilde{\boldsymbol{\theta}}} = \tfrac{1}{\widetilde{\sigma}_\varepsilon^2}\left(\mathbf{Z}(\widetilde{\boldsymbol{\theta}})\right)'\left(\mathbf{y} - \mathbf{f}\left(\mathbf{X}, \widetilde{\boldsymbol{\theta}}\right)\right) = \frac{\left(\mathbf{Z}(\widetilde{\boldsymbol{\theta}})\right)'\widetilde{\boldsymbol{\varepsilon}}}{\widetilde{\sigma}_\varepsilon^2}$$

with the $T \times p$ Jacobian matrix defined under the null as $\mathbf{Z}\left(\widetilde{\boldsymbol{\theta}}\right) := \left.\frac{\partial \mathbf{f}}{\partial \boldsymbol{\theta}}\right|_{\boldsymbol{\theta}=\widetilde{\boldsymbol{\theta}}}$, $\widetilde{\boldsymbol{\varepsilon}} := \hat{\mathbf{u}}$ (OLS) and , $\mathbf{Z}\left(\widetilde{\boldsymbol{\theta}}\right) = \left(\mathbf{y}_{-j} - \mathbf{X}_{-j}\widetilde{\boldsymbol{\beta}} = \widetilde{\mathbf{u}}_{-j} \vdots \mathbf{X} - \widetilde{\rho}_j \mathbf{X}_{-j}\right) = \left(\hat{\mathbf{u}}_{-j}, \mathbf{X}\right)$.

Define now the estimated autocorrelation coefficient as $\hat{r}_j := \left(\hat{\mathbf{u}}'\hat{\mathbf{u}}\right)^{-1}\hat{\mathbf{u}}'_{-j}\hat{\mathbf{u}}$, which is in fact the OLS estimator of $\frac{1}{r_j}$ in $\hat{\mathbf{u}}_{-j} = \frac{1}{r_j}\hat{\mathbf{u}} - \frac{\boldsymbol{\varepsilon}}{r_j}$! Since $\widetilde{\sigma}_\varepsilon^2 = \hat{\sigma}_\varepsilon^2 = \frac{\hat{\mathbf{u}}'\hat{\mathbf{u}}}{T}$ and $\hat{r}_j = \frac{\hat{\mathbf{u}}'_{-j}\hat{\mathbf{u}}}{T\hat{\sigma}_\varepsilon^2}$, the score vector satisfies:

$$\left.\frac{\partial L}{\partial \rho_j}\right|_{\boldsymbol{\theta}=\widetilde{\boldsymbol{\theta}}} = \frac{\hat{\mathbf{u}}'_{-j}\hat{\mathbf{u}}}{\widetilde{\sigma}_\varepsilon^2} = \frac{T\hat{\sigma}_\varepsilon^2 \hat{r}_j}{\hat{\sigma}_\varepsilon^2} = T\hat{r}_j. \tag{3.8}$$

Now the LM test statistic is derived from (3.7) and (3.8) as:

$$\xi_{LM} = T^2\hat{r}_j^2\left[\hat{\sigma}_\varepsilon^{-2}(\hat{\mathbf{u}}'_{-j}\hat{\mathbf{u}}_{-j} - \hat{\mathbf{u}}'_{-j}\mathbf{X}_{-j}(\mathbf{X}'_{-j}\mathbf{X}_{-j})^{-1}\mathbf{X}'_{-j}\hat{\mathbf{u}}'_{-j})\right]^{-1}$$

$$= T\hat{r}_j^2\left[T^{-1}\hat{\sigma}_\varepsilon^{-2}\hat{\mathbf{u}}'_{-j}\hat{\mathbf{u}}_{-j} - \hat{\sigma}_\varepsilon^{-2}(T^{-1}\hat{\mathbf{u}}_{-j}\mathbf{X}_{-j})(T^{-1}\mathbf{X}'_{-j}\mathbf{X}_{-j})^{-1}(T^{-1}\mathbf{X}'_{-j}\hat{\mathbf{u}}_{-j})\right]^{-1},$$

$$\text{or } \xi_{LM} \to T\hat{r}_j^2 \overset{as}{\sim} \chi_1^2, \text{ i.e. } \hat{r}_j\sqrt{T} \overset{as}{\sim} N\left(0,1\right). \tag{3.9}$$

Example 4. Testing for a structural break The detection of a structural break in a linear model is analyzed for a known break point by Chow (1960), Rao (1952) and Kullback and Rosenblatt (1957) leading to a conventional F-test statistic and for an unknown break point e.g. by Quandt (1960) and Brown *et al.* (1975) leading to the Cumulated Sum (or CUSUM) and Moving Sum (or MOSUM) tests.

a. Test with known break points: Chow test with different parameter vectors and equal variances In the economic literature the earliest tests for structural breaks are the tests in Chow (1960) for a single break. The model is a linear regression model with m variables and two regimes with T observations divided over two subperiods, with known break point. Hence, we have the model:

$$\begin{pmatrix} \mathbf{y}_1 \\ \mathbf{y}_2 \end{pmatrix} = \begin{pmatrix} \mathbf{X}_1 & 0 \\ 0 & \mathbf{X}_2 \end{pmatrix} \begin{pmatrix} \boldsymbol{\beta}_1 \\ \boldsymbol{\beta}_2 \end{pmatrix} + \begin{pmatrix} \varepsilon_1 \\ \varepsilon_2 \end{pmatrix},$$

leading to the unrestricted OLS estimator:

$$\hat{\boldsymbol{\beta}} = (\mathbf{X}'\mathbf{X})^{-1}\mathbf{X}'\mathbf{y} = \begin{pmatrix} \mathbf{X}_1'\mathbf{X}_1 & \mathbf{0} \\ \mathbf{0} & \mathbf{X}_2'\mathbf{X}_2 \end{pmatrix} \begin{pmatrix} \mathbf{X}_1'\mathbf{y}_1 \\ \mathbf{X}_2'\mathbf{y}_2 \end{pmatrix} = \begin{pmatrix} \hat{\boldsymbol{\beta}}_1 \\ \hat{\boldsymbol{\beta}}_2 \end{pmatrix}$$

with corresponding residual vector $\hat{\varepsilon} := \binom{\hat{\varepsilon}_1}{\hat{\varepsilon}_2}$.

The restricted coefficient vector can be obtained either indirectly by introducing the restrictions $H_o : \boldsymbol{\beta}_1 = \boldsymbol{\beta}_2$ or $\mathbf{R}\boldsymbol{\beta} = \mathbf{q}$, with $\mathbf{R} := [\mathbf{I}_p, -\mathbf{I}_p], \boldsymbol{\beta} := (\boldsymbol{\beta}_1', \boldsymbol{\beta}_2')'$ and $\mathbf{q} := \mathbf{0}$, or directly by estimating

$$\begin{pmatrix} \mathbf{y}_1 \\ \mathbf{y}_2 \end{pmatrix} = \begin{pmatrix} \mathbf{X}_1 \\ \mathbf{X}_2 \end{pmatrix} \gamma + \begin{pmatrix} \varepsilon_1 \\ \varepsilon_2 \end{pmatrix}.$$

From the resulting restricted residual sum of squares $\tilde{\varepsilon}'\tilde{\varepsilon}$ the appropriate test statistic is the LM test statistic (under the null there is no change of structure and therefore no different parameters), which implies the traditional F-test statistic to test the constancy of the parameter vectors:

$$F_{p,T-2p} = \frac{\left(\tilde{\varepsilon}'\tilde{\varepsilon} - \hat{\varepsilon}_1'\hat{\varepsilon}_1 - \hat{\varepsilon}_2'\hat{\varepsilon}_2\right)/p}{(\hat{\varepsilon}_1'\hat{\varepsilon}_1 + \hat{\varepsilon}_2'\hat{\varepsilon}_2)/(T-2p)}. \tag{3.10}$$

If there are *insufficient observations* to run the two separate regressions, a similar procedure is possible with a specific treatment of the shorter time period.

For example, consider an empirical example involving the years (1973 and 1979) of the oil crises. To study whether the market never actually fundamentally changed or that it only changed temporarily you might consider the same (Chow) test as before, but now only single out the four years 1974, 1975, 1980, and 1981 for special treatment. Fisher (1970) has shown that in such a situation, we can proceed as follows:
i. Estimate the regression with the complete database and compute the restricted residual sum of squares $\tilde{\varepsilon}'\tilde{\varepsilon}$;
ii. Estimate the regression using the longer subperiod, with T_1 observations and compute the unrestricted residual sum of squares $\hat{\varepsilon}'\hat{\varepsilon}$;
iii. Compute the F-statistic as:

$$F_{T_2,T_1-p} = \frac{\left(\tilde{\varepsilon}'\tilde{\varepsilon} - \hat{\varepsilon}'\hat{\varepsilon}\right)/T_2}{\hat{\varepsilon}'\hat{\varepsilon}/(T_1-p)}. \tag{3.11}$$

Comparing (3.11) with (3.10), we observe that the numerator's degrees of freedom in the former test statistic is T_2 and not p (because only $T_2 < p$ coefficients are needed to obtain a perfect fit); otherwise p would be the appropiate number of degrees of freedom.

b. Structural break test with different parameter vectors and unequal variances An important assumption made in using the above Chow test is that the (theoretical) residual variance is the same in both (or all) regressions. With unequal variances for the two subperiods, σ_1^2 and σ_2^2, we have to test under the null whether the residual variances are equal. From (3.5) and $\mathbf{q}(\hat{\boldsymbol{\beta}}) = [\mathbf{I}_p, -\mathbf{I}_p]\binom{\boldsymbol{\beta}_1}{\boldsymbol{\beta}_2} = \binom{0}{0}$, this leads to the Wald test statistic:

$$\xi_W = \left(\hat{\boldsymbol{\beta}}_1 - \hat{\boldsymbol{\beta}}_2\right)' \widehat{\Sigma}_{\hat{\boldsymbol{\beta}}_1 - \hat{\boldsymbol{\beta}}_2} \left(\hat{\boldsymbol{\beta}}_1 - \hat{\boldsymbol{\beta}}_2\right) \overset{as}{\sim} \chi_p^2, \tag{3.12}$$

where $\widehat{\Sigma}_{\hat{\boldsymbol{\beta}}_1 - \hat{\boldsymbol{\beta}}_2} = \widehat{\Sigma}_{\hat{\boldsymbol{\beta}}_1} + \widehat{\Sigma}_{\hat{\boldsymbol{\beta}}_2} - 2\mathbf{cov}(\hat{\boldsymbol{\beta}}_1, \hat{\boldsymbol{\beta}}_2)$ with $\mathbf{cov}(\hat{\boldsymbol{\beta}}_1, \hat{\boldsymbol{\beta}}_2) = \mathbf{0}$ if the two samples are assumed to be independent.

c. Structural break test with unknown break points Quandt (1960) discusses testing the null of constant coefficients against the alternative where a structural change has occurred at some unknown time point and the error variance is also allowed to change. He considered a switching regression where the observations are assumed to be generated by two distinct regression regimes: for some subset of observations $y_i = \mathbf{x}_i'\boldsymbol{\beta}_i + \varepsilon_{1i}$ $(i \in I)$ and for the complementary subset, $y_j = \mathbf{x}_j'\boldsymbol{\beta}_j + \varepsilon_{2j}$ $(j \in J)$. The essence of this formulation is the fact that all first observations up to the (unknown) break point T^* come from one of the regimes, and after the break point, all observations come from the other regime. In order to test the null that there is no change in regimes against the alternative of change in regimes, the appropriate test statistic is an LR test statistic:

$$\xi_{LR} = \frac{1}{2}T^* \log \hat{\sigma}_1^2 + \frac{1}{2}(T - T^*) \log \hat{\sigma}_2^2 - \frac{1}{2}T \log \hat{\sigma}_\varepsilon^2. \tag{3.13}$$

A rough procedure to determine the break point T^* is to take that value for which ξ_{LR} attains its minimum. Quandt (1960) showed empirically that (3.13) for finite samples yields a poor approximation of the (asymptotic) χ^2 distribution.

Brown *et al.* (1975) suggest the CUSUM test based on recursive residuals; this test is for general alternatives, including the case of a single break. The CUSUM test involves considering a plot of the statistic $W_m = \frac{1}{\hat{\sigma}} \sum_{t=p+1}^{m} w_t$ with $m = p+1, \ldots, T$ and w_t the recursive residual. The aim of this test is to detect systematic movements of coefficients.

Example 5. Validity tests of moment restrictions As we saw in Sections 2.2 and 2.3, we can define various types of moment restrictions; these restrictions can be tested by conditional moment tests, e.g. the moment conditions (2.2) can directly be tested from their empirical versions (2.3), with the restricted estimator $\tilde{\boldsymbol{\theta}}$ instead of $\widehat{\boldsymbol{\theta}}$, by an LM test, testing the nullity of the involved moment condition, which under normality boils down to a conventional t-test statistic. More specifically for IVs in Subsection 2.2.2, we have as empirical moment conditions $\frac{1}{n}(\mathbf{Z}'\hat{\varepsilon}) = \mathbf{0}$. Considering Theorem 3.2, this LM test statistic can also be formulated as a Wald test statistic.

Validity test of the IV restrictions Given a consistent estimate of the asymptoptic covariance matrix of $\frac{1}{n}\mathbf{Z}'(\mathbf{y} - \mathbf{X}\hat{\boldsymbol{\beta}})$, the appropiate minimand of (2.5) in the linear IV case in Section 2.2.2, with more instruments than parameters ($k > p$),[1] is a Wald test statistic for the validity of the k moment restrictions. Then testing whether the null that these moment restrictions are valid is made by the Wald test statistic from (2.5) and $\widehat{\boldsymbol{\Sigma}}$ is an estimate of $E(\varepsilon'\varepsilon)$:[2]

$$\xi_W^{IV} = [\mathbf{Z}'(\mathbf{y} - \mathbf{X}\hat{\boldsymbol{\beta}})]'[\mathbf{Z}'\widehat{\boldsymbol{\Sigma}}\mathbf{Z}]^{-1}[\mathbf{Z}'(\mathbf{y} - \mathbf{X}\hat{\boldsymbol{\beta}})] \overset{as}{\sim} \chi^2_{k-p}. \tag{3.14}$$

Note that if the error terms in the underlying linear model under the null are homoskedastic and serially uncorrelated, the Wald test statistic simply boils down to the LM test statistic $\xi_{LM}^{GMM} = nR^2$, where the uncentered R^2 is involved from the regression of the residuals $\hat{\boldsymbol{\varepsilon}} := \mathbf{y} - \mathbf{X}\hat{\boldsymbol{\beta}}$ on the IV matrix $\mathbf{Z}$: $\hat{\boldsymbol{\varepsilon}} = \mathbf{Z}\boldsymbol{\theta} + \boldsymbol{\eta}$ (check this; see also (3.9)). Since $\frac{1}{n}(\mathbf{Z}'\,\hat{\varepsilon}) \approx 0$, the IVs should be suspected to be approximately orthogonal to the IV residuals (the OLS residuals are ortogonal to the explanatory variables by construction). In this case the above uncentered R^2 will be low and we will accept the null that the restrictions (2.4) are valid.

Hansen's test of overidentifying restrictions Referring to the GMM objective in (2.8), where the number of restrictions is larger than the number of parameters ($k > p$), *Hansen*'s (1982) *test of overidentifying restrictions* boils down to the Wald test statistic for the efficient GMM estimator of an arbitrary nonlinear model, or for the weighting matrices approaching the optimal weighting matrix $\hat{\mathbf{W}}$ as in (2.10) we have:

$$\xi_W^{GMM} = n\overline{\mathbf{g}}'\mathbf{W}_n\overline{\mathbf{g}} \overset{as}{\sim} \chi^2_{k-p} \text{ for } \mathbf{W}_n \to \hat{\mathbf{W}} \tag{3.15}$$

where $\overline{\mathbf{g}} = \frac{1}{n}\sum_{i=1}^n \mathbf{g}_i = \frac{1}{n}\sum_{i=1}^n \mathbf{z}_i(y_i - \mathbf{x}_i'\hat{\boldsymbol{\beta}})$.

If Hansen's test in (3.15) leads to rejection, it indicates that $E(\mathbf{g}_i) \neq 0$, but this does not tell us which of these k moment conditions in (2.7) are likely to be violated. In fact any subset of p of these conditions 'hold' for the (non)linear parameter vector $\boldsymbol{\beta}$ it defines.

[1]Hence, the IV estimator is a GMM estimator then.

[2]The appropiate Wald test statistic for nonlinear IV models as NL2SLS in Section B.3 of Appendix B is directly derived from the quadratic objective in (B.2);

$$\xi_W^{IV} = \frac{1}{\hat{\sigma}^2}\left(\mathbf{y} - \mathbf{f}(\mathbf{X};\hat{\boldsymbol{\theta}})\right)'\mathbf{Z}(\mathbf{Z}'\mathbf{Z})^{-1}\mathbf{Z}'\left(\mathbf{y} - \mathbf{f}(\mathbf{X};\hat{\boldsymbol{\theta}})\right) \overset{as}{\sim} \chi^2_{k-p}.$$

However, if a subset of k_1 moment conditions hold beyond any doubt and $k_1 \geqq p$,[3] the remaining $k - k_1$ conditions can be tested. Let $\xi_{1,W}^{GMM}$ and ξ_W^{GMM} be the ξ_W-statistics associated with the k_1 and k moment conditions, respectively. Then, if all k conditions hold, $\xi_W^{GMM} - \xi_{1,W}^{GMM} \overset{as}{\sim} \chi^2_{k-k_1}$.

Example 6. White test on heteroskedasticity Since the null is absence of heteroskedasticity (i.e. homoskedasticity) or $H_o : \sigma_i^2 = \sigma^2, \forall i$, an appropriate test statistic should test under the null whether $E(\varepsilon_i^2 \mathbf{x}_i' \mathbf{x}_i) = \sigma^2 E(\mathbf{x}_i' \mathbf{x}_i)$. Since under the null the best estimation is OLS, this test is of the type of an LM test. This LM test statistic is computed on the basis of a regression of the squared estimated OLS error terms on all possible cross products, that is e.g. for the model $y_i = \beta_o + \beta_1 x_i + \beta_2 z_i + \varepsilon_i$, the test statistic is based on the regression: $\hat{\varepsilon}_i^2 = \alpha_o + \alpha_1 x_i + \alpha_2 x_i^2 + \alpha_3 z_i + \alpha_4 z_i^2 + \alpha_5 x_i z_i + v_i$, with the LM test statistic under the null $\xi_{LM} = nR^2 \overset{as}{\sim} \chi^2_{k-1}$, where k is the number of regressors (excluding the constant term) in the regression of $\hat{\varepsilon}_i^2$ ($k = 5$ in this example) and R^2 is the multiple determination coefficient of this regression (see also Example 5).

Example 7. Hausman test on endogeneity We would like to test the endogeneity of a variable x_i being for example total expenditure in the model:

$$y_i = \beta_o + \beta_1 x_i + \beta_2 \log x_i + \beta_3 \log^2 x_i + \varepsilon_i.$$

The Hausman (1978) test compares the GLS estimator $\hat{\boldsymbol{\beta}}_{GLS}$ (corrected for heteroskedasticity) with the IV estimator $\hat{\boldsymbol{\beta}}_{IV}$ (also corrected for heteroskedasticity). The instruments for this example are e.g. income, logarithmic income, and squared logarithmic income, respectively.

The test is as follows:

H_o: total expenditure is exogenous if and only if $\hat{\boldsymbol{\beta}}_{GLS}$ and $\hat{\boldsymbol{\beta}}_{IV}$ are both consistent, i.e. $\operatorname*{plim}_{n\to\infty} \hat{\boldsymbol{\beta}}_{GLS} = \operatorname*{plim}_{n\to\infty} \hat{\boldsymbol{\beta}}_{IV} = \boldsymbol{\beta}$;

H_1: total expenditure is endogenous if and only if $\hat{\boldsymbol{\beta}}_{IV}$ is consistent and $\hat{\boldsymbol{\beta}}_{GLS}$ is not.

The Hausman test is an example of the Wald test because, under the null, it follows from the central limit theorem that, under certain regularity conditions, asymptotically

$$\hat{\boldsymbol{\beta}}_{IV} - \hat{\boldsymbol{\beta}}_{GLS} \overset{as}{\sim} N(\mathbf{0}, \mathbf{avar}(\hat{\boldsymbol{\beta}}_{IV} - \hat{\boldsymbol{\beta}}_{GLS})), \text{ where } \mathbf{avar}(\hat{\boldsymbol{\beta}}_{IV} - \hat{\boldsymbol{\beta}}_{GLS})$$

is the asymptotic variance-covariance matrix of the difference vector $\hat{\boldsymbol{\beta}}_{IV} - \hat{\boldsymbol{\beta}}_{GLS}$. Hence, from the definition of the Wald test statistic in (3.2), the (Hausman) test statistic under the null for this test statistic is:

$$\xi_H = \left(\hat{\boldsymbol{\beta}}_{IV} - \hat{\boldsymbol{\beta}}_{GLS}\right)' \left(\widehat{\mathbf{avar}}(\hat{\boldsymbol{\beta}}_{IV} - \hat{\boldsymbol{\beta}}_{GLS})\right)^{-1} \left(\hat{\boldsymbol{\beta}}_{IV} - \hat{\boldsymbol{\beta}}_{GLS}\right) \overset{as}{\sim} \chi^2_k, \quad (3.16)$$

[3] The ξ_W^{GMM} test can never test all k moment conditions because we need (at least) p conditions to identify the parameters.

where k is the number of regressors in the regression for y_i. Hausman (1978) works out the inverted asymptotic variance-covariance matrix of the difference between the two estimators (the GLS and the IV estimators), where he selects an asymptotically more efficient estimator. In particular:

$$\mathbf{avar}\left(\hat{\boldsymbol{\beta}}_{IV} - \hat{\boldsymbol{\beta}}_{GLS}\right) = \mathbf{avar}(\hat{\boldsymbol{\beta}}_{IV}) + \mathbf{avar}(\hat{\boldsymbol{\beta}}_{GLS}) - 2\mathbf{acov}(\hat{\boldsymbol{\beta}}_{IV}, \hat{\boldsymbol{\beta}}_{GLS}),$$

but Hausman's (1978) essential result is that the (asymptotic) covariance between an asymptotically efficient estimator and its difference from an asymptotically inefficient estimator is zero, which implies that under the null that GLS is the asymptotically efficient estimator (although both estimators are consistent, the GLS estimator is asymptotically efficient because of the Cramèr-Rao lower bound property):

$$\mathbf{acov}\left[\left(\hat{\boldsymbol{\beta}}_{IV} - \hat{\boldsymbol{\beta}}_{GLS}\right), \hat{\boldsymbol{\beta}}_{GLS}\right] = \mathbf{acov}(\hat{\boldsymbol{\beta}}_{IV}, \hat{\boldsymbol{\beta}}_{GLS}) - \mathbf{avar}(\hat{\boldsymbol{\beta}}_{GLS}) = \mathbf{0},$$

or that $\mathbf{acov}(\hat{\boldsymbol{\beta}}_{IV}, \hat{\boldsymbol{\beta}}_{GLS}) = \mathbf{avar}(\hat{\boldsymbol{\beta}}_{GLS})$, such that:

$$\mathbf{avar}\left(\hat{\boldsymbol{\beta}}_{IV} - \hat{\boldsymbol{\beta}}_{GLS}\right) = \mathbf{avar}(\hat{\boldsymbol{\beta}}_{IV}) - \mathbf{avar}(\hat{\boldsymbol{\beta}}_{GLS}). \tag{3.17}$$

This result is generally proven by the following lemma.

Lemma 3.2 *Let $\hat{\boldsymbol{\beta}}$ be an estimator of $\boldsymbol{\beta}$ that is consistent and asymptotically efficient when H_0 is true, but inconsistent when H_0 is false. Let $\tilde{\boldsymbol{\beta}}$ be an estimator that is consistent under both null and alternative hypotheses. Under H_0:* $\mathbf{avar}(\tilde{\boldsymbol{\beta}} - \hat{\boldsymbol{\beta}}) = \mathbf{avar}(\tilde{\boldsymbol{\beta}}) - \mathbf{avar}(\hat{\boldsymbol{\beta}})$.

Proof. Since both $\tilde{\boldsymbol{\beta}}$ and $\hat{\boldsymbol{\beta}}$ are consistent under H_0, so is $\lambda\tilde{\boldsymbol{\beta}}+(1-\lambda)\hat{\boldsymbol{\beta}}$, for any $\lambda \in [0, 1]$. Asymptotic efficiency of $\hat{\boldsymbol{\beta}}$ means that this composite estimator has maximum asymptotic efficiency when $\lambda = 0$. Its asymptotic variance-covariance matrix is
$\mathbf{avar}(\lambda\tilde{\boldsymbol{\beta}}+(1-\lambda)\hat{\boldsymbol{\beta}}) = \lambda^2\mathbf{avar}(\tilde{\boldsymbol{\beta}})+(1-\lambda)^2\mathbf{avar}(\hat{\boldsymbol{\beta}}) + 2\lambda(1-\lambda)\mathbf{acov}(\tilde{\boldsymbol{\beta}}, \hat{\boldsymbol{\beta}})$
and being a function of one variable we obtain:
$\frac{d}{d\lambda}\mathbf{avar}(\lambda\tilde{\boldsymbol{\beta}}+(1-\lambda)\hat{\boldsymbol{\beta}}) = 2\lambda\mathbf{avar}(\tilde{\boldsymbol{\beta}}) - 2(1-\lambda)\mathbf{avar}(\hat{\boldsymbol{\beta}}) + 2(1-2\lambda)\mathbf{acov}(\tilde{\boldsymbol{\beta}}, \hat{\boldsymbol{\beta}})$.
Minimizing $\mathbf{avar}(\lambda\tilde{\boldsymbol{\beta}}+(1-\lambda)\hat{\boldsymbol{\beta}})$ requires equating the above equation to 0 at $\lambda = 0$ to obtain:
$2\mathbf{avar}(\hat{\boldsymbol{\beta}}) = 2\mathbf{acov}(\tilde{\boldsymbol{\beta}}, \hat{\boldsymbol{\beta}}) = \mathbf{avar}(\tilde{\boldsymbol{\beta}}) + \mathbf{avar}(\hat{\boldsymbol{\beta}}) - \mathbf{avar}(\tilde{\boldsymbol{\beta}} - \hat{\boldsymbol{\beta}})$
and the lemma follows. ■

For $\tilde{\boldsymbol{\beta}} := \hat{\boldsymbol{\beta}}_{IV}$ and $\hat{\boldsymbol{\beta}} := \hat{\boldsymbol{\beta}}_{GLS}$ the above Lemma 3.2 proves the equality (3.17), where the asymptotic variance-covariance matrices of the IV and GLS estimators are $\mathbf{avar}(\hat{\boldsymbol{\beta}}_{IV}) = \sigma_\varepsilon^2[(\mathbf{X}'\mathbf{Z})(\mathbf{Z}'\mathbf{Z})^{-1}(\mathbf{Z}'\mathbf{X})]^{-1}$ and $\mathbf{avar}(\hat{\boldsymbol{\beta}}_{GLS}) = (\mathbf{X}'\boldsymbol{\Sigma}^{-1}\mathbf{X})^{-1}$, respectively.

3.2 Nonnested model tests

If there are separate or non-encompassing families of hypotheses of null and alternative hypotheses the following test situation can be assumed, where it is tested under the null that the simultaneous pdf of the sample equals the joint likelihood function ℓ_0 and under the alternative that another joint likelihood function ℓ_A is valid:

$$\begin{aligned} &H_0 : f(\mathbf{y}, \boldsymbol{\alpha}) =: \ell_0, &&\boldsymbol{\alpha} \in \Omega_{\boldsymbol{\alpha}} \text{ is true} \\ &H_A : g(\mathbf{y}; \boldsymbol{\beta}) =: \ell_A, &&\boldsymbol{\beta} \in \Omega_{\boldsymbol{\beta}} \text{ is true} \end{aligned}$$

with $\Omega_{\boldsymbol{\alpha}} \cap \Omega_{\boldsymbol{\beta}} \neq \Omega_{\boldsymbol{\alpha}}$ or $\Omega_{\boldsymbol{\beta}}$, and $\Omega_{\boldsymbol{\alpha}} \cup \Omega_{\boldsymbol{\beta}} \neq \Omega_{\boldsymbol{\alpha}}$ or $\Omega_{\boldsymbol{\beta}}$. Parameterizing, using a single parameter γ, we can write a new pdf as a (log)linear combination of $f(\cdot)$ and $g(\cdot)$: $h(\mathbf{y}; \boldsymbol{\alpha}, \boldsymbol{\beta}) = c(f(\mathbf{y}; \boldsymbol{\alpha}))^{\gamma} (g(\mathbf{y}; \boldsymbol{\beta}))^{1-\gamma}$ with normalizing constant $c^{-1} := \int_{-\infty}^{+\infty} (f(\mathbf{y}; \boldsymbol{\alpha}))^{\gamma} (g(\mathbf{y}; \boldsymbol{\beta}))^{1-\gamma} dy$. Hence, the simultaneous pdf of $\mathbf{y}$ can be written as $h(\mathbf{y}; \boldsymbol{\alpha}, \boldsymbol{\beta}) = \left(\int_{-\infty}^{+\infty} \ell_0^{\gamma} \ell_A^{1-\gamma} dy\right)^{-1} \ell_0^{\gamma} \ell_A^{1-\gamma}$.

Taking logarithms $\ln h = \ln \ell = \gamma \ln \ell_0 + (1-\gamma) \ln \ell_A - \ln \int_{-\infty 0}^{+\infty} \ell_0^{\gamma} \ell_A^{1-\gamma} dy$ the score for the LM test statistic can be written under the null as:

$$\left.\frac{\partial \ln \ell}{\partial \gamma}\right|_{H_0} = s_0 = s(\tilde{\gamma}; \mathbf{y}) = \ln \ell_0 - \ln \ell_A - E_{H_0}(\ln \ell_0 - \ln \ell_A).$$

Proving this score relation, we can rewrite

$$\begin{aligned} \frac{\partial \ln \ell}{\partial \gamma} &= \ln \ell_0 - \ln \ell_A - \frac{\partial}{\partial \gamma} \ln \int_{-\infty}^{+\infty} \ell_0^{\gamma} \ell_A^{1-\gamma} dy = \ln \ell_0 - \ln \ell_A - \underbrace{\frac{1}{\int_{-\infty}^{+\infty} \ell_0^{\gamma} \ell_A^{1-\gamma} dy}}_{=c} \frac{\partial}{\partial \gamma} \int_{-\infty}^{+\infty} \ell_0^{\gamma} \ell_A^{1-\gamma} dy \\ &= \ln \ell_0 - \ln \ell_A - \frac{\partial}{\partial \gamma} \int_{-\infty}^{+\infty} c \ell_0^{\gamma} \ell_A^{1-\gamma} dy = \ln \ell_0 - \ln \ell_A - \int_{-\infty}^{+\infty} c \frac{\partial}{\partial \gamma} \ell_0^{\gamma} \ell_A^{1-\gamma} dy \\ &= \ln \ell_0 - \ln \ell_A - \int_{-\infty}^{+\infty} \left(c \ell_0^{\gamma} (\ln \ell_0) \ell_A^{1-\gamma} - c \ell_A^{1-\gamma} (\ln \ell_A) \ell_0^{\gamma} \right) dy \\ &= \ln \ell_0 - \ln \ell_A - \int_{-\infty}^{+\infty} (\ln \ell_0 - \ln \ell_A) c \ell_0^{\gamma} \ell_A^{1-\gamma} dy = \ln \ell_0 - \ln \ell_A - (E(\ln \ell_0) - E(\ln \ell_A)), \end{aligned}$$

or $s_0 = \left.\frac{\partial \ln \ell}{\partial \gamma}\right|_{H_0} = \ln \ell_0 - \ln \ell_A - \int_{-\infty}^{+\infty} (\ln \ell_0 - \ln \ell_A) \ell_0 dy = \ln \ell_0 - \ln \ell_A - E_{H_0}(\ln \ell_0 - \ln \ell_A)$.

Now, *two examples* of nonnested tests will be treated:

- Cox test statistic for two linear models;
- Cox test statistic for two nonlinear models.

3.2.1 Cox test for two different linear models

Assume the following testing problem:

$$\begin{array}{ll} H_0 : \mathbf{y} = \mathbf{X}_0\boldsymbol{\beta}_0 + \boldsymbol{\varepsilon}_0 & \boldsymbol{\varepsilon}_0 \sim N_n\left(0, \sigma_0^2\mathbf{I}_n\right) \\ H_A : \mathbf{y} = \mathbf{X}_1\boldsymbol{\beta}_1 + \boldsymbol{\varepsilon}_1 & \boldsymbol{\varepsilon}_1 \sim N_n\left(0, \sigma_1^2\mathbf{I}_n\right) \end{array}$$

with the corresponding logarithmic likelihood functions:

$$\begin{array}{l} L_0\left(\boldsymbol{\beta}_0, \sigma_0^2; \mathbf{y}, \mathbf{X}_0\right) = -\frac{n}{2}\ln 2\pi - \frac{n}{2}\ln\sigma_0^2 - \frac{1}{2\sigma_o^2}\left(\mathbf{y} - \mathbf{X}_0\boldsymbol{\beta}_0\right)'\left(\mathbf{y} - \mathbf{X}_0\boldsymbol{\beta}_0\right) \\ L_1\left(\boldsymbol{\beta}_1, \sigma_1^2; \mathbf{y}, \mathbf{X}_1\right) = -\frac{n}{2}\ln 2\pi - \frac{n}{2}\ln\sigma_1^2 - \frac{1}{2\sigma_1^2}\left(\mathbf{y} - \mathbf{X}_1\boldsymbol{\beta}_1\right)'\left(\mathbf{y} - \mathbf{X}_1\boldsymbol{\beta}_1\right) \end{array}$$

leading to the (unrestricted) ML estimators $\hat{\boldsymbol{\beta}}_0 = (\mathbf{X}_0'\mathbf{X}_0)^{-1}\mathbf{X}_0'\mathbf{y}$ and $\hat{\sigma}_o^2 = \frac{\left(\mathbf{y}\text{-}\mathbf{X}_0\hat{\boldsymbol{\beta}}_0\right)'\left(\mathbf{y}\text{-}\mathbf{X}_0\hat{\boldsymbol{\beta}}_0\right)}{n} = \frac{\left[(\mathbf{I}_n\text{-}\mathbf{X}_0(\mathbf{X}_0'\mathbf{X}_0)^{-1}\mathbf{X}_0')\mathbf{y}\right]'\left[(\mathbf{I}_n\text{-}\mathbf{X}_0(\mathbf{X}_0'\mathbf{X}_0)^{-1}\mathbf{X}_0')\mathbf{y}\right]}{n} = \frac{1}{n}\mathbf{y}'\mathbf{M}_0\mathbf{y}$, with idempotent matrix $\mathbf{M}_0 := \mathbf{I}_n\text{-}\mathbf{X}_0(\mathbf{X}_0'\mathbf{X}_0)^{-1}\mathbf{X}_0'$ and $\hat{\boldsymbol{\beta}}_1 = (\mathbf{X}_1'\mathbf{X}_1)^{-1}\mathbf{X}_1'\mathbf{y}$ with $\hat{\sigma}_1^2 = \frac{1}{n}\mathbf{y}'\mathbf{M}_1\mathbf{y}$ and idempotent projection matrix $\mathbf{M}_1 := \mathbf{I}_n\text{-}\mathbf{X}_1(\mathbf{X}_1'\mathbf{X}_1)^{-1}\mathbf{X}_1'$. The score becomes under the null:

$$\hat{s}_0 = L_0\left(\hat{\boldsymbol{\theta}}_0; \mathbf{y}, \mathbf{X}_0\right)\text{-}L_1\left(\hat{\boldsymbol{\theta}}_1; \mathbf{y}, \mathbf{X}_1\right)\text{-}E_{H_0}\left\{L_0\left(\hat{\boldsymbol{\theta}}_0; \mathbf{y}, \mathbf{X}_0\right)\text{-}L_1\left(\hat{\boldsymbol{\theta}}_1; \mathbf{y}, \mathbf{X}_1\right)\right\}$$

with $\hat{\boldsymbol{\theta}}_0 := (\hat{\boldsymbol{\beta}}_0', \hat{\sigma}_0^2)'$ and $\hat{\boldsymbol{\theta}}_1 := (\hat{\boldsymbol{\beta}}_1', \hat{\sigma}_1^2)'$.

Taking account of the asymptotic properties of the ML estimators:

$$\hat{s}_0 = L_0(\hat{\boldsymbol{\theta}}_0; \mathbf{y}, \mathbf{X}_0)\text{-}L_1(\hat{\boldsymbol{\theta}}_1; \mathbf{y}, \mathbf{X}_1)\text{-}n\left[\underset{n\to\infty}{\text{plim}}_{(H_0)}\left\{\tfrac{1}{n}L_0(\hat{\boldsymbol{\theta}}_0; \mathbf{y}, \mathbf{X}_0)\text{-}L_1(\hat{\boldsymbol{\theta}}_1; \mathbf{y}, \mathbf{X}_1)\right\}\right]$$

At the point $\boldsymbol{\theta}_o = \hat{\boldsymbol{\theta}}_o$, $\left[\underset{n\to\infty}{\text{plim}}_{(H_0)}\left\{\tfrac{1}{n}L_0(\hat{\boldsymbol{\theta}}_0; \mathbf{y}, \mathbf{X}_0)\right\}\right] = L_0(\hat{\boldsymbol{\theta}}_0; \mathbf{y}, \mathbf{X}_0)$, such that the score becomes $\hat{s}_0 = \text{-}L_1(\hat{\boldsymbol{\theta}}_1; \mathbf{y}, \mathbf{X}_1) + n\left[\underset{n\to\infty}{\text{plim}}_{(H_0)}\left\{\tfrac{1}{n}L_1(\hat{\boldsymbol{\theta}}_1; \mathbf{y}, \mathbf{X}_1)\right\}\right]$.

The concentrated loglikelihood function for H_1 becomes:

$$L_1(\hat{\boldsymbol{\theta}}_1; \mathbf{y}, \mathbf{X}_1)\Big|_{H_0} = -\tfrac{n}{2}\ln 2\pi\text{-}\tfrac{n}{2}\ln\hat{\sigma}_1^2\text{-}\tfrac{1}{2\hat{\sigma}_1^2}(\mathbf{y}\text{-}\mathbf{X}_1\hat{\boldsymbol{\beta}}_1)'(\mathbf{y}\text{-}\mathbf{X}_1\hat{\boldsymbol{\beta}}_1)$$

$= \text{-}\frac{n}{2}(\ln 2\pi_1 + 1)\text{-}\underbrace{\frac{n}{2}\ln\hat{\sigma}_1^2\Big|_{H_0}}_{-\frac{n}{2}\ln\hat{\sigma}_{10}^2}$, or $\underset{n\to\infty}{\text{plim}}_{(H_0)}\widetilde{\boldsymbol{\beta}}_1 = \underset{n\to\infty}{\text{plim}}_{(H_0)}\hat{\boldsymbol{\beta}}_1 = \boldsymbol{\beta}_1^o$. Thus $\widetilde{\boldsymbol{\beta}}_1$ can be obtained from reestimating H_1, where $\mathbf{X}_0\hat{\boldsymbol{\beta}}_0$ can be interpreted as a dependent variable or $\widetilde{\boldsymbol{\beta}}_1 = (\mathbf{X}_1'\mathbf{X}_1)\mathbf{X}_1'\mathbf{X}_0\hat{\boldsymbol{\beta}}_0 = (\mathbf{X}_1'\mathbf{X}_1)^{-1}\mathbf{X}_1'\mathbf{X}_0(\mathbf{X}_0'\mathbf{X}_0)^{-1}\mathbf{X}_0\mathbf{y}$.

Substituted in the formula for the residual variance of H_1 under H_0:

$$\begin{aligned} \widetilde{\sigma}_{10}^2 &= \frac{1}{n}\left(\mathbf{y} - \mathbf{X}_1\tilde{\boldsymbol{\beta}}_1\right)'\left(\mathbf{y} - \mathbf{X}_1\tilde{\boldsymbol{\beta}}_1\right) \\ &= \frac{1}{n}\left(\mathbf{y} - \mathbf{X}_0\boldsymbol{\beta}_0 + \mathbf{X}_0\boldsymbol{\beta}_0 - \mathbf{X}_1\tilde{\boldsymbol{\beta}}_1\right)'\left(\mathbf{y} - \mathbf{X}_0\boldsymbol{\beta}_0 + \mathbf{X}_0\boldsymbol{\beta}_0 - \mathbf{X}_1\tilde{\boldsymbol{\beta}}_1\right) \\ &= \frac{1}{n}\Bigg[(\mathbf{y} - \mathbf{X}_0\boldsymbol{\beta}_0)'(\mathbf{y} - \mathbf{X}_0\boldsymbol{\beta}_0) + \left(\mathbf{X}_0\boldsymbol{\beta}_0 - \mathbf{X}_1\widetilde{\boldsymbol{\beta}}_1\right)' \qquad (3.18) \\ &\qquad \left(\mathbf{X}_0\boldsymbol{\beta}_0 - \mathbf{X}_1\tilde{\boldsymbol{\beta}}_1\right) + 2\underbrace{(\mathbf{y} - \mathbf{X}_0\boldsymbol{\beta}_0)'}_{\boldsymbol{\varepsilon}_0}\left(\mathbf{X}_0\boldsymbol{\beta}_0 - \mathbf{X}_1\tilde{\boldsymbol{\beta}}_1\right)\Bigg]. \end{aligned}$$

Substituting in the score:

$$\hat{s}_0 = \frac{n}{2}\ln\hat{\sigma}_1^2 - \frac{n}{2}\left[\operatorname*{plim}_{n\to\infty}\left\{\ln\tilde{\sigma}_{10}^2\right\}\right]_{\boldsymbol{\theta}_0=\hat{\boldsymbol{\theta}}_0} = \frac{n}{2}\ln\frac{\hat{\sigma}_1^2}{\operatorname*{plim}_{n\to\infty}\tilde{\sigma}_{10}^2\Big|_{\boldsymbol{\theta}_0=\hat{\boldsymbol{\theta}}_0}}. \tag{3.19}$$

Since $\operatorname*{plim}_{n\to\infty}\tilde{\sigma}_{10}^2\Big|_{\boldsymbol{\theta}_0=\hat{\boldsymbol{\theta}}_0}$ has to be computed, consider $\mathbf{X}_0\hat{\boldsymbol{\beta}}_0 - \mathbf{X}_1\tilde{\boldsymbol{\beta}}_1$ instead of $\mathbf{X}_0\boldsymbol{\beta}_0 - \mathbf{X}_1\tilde{\boldsymbol{\beta}}_1$, i.e.

$$\begin{gathered}\mathbf{X}_0\hat{\boldsymbol{\beta}}_0\text{-}\mathbf{X}_1(\mathbf{X}_1'\mathbf{X}_1)^{-1}\mathbf{X}_1'\mathbf{X}_0\hat{\boldsymbol{\beta}}_0=(\mathbf{I}_n\text{-}\mathbf{X}_1(\mathbf{X}_1'\mathbf{X}_1)^{-1}\mathbf{X}_1')\mathbf{X}_0\hat{\boldsymbol{\beta}}_0\\ =\mathbf{M}_1\mathbf{X}_0\hat{\boldsymbol{\beta}}_0=\mathbf{M}_1\mathbf{X}_0(\mathbf{X}_0'\mathbf{X}_0)^{-1}\mathbf{X}_0'\underbrace{(\mathbf{X}_0\boldsymbol{\beta}_0+\boldsymbol{\varepsilon}_o)}_{\mathbf{y}}=\mathbf{M}_1\mathbf{X}_0\boldsymbol{\beta}_0+\mathbf{M}_1\mathbf{X}_0(\mathbf{X}_0'\mathbf{X}_0)^{-1}\mathbf{X}_0'\boldsymbol{\varepsilon}_0\end{gathered}$$

so that $\operatorname*{plim}_{n\to\infty}(\mathbf{X}_0\hat{\boldsymbol{\beta}}_0\text{-}\mathbf{X}_1\tilde{\boldsymbol{\beta}}_1)=\mathbf{M}_1\mathbf{X}_0\boldsymbol{\beta}_0$ and $\operatorname*{plim}_{n\to\infty}\{\frac{1}{n}\boldsymbol{\varepsilon}_0'(\mathbf{X}_0\hat{\boldsymbol{\beta}}_0\text{-}\mathbf{X}_1\tilde{\boldsymbol{\beta}}_1)\}=\mathbf{0}$.

Hence, the estimator for the probability limit of $\tilde{\sigma}_{10}^2$ in formula (3.18) can be split in two parts:

- an estimator of the variance of the true model under H_0;
- an estimator of the extra variance according to the difference between H_0 and H_1:

$$\begin{aligned}\hat{\tilde{\sigma}}_{10}^2 &= \operatorname*{plim}_{n\to\infty}\tilde{\sigma}_{10}^2\Big|_{\theta_0=\hat{\theta}_0} = \hat{\sigma}_0^2 + \frac{1}{n}\left(\mathbf{X}_0\hat{\boldsymbol{\beta}}_0 - \mathbf{X}_1\tilde{\boldsymbol{\beta}}_1\right)'\left(\mathbf{X}_0\hat{\boldsymbol{\beta}}_0 - \mathbf{X}_1\tilde{\boldsymbol{\beta}}_1\right)\\ &= \hat{\sigma}_0^2 + \frac{1}{n}\left(\hat{\boldsymbol{\beta}}_0'\mathbf{X}_0'\mathbf{X}_0\hat{\boldsymbol{\beta}}_0\text{-}2\tilde{\boldsymbol{\beta}}_1'\mathbf{X}_1'\mathbf{X}_0\hat{\boldsymbol{\beta}}_0 + \tilde{\boldsymbol{\beta}}_1'\mathbf{X}_1'\mathbf{X}_1\tilde{\boldsymbol{\beta}}_1\right)\end{aligned}$$

$$=\hat{\sigma}_0^2+\frac{1}{n}\Bigg[\underbrace{\mathbf{y}'\mathbf{X}_0(\mathbf{X}_0'\mathbf{X}_0)^{-1}}_{\hat{\boldsymbol{\beta}}_0'}\mathbf{X}_0'\mathbf{X}_0\underbrace{(\mathbf{X}_0'\mathbf{X}_0)^{-1}\mathbf{X}_0'\mathbf{y}}_{\hat{\boldsymbol{\beta}}_0}\text{-}2\underbrace{\mathbf{y}'\mathbf{X}_0(\mathbf{X}_0'\mathbf{X}_0)^{-1}}_{\hat{\boldsymbol{\beta}}_0'}\mathbf{X}_0'\mathbf{X}_1(\mathbf{X}_1'\mathbf{X}_1)^{-1}$$

$$\mathbf{X}_1\mathbf{X}_0\underbrace{(\mathbf{X}_0'\mathbf{X}_0)^{-1}\mathbf{X}_0'\mathbf{y}}_{\hat{\boldsymbol{\beta}}_0}+\mathbf{y}'\mathbf{X}_0(\mathbf{X}_0'\mathbf{X}_0)^{-1}\mathbf{X}_0'\mathbf{X}_1(\mathbf{X}_1'\mathbf{X}_1)\mathbf{X}_1'\mathbf{X}_0(\mathbf{X}_0'\mathbf{X}_0)^{-1}\mathbf{X}_0'\mathbf{y}\Bigg]$$

$$\begin{aligned}&= \hat{\sigma}_0^2 + \frac{1}{n}\left[\hat{\boldsymbol{\beta}}_0'\mathbf{X}_0'\mathbf{X}_0\hat{\boldsymbol{\beta}}_0 - \hat{\boldsymbol{\beta}}_0'\mathbf{X}_0'\mathbf{X}_1(\mathbf{X}_1'\mathbf{X}_1)^{-1}\mathbf{X}_1'\mathbf{X}_0\hat{\boldsymbol{\beta}}_0\right]\\ &= \hat{\sigma}_0^2 + \frac{1}{n}\hat{\boldsymbol{\beta}}_0'\mathbf{X}_0'\mathbf{M}_1\mathbf{X}_0\hat{\boldsymbol{\beta}}_0.\end{aligned}$$

Hence, the score (3.19) becomes:

$$\begin{aligned}\hat{s}_0 &= \frac{n}{2}\ln\left(\frac{\hat{\sigma}_1^2}{\hat{\hat{\sigma}}_{10}^2}\right) = \frac{n}{2}\ln\frac{\hat{\sigma}_1^2}{\hat{\sigma}_0^2 + \frac{1}{n}\hat{\beta}_0'\mathbf{X}_0'\mathbf{M}_1\mathbf{X}_0\hat{\beta}_0} \\ &= \frac{n}{2}\ln\frac{\frac{\hat{\varepsilon}_1'\hat{\varepsilon}_1}{n}}{\frac{\hat{\varepsilon}_0'\hat{\varepsilon}_0}{n} + \frac{\hat{\beta}_0'\mathbf{X}_0'\mathbf{M}_1\mathbf{X}_0\hat{\beta}_0}{n}} = \frac{n}{2}\ln\frac{\mathbf{y}'\mathbf{M}_1\mathbf{y}}{\mathbf{y}'\mathbf{M}_0\mathbf{y} + \hat{\beta}_0'\mathbf{X}_0'\mathbf{M}_1\mathbf{X}_0\hat{\beta}_0}.\end{aligned}$$

The LM test statistic satisfies then from (3.1) that $\xi_{LM} = \hat{s}_0^2\widehat{avar}(\hat{s}_0) \overset{as}{\sim} \chi_1^2$ so that under $H_0 : \hat{s}_0 \overset{as}{\backsim} N(0, avar(\hat{s}_0))$, where $avar(\hat{s}_0)$ is the asymptotic variance of $\hat{s}_0$.

A significant (negative/positive) value means that we have to reject H_0 and an insignificant $\hat{s}_0$ value means that we do not have to reject H_0. Hence, the finite sample distribution of the test statistic $\hat{s}_0$ depends on unknown parameters, and cannot be derived. If, however, $\mathbf{M}_1\mathbf{X}_0 = 0$ ($\mathbf{X}_1$ orthogonal on $\mathbf{X}_0$) then both models are nested and an exact test (statistic) exists, i.e. $\hat{s}_0 = \frac{n}{2}\ln\frac{\mathbf{y}'\mathbf{M}_1\mathbf{y}}{\mathbf{y}'\mathbf{M}_0\mathbf{y}}$, which is equal to an F-test statistic for $H_0 : R^2 = 0$. Given the estimate $\widehat{avar}(\hat{s}_0)$ of the asymptotic variance $avar(\hat{s}_0)$, $\frac{\hat{s}_0}{\sqrt{\widehat{avar}(\hat{s}_0)}} \overset{as}{\backsim} N(0,1)$.

Pesaran (1974) has shown that $\widehat{avar}(\hat{s}_0) = \frac{\hat{\sigma}_0^2}{\hat{\hat{\sigma}}_{10}^4}\hat{\beta}_0'\mathbf{X}_0'\mathbf{M}_1'\mathbf{M}_0\mathbf{M}_1\mathbf{X}_0\hat{\beta}_0$.

In practice, two tests are performed (with $\hat{s}_0$ and $\hat{s}_1$), with both linear models respectively as the H_0 ('true model'). A significant value of the test statistic $\frac{\hat{s}_0}{\sqrt{\widehat{avar}(\hat{s}_0)}}$ points to the rejection of H_0 (if $\hat{s}_0$ is significantly smaller than zero, H_0 is rejected in the direction of $H_A = H_1$, if $\hat{s}_0$ is significantly larger than zero, H_0 is rejected in a direction away from H_1).

The outcome of these two tests points out that:

- either the data is consistent with H_0;
- or the data is consistent with $H_A = H_1$;
- or the data is not consistent with H_0, nor with $H_A = H_1$;
- or the data is consistent with both H_0 and $H_A = H_1$!

The test of the difference between linear models in Subsection 3.2.1 can be extended to nonlinear models

3.2.2 Cox test for two different nonlinear models

Assume the following testing problem for two nonlinear models :

$$\begin{aligned}&H_0 : \mathbf{y} = \mathbf{f}(\mathbf{X}_0, \boldsymbol{\beta}_0) + \varepsilon_0 && \varepsilon_0 \sim N_n\left(0, \sigma_0^2\mathbf{I}_n\right) \\ &H_A : \mathbf{y} = \mathbf{g}(\mathbf{X}_1, \boldsymbol{\beta}_1) + \varepsilon_1 && \varepsilon_1 \sim N_n\left(0, \sigma_1^2\mathbf{I}_n\right).\end{aligned}$$

Testing H_0 against H_A, the underlying LM test statistic for the Cox test leads to the direct analogue of the score (3.19) for the linear case $\hat{s}_0 \overset{as}{\sim} N\left(0, avar(\hat{s}_0)\right)$ under H_0, where, according to Pesaran and Deaton (1978), for the above non-linear models:

$$\widehat{avar}(\hat{s}_0) = \frac{\hat{\sigma}_0^2}{\hat{\sigma}_{10}^4}\left[\mathbf{f}\left(\mathbf{X}_0, \boldsymbol{\beta}_0\right) - \mathbf{g}\left(\mathbf{X}_1, \hat{\boldsymbol{\beta}}_{10}\right)\right]'\left(\mathbf{I}_n - \hat{\mathbf{Z}}(\hat{\mathbf{Z}}'\hat{\mathbf{Z}})^{-1}\hat{\mathbf{Z}}'\right)\left[\mathbf{f}\left(\mathbf{X}_0, \boldsymbol{\beta}_0\right) - \mathbf{g}\left(\mathbf{X}_1, \hat{\boldsymbol{\beta}}_{10}\right)\right],$$

where the $n \times p$ Jacobian matrix $\hat{\mathbf{Z}} := \hat{\mathbf{Z}}(\hat{\boldsymbol{\beta}}_0) = \left.\frac{\partial \mathbf{f}(\mathbf{X}_0, \boldsymbol{\beta}_0)}{\partial \boldsymbol{\beta}_0}\right|_{\boldsymbol{\beta}_0 = \hat{\boldsymbol{\beta}}_0}$ and $\hat{\boldsymbol{\beta}}_{10}$ is the ML estimator of $\boldsymbol{\beta}_1$ if the null were true, so that $\widehat{avar}(\hat{s}_0)$ is calculated by regressing on $\hat{\mathbf{Z}}$ the residuals from the regression of $\mathbf{f}\left(\mathbf{X}_0, \hat{\boldsymbol{\beta}}_0\right)$ on H_A and calculating the residual sum of squares.

Example 3.1 *A linear versus loglinear specification*

H_0 (linear model): $\mathbf{y} = \sum_{i=1}^{p} \alpha_i \mathbf{x}_i + \boldsymbol{\varepsilon}_0$, with $\boldsymbol{\varepsilon}_0 \backsim N_n(0, \sigma_0^2 \mathbf{I}_n)$ against
H_1(loglinear model): $\log \mathbf{y} = \sum_{i=1}^{p} \beta_i \log \mathbf{x}_i + \boldsymbol{\varepsilon}_1$, with $\boldsymbol{\varepsilon}_1 \backsim N_n\left(0, \sigma_1^2 \mathbf{I}_n\right)$;
for linear models there are constant marginal rates $\frac{\partial \mathbf{y}}{\partial \mathbf{x}_i} = \alpha_i$ $(i=1,2,\ldots,k)$ and for loglinear models there are constant elasticities $\frac{\partial \ln \mathbf{y}}{\partial \ln \mathbf{x}_i} = \frac{\partial \mathbf{y}/\mathbf{y}}{\partial \mathbf{x}_i/\mathbf{x}_i} = \frac{\partial \mathbf{y}}{\partial \mathbf{x}_i}\left(\frac{\mathbf{y}}{\mathbf{x}_i}\right)^{-1} = \beta_i$, so that the marginal rates change with the average rates then. A mixed form between loglinear and linear models is the *Box-Cox transformation.*

Example 3.2 *Consider the following model (Box-Cox transformation):*

$$y_j^{(\lambda)} = \sum_{i=1}^{m} \alpha_i x_{ij}^{(\lambda_i)} + \varepsilon_j$$

with $y_j^{(\lambda)} := \frac{y_j^{\lambda} - 1}{\lambda}$ and $x_{ij}^{(\lambda_i)} = \frac{x_{ij}^{\lambda_i} - 1}{\lambda_i}$ $(i = 1, 2, ..., m)$, which implies the following test $H_0 : \lambda = 1$ $(\lambda_i = 1)$ (linear model) and $H_1 : \exists_i\ \lambda_i \neq 1$ or $\lambda \neq 1$.

Note that, if $\lambda = 0$, we have a loglinear model because (according to the Taylor expansion around $\lambda = 0$ $(\lambda_i = 0)$):
$\frac{y_j^{\lambda} - 1}{\lambda} = \log y_j + \frac{\lambda}{2!}(\log y_j)^2 + \frac{\lambda^2}{3!}(\log y_j)^3 + ...$or if $\lambda = 0$, $\frac{y_j^{\lambda} - 1}{\lambda} = \log y_j$ and $\frac{x_{ij}^{\lambda_i} - 1}{\lambda_i} = \log x_{ij}$. Hence, $0 \leqq \lambda \leqq 1$ $(0 \leqq \lambda_i \leqq 1)$. Now, λ and λ_i $(i = 1, 2, ..., m)$ are to be estimated with ML:

$$\max_{\sigma_\varepsilon^2, \alpha_i, \lambda, \lambda_i} L\left(\sigma_\varepsilon^2, \alpha_i, \lambda, \lambda_i; y, X\right) = \max\left\{-\tfrac{n}{2}\ln 2\pi - \tfrac{n}{2}\ln \sigma_\varepsilon^2 \right.$$
$$\left. - \frac{1}{2\sigma_\varepsilon^2}\sum_{j=1}^{n}\left(\frac{y_j^{\lambda} - 1}{\lambda} - \sum_{i=1}^{m}\alpha_i\left(\frac{x_{ij}^{\lambda_i} - 1}{\lambda_i}\right)\right)^2\right\}$$

which yields the ML estimators $\hat{\sigma}_\varepsilon^2, \hat{\alpha}_i, \hat{\lambda}, \hat{\lambda}_i$ $(i = 1, 2, ..., m)$.

If $\hat{\lambda}(\hat{\lambda}_i)$ is close to zero, the corresponding variable occurs multiplicatively; if, on the contrary, $\hat{\lambda}(\hat{\lambda}_i)$ is close to 1, the corresponding variable occurs additively.

Figure 3.1: Demand of rent buildings

Year	Number of corporation houses	Number of private rentbuildings	Total number of rentbuildings
1962			18321
1963			17474
1964			13355
1965			14004
1966			17039
1967	9550	25027	34577
1968	11992	11957	23949
1969	11399	23888	35287
1970	14259	25553	40082
1971	13300	26674	39974
1972	12100	23209	35309
1973	7247	22314	29561
1974	4946	24322	29268
1975	2250	19671	21921
1976	732	16211	16943
1977	358	4842	5200
1978	861	5610	6471
1979	354	6932	7286
1980	1952	8523	20382
1981	4151	15971	20122
1982	5542	14502	20044

Example 3.3 *Dutch rent buildings (empirical example).*

Given the number of rent buildings, which were subsidized in the Netherlands during the period 1962 – 1982 and the following specification for the desired demand of rent buildings:

$$vph^* = \alpha_1 yb + \alpha_2 (\Delta yb)_{-n} + \alpha_3 \left(\frac{hp}{pc}\right) + \alpha_4 \left(\frac{pe}{pc}\right) + \eta_1$$

vph^* = the desired demand of rent buildings
yb =the real available income (National Accounts, CBS)
hp = the rent of rent buildings (Monthly Statistics, CBS)
pe = the cost of heating and lighting (Statistical Pocket, CBS)
pc = the price index of the family consumption (National Accounts, CBS)
n = time that passes by before a change in income is perceived as permanent.
For the investors' supply function:

$$\Delta aph = \boldsymbol{\beta}'\mathbf{x} + \eta_2 = d_1 \Delta vph^* + (1 - d_1)\, \Delta aph_{-1} + d_2' \Delta aww_{-1} + \mu,$$

where the vector $\mathbf{x}$ contains the variables which are important for the investors to invest in rent buildings, aww the supply of social private buildings ("*woningwetwoningen*") and aph the number of private rent buildings in Figure 3.1. We obtain the following function to be estimated:

$$\Delta aph = \beta_0 + \beta_1 \Delta yb + \beta_2 (\Delta\Delta yb)_{-n} + \beta_3 \Delta\left(\frac{hp}{pc}\right) + \beta_4 \left(\frac{pe}{pc}\right) + \beta_5 \Delta aph_{-1} + \beta_6 \Delta aww + \varepsilon_0,$$

with Δaph the increase in supply of rent buildings (Figure 3.1), Δaww the surplus in demand on the market of social private buildings and $\Delta\Delta yb_{-n}$ the presumed permanent real income.

Under H_0 we estimate, with OLS, the following linear relation:

$$\underset{(t-ratio)}{\Delta aph} = \underset{(-1.853)}{-0.459} + \underset{(2.038)}{11.077}\Delta\Delta yb_{-2.5} + \underset{(5.382)}{0.702}\Delta aph_{-1} + \underset{(2.438)}{0.710}\Delta aww_{-1}$$

$$R_0^2 = 0.956, \quad \hat{\sigma}_{\varepsilon_0} = 0.265, \quad \hat{\rho} = -0.076, \quad DW = 2.237.$$

If we look at the above relation it strikes that neither the rent nor the cost of the complementary good energy (lighting, gas and heating) are included in the regression because they are not significant. An explanation for this is presumably the fact that we used real rents and real costs of energy. After all, the nominal rents and nominal cost of energy stick closely to the price-index figure. As a result, both real components imply an approximately constant evolution in time of the two real price variables. The influence of these two variables is captured in the constant term which is significant, despite of the fact that we explain the first order differences in the supply of rent buildings.

Under H_1, we estimate a corresponding relation in logarithms:

$$\Delta \ln(aph) = \underset{(-1.01)}{-0.087} + \underset{(1.757)}{9.409}\Delta\Delta \ln(yb)_{-2.5} + \underset{(4.204)}{0.644}\Delta \ln(aph)_{-1} + \underset{(2.435)}{1.010}\Delta \ln(aww)_{-1}$$

$$R_1^2 = 0.775, \quad \hat{\sigma}_{\varepsilon_1} = 0.63, \quad \hat{\rho} = 0.236, \quad DW = 1.439$$

The linear model is preferred to the loglinear model because $\frac{\hat{s}_0}{\sqrt{\widehat{avar}(\hat{s}_0)}} = 1.255$, which means that H_0 (i.e. the linear model) is not rejected.

3.3 Theoretical and empirical exercises

1. Demonstrate that the score test can be formulated as a conditional moment test.

2. Apply an LM test on the Box-Cox model.

3. Suppose n independent observations $y_1, y_2, ..., y_n$, say, incomes, drawn by simple random sampling from a Pareto distribution which has the following pdf $P(y|\alpha) = \frac{\alpha 10000^{\alpha}}{y^{\alpha+1}}$, $y \geqq 10000$, $\alpha > 0$.
What is the ML estimator of α? Test whether $\alpha = 6$ for the sample $y_1 = 10000, y_2 = 11000, y_3 = 12000, y_4 = 13000, y_5 = 14000$ using LR, LM and Wald tests.

4. Test the validity of the choice for various instruments and of the implied moment conditions when using the data underlying Exercises 2.3.3.

5. Suppose that y follows a Weibull pdf as defined by Greene(2003), pp.522-523:

$$f(y) = \alpha\beta y^{\beta-1}e^{-\alpha y^{\beta}}, \qquad y > 0; \alpha, \beta > 0.$$

(a) Obtain the ML estimators of α and β for a random sample of n observations. Note that $\hat{\beta}_{ML}$ cannot be found analytically.

(b) Obtain the information matrix and the (asymptotic) covariance matrix of the ML estimators.

(c) The following data were generated by the above Weibull pdf: 1.3043, 1.0878, 0.33453, 0.49254, 1.9461, 1.1227, 1.2742, 0.47615, 2.0296, 1.4019, 3.6454, 1.2797, 0.32556, 0.15344, 0.96080, 0.29965, 1.2357, 2.0070, 0.26423, 0.96381.

i. Obtain the ML estimates of α and β based on this sample, and estimate the covariance matrix of the estimates.

ii. Obtain the ML estimate of α under the hypothesis that $\beta = 1$.

iii. Carry out an LM, a Wald, and an LR test for the hypothesis that $\beta = 1$.

Part II

Time Series Analysis

Time plays an important role in economics. It takes for example time for a firm to implement new investment plans in response to a rise in the demand for its product, or for a household to adjust its consumption to a decline in its income. Economic agents may also delay their decisions because of imperfect information. Moreover, it is often fruitful to analyze data over time, e.g. to study business cycle effects. The analysis of time series is an important tool to capture much of these dynamics into economic modeling. The scope of this part is twofold. It serves as a thorough introduction into mostly linear time series analysis for readers with a quantitative background. For more practical purposes, it further provides extensively documented guidelines that incorporate major new developments in the field, including several real world exercises and cases.

Chapter 4 provides a typology of linear dynamic models to become acquainted with terms, such as e.g. autoregressive distributed lag, partial adjustment and error correction, often used in the field. In Chapter 5 univariate ARIMA models and their specific characteristics are discussed. In the same chapter the standard Box-Jenkins procedure is extended to include tests for unit roots and for aberrant observations. Moreover, point and interval forecasting of $ARIMA$ models, but also density forecasting e.g. for underlying asymmetric probability distributions, and theoretical and empirical exercises as well as solved and unsolved (computer) cases based on real-world economic time series databases for $ARIMA$ models are provided. Chapter 6 deals with univariate cointegration and causality which have been proved very useful in modern time series analysis (cf. the shared Nobel prize in economics for Clive W.J. Granger in 2003). After establishing the direction of causality, an overview is given in this chapter of how to implement the identification-estimation-forecasting procedure of transfer function models.

Multivariate modeling including feedback relations between economic variables are introduced in Chapter 7 on multivariate time series (including multivariate cointegration, causality, vector autoregressions and dynamic structural econometric models). Finally, since means, variances and covariances of economic and especially financial data are not time-invariant (cf. the shared Nobel prize in economics for Robert F. Engle in 2003) much attention in the field went to regime switching models in returns, volatility modeling and conditional heteroskedasticity ($(G)ARCH$ modeling). This is the subject of Chapter 8.

Chapter 4

A Typology of Dynamic Models

This introductory chapter is meant to bridge each reader's basic knowledge in time series to an in-depth analysis in time series in the next chapter(s). Therefore, simple linear dynamic models known to most of the readers, as a linear autoregressive distributed lag model, where the dynamics are explained through a sequence of lagged values of an explanatory variable (or of explanatory variables) are discussed. Starting from such a simple model, one can directly verify that there are eight special types of such a simple model. Several of these simple models are (probably) well known such as the partial adjustment model and the error correction model.

4.1 Autoregressive distributed lag models

Define a general linear dynamic transfer function model, where it is assumed that m types of explanatory variables x_i determine a dependent variable y for, in principle, various periods of time without any feedback from y to one or more x_is, or:

$$\alpha(L)\, y_t = \sum_{i=1}^{m} \gamma_i(L)\, x_{it} + \varphi(L)\, \varepsilon_t \quad t = 1, 2, ..., T \tag{4.1}$$

with lag polynomials in the lag (or backward) operator L $(Ly_t := y_{t-1})$:

$$\alpha(L) := 1 - \sum_{i=1}^{p} \alpha_i L^i,\ \gamma_i(L) := \sum_{j=0}^{m_i} \gamma_{ij} L^j,\ \varphi(L) := \sum_{j=0}^{q} \varphi_j L^j,$$

where $\varphi_0 = 1$, and error terms $\varepsilon_t \overset{iid}{\sim} (0, \sigma_\varepsilon^2)$ (*i.e.* white noise or WN with variance σ_ε^2).

For the very simple case that $p = m_2 = 1$, $m_1 = 0$, $q = 0$, $m = 2$ and for x_1 the unity vector:

$$y_t = \beta_0 + \beta_1 x_t + \beta_2 x_{t-1} + \beta_3 y_{t-1} + \varepsilon_t, \tag{4.2}$$

which is called the class of linear *autoregressive* (y_{t-1}) *distributed lag* (x_t, x_{t-1}) or linear AD(1,1) model.

In general one can discern the following (eight!) special types of linear autoregressive distributed lag models from the simple model (4.2):

- the *static* (linear) *regression* model, which is an AD(0,0) model obtained from model (4.2) with $\beta_2 = \beta_3 = 0 : y_t = \beta_0 + \beta_1 x_t + \varepsilon_t$;
- the *univariate* (linear) *time series* model, which is an AD(1,0) model obtained from (4.2) with $\beta_1 = \beta_2 = 0 : y_t = \beta_0 + \beta_3 y_{t-1} + \varepsilon_t$ (with the random walk model as a special case, i.e. $\beta_0 := 0$, $\beta_3 := 1$);
- the (linear) *partial adjustment* model, which is an AD(1,0) model obtained from substituting $\beta_2 = 0$ in (4.2):

$$y_t = \beta_0 + \beta_1 x_t + \beta_3 y_{t-1} + \varepsilon_t; \tag{4.3}$$

- the (linear) *leading indicator* model, which is an AD(0,1) model obtained from substituting $\beta_1 = \beta_3 = 0$ in (4.2) so that $y_t = \beta_0 + \beta_2 x_{t-1} + \varepsilon_t$ (e.g. unemployment in the capital goods sector as a recession indicator);
- the (linear) *finite distributed lag* model, which is (also) an AD(0,1) model derived from model (4.2) with $\beta_3 = 0 : y_t = \beta_0 + \beta_1 x_t + \beta_2 x_{t-1} + \varepsilon_t$;
- the (linear) *difference data/growth rate* model, which is an AD(1,1) model derived from model (4.2) for $\beta := \beta_1 = -\beta_2$ and $\beta_3 = 1 : \Delta y_t = \beta_0 + \beta \Delta x_t + \varepsilon_t$;
- the (linear) '*dead start*' model, being an AD(1,1) model obtained from (4.2) with $\beta_1 = 0 : y_t = \beta_0 + \beta_2 x_{t-1} + \beta_3 y_{t-1} + \varepsilon_t$;
- the (linear) *error correction* model, which also an AD(1,1) model derived from (4.2) with $\sum_{i=1}^{3} \beta_i = 1$ ('homogeneity restriction').

Out of these eight special AD cases, the most popular submodels are the partial adjustment (PA) model and the error correction mechanism (ECM) model, will be discussed more comprehensively in the next sections.

4.2 The partial adjustment model

Given a proportionality *steady state* or *equilibrium* relationship between a dependent variable Y and an independent variable X as in $Y = CX$, where C is a constant of proportionality, logarithms can be taken: $\log Y = \log C + \log X$, so that:

$$y = c + x, \tag{4.4}$$

where $y := \log Y$, $x := \log X$, and $c := \log C$. Now, it could be thought of that economic agents would like to adjust their costs because there are 'disequilibrium costs' (if the static relation (4.4) is not satisfied) and 'adjustment costs'. If both kinds of costs are assumed to be quadratic, the following intertemporal cost function could be minimized with respect to the planned values of the dependent variable:

$$\mathcal{L}_0 = \sum_{t=0}^{\infty} \kappa_t \lambda_1 \underbrace{(y_t^p - c - x_t)^2}_{\text{'disequilibrium cost'}} + \sum_{t=0}^{\infty} \kappa_t \lambda_2 \underbrace{(y_t^p - y_{t-1})^2}_{\text{'adjustment cost'}} \qquad (4.5)$$

with unitary disequilibrium and adjustment costs $\lambda_1, \lambda_2 > 0$ and discount factor:

$$\kappa_t := 1 \ (t = 0) \text{ and } \kappa_t := \prod_{\theta=0}^{t} \left(\frac{1}{1+r_\theta}\right) \ (t > 0).$$

Unconstrained minimization of (4.5) leads to:

$$y_t^p = \beta_0 + \beta_1 x_t + \beta_3 y_{t-1} \qquad (4.6)$$

with $\beta_0 := \frac{c\lambda_1}{\lambda_1+\lambda_2}$; $\beta_1 := \frac{\lambda_1}{\lambda_1+\lambda_2}$ and $\beta_3 := \frac{\lambda_2}{\lambda_1+\lambda_2} = 1 - \beta_1$.

If x_t is unknown at time t, the economic agents have to minimize $E(\mathcal{L}_0)$ instead of $\mathcal{L}_0$. According to the *certainty equivalence property* (4.6) becomes:

$$y_t^p = \beta_0 + \beta_1 E(x_t \,|y_{t-1}) + \beta_3 y_{t-1}, \text{ or:}$$

$$y_t = \beta_0 + \beta_1 x_t + \beta_3 y_{t-1} + \underbrace{(y_t - y_t^p) - \beta_1 (x_t - E(x_t \,|y_{t-1}))}_{\varepsilon_t}, \qquad (4.7)$$

which is precisely the AD(1,0) partial adjustment (PA) model in (4.3).

The mean lag of a PA model

Given a *normalized lag structure* $A := \{a_i | i = 0, 1, 2, ..., a_i \geqq 0, \ \sum_i a_i = 1\}$ and the corresponding *lag generating function* $A(\rho) := \sum_i a_i \rho^i$ with $\rho \in [-1, +1]$, the *mean lag* of an arbitrary linear dynamic model is defined as the first order derivative of the lag generating function at the point $\rho = 1$; in other words, the mean lag is equal to $A'(1)$.

This definition has to be applied on the PA model (4.3). Therefore, a standardized lag structure has first to be derived from solving (4.7) by successive substitution:

$$\begin{aligned} y_t &= \beta_0 + \beta_1 x_t + \beta_3 y_{t-1} + \varepsilon_t \\ &= \beta_0 \sum_{i=0}^{t-1} \beta_3^i + \beta_1 \sum_{i=0}^{t-1} \beta_3^i x_{t-i} + \sum_{i=0}^{t-1} \beta_3^i \varepsilon_{t-i} + \beta_3^t y_0. \end{aligned}$$

If $0 < \beta_3 < 1$ (which is always the case for the PA model since $\beta_3 := \frac{\lambda_2}{\lambda_1+\lambda_2}$ with $\lambda_1, \lambda_2 > 0$) and if $t \to \infty$, then:

$$y_t = \frac{\beta_0}{1 - \beta_3} + \beta_1 \sum_{i=0}^{\infty} \beta_3^i x_{t-i} + \sum_{i=0}^{\infty} \beta_3^i \varepsilon_{t-i}, \qquad (4.8)$$

from which it is clear that $\left\{a_i := \beta_1 \beta_3^i\right\}$ is a normalized lag structure ($a_i > 0$, $\sum_{i=0}^{\infty} a_i = \frac{\beta_1}{1-\beta_3} = 1$). Hence, the mean lag of the ∞ dimensional distributed lag function (4.8) may be derived either *directly* or *indirectly* (by means of a lag generating function).

Explicit or direct derivation of the mean lag of a PA model
Consider the distributed lag function:

$$y_t = \alpha + \beta_0 x_t + \beta_1 x_{t-1} + \beta_2 x_{t-2} + ... + \beta_n x_{t-n} + \varepsilon_t,$$

where:

- the *long run multiplier* is defined as $\beta_0 + \beta_1 + ... + \beta_n$ and should be finite (i.e. $\sum_{j=0}^{\infty} |\beta_j| < 0$); hence $\beta_j \to 0$ as $j \to \infty$ (β_0 = impact multiplier);
- the *mean lag* is a weighted average of the lags in which the j-period lag enters with weight β_j or with relative weight $\frac{\beta_j}{\sum_{i=0}^{\infty} \beta_i}$; hence, the mean lag equals $\frac{\sum_{j=0}^{\infty} j\beta_j}{\sum_{j=0}^{\infty} \beta_j}$;
- the *median lag* is the number of periods required for half of the long run adjustment to be completed.

Now suppose e.g. a geometrically declining lag with coefficients $\beta_j = \lambda^j \beta$, $j = 0, 1, 2, ...,\ 0 < \lambda < 1$
The corresponding ∞ dimensional distributed lag equation becomes:

$$y_t = \alpha + \beta \left(x_t + \lambda x_{t-1} + \lambda^2 x_{t-2} + ... \right),$$

which is similar to (4.8), and where β is called the *impact multiplier.* Then, the long run multiplier can be calculated as $\sum_{j=0}^{\infty} \beta_j = \beta \sum_{j=0}^{\infty} \lambda^j = \frac{\beta}{1-\lambda}$ and the mean lag can be computed as:

$$\frac{\sum_{j=0}^{\infty} j\beta_j}{\sum_{j=0}^{\infty} \beta_j} = \frac{\beta \sum_{j=0}^{\infty} j\lambda^j}{\beta \sum_{j=0}^{\infty} \lambda^j} = \frac{S}{\sum_{j=0}^{\infty} \lambda^j} = \frac{S}{\frac{1}{1-\lambda}}$$

with $S{:=}\lambda + 2\lambda^2 + 3\lambda^3 + ...$ and $\lambda S{=}\lambda^2 + 2\lambda^3 + 3\lambda^4 + ...$, or

$$S - \lambda S {=} \lambda + \lambda^2 + \lambda^3 + ... {=} \tfrac{\lambda}{1-\lambda}, \text{ such that } S = \tfrac{\lambda}{(1-\lambda)^2};$$

hence, the mean lag equals $\frac{\frac{\lambda}{(1-\lambda)^2}}{\frac{1}{1-\lambda}} = \frac{\lambda}{1-\lambda}$.

Indirect derivation of the mean lag (with a lag generating function)
The lag generating function can be derived from the normalized lag structure in (4.8) $A(\rho) = \beta_1 \sum_{i=0}^{\infty} \beta_3^i \rho^i = \frac{1-\beta_3}{1-\beta_3 \rho}$ and the mean lag equals:

$$A'(1) = \beta_3 \frac{1-\beta_3}{(1-\beta_3\rho)^2} \Bigg|_{\rho=1} = \frac{\beta_3}{1-\beta_3}, \tag{4.9}$$

where $(1 - \beta_3)$ is the adjustment coefficient so that the lower the adjustment coefficient the higher the mean lag; if β_3 is large (e.g. 0.9) there is slow adjustment to equilibrium since the mean lag is $\frac{0.9}{1-0.9} = 9$ periods and the adjustment coefficient is low (0.1) then (flexible prices require a large adjustment coefficient).

Rewriting the PA model (4.7) as:

$$\begin{aligned}\Delta y_t &= \beta_0 + \beta_1 (x_t - y_{t-1}) + \varepsilon_t \\ &= \beta_0 + \beta_1 \Delta x_t + \beta_1 (x-y)_{t-1} + \varepsilon_t,\end{aligned}$$

so that the PA model is a special case of the error correction mechanism (ECM) model (*cf. infra*):

$$\Delta y_t = \beta_0 + \underbrace{\beta_1 \Delta x_t}_{\text{impact response}} + \underbrace{(1-\beta_3)(x-y)_{t-1}}_{\text{disequilibium adjustment}} + \varepsilon_t \tag{4.10}$$

According to Friedman (1976), the costs of an increasing number of assets y owing to an extension of the total portfolio x is smaller than the costs of misallocation of stocks within a given portfolio. Hence, for the equilibrium relationship $y = \delta x$ $(0 < \delta < 1)$, a PA model is found from:

$$\min_{\{y_t^p\}} \sum_{t=0}^{\infty} \kappa_t \left[\lambda_1 (y_t^p - \delta x_t)^2 + \lambda_2 (y_t^p - y_{t-1})^2\right],$$

which leads to $y_t^p = \frac{\lambda_1 \delta x_t}{\lambda_1+\lambda_2} + \frac{\lambda_2}{\lambda_1+\lambda_2} y_{t-1}$; or by the certainty equivalence theorem:

$$y_t = \frac{\lambda_1 \delta x_t}{\lambda_1+\lambda_2} + \frac{\lambda_2}{\lambda_1+\lambda_2} y_{t-1} + \varepsilon_t = \beta_1 \delta x_t + (1-\beta_1) y_{t-1} + \varepsilon_t, \text{ i.e.}$$

$$\begin{aligned}\Delta y_t &= \beta_1 (\delta x_t - y_{t-1}) + \varepsilon_t \\ &= \beta_1 \Delta (\delta x_t) + \beta_1 (\delta x_{t-1} - y_{t-1}) + \varepsilon_t \\ &= \underbrace{\beta_1 \delta}_{\alpha_1} \Delta x_t + \underbrace{\beta_1}_{\alpha_2} (\delta x_{t-1} - y_{t-1}) + \varepsilon_t.\end{aligned}$$

Hence, the resulting optimal PA model can be rewritten as an example of an ECM model.

4.3 The error correction mechanism

From the basic linear AD(1,1) model (4.2) with $\beta_0 := 0$ and $\sum_{i=1}^{3} \beta_i = 1$ the ECM model results, or:

$$y_t = \beta_1 x_t + (1 - \beta_1 - \beta_3) x_{t-1} + \beta_3 y_{t-1} + \varepsilon_t, \text{ i.e.}$$

$$\Delta y_t = \beta_1 \Delta x_t + (1-\beta_3)(x-y)_{t-1} + \varepsilon_t, \tag{4.11}$$

which is the simple first order ECM model. The ECM model was first introduced into the econometric literature by Sargan (1964) and it was popularized by Davidson *et al.* (1978).

Remark 4.1 *It is possible to test the homogeneity restriction from:*

$$\Delta y_t = \underbrace{\beta_1}_{\theta_1} \Delta x_t + \underbrace{(1-\beta_3)}_{\theta_2} (x-y)_{t-1} + \underbrace{(\beta_1+\beta_2+\beta_3-1)}_{\theta_3} x_{t-1} + \varepsilon_t$$

with $H_0 : \theta_3 = 0 \Longleftrightarrow \sum_{i=1}^{3} \beta_i = 1.$

The mean lag of an ECM model
Apply successive substitution to (4.2) with $\beta_0 := 0$ and $0 < |\beta_3| < 1$:

$$\begin{aligned} y_t &= \beta_1 x_t + \beta_2 x_{t-1} + \beta_3 y_{t-1} + \varepsilon_t \\ &= \beta_1 x_t + \beta_2 x_{t-1} + \beta_3 \left(\beta_1 x_{t-1} + \beta_2 x_{t-2} + \beta_3 y_{t-2} + \varepsilon_{t-1}\right) + \varepsilon_t \\ &\dots \\ &= \textstyle\sum_{i=0}^{t-1} \left(\beta_3^{i-1}\beta_2 + \beta_3^{i-1}\beta_1\right) x_{t-i} - \frac{\beta_2}{\beta_3} x_t + \sum_{i=0}^{t-1} \beta_3^i \varepsilon_{t-i} + \beta_3^t y_0 \\ &= \textstyle\sum_{i=0}^{\infty} \beta_3^i \left(\frac{\beta_2}{\beta_3} + \beta_1\right) x_{t-i} - \frac{\beta_2}{\beta_3} x_t + \sum_{i=0}^{\infty} \beta_3^i \varepsilon_{t-i}. \end{aligned}$$

Now a normalized lag structure is obtained since all the weights are positive (if $\beta_1 > 0$) and sum to one $\left(\sum_{i=0}^{\infty} \beta_3^i \left(\frac{\beta_2}{\beta_3} + \beta_1\right) - \frac{\beta_2}{\beta_3} = 1\right)$, so that the lag generating function satisfies:

$$\begin{aligned} A(\rho) \quad &: = \textstyle\sum_{i=0}^{\infty} a_i \rho^i \\ &= \left(\frac{\beta_2}{\beta_3} + \beta_1\right) \textstyle\sum_{i=0}^{\infty} \beta_3^i \rho^i - \frac{\beta_2}{\beta_3} \\ &= \frac{\left(\frac{\beta_2}{\beta_3} + \beta_1\right)}{1 - \beta_3 \rho} - \frac{\beta_2}{\beta_3} \end{aligned}$$

and the mean lag of an ECM model amounts to:

$$\begin{aligned} A'(1) &= \left.\frac{\left(\frac{\beta_2}{\beta_3} + \beta_1\right)}{(1 - \beta_3\rho)^2} (-1)(-\beta_3)\right|_{\rho=1} = \left.\frac{(\beta_2 + \beta_1\beta_3)}{(1 - \beta_3\rho)^2}\right|_{\rho=1} \\ &= \frac{(\beta_2 + \beta_1\beta_3)}{(1 - \beta_3)^2} = \frac{1 - \beta_1}{1 - \beta_3} = \frac{\beta_2 + \beta_3}{1 - \beta_3}, \end{aligned} \tag{4.12}$$

so that the mean lag of an ECM model is always higher than that in a PA model (see (4.9)). Hence, the ECM model gives less flexible yet more accurate results.

Chapter 5

Univariate *ARIMA* Models

This chapter investigates the identification, estimation and forecast of autoregressive (AR), moving average (MA), and autoregressive moving average ($ARMA$) time series. If these time series are not stationary, at least one unit root will be present and the underlying time series is called an integrated autoregressive moving average ($ARIMA$) time series model. Finally, a discussion on aberrant observations is included in this chapter.

Only discrete-time time series observations will be considered here and before starting the time series analysis it is of paramount importance to attribute the right dimension to the data.

For example, in Finance, the data y_t may be defined as returns on assets expressed as either ordinary returns, compound returns, or simple returns. We prefer compound returns because aggregate compound returns over several periods are just the sum of individual compound returns and the generalization of discrete-time compound returns to continuous-time ones is straightforward. To illustrate this consider three definitions of returns: 1. ordinary returns $y_{1,t} := z_t + d_t - z_{t-1}$, 2. compound returns $y_{2,t} := \log(z_t + d_t) - \log z_{t-1}$, 3. simple retuns $y_{3,t} := \frac{z_t+d_t-z_{t-1}}{z_{t-1}}$, with z_t the closing price (spot rate) of an asset and d_t the dividend paid at period t. The advantage of using the compound returns definition becomes clear when comparing, without loss of generality for $d_t = 0$, aggregate returns, first over two periods for compound and simple returns, respectively:

$$y^{(2)}_{2,t+1} = \log z_{t+1} - \log z_{t-1} = \log z_{t+1} - \log z_t + \log z_t - \log z_{t-1} = y_{2,t+1} + y_{2,t},$$

$$y^{(2)}_{3,t+1} = \frac{z_{t+1}-z_{t-1}}{z_{t-1}} = \frac{(z_{t+1}-z_t)+(z_t-z_{t-1})}{z_{t-1}} = \frac{z_t-z_{t-1}}{z_{t-1}} + \frac{z_{t+1}-z_t}{z_{t-1}}\frac{z_t}{z_t} = y_{3,t} + y_{3,t+1}(1+y_{3,t})$$
$$= y_{3,t} + y_{3,t+1} + y_{3,t}y_{3,t+1}.$$

In general, for k-period returns:

$$y^{(k)}_{2,t+1} = y_{2,t+k} + y_{2,t+k-1} + \ldots + y_{2,t},$$

$$y^{(k)}_{3,t+1} = y_{3,t} + y_{3,t+1} + \ldots + y_{3,t+k-1} + y_{3,t}y_{3,t+1} + y_{3,t+1}y_{3,t+2} + \cdots$$
$$+ y_{3,t+k-1}y_{3,t+k-2} \cdots y_{3,t},$$

so that nonlinear terms enter in the aggregate of simple returns.

Since the first definition of ordinary returns is not dimensionless, it is clear that the second definition of compound returns is (by far) the most preferable one.

5.1 Stationary processes

After formal definitions of time series and stationary processes, several examples of stationary processes will be presented in this section.

Definition 5.1 *A discrete time series (or stochastic process) is defined as a set* $\{y_t\}$ *of observations, being taken at equidistant time points* ($t = 0, \pm 1, \pm 2, \pm 3, ...$ *or* $\mathbb{Z} := \{0, \pm 1, \pm 2, \pm 3, ...\}$*).*

For linear models, first and second order moments are very characteristic, if they exist:
$E(y_t) = \mu_t$ (expected value)
$E[(y_t - \mu_t)^2] = \sigma_t^2$ (variance)
$E[(y_t - \mu_t)(y_{t+k} - \mu_{t+k})] = \gamma_{t,k}$ ((auto)covariance).

Definition 5.2 *A stochastic process is called strictly ('strongly') stationary if the simultaneous pdf of the vector* $(y_t, y_{t+1}, y_{t+2}, ..., y_{t+k})$ *is independent of* t *for every* $k \epsilon \mathbb{N}$.

Definition 5.3 *A stochastic process* $\{y_t, t \in \mathbb{Z}\}$ *is called (covariance, second order, weak or wide) stationary if* $E(y_t)$ *and* $E(y_t y_{t+k})$ *are finite and do not depend on* t*, for every* $k \epsilon \mathbb{N}$.

It is clear that a strictly stationary time series with finite second order moments is also stationary.

Examples of stationary processes

1. White noise (WN) process describing an *iid* stochastic variable with expectation zero and constant variance, or $\varepsilon_t \overset{iid}{\sim} (0, \sigma_\varepsilon^2)$. For the WN stochastic process $\{\varepsilon_t, t \in \mathbb{Z}\}$: $E(\varepsilon_t) = \mu = 0$ and $E[(\varepsilon_t - \mu)(\varepsilon_s - \mu)] = \delta_{t,s}\sigma_\varepsilon^2$ (with $\delta_{t,s} := 1$ if $t = s$ and $\delta_{t,s} := 0$ if $t \neq s$) $\forall t, s$.

2. Noise is a stationary process with constant expected value (e.g. zero) and $cov(\varepsilon_t, \varepsilon_{t+k}) = \gamma_k$.

3. The deterministic trigonometric series $y_t = A\cos(t\lambda) + B\sin(t\lambda)$, where A and B are known uncorrelated random variables with mean zero and variance σ^2 and where λ is a known number, is a stationary time series. Indeed, $E(y_t) = 0$ and

$$\begin{aligned}\gamma_h &= cov(y_t, y_{t+h}) \\ &= \cos(t\lambda)\cos((t+h)\lambda)var(A) + \sin(t\lambda)\sin((t+h)\lambda)var(B) = \sigma^2\cos(h\lambda)\end{aligned}$$

Even though A and B are random variables, this type of time series is called *deterministic* in time series theory. Once A and B have been determined (at time $-\infty$ say), the process behaves as a deterministic trigonometric function. This type of time series is an important building block when modeling cyclic events in a system, but it is not a typical example of a time series that will be studied in this book. Predicting the future would be too easy in this case.

4. <u>$AR(1)$</u> *process* $y_t = \alpha y_{t-1} + \varepsilon_t$ is stationary if $|\alpha| < 1$, because then:

(a) $E(y_t) = E\left(\sum_{i=0}^{t-1} \alpha^i \varepsilon_{t-i} + \alpha^t y_o\right) = 0$ provided that $E(y_o) = 0$;

(b) $\sigma_y^2 = E(y_t^2) = E\left[\left(\sum_{i=0}^{t-1} \alpha^i \varepsilon_{t-i} + \alpha^t y_o\right)\left(\sum_{j=0}^{t-1} \alpha^j \varepsilon_{t-j} + \alpha^t y_o\right)\right]$ is only finite for increasing t if $|\alpha| < 1$ (then $\sigma_y^2 = \sum_{i=0}^{\infty} \alpha^{2i} E\left(\varepsilon_{t-i}^2\right) = \frac{\sigma_\varepsilon^2}{1-\alpha^2} < \infty$); [1]

(c) the autocovariance only depends on the length of the time period between two observations:

$$\begin{aligned} cov(y_t, y_{t+k}) &= E[y_t(\alpha y_{t+k-1} + \varepsilon_{t+k})] \\ &= E\left[y_t\left(\alpha^k y_t + \textstyle\sum_{j=0}^{k-1} \alpha^j \varepsilon_{t+k-j}\right)\right] \\ &= \alpha^k \sigma_y^2 = \gamma_k = \gamma_{-k} \ (k = 0, \pm 1, \pm 2, ...). \end{aligned}$$

In general, the *random walk* model $\Delta y_t := y_t - y_{t-1} = \varepsilon_t$ with $\varepsilon_t \overset{iid}{\sim} (0, \sigma_\varepsilon^2)$ is not stationary because the belonging variances are not finite:

$$var(y_t) = var\left(\textstyle\sum_{i=0}^{t-1} \varepsilon_{t-i} + y_0\right) = \textstyle\sum_{i=0}^{t-1} var(\varepsilon_t) + var(y_0) = t\sigma_\varepsilon^2 + \sigma_{y_0}^2.$$

Clearly, increases in t lead to exploding values of $var(y_t)$.

Definition 5.4 *The autocovariance function of a stationary process* $\{y_t, t \in \mathbb{Z}\}$ *is a row* $\{\gamma_k\}$, $k = 0, \pm 1, \pm 2, ...$

Definition 5.5 *Autocorrelation is a normalized autocovariance, or* $\rho_k := \frac{\gamma_k}{\gamma_0}$, $k = 0, \pm 1, \pm 2, ...$

E.g., the kth autocorrelation of an $AR(1)$ process is $\rho_k := \frac{\alpha^k \sigma_y^2}{\sigma_y^2} = \alpha^k$.

Definition 5.6 *The autocorrelation function (ACF) of a stationary process* $\{y_t, t \in \mathbb{Z}\}$ *is a row* $\{\rho_k\}$, $k = 0, \pm 1, \pm 2, ...$ *for which* $\rho_k = \rho_{-k}$ *and* $\rho_0 = 1$.

[1] If we substitute forwardly, the appropiate stationarity condition is $|\alpha| > 1$ since the $AR(1)$ model implies that $y_{t-1} = \frac{1}{\alpha} y_t - \frac{1}{\alpha} \varepsilon_t$ then.

5.2 Autoregressive (time series) processes

5.2.1 Stationary autoregressive processes

Definition 5.7 *A stationary process* $\{y_t, t \in \mathbb{Z}\}$ *is called an autoregressive (AR) process of order p if there exists a WN error process* $\{\varepsilon_t, t \in \mathbb{Z}\}$ *such that* $\alpha(L) y_t = y_t - \alpha_1 y_{t-1} - \alpha_2 y_{t-2} - ... - \alpha_p y_{t-p} = \varepsilon_t$, $t \in \mathbb{Z}$.

Stationarity (or stability) condition:

An $AR(p)$ process $\{y_t, t \in \mathbb{Z}\}$ is stationary if all roots z of the equation $\alpha(z) := 1 - \alpha_1 z - \alpha_2 z^2 - ... - \alpha_p z^p = 0$ have a modulus $|z| > 1$, where the modulus of a complex number $z = z_1 + iz_2$ is defined as $z_E := \sqrt{z_1^2 + z_2^2}$ (Euclidean length).

Comparing this definition to Proposition C.1 in Appendix C or $\alpha(\lambda) = (-1)^p (\lambda^p - \alpha_1 \lambda^{p-1} - \alpha_2 \lambda^{p-2} - \ldots - \alpha_p) = 0$, it is logic that the characteristic roots z should lie outside the unit circle because they are just the inverse of the corresponding eigenvalues λ, or $z = \frac{1}{\lambda}$.

Property *Invertibility* of an AR process (with lag or backward operator L ($Ly_t := y_{t-1}$, then $L^i y_t = y_{t-i}$) and identity operator I ($Iy_t := y_t$)).

If $\alpha(L)$ and $\beta(L)$ are two polynomials in L such that $\alpha(L)\beta(L) = I$, where the polynomial orders p and/or q can be infinite, then $\beta(L)$ is called the inverse of $\alpha(L)$ or $\beta(L) = \alpha^{-1}(L)$.

In general, $\alpha(L)$ is said to be *invertible* if the characteristic roots z of $\alpha(z) := \sum_{i=1}^{p} \alpha_i z^i = 0$ are larger than one in absolute value. *An* $AR(p)$ *process* $\{y_t, t \in \mathbb{Z}\}$ *is stationary if and only if* $\alpha(L)$ *is invertible.*

Example: Suppose a stationary $AR(1)$ model $y_t = \alpha y_{t-1} + \varepsilon_t$, with $\varepsilon_t \overset{iid}{\sim} (0, \sigma_\varepsilon^2)$ then $\alpha^{-1}(L) = I + \alpha L + \alpha L^2 + ...$ because

$$\alpha(L)\alpha^{-1}(L) = (I - \alpha L)(I + \alpha L + \alpha^2 L^2 + ...) = 1$$

Computation of the autocorrelations of an AR process: Yule-Walker equations

Premultiplying an $AR(p)$ process with y_{t-k}, taking expectations and dividing by the variance γ_0 of $\{y_t\}$:

$$\frac{E(y_{t-k}y_t)}{\gamma_0} = \alpha_1 \frac{E(y_{t-k}y_{t-1})}{\gamma_0} + ... + \alpha_p \frac{E(y_{t-k}y_{t-p})}{\gamma_0} + \frac{E(y_{t-k}\varepsilon_t)}{\gamma_0}$$

and since $E(y_{t-k}\varepsilon_t) = E(y_t) = 0$ for all $t, k > 0$:

$$\rho_k = \alpha_1 \rho_{k-1} + \alpha_2 \rho_{k-2} + ... + \alpha_p \rho_{k-p} \text{ for } k > 0,$$

or as a system:

$$\begin{aligned} \rho_1 - \alpha_1 \rho_0 - \alpha_2 \rho_1 - ... - \alpha_p \rho_{p-1} &= 0 \\ \rho_2 - \alpha_1 \rho_1 - \alpha_2 \rho_0 - ... - \alpha_p \rho_{p-2} &= 0 \\ \vdots \qquad\qquad &\quad \vdots \\ \rho_p - \alpha_1 \rho_{p-1} - \alpha_2 \rho_{p-2} - ... - \alpha_p \rho_0 &= 0, \end{aligned}$$

Defining:

$$\mathbf{R}:=\begin{pmatrix} \rho_0 & \rho_1 & \rho_2 & \rho_3 & \cdots & \rho_{p-1} \\ \rho_1 & \rho_0 & \rho_1 & \rho_2 & \cdots & \rho_{p-2} \\ \rho_2 & \rho_1 & \rho_0 & \rho_1 & \cdots & \rho_{p-3} \\ \vdots & \vdots & \vdots & \vdots & \ddots & \vdots \\ \rho_{p-1} & \rho_{p-2} & \rho_{p-3} & \rho_{p-4} & \cdots & \rho_0 \end{pmatrix}, \boldsymbol{\alpha}:=\begin{pmatrix} \alpha_1 \\ \alpha_2 \\ \alpha_3 \\ \vdots \\ \alpha_p \end{pmatrix}, \boldsymbol{\rho}:=\begin{pmatrix} \rho_1 \\ \rho_2 \\ \rho_3 \\ \vdots \\ \rho_p \end{pmatrix},$$

the above system, which is called the system of *Yule-Walker equations*, can be written in matrix notation as $\mathbf{R}\,\boldsymbol{\alpha} = \boldsymbol{\rho}$ with general solution:

$$\rho_k = \varphi_1\lambda_1^k + \varphi_2\lambda_2^k + ... + \varphi_p\lambda_p^k \text{ for } k > 0,$$

where the eigenvalues λ_i $(i = 1, 2, \ldots, p)$ are the inverse of the roots z of the characteristic equation $1 - \alpha_1 z - \alpha_2 z^2 - \ldots - \alpha_p z^p = 0$.

The kth component of $\boldsymbol{\alpha}$ is the kth *partial autocorrelation coefficient* (denoted as ρ_{kk}), which is directly computed according to Cramer's rule as $\alpha_k = \rho_{kk} = \frac{|\mathbf{R}^*|}{|\mathbf{R}|}$, where $\mathbf{R}^*$ is the matrix $\mathbf{R}$ (being a nonnegative definite Laurent matrix) with the last column replaced by the vector $\boldsymbol{\rho}$. In general $\boldsymbol{\alpha} = \mathbf{R}^{-1}\boldsymbol{\rho}$, so that the autoregressive parameters can be expressed as a function of the p autocorrelations.

5.2.2 Estimation and identification of *AR* processes

Two alternative measures exist for the identification of an $AR\,(p)$ process, i.e. for the determination of the unknown value of p (and the estimation of the partial autocorrelations). These measures are based on the 'regression' method and the sample Yule-Walker equations, respectively.

The 'regression' method

If a sample is indeed generated by an $AR\,(p)$ process, we have to determine p and the corresponding parameters $\alpha_1, \alpha_2, ..., \alpha_p$ describing the $AR\,(p)$ process. Knowing p, α_i $(i = 1, 2, ..., p)$ can be found from the regression with stochastic regressors $y_t = \alpha_1 y_{t-1} + \alpha_2 y_{t-2} + ... + \alpha_p y_{t-p} + \varepsilon_t$.

Since the first p observations in an $AR(p)$ model are unknown, the $AR(p)$ model is rewritten for the remaining $T - p$ observations:

$$\begin{aligned} y_{p+1} &= \alpha_1 y_p + \alpha_2 y_{p-1} + ... + \alpha_p y_1 + \varepsilon_{p+1} \\ y_{p+2} &= \alpha_1 y_{p+1} + \alpha_2 y_p + ... + \alpha_p y_2 + \varepsilon_{p+2} \\ \vdots & \qquad \vdots \\ y_T &= \alpha_1 y_{T-1} + \alpha_2 y_{T-2} + ... + \alpha_p y_{T-p} + \varepsilon_T \end{aligned}$$

or in matrix notation $\mathbf{y} = \mathbf{Y}\boldsymbol{\alpha} + \boldsymbol{\varepsilon}$ with $\mathbf{y} := (y_{p+1}, y_{p+2}, \ldots, y_T)'$,

$$\mathbf{Y} := \begin{pmatrix} y_p & y_{p-1} & \cdots & y_1 \\ y_{p+1} & y_p & \cdots & y_2 \\ \vdots & \vdots & \ddots & \vdots \\ y_{T-1} & y_{T-2} & \cdots & y_{T-p} \end{pmatrix} \text{ and } \boldsymbol{\varepsilon} := \begin{pmatrix} \varepsilon_{p+1} \\ \varepsilon_{p+2} \\ \vdots \\ \varepsilon_T \end{pmatrix}, \text{ so that the OLS}$$

estimator $\hat{\boldsymbol{\alpha}} = (\mathbf{Y}'\mathbf{Y})^{-1}\mathbf{Y}'\mathbf{y}$ is consistent and asymptotically normally distributed with estimated variance-covariance matrix $\hat{\boldsymbol{\Sigma}}_{\hat{\boldsymbol{\alpha}}} = \hat{\sigma}_{\varepsilon}^2 (\mathbf{Y}'\mathbf{Y})^{-1}$ and estimated variance $\hat{\sigma}_{\hat{\varepsilon}}^2 = \frac{(\mathbf{y}-\mathbf{Y}\hat{\boldsymbol{\alpha}})'(\mathbf{y}-\mathbf{Y}\hat{\boldsymbol{\alpha}})}{T-2p}$.

The order of the $AR(p)$ process is equal to the last significant regression coefficient $\hat{\alpha}_p$ of $\hat{\boldsymbol{\alpha}}$ ($\hat{\alpha}_p = \hat{\rho}_{pp}$).

Definition 5.8 *The row $\{\rho_{kk}\}$, $k = 1, 2, ...$ of partial autocorrelations is called the partial autocorrelation function (PACF).*

Hence we adopt as *identification rule* for the choice of an $AR(p)$ model if $\rho_{kk} \neq 0$ for $k = p$ and $\rho_{kk} = 0$ for $k > p$.

To test the significance of $\hat{\rho}_{kk}$, we have to know the probability distribution of $\hat{\rho}_{kk} = \hat{\alpha}_k$. Hence, under H_0 that $\boldsymbol{\alpha} = 0$, (at least) the variance of the estimators $\hat{\alpha}_k$ is needed.

Example 5.1 *$AR(1)$ versus $AR(2)$ processes:*

$$AR(1): \; y_t = \alpha y_{t-1} + \varepsilon_t; \; \varepsilon_t \overset{iid}{\sim} (0, \sigma_{\varepsilon}^2), t \in \mathbb{Z}$$
$$AR(2): \; y_t = \alpha_1 y_{t-1} + \alpha_2 y_{t-2} + \eta_t \text{ with } \eta_t \overset{iid}{\sim} (0, \sigma_{\eta}^2) \text{ and } t \in \mathbb{Z}.$$

Minimizing $\sum_t \varepsilon_t^2 = \sum_t (y_t - \alpha y_{t-1})^2$ w.r.t. α we find as OLS estimator $\hat{\alpha} = \frac{\sum_t y_t y_{t-1}}{\sum_t y_{t-1}^2}$, with variance

$$\begin{aligned} var(\hat{\alpha}) &= \sigma_{\varepsilon}^2 \left(E\left\{\textstyle\sum_t y_{t-1}^2\right\}\right)^{-1} = \sigma_{\varepsilon}^2 \left(\textstyle\sum_t E\left(y_{t-1}^2\right)\right)^{-1} \\ &= \sigma_{\varepsilon}^2 T^{-1} \left(\frac{\sigma_{\varepsilon}^2}{1-\alpha^2}\right)^{-1} = \frac{1-\alpha^2}{T} \qquad (|\alpha| < 1) \end{aligned}$$

Hence, under $H_0: \alpha = \rho = 0$, $\sqrt{T}\hat{\alpha} \overset{as}{\sim} N(0, 1)$.

Alternatively, when minimizing $\sum_t \eta_t^2 = \sum_t (y_t - \alpha_1 y_{t-1} - \alpha_2 y_{t-2})^2$ w.r.t. α_1 and α_2, we find as OLS estimator:

$$\begin{pmatrix} \hat{\alpha}_1 \\ \hat{\alpha}_2 \end{pmatrix} = \begin{pmatrix} \sum_t y_{t-1}^2 & \sum_t y_{t-1} y_{t-2} \\ \sum_t y_{t-1} y_{t-2} & \sum_t y_{t-2}^2 \end{pmatrix}^{-1} \begin{pmatrix} \sum_t y_t y_{t-1} \\ \sum_t y_t y_{t-2} \end{pmatrix}$$

and corresponding variance-covariance matrix:

$$\begin{aligned} var\begin{pmatrix} \hat{\alpha}_1 \\ \hat{\alpha}_2 \end{pmatrix} &= \sigma_{\eta}^2 \begin{pmatrix} E\left(\sum_t y_{t-1}^2\right) & E\left(\sum_t y_{t-1} y_{t-2}\right) \\ E\left(\sum_t y_{t-1} y_{t-2}\right) & E\left(\sum_t y_{t-2}^2\right) \end{pmatrix}^{-1} \\ &= T^{-1}\sigma_{\eta}^2 \begin{pmatrix} E\left(y_{t-1}^2\right) & E\left(y_{t-1} y_{t-2}\right) \\ E\left(y_{t-1} y_{t-2}\right) & E\left(y_{t-2}^2\right) \end{pmatrix}^{-1} \end{aligned}$$

$$= T^{-1}\sigma_\eta^2 \begin{pmatrix} E\left(y_{t-1}^2\right) & E\left(\left(\alpha_1 y_{t-2}+\alpha_2 y_{t-3}+\eta_{t-1}\right) y_{t-2}\right) \\ E\left(\left(\alpha_1 y_{t-2}+\alpha_2 y_{t-3}+\eta_{t-1}\right) y_{t-2}\right) & E\left(y_{t-2}^2\right) \end{pmatrix}^{-1}$$

$$= T^{-1}\sigma_\eta^2 \begin{pmatrix} E\left(y_{t-1}^2\right) & \alpha_1 E\left(y_{t-2}^2\right) \\ \alpha_1 E\left(y_{t-2}^2\right) & E\left(y_{t-2}^2\right) \end{pmatrix}^{-1}$$

$$= T^{-1}\sigma_\eta^2 \left(\frac{\sigma_\eta^2}{1-\alpha_1^2}\right)^{-1} \begin{pmatrix} 1 & \alpha_1 \\ \alpha_1 & 1 \end{pmatrix}^{-1} = T^{-1} \begin{pmatrix} 1 & -\alpha_1 \\ -\alpha_1 & 1 \end{pmatrix}.$$

Hence, under $H_0 : \alpha_1 = \alpha_2 = 0$,

$$\sqrt{T} \begin{pmatrix} \hat{\alpha}_1 \\ \hat{\alpha}_2 \end{pmatrix} \overset{as}{\sim} N_2 \left(\begin{pmatrix} 0 \\ 0 \end{pmatrix}, \begin{pmatrix} 1 & 0 \\ 0 & 1 \end{pmatrix} \right).$$

By induction, we have for a general $AR(p)$ model under H_0 $(\alpha_1 = \alpha_2 = ... = \alpha_p = 0)$:

$$\sqrt{T} \begin{pmatrix} \hat{\alpha}_1 \\ \hat{\alpha}_2 \\ \vdots \\ \hat{\alpha}_p \end{pmatrix} \overset{as}{\sim} N_p \left(\begin{pmatrix} 0 \\ 0 \\ \vdots \\ 0 \end{pmatrix}, \begin{pmatrix} 1 & 0 & \cdots & 0 \\ 0 & 1 & \cdots & 0 \\ \vdots & \vdots & \ddots & \vdots \\ 0 & \cdots & \cdots & 1 \end{pmatrix} \right),$$

or, in general, the (asymptotic) variance of $\hat{\alpha}_k = \hat{\rho}_{kk}$ (partial autocorrelation coefficient) is estimated as $\widehat{var}\left(\hat{\rho}_{kk}\right) = \frac{1}{T}$ $(\forall k > 0)$ and $\sqrt{T}\hat{\rho}_{kk} \overset{as}{\sim} N(0,1)$ under H_0.

Sample Yule-Walker equations

Given a consistent estimator $r_k = \frac{s_k}{s_0}$ $(k = 0, 1, 2, ..., K)$ of the autocorrelation coefficients with $s_k := \frac{1}{T}\sum_{t=1}^{T-k} \left(x_{t+k} - \overline{x}\right)\left(x_t - \overline{x}\right)$, the partial autocorrelation coefficients ρ_{kk} can consistently be estimated by Cramer's Rule as:

$$\hat{\alpha}_k = r_{kk} = \frac{|\hat{\mathbf{R}}^*|}{|\hat{\mathbf{R}}|}, \quad k = 1, 2, ...$$

with $\hat{\mathbf{R}}$ the estimated nonnegative definite Laurent matrix:

$$\hat{\mathbf{R}} := \begin{pmatrix} 1 & r_1 & r_2 & \cdots & r_{k-1} \\ r_1 & 1 & r_1 & \cdots & r_{k-2} \\ \vdots & \vdots & \vdots & \ddots & \vdots \\ r_{k-1} & r_{k-2} & r_{k-3} & \cdots & 1 \end{pmatrix}$$

and $\hat{\mathbf{R}}^*$ the matrix $\hat{\mathbf{R}}$ with the last column replaced by $\hat{\boldsymbol{\rho}} := \left(r_1, r_2, \ldots, r_k\right)'$.

Notice that both methods (i.e. the regression method and the sample Yule-Walker equations) are asymptotically equivalent.

Example 5.2 *The following estimated autocorrelations are obtained from a row of T=200 observations, originating from a stochastic process for which*

$\rho_1 = -0.4$ *and* $\rho_k = 0,\ k \geqq 2.$

k	1	2	3	4	5	6	7	8	9	10
r_k	-0.38	-0.08	0.11	0.08	0.02	0.00	0.00	0.00	0.07	-0.08

Since $\hat{\sigma}^2_{r_k} = \frac{1}{T} = \frac{1}{200} = 0.005$, the standard deviation of the autocorrelations is 0.07 and since $r_1 = -0.38$, being in modulus more than 5 times the standard deviation, it can directly be concluded that $\rho_1 \neq 0$; $\rho_k = 0$ for $k \geqq 2$.

Example 5.3 *Suppose a row of T=50 observations with as estimated autocorrelations* $r_k = \hat{\rho}_k$*:*

k	1	2	3	4	5	6	7	8	9	10	11	12
r_k	0.84	0.73	0.61	0.54	0.47	0.46	0.38	0.29	0.17	0.05	0.04	-0.01

According to Cramer's rule solutions of the Yule-Walker equations, the estimated partial autocorrelations satisfy:

$$r_{11} = 0.84,\ r_{22} = \frac{\begin{vmatrix} 1 & 0.84 \\ 0.84 & 0.73 \end{vmatrix}}{\begin{vmatrix} 1 & 0.84 \\ 0.84 & 1 \end{vmatrix}} = 0.08,\ r_{33} = \frac{\begin{vmatrix} 1 & 0.84 & 0.84 \\ 0.84 & 1 & 0.73 \\ 0.73 & 0.84 & 0.61 \end{vmatrix}}{\begin{vmatrix} 1 & 0.84 & 0.73 \\ 0.84 & 1 & 0.84 \\ 0.73 & 0.84 & 1 \end{vmatrix}} = -0.075.$$

Since, under $H_0 : \rho_{kk} = 0\ (k = 1, 2, ...)$:

$$\sqrt{T}\begin{pmatrix} r_{11} \\ r_{22} \\ \vdots \\ r_{kk} \end{pmatrix} \overset{as}{\sim} N_k\left(\begin{pmatrix} 0 \\ 0 \\ \vdots \\ 0 \end{pmatrix}, \begin{pmatrix} 1 & 0 & \cdots & 0 \\ 0 & 1 & \cdots & 0 \\ \vdots & \vdots & \ddots & \vdots \\ 0 & 0 & \cdots & 1 \end{pmatrix}\right),$$

$\hat{var}(r_{kk}) = T^{-1} = 0.02$, or $2 * \hat{\sigma}_{r_{kk}} = 0.28$ and r_{22} and r_{33} are *not* significant.

5.3 Moving average (time series) processes

Suppose the following $AR(\infty)$ time series process:

$$y_t = -\alpha y_{t-1} - \alpha^2 y_{t-2} - \alpha^3 y_{t-3} - ... + \varepsilon_t$$

with $\varepsilon_t \overset{iid}{\sim} (0, \sigma^2_\varepsilon)$, or with lag operator L:

$$\left(I + \alpha L + \alpha^2 L^2 + \alpha^3 L^3 + ...\right) y_t = \frac{y_t}{1-\alpha L} = \varepsilon_t,$$

y_t is stationary if $|\alpha| < 1$. Then, $y_t = (I - \alpha L)\,\varepsilon_t$.

Hence, the above stationary $AR(\infty)$ stochastic process for y_t can be rewritten as a weighted sum of a WN process, and so as a moving average of order one ($MA(1)$).

5.3.1 Definition

An MA stochastic process can be interpreted as the output of a linear filter on an input process of WN terms (see Wold's decomposition theorem,[2] Wold (1938)), or, more precisely:

Definition 5.9 *A stochastic process $\{y_t, t \in \mathbb{Z}\}$ is called an MA process of order q, if there exists a WN process $\{\varepsilon_t, t \in \mathbb{Z}\}$ such that:*

$$y_t = \sum_{j=0}^{q} \varphi_j \varepsilon_{t-j}, \quad t \in \mathbb{Z},$$

with $F(z) := \sum_{j=0}^{q} \varphi_j z^j \neq 0$ and $|z| < 1$, $\varphi_q \neq 0$ and $\varphi_0 := 1$.

Proposition 5.1 *Any stationary AR process can be written as an MA process*

In general, we have an $MA(q)$ process:

$$\begin{aligned} y_t &= \varepsilon_t + \varphi_1 \varepsilon_{t-1} + \varphi_2 \varepsilon_{t-2} + ... + \varphi_q \varepsilon_{t-q} \\ &= \left(1 + \varphi_1 L + \varphi_2 L^2 + ... + \varphi_q L^q\right) \varepsilon_t = \varphi(L) \varepsilon_t. \end{aligned}$$

Such an $MA(q)$ process can be written as an $AR(\infty)$ process if all characteristic roots z of $\varphi(z) = 1 + \varphi_1 z + \varphi_2 z^2 + ... + \varphi_q z^q = 0$ lie outside the (complex) unit circle ($|z| > 1$). Then the polynomial $\varphi(z)$ is said to be *invertible*. In this case the $MA(q)$ process can be written as an $AR(\infty)$ process *uniquely*.

Remark that, alternatively formulated, the *invertibility condition* on $\varepsilon_t = [\varphi(L)]^{-1} y_t$ can be argued from:

$$y_t = \left(1 + \varphi_1 L + \varphi_2 L^2 + ... + \varphi_q L^q\right) \varepsilon_t = \prod_{i=1}^{q} (I - \pi_i L) \varepsilon_t,$$

where $\pi_1, \pi_2, \ldots, \pi_q$ are the roots of the equation $\sum_{i=0}^{q} \varphi_i z^{q-i} = 0$ ($\varphi_0 := 1$) then, the condition for invertibility is that $|\pi_i| < 1$ for all $i = 1, 2, \ldots, q$.

Example 5.4 *For the $MA(2)$ process $y_t = \left(1 + \varphi_1 L + \varphi_2 L^2\right) \varepsilon_t$, π_1 and π_2 are roots of the quadratic equation $z^2 + \varphi_1 z + \varphi_2 = 0$ and the invertibility condition $|\pi_i| < 1$ implies $\frac{1}{2}\left|-\varphi_1 \pm \sqrt{\varphi_1^2 - 4\varphi_2}\right| < 1$, which yields the result that φ_1 and φ_2 must satisfy $\varphi_1 + \varphi_2 > -1$, $\varphi_2 - \varphi_1 > -1$, and $|\varphi_2| < 1$.*

The last condition in this example is derived from the fact that φ_2 is equal to the product of the roots, or $\varphi_2 = \pi_1 \pi_2$, and the first two conditions are derived from the fact that if $\varphi_1^2 - 4\varphi_2 > 0$, then $\varphi_1^2 - 4\varphi_2 < (2 + \varphi_1)^2$ or $\varphi_1^2 - 4\varphi_2 < (2 - \varphi_1)^2$. Under the above three conditions, the example $MA(2)$ model can uniquely be written as an $AR(\infty)$ model.

[2] This theorem states that for any stationary stochastic process, for a given choice of p, there is a Wold representation of the stationary series:

$$y_t = \sum_{i=1}^{p} \alpha_i y_{t-i} + \sum_{i=0}^{\infty} \varphi_i \varepsilon_{t-i}$$

with $\sum_{i=1}^{p} \alpha_i y_{t-i}$ a *linearly 'deterministic' component* and $\sum_{i=0}^{\infty} \varphi_i \varepsilon_{t-i}$ a *linearly nondeterministic (stochastic) component.*

5.3.2 Identification of an MA process

Identifying an MA process is equivalent to determining the appropriate number q of lags.

Proposition 5.2 *The order q of an MA process is the maximum value of k for which the autocorrelation coefficient ρ_k is different from zero*

Proof

In the case of the $MA(1)$ process $y_t = \varepsilon_t - \alpha\varepsilon_{t-1}$ with $\{y_t, t \in \mathbb{Z}\}$ (equivalent to a stationary $AR(1)$ process), then:

$$\begin{aligned}
E(y_t) &= E(\varepsilon_t) - \alpha E(\varepsilon_{t-1}) = 0 \\
cov(y_t, y_{t+k}) &= E(y_t y_{t+k}) = E(\varepsilon_t \varepsilon_{t+k}) - \alpha E(\varepsilon_{t-1}\varepsilon_{t+k}) \\
&\quad -\alpha E(\varepsilon_t \varepsilon_{t+k-1}) + \alpha^2 E(\varepsilon_{t-1}\varepsilon_{t+k-1}) \\
&= \begin{cases} \sigma_\varepsilon^2 + \alpha^2\sigma_\varepsilon^2 & if\ k = 0 \\ -\alpha\sigma_\varepsilon^2 & if\ k = \pm 1 \\ 0 & \text{elsewhere} \end{cases}
\end{aligned}$$

Hence, $cov(y_t, y_{t+k})$ only depends on k, or the process $\{y_t, t \in \mathbb{Z}\}$ is stationary: $cov(y_t, y_{t+k}) = \gamma_k$.

For a general $MA(q)$ process:

$$\begin{aligned}
E(y_t) &= E(\varepsilon_t) + \varphi_1 E(\varepsilon_{t-1}) + ... + \varphi_q E(\varepsilon_{t-q}) = 0 \\
cov(y_t, y_{t+k}) &= E(y_t y_{t+k}) \\
&= E\left\{(\varepsilon_t + \varphi_1\varepsilon_{t-1} + ... + \varphi_q\varepsilon_{t-q})\right. \\
&\quad \left.(\varepsilon_{t+k} + \varphi_1\varepsilon_{t+k-1} + ... + \varphi_q\varepsilon_{t+k-q})\right\} \\
&= E\left\{\left(\textstyle\sum_{i=0}^q \varphi_i\varepsilon_{t-i}\right)\left(\textstyle\sum_{j=0}^q \varphi_j\varepsilon_{t+k-j}\right)\right\} \\
&= E\left\{\textstyle\sum_{i=0}^q \sum_{j=0}^q \varphi_i\varphi_j\varepsilon_{t-i}\varepsilon_{t+k-j}\right\} \\
&= \textstyle\sum_{i=0}^q \sum_{j=0}^q \varphi_i\varphi_j E\left\{\varepsilon_{t-i}\varepsilon_{t+k-j}\right\} \\
&= \begin{cases} \sigma_\varepsilon^2 \sum_{i=0}^{q-k} \varphi_i\varphi_{i+k} & for\ k = 0, 1, .., q \\ 0 & for\ k > q, \end{cases}
\end{aligned}$$

which illustrates the stationarity of an $MA(q)$ process.

Thus, in general, the *autocovariances* of an $MA(q)$ process are:

$$\gamma_k = \begin{cases} \sigma_\varepsilon^2 \sum_{i=0}^{q-k} \varphi_i\varphi_{i+k} & for\ k = 0, 1, .., q \text{ with } \varphi_0 := 1 \\ 0 & \text{for } k > q \end{cases} \tag{5.1}$$

and its *autocorrelations* are:

$$\rho_k = \frac{\gamma_k}{\gamma_0} = \begin{cases} \frac{\sum_{i=0}^{q-k} \varphi_i\varphi_{i+k}}{\sum_{i=0}^{q-k} \varphi_i^2} & for\ k = 0, 1, .., q \text{ with } \varphi_0 := 1 \\ 0 & \text{elsewhere} \end{cases} \tag{5.2}$$

If a sample is available $\{y_t, t = 1, 2, ..., T\}$, consistent and unbiased estimators of the autocovariances (5.1) and the autocorrelations (5.2) can directly be provided.■

5.3.3 Parameter estimation of an *MA*(q) process

If a data generating process (DGP) of a time series $\{y_t, t = 1, 2, ..., T\}$ is $MA(q)$, then the MA parameters are estimated by OLS, minimizing the residual sum of squares:

$$S(\varphi) = \sum_{t=1}^{T} \varepsilon_t^2 = \sum_{t=1}^{T} \left(y_t - \varphi_1 \varepsilon_{t-1} - \varphi_2 \varepsilon_{t-2} - ... - \varphi_q \varepsilon_{t-q}\right)^2$$

with respect to $\boldsymbol{\varphi} := \left(\varphi_1, \varphi_2, ..., \varphi_q\right)'$. Since the WN process $\{\varepsilon_t, t \in \mathbb{Z}\}$ is not observable, we have to replace it by observations $\{y_1, y_2, ..., y_T\}$ (AR process).

Example 5.5 *Assume an* $MA(1)$ *process* $y_t = \varepsilon_t - \varphi\varepsilon_{t-1}$ $(|\varphi| < 1)$, *which can be rewritten as an* $AR(\infty)$ *process:*

$$y_t + \varphi y_{t-1} + \varphi^2 y_{t-2} + ... = \varepsilon_t,$$

so that the residual sum of squares is rewritten as:

$$S(\varphi) = \sum_{t=1}^{T} \varepsilon_t^2 = \sum_{t=1}^{T} \left(y_t + \varphi y_{t-1} + \varphi^2 y_{t-2} + ... + \varphi^{t-1} y_1\right)^2,$$

which can be minimized by procedures as quasi-Newton (e.g. Davidon-Fletcher-Powell), Marquardt, etc. from Chapter 1 and Appendix A or by a simple grid search for $\varphi \in (-1, 1)$.

5.4 *ARMA* models

Combine an MA process of order q with a linear difference equation of order p to obtain:

$$y_t = a_0 + \sum_{i=1}^{p} a_i y_{t-i} + \sum_{i=0}^{q} \varphi_i \varepsilon_{t-i}, \tag{5.3}$$

which is an *autoregressive moving average* model of order (p, q) or an $ARMA(p, q)$ model (typically $\varphi_0 := 1$). If $q = 0$ we have a pure $AR(p)$ model and if $p = 0$ we have a pure $MA(q)$ process. Both p and q can be infinite.

In the sequel of this section, the impact of stationarity restrictions for $ARMA$ models will be analyzed.

5.4.1 Stationarity restrictions

In general, a stationary finite order process $AR(p)$ can be written as an infinite order MA with restricted parameters and a finite order $MA(q)$ can be represented as a restricted infinite order AR. Hence, if the AR polynomial $\alpha(L)$ is invertible, the process $\{y_t, t \in \mathbb{Z}\}$ is stationary and can be written as an $MA(\infty)$ process. On the other hand, if the MA polynomial $\varphi(L)$ is invertible, the $\{y_t, t \in \mathbb{Z}\}$ process can be written as an $AR(\infty)$ process.

Definition 5.10 *A stationary process $\{y_t, t \in \mathbb{Z}\}$ is called a (finite) ARMA process if there exists a WN process $\{\varepsilon_t, t \in \mathbb{Z}\}$ such that for $p, q \geqq 1$:*

$$y_t = \sum_{i=1}^{p} \alpha_i y_{t-i} + \sum_{j=0}^{q} \varphi_j \varepsilon_{t-j} \;\; t \in \mathbb{Z}, \; \varphi_0 := 1, \tag{5.4}$$

where for the polynomials $\alpha(z) := 1 - \sum_{i=1}^{p} \alpha_i z^i$ and $\varphi(z) := 1 + \sum_{j=1}^{q} \varphi_j z^j$, with $\alpha(z) \neq 0, \varphi(z) \neq 0$ for $|z| \leqq 1$, and $(\alpha(z), \varphi(z)) \neq (0,0) \; \forall z \in \mathbb{C}$. This process $\{y_t, t \in \mathbb{Z}\}$ is of order (p, q), i.e. $ARMA(p,q)$, if:

$$\alpha_p \neq 0, \alpha_{p+i} = 0 \;\; \forall i > 0, \varphi_q \neq 0, \varphi_{q+j} = 0 \; \forall j > 0,$$

which can be shortly written as: $\alpha(L) y_t = \varphi(L) \varepsilon_t$ and is called *invertible* if there exists a row of constants $\{\Upsilon_j, \; j = 0, 1, 2, ...\}$ such that $\sum_{j=0}^{\infty} |\Upsilon_j| < \infty$ and

$$y_t = \alpha^{-1}(L) \varphi(L) \varepsilon_t = \Upsilon(L) \varepsilon_t = \sum_{j=0}^{\infty} \Upsilon_j \varepsilon_{t-j}, \; t \in \mathbb{Z} \tag{5.5}$$

or, alternatively, a row of constants $\{\pi_i, \; i = 0, 1, 2, ...\}$ such that $\sum_{i=0}^{\infty} |\pi_i| < \infty$ and

$$\varepsilon_t = \varphi^{-1}(L) \alpha(L) y_t = \pi(L) y_t = y_t - \sum_{i=1}^{\infty} \pi_i y_{t-i}, \; t \in \mathbb{Z}. \tag{5.6}$$

Remark 5.1 ***Pay attention to the common roots problem***

In $ARMA(p,q)$ models, MA and AR parts might cancel out. This is known as the *common roots* problem.
For example, the $ARMA(2,1)$ model $y_t = y_{t-1} - 0.25 y_{t-2} + \varepsilon_t - 0.5 \varepsilon_{t-1}$ or

$$(I - 0.5L)(I - 0.5L) y_t = (I - 0.5L) \varepsilon_t$$

is the same as the $AR(1)$ model $y_t = 0.5 y_{t-1} + \varepsilon_t$. This implies that it will sometimes be difficult to estimate higher order $ARMA(p,q)$ models when the AR and MA roots are almost identical.

5.4.2 The $ARMA(1,1)$ model

Consider the process for $\{y_t - \mu, \; t \in \mathbb{Z}\} : \alpha(L)(y_t - \mu) = \varphi(L) \varepsilon_t$, or:

$$y_t - \mu = \alpha(y_{t-1} - \mu) + \varepsilon_t + \varphi \varepsilon_{t-1} \;\; \text{with} \tag{5.7}$$

$|\alpha| < 1$: *stationarity or invertibility condition* (Υ_j-weights convergence); if $|\alpha| < 1$, then $(I - \alpha L)$ is invertible and y_t is stationary

[3] If the mean of y_t is μ, $\{y_t - \mu, t \in \mathbb{Z}\}$ is (also) an $ARMA(p,q)$ process.

$|\varphi| < 1$: *invertibility condition* (π_i-weights convergence); if the invertibility condition $|\varphi| < 1$ is satisfied, the $MA(1)$ process can be written as an $AR(\infty)$ process.

Hence, the stationarity condition for $ARMA(1,1)$ is the same as for $AR(1)$ and the invertibility condition for $ARMA(1,1)$ is the same as for $MA(1)$. For financial series we have, typically, $\alpha > 0$, $\varphi < 0$, and $\alpha > |\varphi|$ (see Mills (1993), p. 22)

A pure MA model ($MA(\infty)$) is given by:

$$\begin{aligned} y_t - \mu &= \frac{I+\varphi L}{I-\alpha L}\varepsilon_t = \textstyle\sum_{i=0}^{\infty} (\alpha L)^i (I+\varphi L)\varepsilon_t \\ &= \textstyle\sum_{i=0}^{\infty} (\alpha L)^i (\varepsilon_t + \varphi\varepsilon_{t-1}) \\ &= \varepsilon_t + \varphi\varepsilon_{t-1} + \alpha\varepsilon_{t-1} + \alpha\varphi\varepsilon_{t-2} + \alpha^2\varepsilon_{t-2} + \alpha^2\varphi\varepsilon_{t-3} + \ldots \\ &= \varepsilon_t + (\alpha+\varphi)\textstyle\sum_{i=1}^{\infty} \alpha^{i-1}\varepsilon_{t-i}. \end{aligned} \tag{5.8}$$

A pure AR model ($AR(\infty)$) can be obtained as:

$$\begin{aligned} \frac{I-\alpha L}{I+\varphi L}(y_t-\mu) &= (I-\alpha L)\left\{\textstyle\sum_{i=0}^{\infty} (-\varphi L)^i\right\}(y_t-\mu) \\ &= (y_t-\mu) - \alpha(y_{t-1}-\mu) - \varphi(y_{t-1}-\mu) \\ &\quad +\alpha\varphi(y_{t-2}-\mu) + \ldots, \end{aligned}$$

which yields:

$$y_t - \mu = (\alpha+\varphi)\textstyle\sum_{i=1}^{\infty} (-\varphi)^{i-1}(y_{t-i}-\mu) + \varepsilon_t. \tag{5.9}$$

If we (post)multiply $(y_t-\mu)$ in (5.7) by $(y_{t-k}-\mu)$ and take expectations (for stationary processes $\gamma_k \to \rho_k$) ($k = 0, 1, 2, ..., K$):

$$\begin{aligned} E(y_t-\mu)(y_{t-k}-\mu) &= \gamma_k \\ &= \underbrace{\alpha E(y_{t-1}-\mu)(y_{t-k}-\mu)}_{\gamma_{k-1}} + E(\varepsilon_t(y_{t-k}-\mu)) \\ &\quad +\varphi E(\varepsilon_{t-1}(y_{t-k}-\mu)), \end{aligned} \tag{5.10}$$

- for $k > 1$: $\gamma_k = \alpha\gamma_{k-1} = \alpha^k\gamma_0$, or $\rho_k = \frac{\gamma_k}{\gamma_0} = \alpha^k$;
- for k=1: $\gamma_1 = \alpha\gamma_0 + \varphi E\left(\varepsilon_{t-1}\left(\varepsilon_{t-1} + (\alpha+\varphi)\sum_{i=1}^{\infty}\alpha^{i-1}\varepsilon_{t-1-i}\right)\right) = \alpha\gamma_0 + \varphi\sigma_\varepsilon^2$;
- for k=0: $\gamma_0 = \alpha\gamma_{-1} + E\left(\varepsilon_t\left(\varepsilon_t + (\alpha+\varphi)\sum_{i=1}^{\infty}\alpha^{i-1}\varepsilon_{t-i}\right)\right)$

$$+\varphi E\left(\varepsilon_{t-1}\left(\varepsilon_t + (\alpha+\varphi)\sum_{i=1}^{\infty}\alpha^{i-1}\varepsilon_{t-i}\right)\right) = \alpha\gamma_1 + \sigma_\varepsilon^2 + \varphi(\alpha+\varphi)\sigma_\varepsilon^2.$$

Solving for γ_0 and γ_1 in the equations for $k = 1$ and $k = 0$ leads to:

$$\gamma_0 = \frac{(1+2\alpha\varphi+\varphi^2)}{1-\alpha^2}\sigma_\varepsilon^2 \text{ and } \gamma_1 = \frac{(1+\alpha\varphi)(\alpha+\varphi)}{1-\alpha^2}\sigma_\varepsilon^2.$$

Hence, $\rho_1 = \frac{\gamma_1}{\gamma_0} = \frac{(1+\alpha\varphi)(\alpha+\varphi)}{(1+2\alpha\varphi+\varphi^2)}$ and $\rho_k = \alpha\rho_{k-1} = \alpha^k\rho_0$ for $k > 1$. Thus the ACF of an $ARMA\,(1,1)$ process is therefore similar to that of an $AR\,(1)$ process, being characterized by an exponential decay but this decay starts from the above value ρ_1 and not from $\rho_1 = \alpha$ as in the $AR\,(1)$ case. The PACF behaves like that of an $MA\,(1)$ process after an initial value $\phi_{11} = \rho_1$ and will therefore be dominated by an exponential decay.

For many financial series, α will be positive and φ will be negative; with $\alpha > |\varphi|$, ρ can be much smaller than α if $\alpha - |\varphi|$ is small. However, unlike pure AR or MA models, the mixed $ARMA$ model is characterized by both an ACF and a PACF that tail off to infinity, rather than that they are cut off at a particular lag.

The general $ARMA\,(p,q)$ model:

$$(y_t - \mu) = \alpha_1 (y_{t-1} - \mu) + ... + \alpha_p (y_{t-p} - \mu) + \varepsilon_t + \varphi_1\varepsilon_{t-1} + ... + \varphi_q\varepsilon_{t-q}$$

can be multiplied by $(y_{t-k} - \mu)$ and expectations can be taken so that the autocovariances and autocorrelations are obtained from the pth order difference equations (Yule-Walker equations) $\gamma_k - \alpha_1\gamma_{k-1} - \alpha_2\gamma_{k-2} - ... - \alpha_p\gamma_{k-p} = 0, \;\; k > q$, or:

$$\rho_k - \alpha_1\rho_{k-1} - \alpha_2\rho_{k-2} - ... - \alpha_p\rho_{k-p} = 0, \;\; k > q. \tag{5.11}$$

Initial values Comparing (5.11) to the Yule-Walker equations for AR processes in Subsection 5.2.1, it is observed that the ACF of an $ARMA\,(p,q)$ process will follow the same pattern as that of an $AR\,(p)$ process, being described by combinations of damped exponentials and/or damped sine waves. Since the p starting values $(\rho_q, \rho_{q-1}, ..., \rho_{q-p+1})$ have to lie on the solution of (5.11), the solution itself holds for $k > q - p$. The initial $(q - p + 1)$ starting values $\rho_0, \rho_1, ..., \rho_{q-p}$, however, will not follow this pattern. For $k > q - p$ the ACF of $ARMA\,(p,q)$ model follows the same pattern as that of an $AR\,(p)$ process after $q - p + 1$ initial values while the PACF for $k > p - q$ behaves like that of an $MA\,(q)$ process. Some special cases are the following:

- $ARMA\,(1,1) : q - p + 1 = 1 \Longrightarrow$ one starting value, ρ_1; exponential decay from $k = 1$;
- $ARMA\,(2,2) : q - p + 1 = 1 \Longrightarrow$ one starting value, ρ_1; exponential decay or damped sine wave from $k = 1$;
- $ARMA\,(1,2) : q - p + 1 = 2 \Longrightarrow$ two starting values, ρ_1, ρ_2; exponential decay from $k = 2$;
- $ARMA\,(2,1) : q - p + 1 = 0 \Longrightarrow$ no different starting values; all autocorrelations (for $k \geqq 0$) are described by either a sum of two damped exponentials or a damped sine wave.

More generally, the PACF of an $ARMA(p,q)$ process eventually behaves like that of an $MA(q)$ process for $k > p-q$. For $k \leqq p-q$ the PACF does not follow this general pattern.

Summary 5.1 *Both the ACF and the PACF of a mixed ARMA process will be infinite in extent and tail off as k increases. Eventually (for $k > q-p$), the ACF is determined by the AR part of the model, while for $k > p-q$ the PACF is determined by the MA part of the model.*

5.5 Testing for unit roots

In order to avoid *spurious regressions* we have to get rid of *unit roots.* To introduce testing procedures whether a unit root is present we concentrate on AR models. The statistical theory related to $AR(1)$ processes where the AR parameter is equal to one (unstable process) and greater than one (explosive process) was developed by Anderson (1959), White (1958 and 1959) and Rao (1961). Fuller (1976) and Dickey and Fuller (1979) derived the asymptotic distribution of the estimated $AR(1)$ parameters under the assumption of WN errors and the initial condition of the dependent variable equal to zero. Phillips (1987) derived this asymptotic distribution under more general assumptions about the error terms and the initial condition. However, as indicated by Maddala and Kim (1998, p. 3) "the econometric literature on unit roots took off after the publication of the paper by Nelson and Plosser (1982) that argued that most macroeconomic series have unit roots and that this is important for the analysis of macroeconomic policies". Although an $AR(1)$ model is restrictive, an arbitrary AR model is not particularly restrictive since any $ARMA$ model will always have an AR representation, provided that the MA polynomial $\phi(L)$ is invertible.

5.5.1 Testing for unit roots in a first order autoregressive model

It is said that a series is integrated of order 1, or $y_t \sim I(1)$, if y_t has a *unit root* or $\Delta y_t = \varepsilon_t$ ($\varepsilon_t \in WN$).

Following Verbeek (2000), let us consider, first of all, the $AR(1)$ process:

$$y_t = \alpha y_{t-1} + \varepsilon_t \text{ with } \varepsilon_t \in WN,\ t \in \mathbb{Z}.$$

A test for a unit root is a test for $\alpha = 1$ and it seems obvious to use the estimate $\hat{\alpha}$ for α from an OLS procedure (which is consistent, irrespective of the true value of α) and the corresponding standard error to test the null hypothesis. However, as was shown in the seminal paper by Dickey and Fuller (1979), under the null that $\alpha = 1$ the standard t-ratio does not have a t-distribution, not even asymptotically. The reason for this is the nonstationarity of the process, invalidating standard results on the distribution of the OLS estimator $\hat{\alpha}$. For example, if $\alpha = 1$ the variance of y_t, denoted by γ_0, is not finite. For any finite sample size T, however, a finite estimate of the variance for y_t, should be obtained. To test the null that $\alpha = 1$ it is possible to use either the test statistic $K := T(\hat{\alpha} - 1)$ or the normalized test statistic for $\hat{\alpha}$, i.e. the t-ratio $\hat{\tau} := \frac{\hat{\alpha}-1}{\hat{\sigma}_{\hat{\alpha}}}$, where $\hat{\sigma}_{\hat{\alpha}}$ denotes the standard error of $\hat{\alpha}$. Critical values, however, have to be taken from the appropriate distribution, which under the null hypothesis of nonstationarity, is nonstandard. In particular, the distribution is skew to the right so that critical values are smaller than those for (the normal approximation of) the t-distribution. Using a 5 per cent significance level in a one-tailed test of $H_0 : \alpha = 1$ (a unit root) against $H_1 : |\alpha| < 1$ (stationarity) the correct critical value is -1.95, rather than -1.65 for the normal approximation to the Student t-distribution. Consequently, if you use the standard t-tables you may reject a unit root too often. Selected percentiles of the appropriate distribution are published in several works by Dickey and Fuller (see Fuller (1976 and 1996), and Dickey and Fuller (1979 and 1981)). In Figure 5.1,[4] reproduced from Maddala and Kim (1998), p.64, we present 1 per cent and 5 per cent critical values for this test (under the heading "$AR(1)$ without a drift"), usually referred to as the *Dickey-Fuller (DF)* test, for a range of different sample sizes T.

Usually, a slightly more convenient regression procedure is used:

$$\Delta y_t = (\alpha - 1)\, y_{t-1} + \varepsilon_t \tag{5.12}$$

In this case the model is rewritten as from which the DF test statistic for the null: $\alpha - 1 = 0$, $\frac{\widehat{\alpha-1}}{\hat{\sigma}_{\hat{\alpha}}}$, is equivalent to $\hat{\tau}$ above. The reason for this is that the OLS method is invariant to linear transformations of the model. Under the null hypothesis applied on (5.12), y_t is described by a random walk, while under the alternative y_t is an $AR(1)$ model with zero mean. If we consider the series y_t that may have nonzero mean, it is appropriate to include a constant term. As the constant in a stationary $AR(1)$ model satisfies $\alpha_0 := (1 - \alpha)\,\mu$ (see Section 5.2 on AR processes), where μ is the mean of the series, the null hypothesis of a unit root also implies that the intercept term should be zero. The testing regression is thus:

$$\Delta y_t = \alpha_0 + (\alpha - 1)\, y_{t-1} + \varepsilon_t \text{ with } \varepsilon_t \in WN,\ t \in \mathbb{Z}, \tag{5.13}$$

where the null hypothesis is the joint hypothesis $H_0 : \alpha_0 = 0, \alpha - 1 = 0$. Although it is possible to test these two restrictions jointly (*cf. infra*), it is easier (and more common) to test only that $\alpha - 1 = 0$. The distribution of the t-ratio for this hypothesis, denoted $\hat{\tau}_\mu$, under the assumption that $H_0 : \alpha - 1 = 0$

[4] Notes to Figure 5.1
F-test a. is for $H_o : \alpha = 0, \rho = 1$ in $y_t = \alpha + \rho y_{t-1} + \varepsilon_t$ and $H_o : \delta = 0, \rho = 1$ in $y_t = \alpha + \delta t + \rho y_{t-1} + \varepsilon_t$; F^*-test (b) is for $H_o : \alpha = 0, \delta = 0, \rho = 1$ in $y_t = \alpha + \delta t + \rho y_{t-1} + \varepsilon_t$
- sources Fuller (1976) for K-test and t-test, Dickey and Fuller (1981) for F^*-test with 2 and $T - p - 3$ degrees of freedom
- the traditional F-values are much lower than the F^* or Dickey-Fuller test statitics in the figure, e.g. in the case of a drift and time trend under the 5% level for a sample size of 25: 3.42 compared to 7.24, for a sample size of 50: 3.20 compared to 6.73 and for a sample size of ∞: 3.00 compared to 6.25 (see Hasza and Fuller (1979)).

Figure 5.1: Dickey - Fuller critical values

Sample size	K-test		t-test		F*-test (a)		F*-test(b)	
	1%	5%	1%	5%	1%	5%	1%	5%
AR(1) without a drift								
25	-11.9	-7.7	-2.62	-1.95				
100	-13.3	-7.9	-2.60	-1.95				
250	-13.6	-8.0	-2.58	-1.95				
500	-13.7	-8.0	-2.58	-1.95				
infinity	-13.8	-8.1	-2.58	-1.95				
AR(1) with a drift								
25	-17.2	-12.5	-3.75	-3.00	7.88	5.18		
50	-18.9	-13.3	-3.58	-2.93	7.06	4.86		
100	-19.8	-13.7	-3.51	-2.89	6.7	4.71		
250	-20.3	-14.0	-3.46	-2.88	6.52	4.63		
500	-20.5	-14.0	-3.44	-2.87	6.47	4.61		
infinity	-20.7	14.1	-3.43	-2.86	6.43	4.59		
AR(1) with drift and time trend								
25	-22.5	-17.9	-4.38	-3.60	10.61	7.24	8.21	5.68
50	-25.7	-19.8	-4.15	-3.50	9.31	6.73	7.02	5.13
100	-27.4	-20.7	-4.04	-3.45	8.73	6.43	6.50	4.88
250	-28.4	-21.3	-3.99	-3.43	8.43	6.34	6.22	4.75
500	-28.9	-21.5	-3.98	-3.42	8.34	6.30	6.15	4.71
infinity	29.5	-21.8	-3.96	-3.41	8.27	6.25	6.09	4.68

is correct, is also nonstandard. The critical values for $\hat{\tau}_\mu := \frac{\hat{\alpha}-1}{\hat{\sigma}_{\hat{\alpha}}}$, also presented in Figure 5.1 (under the heading "$AR(1)$ with a drift"), are smaller than those for $\hat{\tau}$. For large samples sizes, the unit root hypothesis is rejected at the 5 per cent level if $\hat{\tau}_\mu < -2.86$.

It is possible that (5.13) holds with $\alpha = 1$ and a nonzero intercept $\alpha_0 \neq 0$. Because in this case α_0 cannot equal $(1-\alpha)\mu$, (5.13) cannot be derived from a pure $AR(1)$ model. This is seen by considering the resulting process:

$$\Delta y_t = \alpha_0 + \varepsilon_t, \tag{5.14}$$

which is known as a *random walk with drift*, where α_0 is the drift parameter. In the model for the level variable y_t, α_0 corresponds to a linear time trend, since

$$\begin{aligned} y_t &= y_{t-1} + \alpha_0 + \varepsilon_t \\ &= y_{t-2} + \alpha_0 + \varepsilon_{t-1} + \alpha_0 + \varepsilon_t \\ &= y_{t-3} + \alpha_0 + \varepsilon_{t-2} + \alpha_0 + \varepsilon_{t-1} + \alpha_0 + \varepsilon_t \\ &= \ldots \\ &= y_0 + t\alpha_0 + \textstyle\sum_{i=0}^{t-1} \varepsilon_{t-i}. \end{aligned}$$

Hence, for some given starting value y_0, $E(y_t) = y_0 + t\alpha_0$ is found. This shows that the interpretation of the intercept term in (5.13) heavily depends upon the presence of a unit root. In the stationary case, α_0 reflects the nonzero mean of the series; in the unit root case, it reflects a *deterministic trend* in y_t. Because in the latter case first differencing produces a stationary time series, the process for y_t is referred to as *difference stationary* then. In general, a difference stationary process is a process that can be made stationary by differencing.

It is also possible that nonstationarity is caused by the presence of a deterministic time trend in the process, rather than by the presence of a unit root. It is said that a process is driven by a deterministic trend rather than by a stochastic trend then. This happens when the $AR(1)$ model (5.13) is extended to:

$$y_t = \alpha_0 + \alpha y_{t-1} + \delta t + \varepsilon_t \tag{5.15}$$

with $|\alpha| < 1$ and $\delta \neq 0$. In this case, we have a nonstationary process because of the linear trend δt. This deterministic nonstationarity can be removed by regressing y_t upon a constant and t, and then considering the residuals of this regression, or by simply including t as an additional variable in the model. The process for y_t in this case is referred to as being *trend stationary*. Nonstationary processes may thus be characterized by the presence of a deterministic trend, like δt, a stochastic trend implied by the presence of a unit root, or both.[5]

[5] In this respect, the DGPs for a trend and difference stationary process can be written as $y_t = \gamma_0 + \delta t + \varepsilon_t$ and $y_t = \alpha_0 + \alpha y_{t-1} + \varepsilon_t$, respectively (with ε_t a stationary error process, say $ARMA$ or WN). We can nest these two models in the following model: $y_t = \gamma_0 + \delta t + u_t$ with $u_t = \rho u_{t-1} + \varepsilon_t$, so that $y_t = \gamma_0 + \delta t + \rho\left[y_{t-1} - \gamma_0 - \delta(t-1)\right] + \varepsilon_t$.
If $|\rho| < 1$, y_t is trend stationary and if $|\rho| = 1$, y_t is difference stationary.

It is possible to test whether y_t follows a random walk against the alternative that it follows the trend stationary process in (5.15). This can be tested by subtracting y_{t-1} from both sides of (5.15) and applying OLS on:

$$\Delta y_t = \alpha_0 + (\alpha - 1)\, y_{t-1} + \delta t + \varepsilon_t. \tag{5.16}$$

The null hypothesis to be tested is that the process is a random walk rather than a trend stationary one and corresponds to $H_0 : \alpha_0 = \delta = \alpha - 1 = 0$. Instead of testing this joint hypothesis, it is quite common to use the t-ratio corresponding to $\hat{\alpha} - 1$ denoted $\hat{\tau}_\tau := \frac{\hat{\alpha}-1}{\hat{\sigma}_{\hat{\alpha}}}$, assuming that the other restrictions in the null hypotheses are satisfied (*cf. infra* for joint test statistics). Although the null hypothesis is still the same as in the previous two unit root tests, the testing regression is different and thus we have, again, a different distribution of the test statistic. The critical values for $\hat{\tau}_\tau$, given in Figure 5.1 under the heading "$AR(1)$ with a drift and a time trend" are still smaller than those for $\hat{\tau}_\mu$ (i.e. still more skewed to the right). In fact, with an intercept and a deterministic trend included the probability that $\hat{\alpha} - 1$ is positive (given that the true value $\alpha - 1$ equals zero) is in many cases negligibly small. It should be noted, however, that if the unit root hypothesis $\alpha - 1 = 0$ is rejected, we cannot conclude that the process for y_t is likely to be stationary. Under the alternative hypothesis δ may be nonzero so that the process for y_t is not stationary (but only trend stationary).

Notice that if we reject the null hypothesis of a single unit root, we do not yet know whether the model is stationary or has more than one unit root. This has to do with the possible event that two (or even three) unit roots might exist. This will be investigated later in this section. To proceed consider first the occurrence of unit roots in higher order autoregressive models.

5.5.2 Testing for unit roots in higher order AR models

A test for a single unit root in higher order AR processes can easily be obtained by extending the Dickey-Fuller test procedure to the so-called *augmented Dickey-Fuller tests* (ADF tests). Assume that we consider the following $AR(2)$ model:

$$y_t = \alpha_1 y_{t-1} + \alpha_2 y_{t-2} + \varepsilon_t \tag{5.17}$$

which can be rewritten in factorized form and using lag and identity operators as

$$(I - \theta_1 L)\,(I - \theta_2 L)\, y_t = \varepsilon_t.$$

The stationarity condition requires that θ_1 and θ_2 are both less than one in absolute value, but if $\theta_1 = 1$ and $|\theta_2| < 1$, we have a single unit root, $\alpha_1 + \alpha_2 = 1$ and $\alpha_2 = -\theta_2$. Equation (5.17) can be used to test the unit root hypothesis by testing $\alpha_1 + \alpha_2 = 1$, given $|\alpha_2| < 1$. This is conveniently done be rewriting (5.17) as:

$$\Delta y_t = (\alpha_1 + \alpha_2 - 1)\, y_{t-1} - \alpha_2 \Delta y_{t-1} + \varepsilon_t. \tag{5.18}$$

The coefficients in (5.18) can consistently be estimated by OLS and the estimate of the coefficient for y_{t-1} serves for testing the null hypothesis $\pi := \alpha_1 + \alpha_2 - 1 = 0$. The resulting t-ratio, $\hat{\pi}/\hat{\sigma}_{\hat{\pi}}$, has the same distribution as $\hat{\tau} := \frac{\hat{\alpha}-1}{\hat{\sigma}_{\hat{\alpha}}}$ above. In the spirit of the Dickey-Fuller procedure, one might add an intercept term or an intercept and a time trend to equation (5.18). Depending on which variant is used, the resulting test statistic has to be compared with a critical value taken from the appropriate row of Figure 5.1.

This procedure can easily be generalized to the testing of a *single* unit root in an $AR(p)$ process by rewriting any $AR(p)$ process $y_t = \alpha_0 + \delta t + \sum_{i=1}^{p} \alpha_i y_{t-i} + \varepsilon_t$ for $\{y_t, t \in \mathbb{Z}\}$ with $p \geqq 2$, $\varepsilon_t \in WN$ as an ECM model. For example, from (5.18) where a drift α_0 and a time trend δt have to be added and repeating it for $p = 3, 4, \ldots$the following ECM model is derived for general p:

$$\Delta y_t = \gamma_0 y_{t-1} + \sum_{i=1}^{p-1} \gamma_i \Delta y_{t-i} + \alpha_0 + \delta t + \varepsilon_t, \tag{5.19}$$

where $\gamma_0 := \sum_{i=1}^{p} \alpha_i - 1 = -\alpha(1)$ and $\gamma_i := -\sum_{j=i+1}^{p} \alpha_j$.

If $\gamma_0 = \alpha(1) = 0$ there is one unit root in (5.19). Thus, as before, the null that $\gamma_0 = 0$ corresponds to a unit root, which can again be tested using t-ratios. If the $AR(p)$ assumption is correct and under the null of a unit root, the asymptotic distributions of the $\hat{\tau}, \hat{\tau}_\mu$ and $\hat{\tau}_\tau$ statistics, calculated from (5.19) are the same as in Figure 5.1. The small sample critical values are somewhat different from the tabulated ones and are provided by, for example, MacKinnon (1991).

As mentioned before, the joint F-type (or F^*-) statistics should be applied for testing joint null hypotheses in the drift case (5.13) and the drift and trend case (5.16). The joint null hypotheses are then $H_0 : (\alpha_0, \gamma_0)' = (0,0)'$ and $H_0 : (\alpha_0, \delta, \gamma_0)' = (0,0,0)'$, respectively. Critical values of F^*-statistics can be found in the last four columns of Figure 5.1.

Phillips and Perron (1988) have suggested an alternative to the augmented Dickey-Fuller tests. Instead of adding additional lags in the regressions to obtain an error term that has no autocorrelation, they stick to the original Dickey-Fuller regressions, but adjust the $\hat{\tau}$-statistics to take into account the (potential) autocorrelation pattern in the errors. The (asymptotic) critical values are again the same as those reported in Figure 5.1. The Phillips-Perron (PP) test, sometimes referred to as a nonparametric test for a unit root, is, like the ADF test, applicable for general $ARMA$ models (see Hamilton (1994), pp.506-515, and Maddala and Kim (1998), pp.78-81 for more details), Monte Carlo studies do not show a clear ranking of the two tests regarding their power (probability to reject the null if it is false) in finite samples.

To circumvent the problem that unit root tests often have low power, Kwiatkowski, Philips, Schmidt, and Shin (1992) propose an alternative test where stationarity is the null hypothesis and the existence of a unit root is the alternative. This test is usually referred to as the *KPSS test*. The basic idea is that a time series is decomposed into the sum of a deterministic time trend, a random walk and a stationary error term (not necessarily WN). The null (of trend stationarity) specifies that the variance of the random walk component is zero. The test is actually an LM test and computation of the test statistic is fairly simple and runs as follows. Kwiatkowski *et al.* (1992) start from the model $y_t = \delta t + \zeta_t + \varepsilon_t$,where ε_t is a stationary process and ζ_t is a random walk process given by $\zeta_t = \zeta_{t-1} + u_t$ with $u_t \in WN$ with variance σ_u^2. The null of stationarity is then formulated as $H_0 : \sigma_u^2 = 0$ (or ζ_t is a contant). The $KPSS$ test is a special case of a test for parameter constancy, against H_1 that the parameters follow a random walk. Nabeya and Tanaka (1988) constructed a similar test for the regression model $y_t = \beta_t x_t + \gamma' \mathbf{z}_t + \varepsilon_t$ with $\beta_t = \beta_{t-1} + u_t$ so that the $KPSS$ test is a special case with $x_t = 1$ and $z_t = t$. The Nabeya and Tanaka LM test statistic for the $KPSS$ case is $\xi_{LM} = \frac{\sum_{t=1}^{T} S_t^2}{\hat{\sigma}_{\hat{e}}^2}$, where $\hat{e}_t$ are the residuals from the auxiliary regression of y_t on a constant and a time trend, $\hat{\sigma}_{\hat{e}}^2$ is the estimated residual variance from this regression ($\hat{\sigma}_{\hat{e}}^2 = \frac{1}{T} \sum_{t=1}^{T} \hat{e}_t^2$)[6] and S_t is the partial sum of $\hat{e}_t$ defined as $S_t := \sum_{i=1}^{t} \hat{e}_i$ for $t = 1, 2, \ldots, T$. If $\varepsilon_t \in WN$, the asymptotic distribution of ξ_{LM} has been derived in Nabeya and Tanaka (1988). However, Kwiatkowski *et al.* (1992) consider the case of a general stationary error process and tabulate the critical values of the asymptotic distribution by simulation. For example, the 5 per cent critical value of ξ_{LM} is 0.146. If the null is stationarity rather than trend stationary, the trend term should be omitted from the auxiliary regression. The test statistic is then computed in the same fashion, but the 5 per cent critical value is 0.463.

5.5.3 Multiple unit roots

Double unit roots

In practice, some economic variables are $I(2)$, that is, they have two or *double unit roots.* Known examples are money supply and prices. Hence questions arise as to how to test for more than one unit root. There are two approaches: the first, a *bottom-up approach* is the one discussed in Choi and Yu (1997) and suggests testing sequentially: $I(0)$ versus $I(1)$ and $I(1)$ versus $I(2)$ if the first hypothesis is rejected. The second approach, suggested in Dickey and Pantula (1987), is a *top-down approach.* Test first $I(2)$ versus $I(1)$ and if the hypothesis of $I(2)$ is rejected, then test $I(1)$ versus $I(0)$. This procedure is appropriate for tests of nonstationarity as the null, whereas the bottom-up approach is appropriate for tests of stationarity as the null.

[6] The latter estimator $\hat{\sigma}_{\hat{e}}^2$ may involve corrections for autocorrelation based on the Newey-West formula.

How to test the occurrence of two unit roots?

Disentangling (5.19) once more, the $\alpha(L)$ polynomial satisfies:

$$\begin{aligned} \alpha(L) \quad &: \quad = I - \sum_{i=1}^{p} \alpha_i L^i = \alpha(1) L + \gamma(L)(I-L) \\ &= \quad \alpha(1) L + \gamma(1) L (I-L) + \xi(L)(I-L)^2, \end{aligned} \tag{5.20}$$

so that a second order ECM model emerges:

$$\Delta^2 y_t = \beta_1 y_{t-1} + \beta_2 \Delta y_{t-1} + \sum_{i=1}^{p-2} \xi_i \Delta^2 y_{t-i} + \alpha_0 + \delta t + \varepsilon_t, \tag{5.21}$$

with $\beta_1 := -\alpha(1)$ and $\beta_2 := -\gamma(1)$. The null hypothesis of two unit roots ($\beta_1 = \beta_2 = 0$) can be tested with an F^*-distribution as in Figure 5.1.

Hence, if the ADF test does not allow rejection of the null of one unit root, the presence of a second unit root may be tested by estimating the regression $\Delta^2 y_t$ on $\Delta y_{t-1}, \Delta^2 y_{t-1}, ..., \Delta^2 y_{t-p+1}$ and comparing the t-ratio of the coefficient on Δy_{t-1} with the appropriate critical value from the above table. Alternatively, the presence of two unit roots may be tested *jointly*, by estimating the regression of $\Delta^2 y_t$ on $y_{t-1}, \Delta y_{t-1}, \Delta^2 y_{t-1}, ..., \Delta^2 y_{t-p+1}$ and computing the usual F-ratio for testing the joint significance of y_{t-1} and Δy_{t-1}. Again, though, this test statistic has a distribution under the null hypothesis of a double unit root that is not the usual F-distribution; critical values of this distribution are given in Figure 5.1.

Triple unit roots

Dickey and Pantula (1987) consider the case of three unit roots which is pretty rarely the case with economic variables. For completeness, we shall review their procedure here. The standard (A)DF unit root tests assume that there is at most one unit root in the series. Dickey and Pantula (1987) proposed t^*- and F^*-tests that compare a null hypothesis of k unit roots with an alternative of $k-1$ unit roots.

Consider $y_t = \sum_{j=1}^{p} \alpha_j y_{t-j} + \varepsilon_t$, where $\{\varepsilon_t\}$ is a sequence of *iid* random variables with mean 0 and variance $\sigma_\varepsilon^2 = 1$ and $y_{-p+1} = ... = y_0 = 0$ for $p = 3$ and three possible unit roots. By disentangling (5.20) one step further:

$$\begin{aligned} \alpha(L) &= \alpha(1) L + \gamma(1) L (I-L) \\ &\quad + \xi(1) L (I-L)^2 + \vartheta(L)(I-L)^3. \end{aligned} \tag{5.22}$$

The $AR(3)$ model can then be rewritten according to (5.21):

$$\Delta^3 y_t = \theta_1 y_{t-1} + \theta_2 \Delta y_{t-1} + \theta_3 \Delta^2 y_{t-1} + \alpha_0 + \delta t + \varepsilon_t$$

with $\theta_1 := -\alpha(1), \theta_2 := -\gamma(1)$ and $\theta_3 := -\xi(1)$.

Let z_1, z_2 and z_3 denote the roots of the corresponding characteristic equation:

$$z^3 - \rho_1 z^2 - \rho_2 z - \rho_3 = 0$$

where $|z_i| \leq 1 \Leftrightarrow |\rho_i| \geq 1$

Following Maddala and Kim (1998), pp. 131-133, consider the hypotheses:

- *no unit root* $H_0 : |z_1| < 1$ or $H_0 : \theta_1 < 0$ and some restrictions on θ_2 and θ_3, e.g., the restrictions on θ_2 and θ_3 are $-12 < \theta_2 + 2\theta_1 < 0$ and $-2 < \theta_3 < 0$;

- *one unit root* $H_1 : z_1 = 1,\ |z_2| < 1$ or $H_0 : \theta_1 = 0,\ \theta_2 < 0$ and some restrictions on θ_2 and θ_3; the restrictions on θ_2 and θ_3 are: $0 < 4 + \theta_2 + 2\theta_3$, $-2 < \theta_3 < 0$;

- *two unit roots* $H_2 : z_1 = z_2 = 1,\ |z_3| < 1$ or $H_0 : \theta_1 = \theta_2 = 0,\ \theta_3 < 0$;

- *three unit roots* $H_3 : z_1 = z_2 = z_3 = 1$ or $H_0 : \theta_1 = \theta_2 = \theta_3 = 0$.

Dickey and Pantula (1987) showed that applying F^*-statistics for testing from the higher number of unit roots to the lower number of unit roots is valid, while the reverse order of applying F^*-statistics is not valid. Based on the asymptotic distributions of F^*-statistics derived by Dickey and Pantula, it is suggested to test the hypotheses sequentially in the order H_3, H_2, and H_1:

1. if H_3 is rejected by the F^*-test, then go to step 2; otherwise conclude that H_3 is not rejected;

2. if H_2 is rejected by the F^*-test, then go to step 3; otherwise conclude that H_2 is not rejected;

3. if H_1 is rejected by the F^*-test, then conclude that H_0 is true; otherwise conclude that H_1 is not rejected.

Dickey and Pantula (1987) proposed alternative testing procedures based on t-ratios. However, the t-ratios from the regression of x_t on y_{t-1}, z_{t-1}, and w_{t-1} have different asymptotic distributions depending on the number of unit roots. And thus they argue that a sequential procedure based on these statistics is not consistent. They suggest to use alternative t^*-statistics:

1. for H_3 against H_2, obtain the t^*-statistic from the regression x_t on w_{t-1};

2. for H_2 against H_1, use the t^*-statistic in the regression x_t on z_{t-1} and w_{t-1} for testing the coefficient of z_{t-1} is 0;

3. for H_1 against H_0, use the t^*-statistic in the regression x_t on y_{t-1}, z_{t-1} and w_{t-1} for testing the coefficient of y_{t-1} is 0.

Their results of a Monte Carlo power study show that the procedure based on t^*-statistics is more powerful than that based on F^*-statistics in most cases.

5.5.4 Seasonal unit roots

If seasonal data are available, a *seasonal AR* (*SAR*) model can be defined as e.g. in Osborn *et al.* (1988)

$$\Delta_1 \Delta_S y_t = \gamma_0 + \sum_{i=1}^{S-1} \gamma_i D_{it} + \beta_1 \Delta_S \, y_{t-1} + \beta_2 \Delta_1 \, y_{t-S} + \sum_{j=1}^{p} \alpha_i \Delta_1 \Delta_S \, y_{t-j} + u_t,$$

where D_{it} $i = 1, 2, \ldots, S$ are seasonal dummies (e.g. $S = 4$ for quarterly effects, $S = 5$ for working days effects, $S = 12$ for monthly effects and so on); hence, the deterministic seasonal part of this equation implies that the seasonal coefficients satisfy $\sum_{i=1}^{S-1} \gamma_i = -\gamma_S$. However, dummy variables should not (necessarily) be used to correct for seasonality. This can better be done by considering either seasonal unit roots testing and/or direct seasonal autoregressive modeling.

A time series is said to be *seasonally integrated* if the *seasonal difference* Δ_S filter is needed to make the series stationary and the time series can be said to have *seasonal unit roots* then. Known examples of seasonal unit root tests are the DHF test of Dickey, Hasza, and Fuller (1984), the HEGY test of Hylleberg, Engle, Granger, and Yoo (1990) and the CH test of Canova and Hansen (1995). The DHF test is based on the $SAR(1)$ model $y_t = \alpha_S y_{t-S} + u_t$ where u_t is some stationary invertible $ARMA(p, q)$ model. Consider testing the null hypothesis $\alpha_S = 1$ against the alternative $\alpha_S < 1$. The filter Δ_S has S roots, one of which is a unit root. The DHF test rules out the possibility that the time series y_t can be made stationary by using the simple $\Delta = \Delta_1$ filter to take care of the unit root and seasonality can be taken care of by seasonal dummies. The DHF test tests, as in Subsection 5.5.1 for a single unit root, the null hypothesis that all the S roots implied by the filter Δ_S are unity (one nonseasonal and $S - 1$ seasonal unit roots) against the alternative of no unit roots.

Explanation of the HEGY and CH tests can be found in Maddala and Kim (1998), pp. 367 – 376.

Example 5.6 *Test for a unit root*

In their 1981 paper, Dickey and Fuller apply their methodology to a model for the logarithms of a quarterly series on output, the Federal Reserve Board Production Index. The model used is:

$$y_t = \mu + \delta t + \alpha y_{t-1} + \gamma \Delta y_{t-1} + \varepsilon_t.$$

The test is carried out by testing the joint hypothesis that both δ and γ are zero in model (5.19) with $p = 2$. The example in Greene (2000), pp.784-785, uses quarterly US real GNP from 1950^{III} to 1983^{IV}. Normally one would test for seasonal integration, but the simple $AR(1)$ regression is directly analyzed here (estiamtes of asymptotic t-ratios in brackets):

$$\Delta \hat{y}_t = \underset{(3.60)}{0.0256} - \underset{(-2.40)}{0.0075} y_{t-1} \text{ with } y_t := \ln \frac{GNP_t}{P_t}.$$

Based on the five per cent conventional critical point of the Student t-distribution (-1.645), the null of nonstationarity (a unit root) is rejected. But the interpolated critical value for 134 observations with a drift from Figure 5.1 would be -2.89 (5% significance level). So the hypothesis of a unit root is not rejected. The augmented regressions (5.18) with drift and with drift and trend term, respectively produce,

$$\Delta\hat{y}_t = \underset{(3.90)}{0.0043} + \underset{(6.30)}{0.4918}\Delta y_{t-1},\ R^2 = 0.24829$$

and $\Delta\hat{y}_t = \underset{(3.63)}{0.1420} + \underset{(3.33)}{0.00065}t - \underset{(-3.48)}{0.0813}y_{t-1} + \underset{(6.72)}{0.4926}\Delta y_{t-1},\ R^2 = 0.3179.$

For a sample of 134 observations, the following F^*-statistic is obtained for the first equation:

$$F^* = \frac{(0.3179-0.24829)/2}{(1-0.3179)/(134-4)} = 6.633,$$

which is (slightly) larger than the Dickey-Fuller F^*-value of 6.49 in Figure 5.1. Therefore the null of a unit root in the DGP for the quarterly logs of US real GNP is rejected.

5.6 *ARIMA* models

In this section, some guidelines for 'how to work in practice?' for a nonstationary DGP $\{y_t\}$, $t = 1, 2, ..., T$ will be summarized.

1. First determine a possible deterministic (time) trend from $\{y_t\}$ and remove this trend by considering the residuals $y_t - \hat{y}_t =: y_t^*$, where e.g. in the case of a third order deterministic trend $\hat{y}_t = \hat{\alpha}_0 + \hat{\alpha}_1 t + \hat{\alpha}_2 t^2 + \hat{\alpha}_3 t^3$.

2. Verify whether there is a possible stochastic trend, i.e. determine the number of unit roots, d, by following the top-down procedure mentioned above, i.e. determine d in $\Delta^d y_t^* =: x_t$ by testing stationarity using a(n) (A)DF, PP or KPSS test; if the series is not stationary, take an appropriate number of differences d (take an appropriate number of seasonal differences if seasonal unit roots are present). An $ARMA(p, q)$ model that is nonstationary is called an $ARIMA(p, d, q)$ model (in the case of seasonal unit roots a nonstationary $ARMA(P, Q)$ model is a seasonal $ARIMA(P, D, Q)$ (or $SARIMA(P, D, Q)$) model where D is the appropriate number of seasonal differences).

3. Apply an $ARMA$ or possibly a seasonal $ARMA$ (model on the -now stationary- series x_t in order to arrive at a final estimate of the time series model. The specific (seasonal) $ARMA$ model is found from the ACF and PACF, which gives in theoretical notation $\{\rho_k\}$ and $\{\rho_{kk}\}$:

	ACF $\{\rho_k\}$	PACF $\{\rho_{kk}\}$
$AR\,(p)$	$\lvert\rho_k\rvert$ exponentially decaying	H_0: $\rho_{kk} = 0, k > p$
$MA\,(q)$	$H_0 : \rho_k = 0, k > q$	$\lvert\rho_{kk}\rvert$ exponentially decaying
$ARMA\,(p,q)$	$\lvert\rho_k\rvert$ exponentially decaying	$\lvert\rho_{kk}\rvert$ exponentially decaying (or damped sine waves)

In the presence of seasonal effects (with S the seasonal length, e.g. $S = 4, S = 12, ...$):

	ACF $\{\rho_k\}$	PACF $\{\rho_{kk}\}$
$SAR(T)\,(P)$	autocorrelations $\rho_k$on $S, 2S, 3S, ...$ are exponentially decaying	ρ_{kk} on $S, 2S, 3S, ...$ break down after $S \times P$
$SMA(T)\,(Q)$	autocorrelations ρ_k on $S, 2S, 3S, ...$ break down after $S \times Q$	ρ_{kk} on $S, 2S, 3S, ...$ slowly go to zero

In the next section the identification and estimation steps of $ARIMA$ models is further documented.

5.7 Box-Jenkins approach for $ARIMA$ models

The Box-Jenkins (BJ) approach for a time series model is an empirical procedure to compute an $ARIMA$ model for a certain database, which describes this database as good as possible (see e.g. Box and Jenkins (1970)). In general, there are three steps in the BJ analysis: identification, estimation, and diagnostic checking.

5.7.1 Identification of a tentative time series model

The correlogram of a stationary series cuts off (at a certain significance level) after a certain (finite) number of lags. This is not the case with nonstationary series. To decide upon the number of differences one plots the correlogram of a series y, Δy, $\Delta^2 y$, ... and evaluates the correlogram at each stage. The series should be differenced until the correlogram dampens (usually not more than once or twice).

Identification is based on the estimated autocorrelations and the estimated partial autocorrelations.

- If the autocorrelations have a slow decay or do not die out at all, the time series is nonstationary and it is suggested to take differences until stationarity is obtained; then an $ARMA$ model is identified for the 'differentiated' series ($ARIMA$ model of the original series).

- If the autocorrelations $\rho_k = 0$ for $k > q$ and the partial autocorrelations dampen slowly, an $MA\,(q)$ model is obtained; hence q is the largest value of k for which the estimated autocorrelations r_k (or $\hat{\rho}_k$) lie outside the 99% confidence interval $\left(\pm 1.96/\sqrt{T}\right)$.

- If the partial autocorrelations $\rho_{kk} = 0$ for $k > p$ and the autocorrelations die out slowly, an $AR(p)$ model is attained; hence p is the largest value of k for which the estimated partial autocorrelations r_{kk} (or $\hat{\rho}_{kk}$) lie outside the 99% confidence interval $\left(\pm 1.96/\sqrt{T}\right)$.

- If both the partial autocorrelations and the autocorrelations die out slowly, an $ARMA(p,q)$ model is obtained, and the identification procedure is started according to the highest absolute values of the estimated autocorrelations and partial autocorrelations.

5.7.2 Estimation of a time series model

The parameters of a pure AR process can be estimated by the regression method or by the sample Yule-Walker equations, as described in Section 5.2. If there are also MA terms, the minimization of the sum of squared errors or the maximization of the sample likelihood function requires nonlinear estimation methods as e.g. Newton, Davidon-Fletcher-Powell, Marquardt, etc.

An OLS estimate for the $ARMA$ model $\alpha(L) y_t = \varphi(L) \varepsilon_t + \mu$ is found from the minimization of $\sum_{t=p+1}^{T} \varepsilon_t^2$ with:

$$\varepsilon_t := \alpha(L) \varphi^{-1}(L) y_t - \left(\varphi_1 + \varphi_2 + ... + \varphi_q\right)^{-1} \mu. \tag{5.23}$$

Application of OLS on the $AR(\infty)$ model (5.23) yields consistent estimators since $E(y_{t-j}\varepsilon_t) = 0$ for $j > 0$. The OLS residuals are used as approximations to the error terms and, either the q starting values $\{\hat{\varepsilon}_p, \hat{\varepsilon}_{p-1}, ..., \hat{\varepsilon}_{p-q+1}\}$ are set equal to the unconditional expectations of $\{\varepsilon_p, \varepsilon_{p-1}, ..., \varepsilon_{p-q+1}\}$ which are equal to zero ('conditional estimation') or these residuals are *backforecasted* (yielding the generally better 'unconditional estimation'). Backforecasting and its impact on parameter estimations are illustrated as follows:

Assume the $ARMA(p,q)$ model with $\varepsilon_t \overset{iid}{\sim} (0, \sigma_\varepsilon^2)$:

$$y_t = \alpha_1 y_{t-1} + \alpha_2 y_{t-2} + ... + \alpha_p y_{t-p} + \varepsilon_t + \varphi_1 \varepsilon_{t-1} + ... + \varphi_q \varepsilon_{t-q}.$$

Assume also that $\bar{\varphi}_1, \bar{\varphi}_2, ..., \bar{\varphi}_q$ and $\bar{\alpha}_1, \bar{\alpha}_2, ..., \bar{\alpha}_p$ are initial estimates of the corresponding parameters. Rewriting the $ARMA(p,q)$ model as $\varepsilon_t = y_t - \sum_{i=1}^{p} \alpha_i y_{t-i} - \sum_{j=1}^{q} \varphi_j \varepsilon_{t-j}$, we observe that we need values for $y_0, y_1, ..., y_{1-p}$, $\varepsilon_0, \varepsilon_{-1}, ..., \varepsilon_{1-q}$ in order to compute ε. The problem of how to determine the initial values as a function of $y_1, y_2, ..., y_T$ is known as the starting value problem. 'Backward forecasting' is a procedure that gives a solution to this problem.

Assume that y_t satisfies the above $ARMA$ model, then there exists a WN process $\{\varepsilon_t^*\}$ such that y_t also satisfies the forward equation:

$$y_t - \alpha_1 y_{t+1} - ... - \alpha_p y_{t+p} = \varepsilon_0^* + \textstyle\sum_{j=1}^{q} \varphi_j \varepsilon_{t+j}^*,$$

where ε_t^* can be expressed in terms of $y_t, y_{t+1}, \ldots$ in the following way:

$$\varepsilon_t^* = \sum_{j=0}^{\infty} \theta_j^* y_{t+j} \quad (\theta_0^* := 1), \tag{5.24}$$

where the parameters θ_j^* can be determined from the power series of

$$\theta^*(z) := \varphi^{-1}(z) A(z) \text{ with } |z| \leqq 1 \text{ and } \alpha(z) := 1 - \alpha_1 z - \ldots - \alpha_p z^p \neq 0$$
$$\text{and } \varphi(z) := 1 + \varphi_1 z + \ldots + \varphi_q z^q \neq 0.$$

From (5.24) $y_t = \varepsilon_t^* - \sum_{j=1}^{\infty} \theta_j^* y_{t+j} \cong \varepsilon_t^* - \sum_{j=1}^{p} \theta_j^* y_{t+j}$ and by setting $t = 0$, $y_0 \cong \varepsilon_0^* - \sum_{j=1}^{p} \theta_j^* y_j$ and if $\varepsilon_t^* \overset{iid}{\sim} (0, \sigma_{\varepsilon^*}^2)$, then ε_0^* is independent of $y_1, y_2, \ldots, y_T$ and the best linear estimator for y_0, using $y_1, y_2, \ldots, y_T$, is the conditional expectation:

$$E(y_0 | y_1, y_2, \ldots, y_T) \approx -\sum_{j=1}^{p} \theta_j^* y_j.$$

Setting t=-1,-2, ... forecasts of $y_{-1}, y_{-2}, \ldots, y_{-p+1}$ can be obtained. Problems arise, however, when p is large with respect to the sample size T. E.g. $y_t = (1 - 0.5L^{24}) \varepsilon_t$ is to be represented as:

$$\begin{aligned} \varepsilon_t^* &= y_t + 0.5y_{t+24} + 0.25y_{t+48} + 0.125y_{t+72} + 0.0625y_{t+96} + \ldots \\ &= y_t + 0.5y_{t+24} + 0.25y_{t+48} + 0.125y_{t+72}, \end{aligned} \tag{5.25}$$

which means that many observations are necessary for the *AR* representation to be useful.

It is also possible to use (approximate) ML besides OLS to estimate an $ARMA(p, q)$ model. Recall that the intuition of the ML principle is as follows. First, the likelihood of observing the sample under consideration is determined as a function of some unknown parameters that characterize the (assumed) distribution of the data. There are various ways to express the joint likelihood function, but typically some nonlinear function of the unknown parameters (that need to be optimized) will be obtained. Next, the values for the unknown parameters that maximize this function (i.e. that give us the highest likelihood) are retained as our ML estimates. Consider an $ARMA(p, q)$ model, with *AR* parameters in a p variate vector $\boldsymbol{\alpha}$ and *MA* parameters in a q variate vector $\boldsymbol{\varphi}$ and assume $\varepsilon_t \overset{iid}{\sim} N(0, \sigma_\varepsilon^2)$. In order to estimate the parameters we have to solve the following maximization problem:[7]

$$\begin{aligned} \max \ln \ell(\boldsymbol{\alpha}, \boldsymbol{\varphi}, \sigma^2) &= \max \ln \{p(y_1, y_2, \ldots, y_T | y_0, y_{-1}, \ldots; \boldsymbol{\alpha}, \boldsymbol{\varphi}, \sigma_\varepsilon^2)\} \\ &= \max \left[\ln p(y_1) + \sum_{t=2}^{T} \ln \{p(y_t | y_{t-1}, \ldots, y_0, y_{-1}, \ldots; \boldsymbol{\alpha}, \boldsymbol{\varphi}, \sigma_\varepsilon^2)\} \right] \end{aligned} \tag{5.26}$$

conditional ML (CML) is obtained if only the summation term in (5.26) with the parameters and the initial conditions is maximized; an exact ML (EML) is

[7] For computational convenience, the (natural) logarithm of the likelihood function is maximized.

obtained if the complete loglikelihood function is maximized. Asymptotically, CML and EML are equivalent in this case.

Consider now four models in turn:

- <u>*AR*(1) **model**:</u>
  $y_t = \alpha y_{t-1} + \varepsilon_t$, with $\varepsilon_t \overset{iid}{\sim} N(0, \sigma_\varepsilon^2)$, $t = 1, 2, ..., T$ or its pdf satisfies:
  $p(y_t | y_{t-1}, ..., y_0, y_{-1}, ...; \alpha, \sigma_\varepsilon^2) = f_N(y_t | y_{t-1}; \alpha, \sigma_\varepsilon^2)$ for $t = 1, 2, ..., T$.

  Precise values are not known, but assume that the time series process has been going on long enough to be in stationary state; i.e. y_1 will have the stationary distribution. This implies that
  $E(y_1){=}0$ and $var(y_1){=}var\left(\sum_{i=0}^{\infty} \alpha^i \varepsilon_{1-i}\right){=}\sigma_\varepsilon^2 \sum_{i=0}^{\infty} \alpha^{2i}{=}\sigma_\varepsilon^2 / \left(1 - \alpha^2\right)$,

where for normally distributed y_1 the loglikelihood function to be maximized satisfies:

$$\begin{aligned} L(\alpha, \sigma_\varepsilon^2) &= \ln p(y_1) + \textstyle\sum_{t=2}^{T} \ln p(y_t | I_{t-1}) \\ &= -\frac{T}{2} \ln(2\pi\sigma_\varepsilon^2) + \frac{1}{2} \ln(1-\alpha^2) - \frac{(1-\alpha^2)}{2\sigma_\varepsilon^2} y_1^2 - \frac{1}{2\sigma_\varepsilon^2} \sum_{t=2}^{T} (y_t - \alpha y_{t-1})^2 . \end{aligned} \tag{5.27}$$

- This is often simplified by assuming y_1 as *known*, in which case (5.27) becomes:

  $$L(\alpha, \sigma_\varepsilon^2) = -\frac{T-1}{2} \ln(2\pi\sigma_\varepsilon^2) - \frac{1}{2\sigma_\varepsilon^2} \textstyle\sum_{t=2}^{T} (y_t - \alpha y_{t-1})^2, \tag{5.28}$$

  which is simply maximized by minimizing the sum of squares (OLS of y_t on y_{t-1}). Note that the difference between (5.27) and (5.28) tends to zero as T grows.

- <u>*MA*(1) **model**:</u>
  $y_t = \varepsilon_t + \varphi \varepsilon_{t-1}$ with $\varepsilon_t \overset{iid}{\sim} N(0, \sigma_\varepsilon^2)$, $t = 1, 2, ..., T$.
  If ε_0 is known, $\{\varepsilon_t\}$ can be constructed from $\{y_t\}$ and φ.
  So then $p(y_t | y_{t-1}, ..., y_0, y_{-1}; \varphi, \sigma_\varepsilon^2) = f_N(y_t | \varepsilon_{t-1}; \varphi\sigma_\varepsilon^2)$ and:

  $$L(\varphi, \sigma_\varepsilon^2) = -\frac{T}{2} \ln 2\pi\sigma_\varepsilon^2 - \frac{1}{2\sigma_\varepsilon^2} \textstyle\sum_{t=1}^{T} (y_t - \varphi \varepsilon_{t-1})^2 . \tag{5.29}$$

  Again L is maximized by minimizing the sum of squares. Note that now ε_{t-1} is in the sum of squares, which depends on φ and one obtains a nonlinear function to be minimized. This NLLS estimator will be the CML estimator, since $\varepsilon_t = \sum_{j=0}^{t-1} (-\varphi)^j y_{t-j} - \varphi^t \varepsilon_0$ with $\varepsilon_0 = 0$.

  To obtain the EML estimator, remember that e.g. (5.29) with $\varepsilon_0 = 0$ implies $var(y_1) = \sigma_\varepsilon^2$, whereas the actual stationary variance (not fixing

ε_0) is equal to $var(y_1) = \sigma_\varepsilon^2(1+\varphi^2)$. Including all this makes it more complicated; however, the importance of the difference between CML and EML vanishes with T (CML and EML are asymptotically equivalent).

- ***ARMA* (1, 1) model:**
  $y_t = \alpha y_{t-1} + \varepsilon_t + \varphi\varepsilon_{t-1}$ with $\varepsilon_t \overset{iid}{\sim} N(0,\sigma_\varepsilon^2)$, $t = 1, 2, ..., T$.

Exercise 5.1 *Show that an ARMA (1, 1) model has the following loglikelihood function:*

$$L(\alpha,\varphi,\sigma_\varepsilon^2) = -\frac{T}{2}\ln 2\pi\sigma_\varepsilon^2 - \frac{1}{2}\ln|\mathbf{Z}'\mathbf{Z}| - \frac{1}{2\sigma_\varepsilon^2}S(\alpha,\varphi) \qquad (5.30)$$

with $|\mathbf{Z}'\mathbf{Z}| = \frac{(1-\alpha^2)(1-\varphi^2)+(1-\varphi^{2T})(\varphi-\alpha)}{(1-\varphi^2)(1-\alpha^2)}$ and
$S(\alpha,\varphi) = E(\varepsilon_0^2|y) + \frac{(1-\alpha^2)}{(\varphi-\alpha)^2}[E(y_0|y) - E(\varepsilon_0|y)]^2 + \sum_{t=1}^T E(\varepsilon_t^2|y)$.

- ***ARIMA* (p, d, q) model:**
  $y_t = \alpha_1 y_{t-1} + \alpha_2 y_{t-2} + ... + \varepsilon_t + \varphi_1\varepsilon_{t-1} + \varphi_2\varepsilon_{t-2} + ...$, with $\varepsilon_t \overset{iid}{\sim} N(0,\sigma_\varepsilon^2)$, $t = 1, 2, ..., T$, has a joint likelihood function which is of the following form:
  $\ell(\Omega,\Psi,\sigma_\varepsilon^2) = g_1(\Omega,\Psi,\sigma_\varepsilon^2)\exp\left\{-\frac{1}{2\sigma_\varepsilon^2}V(\alpha,\varphi)\right\}$ with $V(\alpha,\varphi) := \sum_{t=1-p-q}^T E(\varepsilon_t^2|y)$.

5.7.3 Diagnostic Checking

A fundamental issue is whether you want a model that gives the most adequate description of the data or a model that gives you the most accurate forecasts. For forecasting the idea of *parsimony* is important, i.e. one opts for a relatively small model because such a model usually forecasts better in the sense that it has smaller confidence intervals around the forecasts. The aim is in fact to approximate the true DGP, but not to pin down the exact process. Forecasting often becomes worse if supplementary statistical significant regressors are included that are economically unimportant and are thus not really needed for forecasting. Since these parameters are estimated with an error, confidence intervals will become wider. Most of the time it is not difficult to find statistical significant regressors of high(er) lag order.[8] Although the inclusion of these regressors in the model might give you a more adequate statistical description of the fit of data, think e.g. of the R^2 which usually goes up if more regressors are included, a large and complicated lag structure often hampers the economic interpretation (in addition to forecasting). Such *overfitting* can occur when one tries to obtain a correlogram of the residuals that displays a 'perfect' WN pattern. The following model selection criteria were developed that incorporate

[8] Clearly, the interpretation of high depends on the nature and frequency of the data and on the researcher's judgement.

this idea:

Akaike Information Criterion (AIC):

$$AIC\,(p,q) = \ln\left(\hat{\sigma}^2_{\hat{\varepsilon}}\right) + 2\,(p+q)\,/T, \tag{5.31}$$

Schwartz (Bayesian) Information Criterion (BIC, SC, SBC):

$$BIC\,(p,q) = \ln\left(\hat{\sigma}^2_{\hat{\varepsilon}}\right) + \frac{(p+q)}{T}\ln\left(T\right), \tag{5.32}$$

where $\hat{\sigma}^2_{\hat{\varepsilon}}$ is the estimated error variance, i.e. $\hat{\sigma}^2_{\hat{\varepsilon}} = \frac{\hat{\varepsilon}'\hat{\varepsilon}}{T}$ for ML and $\hat{\sigma}^2_{\hat{\varepsilon}} = \frac{\hat{\varepsilon}'\hat{\varepsilon}}{T-p-q}$ for LS, T is the sample size and $p+q$ is the total number of estimated parameters ($ARMA(p,q)$ example in this case).

Increasing p or q by one the fit (measured by the lack of fit in $\hat{\sigma}^2_{\varepsilon}$) will increase, but the penalty for lack of parsimony is equal to $2/T$ for AIC and $\ln(T)/T$ for BIC. BIC has a higher penalty, and is consistent as $T \to \infty$. AIC will tend to result in models that are asymptotically overparameterized This suggests to prefer the model for which AIC and BIC are smallest. Usually all models up to a certain order are compared. Note that it is possible that the two criteria lead to a different conclusion. The latter, in combination with the idea of parsimony, implies that the researcher will have to make a judgement at some point and that two equally skilled econometricians might come up with a different model.

The analysis of the residuals is mostly based on the hypothesis that the resulting model should have WN residuals. Assume the $ARMA\,(p,q)$ process:

$$\breve{y}_t - \textstyle\sum_{i=1}^{p} \alpha_i \breve{y}_{t-i} = 1 + \sum_{j=1}^{q} \varphi_j \varepsilon_{t-j}$$

with $\breve{y}_t := y_t - \mu$; then $\hat{\varepsilon}_t = \left(\hat{\varphi}\,(L)\right)^{-1} \hat{\alpha}\,(L)\,\widetilde{y}_t$ ($\widetilde{y}_t := y_t - \hat{\mu}$) are the residuals.

Assume now that the model is correct and that the real parameters are known; then it can be shown that the estimated autocorrelations $r_k(\varepsilon) = \frac{\sum_{t=k+1}^{T} \varepsilon_t \varepsilon_{t-k}}{\sum_{t=1}^{T} \varepsilon_t^2}$ are not significantly different from zero, and are asymptotically normally distributed with variance $\frac{1}{T}$ (see Example 5.1 in Section 5.2.2). In practice the parameters are not known, but the estimates $\left(\hat{\alpha}_1, \hat{\alpha}_2, ..., \hat{\alpha}_p, \hat{\varphi}_1, \hat{\varphi}_2, ..., \hat{\varphi}_q\right)$ can be obtained, and the resulting residuals $\hat{\varepsilon}_t$ can be computed. Since the estimated autocorrelation $r_k\left(\hat{\varepsilon}_t\right)$ can give information on the model adjustment, a test statistic has been developed to test whether the residuals still display autocorrelation or can be classified as WN. Under H_0 $r_k\left(\hat{\varepsilon}\right) \overset{iid}{\sim} N\left(0, \frac{1}{T}\right)$, the *Box-Pierce portmanteau* test statistic (see Box and Jenkins (1970)) is ($T^* := T - d$, d is the number of differences):

$$Q = T^* \textstyle\sum_{k=1}^{K} r_k^2\left(\hat{\varepsilon}\right) \overset{as}{\sim} \chi^2_{K-p-q}$$

presents the basis for the decision on the model. Davies *et al.* (1977) have shown from asymptotic theory, however, that even for reasonably large sample sizes, the expectation and the variance of Q can significantly differ from $K - p - q$.

It seems that $E(Q) << K - p - q$. Therefore, Ljung and Box (1978) have proposed an adjusted portmanteau test statistic:

$$Q^* = T^* (T^* + 2) \sum_{k=1}^{K} (T^* - k)^{-1} r_k^2 (\hat{\varepsilon}), \tag{5.33}$$

which is asymptotically χ^2-distributed with $K - p - q$ degrees of freedom for not too large T. However, the low power of the Q- and Q^*-statistics has been widely documented in the literature. In addition, Maddala and Kim (1998, p. 19) argue that the Q-statistics are even not appropriate in autoregressive models because the *OLS* estimates of the coefficients of lagged dependent variables are inconsistent under the presence of serially correlated errors (cf. the arguments against the use of the Durbin-Watson statistic with lagged dependent variables ;[9] note that this does not imply that a correlogram should not be used for identification).

The Box-Jenkins approach is summarized in Figure 5.2.

[9] The Durbin-Watson test is a two-sided test that rejects the null of no first order autocorrelation if the DW-statistic is outside of the range $(DW_\ell, 4-DW_\ell)$, where the interval bounds are available in DW tables (see Durbin and Watson (1950 and 1951)). More specifically, the statistic is given by $DW := \frac{\sum_{t=2}^{T}(\hat{\varepsilon}_t - \hat{\varepsilon}_{t-1})^2}{\sum_{t=1}^{T}\hat{\varepsilon}_t^2}$, where $\hat{\varepsilon}_t$ is the estimated OLS residual. If the number of observations is large, $DW \approx 1 + \sum_{t=2}^{T}\hat{\varepsilon}_t^2 / \sum_{t=1}^{T}\hat{\varepsilon}_t^2 - 2\sum_{t=2}^{T}\hat{\varepsilon}_t\hat{\varepsilon}_{t-1} / \sum_{t=1}^{T}\hat{\varepsilon}_t^2$ $\approx 2(1 - \hat{\rho})$.

If there are lagged dependent variables among the explanatory variables, the DW-statistic is no longer valid and is given by Durbin (1970).

Figure 5.2: Summarizing flow chart of the Box-Jenkins approach

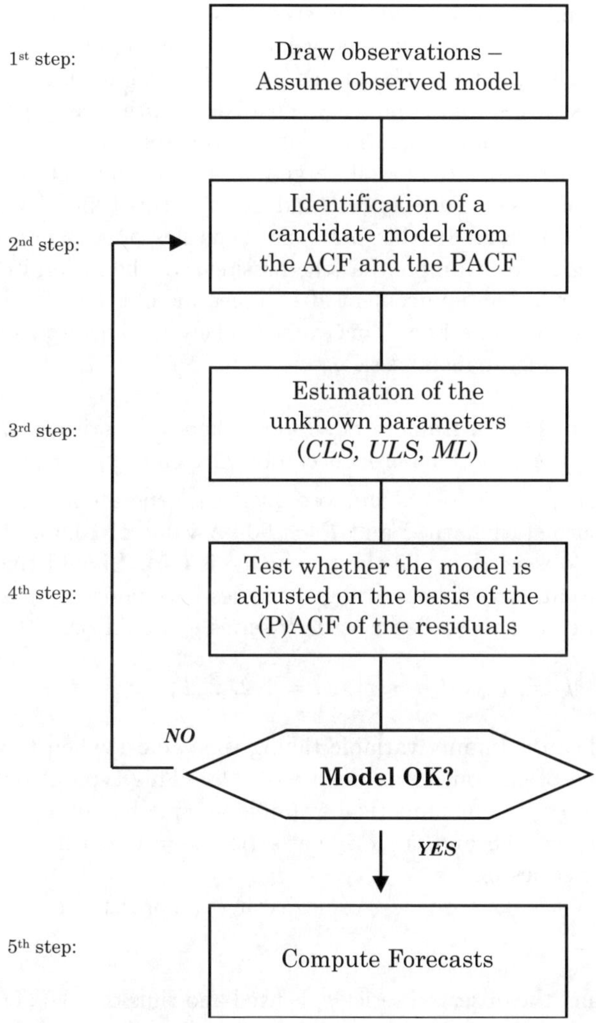
1st step:
Draw observations – Assume observed model
2nd step:
Identification of a candidate model from the ACF and the PACF
3rd step:
Estimation of the unknown parameters (CLS, ULS, ML)
4th step:
Test whether the model is adjusted on the basis of the (P)ACF of the residuals
NO
Model OK?
YES
5th step:
Compute Forecasts

5.8 Time series with aberrant observations

5.8.1 Definition and problems

Sometimes, when a time series is considered, one or more observations can be strikingly different from the other observations. Such observations are called *aberrant*. Inspecting for example a graph of the series, it seems as if these aberrant observations do not fit the general pattern observed. A value that shows an unexpectedly large (positive or negative) difference is also called an *outlier*. The more general terminology *aberrant observations* refers to different types of outliers. In general one or more extreme value(s) in economic time series are not expected. There are important exceptions, however, such as financial data and micro data. Stocks are for example very sensitive to (principally negative) news or to the external economic situation. Since more volatile periods imply more uncertainty about the future as well, large fluctuations in stock prices are likely to be clustered in time. To examine (and exploit) this type of volatility clustering consult Section 8.2. In this part the focus lies on the presence of isolated outliers. Imagine for example that the president of the Federal Reserve Bank faints at an important announcement after a meeting of the Federal Funds Board, but later turns out to be fine. This event is likely to show up as a spike or one time outlier in some financial series.

To put things in an $ARMA$ framework suppose there is a series x_t which is generated by an $AR(1)$ process.This process can be written as $x_t = \alpha_1 x_{t-1} + \varepsilon_t$, for $t = 1, 2, ..., T$, where $\varepsilon_t \overset{iid}{\sim} N(0,1)$. Suppose a series is generated with $x_0 = 0$, $\alpha_1 = 0.7$ (< 1 to assure stationarity) and $T = 200$. A value of 10.5 at $t = 100$ is not expected to be observed (while the true value is 0.5). Should this value occur (possibly due to an error when noting the series for example), it is called an outlier. Then, instead of the series x_t, a new series y_t would be:

$$y_t = x_t + \omega I_t[t = \tau] \qquad t = 1, 2, ..., T, \tag{5.34}$$

with $I_t[t = \tau]$ an indicator dummy variable taking the value 1 when $t = \tau$ and 0 otherwise. The size of the outlier is denoted by ω. This type of outlier is called an *additive outlier*. In our numerical example y_t equals x_t except at time $t = \tau = 100$ when $y_t = 10.5$ and the original series $x_t = 0.5$ (thus $\omega = 10$); Figure (5.3) plots the series y_t.

Using this example the consequences of neglecting the presence of an outlier can be illustrated:

1. When estimating, the observed series y_t is used and thus the $AR(1)$ model estimated looks like $y_t = \hat{\alpha}_1 y_{t-1} + \hat{\varepsilon}_t$.

   Using y_t instead of the unobserved x_t a value that can significantly differ from the true value α_1^* is obtained. Table 5.1 illustrates an example for different values of ω. The larger the value of ω, the more the estimate of α_1 deviates towards zero, the estimated standard error is increasing in ω.

Figure 5.3: Series y_t with outlier at time $t = 100$

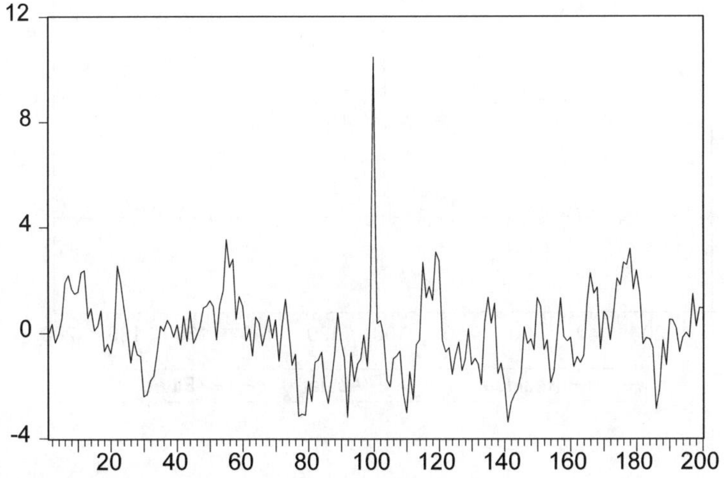

Table 5.1: Effect of an additive outlier of magnitude ω on the estimated $\hat{\alpha}_1$

ω	$\hat{\alpha}_1$	*SK*	*K*	*JB*
0	0.707 (0.050)	-0.024	2.997	0.019
5	0.673 (0.053)	0.269	4.600	23.637
10	0.581 (0.058)	1.761	19.834	2452.562
20	0.378 (0.066)	5.799	67.118	35203.48

Standard errors within brackets, SK=skewness, K=kurtosis, JB=Jarque-Bera.

Figure 5.4: Effect of neglecting a single additive outlier on residuals of an $AR(1)$ model ($\omega = 10$)

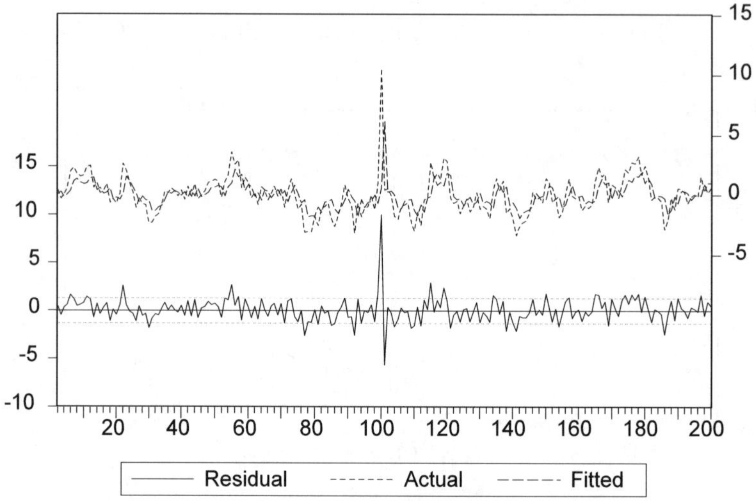

Moreover, observe that additive outliers generate large values of skewness and kurtosis and hence, significant values for the *Jarque-Bera* (JB) test for normality.[10] Large values of the JB statistic can be an indication of the presence of one or more outliers (rejection of the null hypothesis is not sufficient, the value of the JB statistic must be large). Figure 5.4 indicates that due to the additive outlier a large positive error at $t = \tau = 100$ and a large negative error at $t = \tau + 1 = 101$ (due to the $AR(1)$) are obtained.

2. To use the model for forecasting, our forecast clearly suffers from the fact that $\hat{\alpha}_1$ is not accurate. If the aberrant observation is close to our forecasting period, there is an additional problem. Suppose one wants to forecast y_{101} in the example, using y_{100} -the aberrant observation- $\hat{y}_{101}$ will be highly inaccurate.

3. Estimating the $AR(1)$ model using y_t, the estimated variance of the error terms $\hat{\sigma}^2_\varepsilon$ can be expected to be larger than the true value σ^2_ε (based on x_t). This implies that forecast confidence intervals will be too wide.

Next to the additive outlier, there are also innovational outliers, temporary changes and level shifts.

[10] The JB statistic is defined as $JB := T\left(\frac{SK^2}{6} + \frac{(K-3)^2}{24}\right)$ with $SK := \frac{E[(y_t-\mu)^3]}{(var[y_t])^{3/2}}$, the skewness coefficient, and $K := \frac{E[(y_t-\mu)^4]}{(var[y_t])^2}$, the kurtosis coefficient.

To illustrate these different types consider the following general expression for $x_t = \frac{\varphi(L)}{\alpha(L)}\varepsilon_t$ for $t = 1, 2, ..., T$ where $\varphi(L)$ and $\alpha(L)$ are polynomials in L, with all roots of $\varphi(z) = 0$ and those of $\alpha(z) = 0$ outside the unit circle. The time series subject to the influence of an aberrant observation y_t can in general terms be described as follows:

$$y_t = x_t + \omega\frac{G(L)}{H(L)}I_t[t = \tau] \tag{5.35}$$

with $I_t = 1$ if $t = \tau$. An aberrant observation impact can now be classified by imposing a special structure on $G(L)/H(L)$. The different types are: an *innovational outlier*, represented by an $ARMA$ model written in the form of an infinite dimensional MA model $\left(\frac{G(L)}{H(L)} = \frac{\varphi(L)}{\alpha(L)}\right)$, an *additive outlier* $\left(\frac{G(L)}{H(L)} = 1\right)$, a *level shift* with a unit root in the AR polynomial $H(L)$, $\left(\frac{G(L)}{H(L)} = \frac{1}{(1-L)}\right)$, and a *temporary change* $\left(\frac{G(L)}{H(L)} = \frac{1}{(1-\alpha L)}, \alpha \in (0,1)\right)$.

5.8.2 Testing for and dealing with aberrant observations

Suppose no information is available on external events so that there is no idea about whether there are aberrant observations and when they occur.

As already noted before, the presence of an outlier causes a violation of the normality assumption for the residuals. Therefore, the JB statistic is used to test for normality. A rejection of the null might be an indication of aberrant observations. Note that -in practice- the JB statistic should have a large value (cf. Table 5.1). If this is the case a closer look at the residuals is useful and residuals exceeding four times the standard deviation are considered as outliers then. Once the extreme value is located, several statistical methods can be used. Here the method of Chen and Liu (1993) is discussed.

Starting from a univariate time series $\{x_t\}$ generated by an $ARMA(p, q)$ process $(I - \alpha_1 L - ... - \alpha_p L^p)\, x_t = (1 + \varphi_1 L + ... + \varphi_q L^q)\, \varepsilon_t$ where ε_t is WN with variance σ_ε^2. The actually observed series (*containing an additive outlier*) can again be described as:

$$y_t = x_t + \omega I_t[t = \tau] \tag{5.36}$$

with I_t a dummy variable equal to one if $t = \tau$ and zero otherwise.

To examine the effects of aberrant observations (the case of additive outliers here) on the estimated residuals, the time series parameters are assumed to be known and the series is observed from $t = -J$ to $t = T$, where J is an integer larger than $p + q$ and that $1 \leqq \tau \leqq T$. Define the polynomial $\pi(L)$ as $\pi(L) := \frac{\alpha(L)}{\varphi(L)} = I - \pi_1 L - \pi_2 L^2 - ...$, so that the $ARMA(p, q)$ model can be written from (5.6) as an infinite dimensional autoregressive model, or for $t = 1, 2, ...$:

$$e_t := \varepsilon_t + \pi(L)\,\omega I_t[t = \tau] = \pi(L)\, y_t, \tag{5.37}$$

which, using (5.36), can be interpreted as a regression model for e_t on z_t, given by:

$$e_t = \omega z_t + \varepsilon_t \tag{5.38}$$

where (for our case of an additive outlier):

$$\begin{aligned} z_t &= 0 \Leftrightarrow t \neq \tau, \text{ except for } z_{t+k} = -\pi_k \Leftrightarrow t > \tau \\ z_t &= 1 \Leftrightarrow t = \tau \quad . \end{aligned} \tag{5.39}$$

Using possibly contaminated estimated residuals, $\hat{e}_t$, the $AR(\infty)$ model (5.37) can be expressed as:

$$\hat{e}_t = \hat{\pi}(L)\, y_t \quad \text{for } t = 1, 2, ..,$$

so that in that case the OLS estimator for the effect of a single outlier at $t = \tau$ can simply be computed as:

$$\hat{\omega} := \hat{\omega}_{AO}(\tau) = \frac{\sum_{t=\tau}^{T} \hat{e}_t \hat{z}_t}{\sum_{t=\tau}^{T} \hat{z}_t^2}, \tag{5.40}$$

where the 'regressor' $\hat{z}_t$ is given by (5.39) with π_k replaced by $\hat{\pi}_k$.

Chang *et al.* (1988) show that a possible approach for detecting additive outliers is to examine the maximum value of the standardized statistic of the outlier effect:

$$\hat{\zeta}_{AO}(\tau) = \hat{\omega}_{AO}(\tau) / SE(\hat{\omega}_{AO}(\tau)) \tag{5.41}$$

where $SE(\hat{\omega}_{AO}(\tau)) := \hat{\sigma}_{\hat{\varepsilon}} \left(\sum_{t=\tau}^{T} \hat{z}_t^2 \right)^{-1/2}$; this test statistic is approximately normally distributed. To calculate (5.41) $\hat{\sigma}_{\hat{\varepsilon}}$ has to be known. The residual standard error of the estimated $ARMA(p,q)$ process cannot be used, because in the presence of outliers the residuals are contaminated. Hence $\sigma_{\hat{\varepsilon}}$ may be overestimated if the sample standard deviation is used. Chen and Liu (1993) propose three different methods to calculate a better $\sigma_{\hat{\varepsilon}}$:

1. The Mean Absolute Deviation method. The standard deviation can then be calculated as $\hat{\sigma}_{\hat{\varepsilon}} = 1.483 \times median\{|\hat{e}_t - \widetilde{e}_t|\}$, where $\widetilde{e}_t$ is the median of the estimated residuals.

2. The $(100 \times \alpha)$ per cent 'trimmed' method. To compute the $(100 \times \alpha)$ per cent trimmed standard deviation, first take absolute values of the residuals and then remove the $(100 \times \alpha)$ per cent largest values. Compute then the standard deviation based on this trimmed sample.

3. The last method is the 'omit-one' method. This method implies calculating the standard deviation of the estimated residuals, leaving out the residual at period $t = \tau$.

If $\hat{\zeta}_{AO}(\tau)$ exceeds a critical value C (since $\hat{\zeta}_{AO}(\tau)$ follows an asymptotically normal distribution, the usual critical value of 4 can be applied), we conclude that the impact of the AO at period $t = \tau$ is significant. The observed series y_t is then adjusted as follows:

$$y_t^{adj} = y_t - \hat{\omega}_{AO}(\tau) I_t[t = \tau] \tag{5.42}$$

5.8.3 Detecting and estimating aberrant observations

Chen and Liu (1993) offer the following joint estimation procedure in the presence of outliers. There are three steps in the estimation. First, we estimate an $ARMA$ model and look for outliers. Afterwards the parameters of the model and the outliers are estimated simultaneously. The final step consists in detecting extreme values based on the final model obtained in step two. Every step contains several substeps. The detailed procedure for AOs goes as follows.

1. **Stage I: Initial parameter estimation and detection of outliers**

   (a) Estimate the model and obtain the residuals.
   What follows is an iterative procedure. Use the observed series $\{y_t\}$ for the first iteration. If you already went through an iteration, use the series that is adjusted based on the findings in this iteration.

   (b) Calculate $\hat{\zeta}_{AO}(\tau)$ using the residuals obtained in step (a). If $\hat{\zeta}_{AO}(\tau) > C = 4$, the outlier is significant.

   (c) If there is no significant outlier, go to (d). If there is a significant outlier remove its effect from the residuals and the observations according to (5.42). Return to (b) to test whether there are more outliers present.

   (d) If no outliers are found during the first iteration, stop here. The series is not affected by aberrant observations. If outliers are found under the given parameter estimates, go back to (a) to revise our estimates. Proceed with (b) to test whether any outlier is present under the revised parameter estimates. Go to stage II if no additional significant outliers are detected after parameter revising.

2. **Stage II: Simultaneous estimation of outlier effects and model parameters**

   (a) Suppose m time points $\tau_1, \tau_2, ..., \tau_m$ that are possible outliers were detected. The outlier effects $\omega_j := \omega_{AO}(\tau_j)$ can be estimated as follows. If the series $\{y_t\}$ is affected by m outliers $\tau_1, \tau_2, ..., \tau_m$, (5.36) can be rewritten as $y_t = x_t + \sum_{j=1}^{m} \omega_j I_t[t = \tau_j]$.
   The residuals (5.37) then become:

   $$e_t = \textstyle\sum_{j=1}^{m} \omega_j \pi(L) I_t[t = \tau_j] + \varepsilon_t. \tag{5.43}$$

   (b) Calculate $\hat{\zeta}_{AO}(\tau_j)$ for the estimated ω_js where:

   $$\hat{\zeta}_{AO}(\tau_j) = \frac{\hat{\omega}_j}{SE(\hat{\omega}_j)} \quad j = 1, ..., n \tag{5.44}$$

   If $\min_j |\hat{\zeta}_{AO}(\tau_j)| = \hat{\zeta}_{AO}(\tau_v) \leqq C$ (same critical value as at the end of the previous paragraph and in step (b) of stage I) then remove the outlier at time point τ_v from the series of outliers $\tau_1, \tau_2, ..., \tau_m$ detected in stage I. Return to step (a) of stage II with the remaining $m-1$ outliers. If $\min_j |\hat{\zeta}_{AO}(\tau_j)| = \hat{\zeta}_{AO}(\tau_v) > C$, proceed to step c.

(c) Obtain the adjusted series by removing the outlier effects according to (5.42), using the most recent estimates of ω_j at step (a) of stage II. Thus adjust only for the outlier effects that are significant based on the iterations in steps (a) and (b) of stage II.

(d) Estimate the model parameters based on the adjusted series obtained at step (c) of stage II.
If the relative change of the standard error of the residuals from the previous step is greater than ζ, go to step (a) of stage II for further iterations, otherwise go to stage III. ζ, a predetermined constant, is the tolerance level chosen by the user to control the accuracy of the parameter estimates. An appropriate tolerance value could e.g. be 0.001.

3. **Stage III: Detection of outliers based on the final parameter estimates**

   (a) Compute the residuals by filtering the originally observed series $\{y_t\}$ based on the parameters estimates obtained in stage II, step (d).

   (b) Use the residuals obtained in step (a) and iterate through stages I and II with the following modifications:

      i. the parameter estimates used in the 'inner loop' of stage I are fixed to those obtained in step (d) of stage II;

      ii. steps (c) and (d) are omitted in stage II; the estimated $\hat{\omega}_j$s of the last iteration at step a. of stage II are the final estimates of the effects of the detected outliers.

5.9 Forecasting of $ARIMA$ models

One of the most important uses of AR, MA and $AR(I)MA$ models is to forecast future values of the $\{y_t\}$ series. The objective is e.g. to predict the value of y for the period $t+h$ based on the information assembled until period t; t can then be called the *origin* and h the *lead time* or *forecast horizon* and $y_t(h)$ is then the *h-period* or *h-step (ahead)* forecast of y_{t+h}. Most of the time one is interested in *out-of-sample* forecasts, i.e. y_{t+h} is not used (or even not observed) to estimate the parameters of our $ARIMA$ model. An in-sample forecast, $y_{t,h}$ is obtained when y_{t+h} is used to estimate the parameters of the model. If not mentioned otherwise, throughout this chapter the focus will be on out-of-sample forecasting. To simplify the discussion, it is assumed in this chapter that the actual DGP is known. In this case, there is no uncertainty about the model choice, the parameter values or the errors $\{\varepsilon_t\}$ (note that by doing this the uncertainty of the forecasts is underestimated). Furthermore, the prime focus is on stationary (and invertible) processes. AR, MA and $ARMA$ forecasts will be dealt with separately and the issue of forecasting $ARIMA$ models will be included as an extension. As forecast criterion the minimization of the conditional expected (mean) squared forecast error $f_t(h)$ or the (conditional) forecast variance (the variance of the forecast error), given the information available up to period t is chosen, or

$$\min_{y_{t,h}} E\left\{ \underbrace{(y_{t+h} - y_t(h))^2}_{f_t(h)} \middle| \underbrace{y_t, y_{t-1}, y_{t-2}, ..., y_1}_{\Im_t} \right\}, \tag{5.45}$$

i.e. $y_t(h)$ is chosen as to minimize the expected value of $(y_{t+h} - y_t(h))^2$, given the information available up to time t gathered in the information set $\Im_t$. In general using $E_t(y_{t+h})$ as a shorthand notation to write the conditional expectation of y_{t+h} given the information available at period t, the solution to (5.45) is:

$$y_t(h) = E(y_{t+h}|\Im_t) = E_t(y_{t+h}).$$

5.9.1 Forecasting AR processes

Iterative forecasts

Assume one wants to forecast from the $AR(1)$ model $y_t = \alpha_0 + \alpha_1 y_{t-1} + \varepsilon_t$ with $\varepsilon_t \in WN$. Updating one period: $y_{t+1} = \alpha_0 + \alpha_1 y_t + \varepsilon_{t+1}$.

Since by assumption the parameters α_0 and α_1 are known, y_{t+1} can be forecast, conditional upon the available information at period t, as $E_t(y_{t+1}) = \alpha_0 + \alpha_1 y_t$ or formally $E_t(y_{t+i}) = E(y_{t+i} | y_t, y_{t-1}, ..., \varepsilon_t, \varepsilon_{t-1}, ...)$.

Since $y_{t+2} = \alpha_0 + \alpha_1 y_{t+1} + \varepsilon_{t+2}$, the forecast of y_{t+2} conditional upon the information available at period t is $E_t(y_{t+2}) = \alpha_0 + \alpha_1 E_t(y_{t+1})$. Substituting $E_t(y_{t+1})$ by the above forecast, $E_t(y_{t+2}) = \alpha_0 + \alpha_1\alpha_0 + \alpha_1^2 y_t$ is obtained. By continuing this forward substitution, one can see that in general the forecast function is:

$$\begin{aligned} E_t(y_{t+h}) &= \alpha_0 + \alpha_1 y_t & \text{for } h = 1 \\ &= \alpha_0\left(1 + \alpha_1 + \alpha_1^2 + ... + \alpha_1^{h-1}\right) + \alpha_1^h y_t & \text{for } h \geqq 2. \end{aligned} \tag{5.46}$$

Expression (5.46) is called the *forecast function* and yields the h-step ahead forecast for each y_{t+h}. Introducing now the assumption of stationarity of the $AR(1)$ model, i.e. $|\alpha_1| < 1$, (5.46) yields a converging sequence of forecasts. Taking the limit of $E_t(y_{t+h})$ as $h \to \infty$, one obtains:

$$\lim_{h \to \infty} E_t(y_{t+h}) = \frac{\alpha_0}{1 - \alpha_1}, \tag{5.47}$$

which is the unconditional mean (forecast).

Forecasting using the $MA(\infty)$ representation

For a *stationary AR* process, optimal forecasts can also be obtained by using the MA representation. Consider the following $AR(1)$ process:

$$y_t - \mu = \alpha_1 (y_{t-1} - \mu) + \varepsilon_t$$

with $\mu := \frac{\alpha_0}{1-\alpha_1}$ the mean of the series, $\varepsilon_t \in WN$, and $|\alpha_1| < 1$. Some rewriting yields (with L and I as lag and identity operators):

$$(y_t - \mu) - \alpha_1 (y_{t-1} - \mu) = (I - \alpha_1 L)(y_t - \mu) = \varepsilon_t.$$

The pure $MA(\infty)$ representation is then (see (5.5)):

$$\begin{aligned} y_t - \mu &= (I - \alpha_1 L)^{-1} \varepsilon_t = \textstyle\sum_{i=0}^{\infty} (\alpha_1 L)^i \varepsilon_t \\ &= \varepsilon_t + \alpha_1 \varepsilon_{t-1} + \alpha_1^2 \varepsilon_{t-2} + \alpha_1^3 \varepsilon_{t-3} + \ldots \end{aligned} \tag{5.48}$$

Based on (5.48) the one-period forecast is:

$$E_t(y_{t+1}) = \mu + E_t(\varepsilon_{t+1}) + \alpha_1 \varepsilon_t + \alpha_1^2 \varepsilon_{t-1} + \alpha_1^3 \varepsilon_{t-2} + \ldots \tag{5.49}$$

Except for $E_t(\varepsilon_{t+1})$, the error terms prevailing in (5.49) are known at time t and can be replaced by the observed values; ε_{t+1} is however not known or observed at time t, therefore the expected error $E_t(\varepsilon_{t+1})$ is used. Because of the WN assumption, the best expectation of ε_{t+1} is simply zero.

Remark 5.2 *Nearly always, only y_t is observed in practice. Therefore consistent estimators for the parameters α_i and the error terms ε_t have to be found.*

Equation (5.49) can be rewritten as $E_t(y_{t+1}) = \mu + \alpha_1 \underbrace{(\varepsilon_t + \alpha_1 \varepsilon_{t-1} + \alpha_1^2 \varepsilon_{t-2} + \ldots)}_{y_t - \mu}$; similarly, the two-period forecast is

$$\begin{aligned} E_t(y_{t+2}) &= \mu + E_t(\varepsilon_{t+2}) + \alpha_1 E_t(\varepsilon_{t+1}) + \alpha_1^2 \varepsilon_t + \alpha_1^3 \varepsilon_{t-1} + \ldots \\ &= \mu + \alpha_1^2 (\varepsilon_t + \alpha_1 \varepsilon_{t-1} + \alpha_1^2 \varepsilon_{t-2} + \ldots). \end{aligned}$$

For a general h-period (ahead) forecast:

$$\begin{aligned} E_t(y_{t+h}) &= \mu + E_t(\varepsilon_{t+h}) + \alpha_1 E_t(\varepsilon_{t+h-1}) + \alpha_1^2 E_t(\varepsilon_{t+h-2}) + \ldots \\ &= \mu + \alpha_1^h (\varepsilon_t + \alpha_1 \varepsilon_{t-1} + \alpha_1^2 \varepsilon_{t-2} + \ldots) = (1 - \alpha_1^h)\mu + \alpha_1^h y_t. \end{aligned} \tag{5.50}$$

Notice again that for $h \to \infty$, $E_t(y_{t+h})$ converges to the series' (unconditional) mean μ.

Proposition 5.3 *For any* ***stationary*** *AR model, the conditional forecast of y_{t+h}, $E_t(y_{t+h})$, converges to the unconditional mean as $h \to \infty$.*

The terms $y_t - \mu = \varepsilon_t + \alpha_1 \varepsilon_{t-1} + \alpha_1^2 \varepsilon_{t-2} + \ldots$ are also called *impulse responses*. The *impulse response function* is the path that y_t follows if it is kicked by a single unit shock ε_t, i.e. $\varepsilon_{t-j} = 0$, $\varepsilon_t = 1$, $\varepsilon_{t+j} = 0$ $(\forall\ j \neq 0)$ (Cochrane (1997)).[11] This is interesting because it allows to start thinking about 'causes' and 'effects'.

[11] See *http://www-gsb.uchicago.edu/fac/john.cochrane/research/Papers/timeser1.pdf*

Forecast errors

Forecasts are unfortunately not perfect. Define the h-period (ahead) forecast error (for an $AR(1)$ model) based on information available at period t, $f_t(h)$, as the difference between the realized value of y_{t+h} and the forecast value or $f_t(h) := y_{t+h} - E_t(y_{t+h})$. Hence, the one-step forecast error is

$$f_t(1) = y_{t+1} - E_t(y_{t+1}) = \varepsilon_{t+1},$$

so that ε_{t+1} is the *unforecastable* portion of y_{t+1} given the information available at time t. By substituting y_{t+1} in $y_{t+2} = \alpha_0 + \alpha_1 y_{t+1} + \varepsilon_{t+2}$,

$$y_{t+2} = \alpha_0 + \alpha_1\alpha_0 + \alpha_1^2 y_t + \varepsilon_{t+2} + \alpha_1\varepsilon_{t+1}.$$

Recalling $E_t y_{t+2}$ from this subsection:

$$\begin{aligned} f_t(2) &= y_{t+2} - E_t(y_{t+2}) \\ &= \left(\alpha_0 + \alpha_1\alpha_0 + \alpha_1^2 y_t + \varepsilon_{t+2} + \alpha_1\varepsilon_{t+1}\right) \\ &\quad - \left(\alpha_0 + \alpha_1\alpha_0 + \alpha_1^2 y_t\right) = \varepsilon_{t+2} + \alpha_1\varepsilon_{t+1}. \end{aligned} \tag{5.51}$$

In general, for an $AR(1)$ model:

$$f_t(h) = \varepsilon_{t+h} + \alpha_1\varepsilon_{t+h-1} + \alpha_1^2\varepsilon_{t+h-2} + \alpha_1^3\varepsilon_{t+h-3} + ... + \alpha_1^{h-1}\varepsilon_{t+1}. \tag{5.52}$$

Note that exactly the same result is obtained by subtracting the second line in (5.50) from (5.48) (for $t+h$ instead of t!):

$$\begin{aligned} y_{t+h} - E_t(y_{t+h}) &= \mu + \varepsilon_{t+h} + \alpha_1\varepsilon_{t+h-1} + \alpha_1^2\varepsilon_{t+h-2} + \alpha_1^3\varepsilon_{t+h-3} + ... \\ &\quad -\mu - \alpha_1^h\left(\varepsilon_t + \alpha_1\varepsilon_{t-1} + \alpha_1^2\varepsilon_{t-2} + ...\right) \\ &= \varepsilon_{t+h} + \alpha_1\varepsilon_{t+h-1} + \alpha_1^2\varepsilon_{t+h-2} + \alpha_1^3\varepsilon_{t+h-3} + ... + \alpha_1^{h-1}\varepsilon_{t+1}. \end{aligned}$$

Since $E_t(\varepsilon_{t+h}) = E_t(\varepsilon_{t+h-1}) = ... = E_t(\varepsilon_{t+1}) = 0$, the conditional expectation of (5.52) is $E_t(f_t(h)) = 0$. Since the expected value of the forecast error is zero, the forecasts based on (5.46) are unbiased.

To measure the inaccuracy of our forecasts the variance of the forecast error can be calculated. Under the WN assumption, the variance of the one-step forecast error is simply σ_ε^2. From (5.51) the two-step forecast variance is $\sigma_\varepsilon^2\left(1+\alpha_1^2\right)$. Based on (5.52) the following general expression for the variance of the h-step (ahead) forecast error is:

$$\begin{aligned} var\left[f_t(h)\right] &= \sigma_\varepsilon^2 \text{ (for } h=1) \\ &= \sigma_\varepsilon^2\left[1+\alpha_1^2+\alpha_1^4+...+\alpha_1^{2(h-1)}\right] = \sigma_\varepsilon^2\frac{\left(1-\alpha_1^{2h}\right)}{\left(1-\alpha_1^2\right)}, \end{aligned} \tag{5.53}$$

where the property of stationarity is only used at the very end.[12]

[12] It is obvious from (5.46) and (5.53) that the forecast variances for nonstationary AR models can be computed.

Notice from (5.53) that the variance of the forecast error is an increasing function of h. This implies that one can have more confidence in short-term rather than long-term forecasts, because the variance of the former is smaller. Taking the limit as $h \to \infty$:

$$\lim_{h \to \infty} var\left[f_t\left(h\right)\right] = \frac{\sigma_\varepsilon^2}{1 - \alpha_1^2},$$

the forecast error variance converges to the unconditional variance of the $\{y_t, t \in \mathbb{Z}\}$ series. Under the WN assumption and the application of the central limit theorem or under normally distributed forecast errors, (asymptotic) confidence intervals around the forecasts can be calculated. For a general forecast horizon h, the following 95% confidence interval is found $\alpha_0 \sum_{i=0}^{h-1} \alpha_1^i + \alpha_1^h y_t \pm 1.96\, \sigma_\varepsilon \sqrt{\sum_{i=0}^{h-1} \alpha_1^{2i}}$.

Note that, if there is any uncertainty concerning the parameters (which is generally the case since they have to be estimated) the confidence intervals will be wider than those reported here.

5.9.2 Forecasting MA processes

The ideas behind forecasting AR models and MA models are very similar. Consider the following invertible $MA\left(1\right)$ representation $y_t - \mu = \left(I + \varphi L\right) \varepsilon_t$ with $|\varphi| < 1$ and $\varepsilon_t \overset{iid}{\sim} \left(0, \sigma_\varepsilon^2\right)$. Rewriting this $MA\left(1\right)$ process as a pure $AR\left(\infty\right)$ process:

$$\begin{aligned} \varepsilon_t &= \frac{1}{I + \varphi L}\left(y_t - \mu\right) = \sum_{i=0}^{\infty} \left(-\varphi L\right)^i \left(y_t - \mu\right) \\ &= \left(y_t - \mu\right) \text{-} \varphi\left(y_{t-1}\text{-}\mu\right) + \varphi^2\left(y_{t-2}\text{-}\mu\right) \text{-} \varphi^3\left(y_{t-3} - \mu\right) + ... \end{aligned} \tag{5.54}$$

The formula for the one-period forecast is:

$$E_t(y_{t+1}) = \mu + E_t(\varepsilon_{t+1}) + \varphi\varepsilon_t. \tag{5.55}$$

Again the best guess for the conditional expectation of ε_{t+1}, $E_t(\varepsilon_{t+1})$, is simply zero because of the WN assumption; the forecast is then calculated as (based on (5.54)):

$$\begin{aligned} E_t(y_{t+1}) = \mu + \varphi\left(y_t - \mu\right) - \varphi^2\left(y_{t-1} - \mu\right) \\ + \varphi^3\left(y_{t-2} - \mu\right) - \varphi^4\left(y_{t-3} - \mu\right) + ... \end{aligned}$$

From (5.55) one immediately sees that a two (and more) period forecast is the (unconditional) mean μ of the series $\{y_t, t \in \mathbb{Z}\}$.

Alternatively the WN error term can be seen as the outcome of the infinite recursion $\varepsilon_t = \left(y_t - \mu\right) - \varphi\varepsilon_{t-1}$, which eventually brings us to the same result.

Forecasting from a more general invertible $MA\left(q\right)$ process,

$$(y_t - \mu) = \varphi(L)\varepsilon_t = \left(I + \varphi_1 L + \varphi_2 L^2 + ... + \varphi_q L^q\right)\varepsilon_t,$$

goes by the same principle. This results in the h-period forecast (for $h \leqq q$):

$$E_t(y_{t+h}) = \mu + \left(\varphi_h + \varphi_{h+1}L + \varphi_{h+2}L^2 + ... + \varphi_q L^{q-h}\right)\varepsilon_t \tag{5.56}$$

with $\varepsilon_t = (y_t - \mu) - \varphi_1\varepsilon_{t-1} + \varphi_2\varepsilon_{t-2} + ... + \varphi_q\varepsilon_{t-q}$. For $h > q$, $E_t(y_{t+h})$ equals the (unconditional) mean μ.

5.9.3 Forecasting *ARMA* processes

From the above, it is possible to use forward iterations to derive the forecast function. This can directly be extended to a general $ARMA(p,q)$ process. To keep the algebra as simple as possible, consider an $ARMA(2,1)$ model $y_t = \alpha_0 + \alpha_1 y_{t-1} + \alpha_2 y_{t-2} + \varepsilon_t + \varphi_1\varepsilon_{t-1}$. Updating one period yields $y_{t+1} = \alpha_0 + \alpha_1 y_t + \alpha_2 y_{t-1} + \varepsilon_{t+1} + \varphi_1\varepsilon_t$. Under the assumptions that *i*) all coefficients are known, *ii*) all variables are known up to period t, and *iii*) $E_t(\varepsilon_{t+j}) = 0$ for $j > 0$, the one-step (ahead) forecast is

$$E_t(y_{t+1}) = \alpha_0 + \alpha_1 y_t + \alpha_2 y_{t-1} + \varphi_1\varepsilon_t$$

and since $y_{t+2} = \alpha_0 + \alpha_1 y_{t+1} + \alpha_2 y_t + \varepsilon_{t+2} + \varphi_1\varepsilon_{t+1}$, the two-step forecast is

$$\begin{aligned} E_t(y_{t+2}) &= \alpha_0 + \alpha_1 E_t(y_{t+1}) + \alpha_2 y_t \\ &= \alpha_0 + \alpha_1\alpha_0 + \alpha_1^2 y_t + \alpha_1\alpha_2 y_{t-1} + \alpha_1\varphi_1\varepsilon_t + \alpha_2 y_t \\ &= \alpha_0(1+\alpha_1) + \left(\alpha_1^2 + \alpha_2\right)y_t + \alpha_1\alpha_2 y_{t-1} + \alpha_1\varphi_1\varepsilon_t \end{aligned}$$

Exercise 5.2 *Show that the three-period forecast is*

$$\begin{aligned} E_t(y_{t+3}) &= \alpha_0\left(1 + \alpha_1 + \alpha_1^2 + \alpha_2\right) + \left(\alpha_1^3 + 2\alpha_1\alpha_2\right)y_t \\ &\quad + \left(\alpha_1^2\alpha_2 + \alpha_2^2\right)y_{t-1} + \varphi_1\left(\alpha_1^2 + \alpha_2\right)\varepsilon_t. \end{aligned}$$

A general h-step (ahead) forecast (for $h \geqq 2$) can be found from

$$E_t(y_{t+h}) = \alpha_0 + \alpha_1 E_t(y_{t+h-1}) + \alpha_2 E_t(y_{t+h-2}). \tag{5.57}$$

This forecast rule (5.57) and the solution for the three-step forecast indicate that the forecasts will satisfy a second order difference equation; as long as the characteristic roots of (5.57) lie within the unit circle, the forecasts will converge to the unconditional mean $\frac{\alpha_0}{1-\alpha_1-\alpha_2}$.[13]

Example 5.7 *Optimal forecasting from a random walk and noise model*

[13] Alternatively, one can find the general solution by combining homogeneous solutions and the particular solution and imposing the initial conditions.

First, consider an ordinary random walk $y_t - y_{t-1} = \varepsilon_t$. Given initial condition y_0 recursive substitution yields:

$$\begin{aligned} y_t &= y_{t-1} + \varepsilon_t = y_{t-2} + \varepsilon_{t-1} + \varepsilon_t = ... \\ &= y_0 + \sum_{i=1}^{t} \varepsilon_i. \end{aligned}$$

The unconditional expected values or means of a random walk are constant: $E(y_{t+1}) = ... = E(y_{t+s}) = y_0$. Conditional expectations are formed as: $E(y_{t+1}|\Im_t) = E_t(y_{t+1}) = E(y_t + \varepsilon_{t+1}|\Im_t) = E_t(y_t + \varepsilon_{t+1}) = y_t$ and, similarly, $E(y_{t+s}|\Im_t) = E_t(y_{t+s}) = y_t + E_t\left(\sum_{i=1}^{s} \varepsilon_{t+i}\right) = y_t$ and $var(y_t) = t\sigma_\varepsilon^2$ and $\rho_s = \left(\frac{t-s}{t}\right)^{1/2}$ (the autocorrelation function of a random walk will thus show a *slight* tendency to decay)!

Consider now a random walk and noise (η): $y_t = \sum_{i=1}^{t} \varepsilon_i + \eta_t$. Let the forecast of y_t be a linear function of the past values or $E(y_t|\Im_t) = E_t(y_t) = \sum_{i=1}^{\infty} v_i y_{t-i}$, with v_i selected such that the mean squared forecast error is minimized. Substituting gives:

$$E_t(y_t) = v_1\left(\sum_{i=1}^{t-1} \varepsilon_i + \eta_{t-1}\right) + v_2\left(\sum_{i=1}^{t-2} \varepsilon_i + \eta_{t-2}\right) + ...$$

The optimization problem is to select the v_js so as to minimize the mean squared error:

$$\begin{aligned} E(y_t - E_t(y_t))^2 &= E\left[\sum_{i=1}^{t} \varepsilon_i + \eta_t - v_1\left(\sum_{i=1}^{t-1} \varepsilon_i + \eta_{t-1}\right) - ...\right]^2 \\ &= \sigma_\varepsilon^2 + \sigma_\eta^2 + \sigma_\varepsilon^2 \sum_{i=1}^{\infty}\left(1 - \sum_{j=1}^{i} v_j\right)^2 + \sigma_\eta^2 \sum_{j=1}^{i} v_j^2. \end{aligned}$$

The first order condition for minimizing the mean squared error is (for $k = 1, 2, ...$):

$$\frac{\partial E(y_t - E_t(y_t))^2}{\partial v_k} = 2\sigma_\eta^2 v_k - 2\sigma_\varepsilon^2 \sum_{j=k}^{\infty}\left(1 - \sum_{i=1}^{j} v_i\right) = 0 \qquad (i)$$

$$\left.\begin{aligned} &\text{For } k = 1: 2\sigma_\eta^2 v_1 - 2\sigma_\varepsilon^2 \sum_{j=1}^{\infty}\left(1 - \sum_{i=1}^{j} v_i\right) = 0 \\ &\text{For } k = 2: 2\sigma_\eta^2 v_2 - 2\sigma_\varepsilon^2 \sum_{j=2}^{\infty}\left(1 - \sum_{i=1}^{j} v_i\right) = 0 \end{aligned}\right\}$$

Subtracting yields:

$$\sigma_\varepsilon^2(1 - v_1) + \sigma_\eta^2(v_2 - v_1) = 0 \qquad (ii)$$

The second difference of (i) yields (for $k = 2, 3, ...$): $-v_{k-1} + \left(2 + \frac{\sigma_\varepsilon^2}{\sigma_\eta^2}\right) v_k - v_{k+1}$ which is a homogeneous second order difference equation with a solution of the following form: $v_k = A_1\lambda_1^k + A_2\lambda_2^k$, where λ_1 and λ_2 are the characteristic roots. The smaller root satisfies: $\lambda_1^2 - \left(2 + \frac{\sigma_\varepsilon^2}{\sigma_\eta^2}\right)\lambda_1 + 1 = 0$. Substituting $v_1 = A_1\lambda_1$ and $v_2 = A_1\lambda_1^2$ into (ii):

$$\sigma_\varepsilon^2(1 - A_1\lambda_1) + \sigma_\eta^2 A_1\left(\lambda_1^2 - \lambda_1\right) = 0 \quad (iii)$$

Solving (iii) we find $A_1 = \frac{1-\lambda_1}{\lambda_1}$, or the v_h are determined by $v_h = (1-\lambda_1)\lambda_1^{k-1}$. Hence, the one-step ahead forecast of y_t is $E_t(y_t) = (1-\lambda_1)\sum_{j=1}^{\infty}\lambda_1^{j-1}y_{t-j}$. With $|\lambda_1| < 1$, $(1-\lambda_1)\sum_{j=1}^{\infty}\lambda_1^{j-1} = 1$.

Thus the optimal linear forecast is the geometrically weighted average of the past realizations of y.

5.9.4 Forecasting an *ARIMA* process

Suppose the observed time series $\{y_t | t = 1, 2, ..., T\}$ is regarded as a realization of the general $ARIMA(p, d, q)$ process $\alpha(L)\Delta^d y_t = \varphi(L)\varepsilon_t$ and the future value y_{T+h} has to be forecast. The linear filter representation of y_{T+h} is $y_{T+h} = \varepsilon_{T+h} + \tau_1\varepsilon_{T+h-1} + ... + \tau_{h-1}\varepsilon_{T+1} + \tau_h\varepsilon_T + \tau_{h+1}\varepsilon_{T-1} + ...$, where $\tau(L) := \alpha^{-1}(L)\Delta^{-d}\varphi(L)$. Again the forecast of y_{T+h} will be a linear combination of the past and present values of y so that it can be regarded as the conditional expectation of y_{T+h} given $y_T, y_{T-1}, ..., y_1$. Hence the h-step (ahead) forecast made at origin T is denoted as $E_T(y_{T+h})$:

$$\begin{aligned} E_T(y_{T+h}) &= E(y_{T+h} | y_T, y_{T-1}, ..., y_1) \\ &= E(\varepsilon_{T+h} + \tau_1\varepsilon_{T+h-1} + ... + \tau_{h-1}\varepsilon_{T+1} + \tau_h\varepsilon_T + \tau_{h+1}\varepsilon_{T-1} + ... | y_T, y_{T-1}, ..., y_1). \end{aligned}$$

Now, $E(\varepsilon_{T+j} | y_T, y_{T-1}, ..., y_1) = \varepsilon_{T+j}$, for $j \leqq 0$ and equal to 0 if $j > 0$, i.e. past values of ε_{T+j} are known and future values, although unknown, have zero expectation. Hence, $E_T(y_{T+h}) = \tau_h\varepsilon_T + \tau_{h+1}\varepsilon_{T-1} + ...$, which is the minimum mean squared error forecast of y_{T+h} made at origin T. The h-step (ahead) forecast *error* for origin T is:

$$\begin{aligned} f_T(h) &= y_{T+h} - E_T(y_{T+h}) \\ &= \varepsilon_{T+h} + \tau_1\varepsilon_{T+h-1} + ... + \tau_{h-1}\varepsilon_{T+1}, \end{aligned} \tag{5.58}$$

so that the forecast error is a linear combination of the unobservable future shocks entering the system after time T; in particular, the one-step (ahead) forecast error ($h = 1$) is $f_T(1) = y_{T+1} - y_T = \varepsilon_{T+1}$. From (5.58) again an unbiased forecast is obtained since $E(f_T(h) | y_T, y_{T-1}, ..., y_1) = 0$ and the variance of the forecast error satisfies: $var(f_T(h)) = \sigma_\varepsilon^2(1 + \tau_1^2 + \tau_2^2 + ... + \tau_{h-1}^2)$.

For a minimum mean squared forecast error, the one-step (ahead) forecast errors at different origins must be uncorrelated. However, it is not the case that h-step (ahead) forecast errors, made at different origins, will be uncorrelated, nor will be forecasts for different lead times made at the same origin. According to Box and Jenkins (1970) the autocorrelations between h-step (ahead) forecasts made at origins T and $T - j$ ($j > 0$) are given by:

$$\begin{aligned} \rho(f_T(h), f_{T-j}(h)) &= \frac{\sum_{i=j}^{h-1}\tau_i\tau_{i-j}}{\sum_{i=0}^{h-1}\tau_i^2}, \quad 0 \leqq j \leqq h \\ &= 0 \text{ elsewhere.} \end{aligned} \tag{5.59}$$

The correlation coefficient between the T-origin forecast errors at lead times h and $h+j$ is:

$$\rho\left(f_T\left(h\right), f_T\left(h+j\right)\right) = \frac{\sum_{i=j}^{h-1} \tau_i \tau_{h+i}}{\left[\left(\sum_{k=0}^{h-1} \tau_k^2\right)\left(\sum_{l=0}^{h+j-1} \tau_l^2\right)\right]^{1/2}}. \tag{5.60}$$

Further examples of forecasting with *ARIMA* models

1. Forecast $ARIMA(0,1,1)$ model $\Delta y_t = (1-\varphi L)\,\varepsilon_t$. From this model $y_{T+h} = y_{T+h-1} + \varepsilon_{T+h} - \varphi\varepsilon_{T+h-1}$, for $h=1$: $E_T(y_{T+h}) = E_T(y_{T+1}) = y_T - \varphi\varepsilon_T$ and for $h=2$: $E_T(y_{T+2}) = E_T(y_{T+1}) = y_T$, or in general: $E_T(y_{T+h}) = E_T(y_{T+h-1}) = y_T$ $(h>1)$. Hence, for all lead times, the forecasts from origin T will follow a straight line parallel to the time axis passing through $E_T(y_{T+1})$. Note that, since $E_T\left(y_{T+h}\right) = y_T - \varphi\varepsilon_T$ and $\varepsilon_T = (1-L)\,(1-\varphi L)^{-1}\,y_T$, the h-step forecast can be written as:

$$\begin{aligned} E_T(y_{T+h}) &= (1-\varphi)\,(1-\varphi L)^{-1}\,y_T \\ &= (1-\varphi)\,(y_T + \varphi y_{T-1} + \varphi y_{T-2} + ...), \end{aligned}$$

i.e. the forecast of all future values of y is an exponentially weighted moving average of current and past values. Since for this model $\tau_j := 1-\varphi$ $(j=1,2,...)$, the expression for the variance of the h-step (ahead) forecast error is: $var\left(f_T\left(h\right)\right) = \sigma_\varepsilon^2\left(1+(h-1)\,(1-\varphi)^2\right)$, which is increasing in the forecast horizon h.

2. Forecast $ARMA(p,q)$ model: $\left(1-\alpha_1 L - \alpha_2 L^2 - ... - \alpha_p L^p\right) y_t = \left(1+\varphi_1 L + \varphi_2 L^2 + ... + \varphi_q L^q\right)\varepsilon_t$ with $\varepsilon_t \overset{iid}{\sim} \left(0,\sigma_\varepsilon^2\right)$.

   - If all characteristic roots z satisfying $1-\alpha_1 z - \alpha_2 z^2 - ... - \alpha_p z^p = 0$ lie outside the unit circle the $MA\left(\infty\right)$ representation (5.5) has coefficients

$$\begin{aligned} \Upsilon_1 &: = \varphi_1 + \alpha_1 \\ \Upsilon_2 &: = \varphi_2 + \alpha_2 + \alpha_1\tau_1 \\ &\vdots \\ \Upsilon_n &: = \varphi_n + \sum_{i=1}^{\min(n,p)} \alpha_i \tau_{n-i} \quad \text{for } n = 1,2,...,p \\ &: = \sum_{i=1}^{\min(n,p)} \alpha_i \tau_{n-i} \quad \text{for } n > p. \end{aligned}$$

   - Moreover, if the process is invertible, i.e., if all characteristic roots w satisfying $1+\varphi_1 w + \varphi_2 w^2 + ... + \varphi_q w^q = 0$ lie outside the unit circle, the $AR\left(\infty\right)$ representation is (5.6). Forecasts for the $\{y_t\}$ process can now be derived:

$$\begin{aligned} E_t(y_{t+1}) &= \textstyle\sum_{j=1}^{\infty} \pi_j y_{t+1-j} \\ E_t(y_{t+2}) &= \pi_1 E_t(y_{t+1}) + \textstyle\sum_{j=2}^{\infty} \pi_j y_{t+2-j} \\ E_t(y_{t+3}) &= \pi_1 E_t(y_{t+2}) + \pi_2 E_t(y_{t+1}) + \textstyle\sum_{j=3}^{\infty} \pi_j y_{t+3-j}, \text{ etc.} \end{aligned}$$

3. This result can also be used for forecasting with an $ARIMA\,(p,d,q)$ model; in operator form:

$$\alpha\,(L)\,\chi\,(L)\,y_t = \varphi\,(L)\,\varepsilon_t,$$

with $\chi\,(L)\,y_t = \Delta^d y_t = \sum_{j=0}^{d} \binom{d}{j} (-1)^j\, y_{t-j} := x_t$ such that:

$$\begin{aligned} y_t &= (1-\chi\,(L))\,y_t + x_t \\ &= -\textstyle\sum_{j=1}^{d} \dbinom{d}{j} (-1)^j\, y_{t-j} + x_t \end{aligned}$$

or $y_{t+1} = -\sum_{j=1}^{d} \binom{d}{j} (-1)^j\, y_{t+1-j} + x_{t+1}$.

Since $E_t(y_{t+1}) = E\,(y_{t+1}|y_t, y_{t-1}, ..., x_t, x_{t-1}, ...)$,

$$E_t(y_{t+1}) = -\textstyle\sum_{j=1}^{d} \dbinom{d}{j} (-1)^j\, y_{t+1-j} + E_t x_{t+1}.$$

As $\{x_t, t \in \mathbb{Z}\}$ follows an $ARMA\,(p,q)$ process, the forecasts $E_t(x_{t+i})$ can be obtained as discussed above. This eventually yields the forecasts for the $ARIMA\,(p,d,q)$ model.

5.9.5 *Ad hoc* forecasting methods

This section briefly deals with a number of forecasting methods which are not explicitly based on a probability model (see also Chatfield (2001)). For this reason they might be regarded as being of *ad hoc* nature. Generally, they will, however, be based on a preliminary examination of the data, e.g. for answering the question: "Is it necessary to use a seasonal method or not?"

Simple exponential smoothing

Simple exponential smoothing (SES) is the simplest version of a variety of methods that produce forecasts by simply updating formulae (with newly available data). SES computes 1-step (ahead) forecasts by a formula that is equivalent to computing a geometric sum of past observations:

$$y_T\,(1) = \alpha y_T + \alpha\,(1-\alpha)\,y_{T-1} + \alpha\,(1-\alpha)^2\,y_{T-2} + ... \tag{5.61}$$

with α denoting a smoothing parameter $(0 < \alpha < 1)$ and $y_T\,(1)$ the one-step forecast of y_{T+1}.

In practice the above equation is rewritten in an updating format. There are two equivalent updating formats: the *recurrence* form and the *error correction* form:

- Recurrence form:

$$y_T(1) = \alpha y_T + (1-\alpha) y_{T-1}(1), \tag{5.62}$$

which can be illustrated by rewriting (5.62) as a first order moving average model in growth rates (verify!);

- Error correction form:

$$y_T(1) = y_{T-1}(1) + \alpha e_T \tag{5.63}$$

with $e_T := y_T - y_{T-1}(1)$, i.e. the prediction error at time T.

The original motivation for SES (Brown (1962)) comes from applying discounted least squares to a model with constant local mean μ and WN errors $\{\varepsilon_t\}$ such that $y_t = \mu + \varepsilon_t$, where μ was estimated, not by least squares, but by applying geometrically decaying weights to the squared prediction errors. In other words, μ is estimated by minimizing $S := \sum_{j=0}^{T-1} \beta^j (y_{T-j} - \mu)^2$. It turns out that the resulting estimate of μ -using data up to period T- is such that it can be updated by SES (e.g. Kendall and Ord (1990), Section 8.19). This procedure is intuitively reasonable, but theoretically dubious in that if one really believes that the underlying mean is *constant* then OLS should be used. In fact, it can be shown that SES is optimal for several underlying models when the form of the 'drift' is mathematically specified, e.g. for an $ARIMA(0,1,1)$ model.

Holt's linear trend method (local linear trend)

This method is a generalization of the SES method. It is helpful to first recast the SES by regarding it as a method for producing an estimate of the local level at time t ($=: L_t$). Then (5.62) can be written as:

$$y_T(1) = L_T = \alpha y_T + (1-\alpha) L_{T-1} \tag{5.64}$$

where L_T is an estimate of the 'local level' at time T.

The generalization with respect to SES is to introduce a local trend term T_t, which measures the expected increase or decrease per unit time period in the local mean level. The updating equation for the local level is obtained as a generalization of (5.64): $L_T = \alpha y_T + (1-\alpha)(L_{T-1} + T_{T-1})$. In addition we update the local estimate of the growth rate by: $T_T = \gamma (L_T - L_{T-1}) + (1-\gamma) T_{T-1}$, where γ is another smoothing parameter. The h-step ahead forecast is then given by: $y_T(h) = L_T + hT_T$.

Holt's linear trend procedure is optimal when the underlying process is either $ARIMA(0,2,2)$ or a linear growth state-space model (there are two smoothing parameters now). A useful variant of Holt's linear trend method is to include a damping parameter, say θ ($0 < \theta < 1$), such that the estimate of the trend or growth rate at time t, say T_t, is damped to θT_t in the subsequent time period. The h-step ahead forecast at time T is then:

$$y_T(h) = L_T + \left(\sum_{i=1}^{h} \theta^i\right) T_T \tag{5.65}$$

This method involves estimating 1 more parameter θ and is optimal for a particular $ARIMA\,(1,1,2)$ model.

Double exponential smoothing

In the previous subsections the SES was shown to be suitable for a time series which could be approximated by a horizontal line near the forecast origin T. A straightforward generalization can be obtained if the time series y_t can be approximated by a line in the neighborhood of $T: y_t = \mu_1 + (t-T)\mu_2$, which suggests that the forecast function $y_T(h) = \mu_1(T) + h\mu_2(T)$ can be used, where $\mu_1(T)$ and $\mu_2(T)$ are the values of μ_1 and μ_2 minimizing the expression $S := \sum_{j=0}^{T-1} \beta^j (y_{T-j} - \mu_1 + j\mu_2)^2$. Necessary conditions for this minimum are

$$\frac{\partial S}{\partial \mu_1} = -2\sum_{j=0}^{T-1} \beta^j (y_{T-j} - \mu_1 + j\mu_2)^2 = 0$$
$$\frac{\partial S}{\partial \mu_2} = 2\sum_{j=0}^{T-1} j\beta^j (y_{T-j} - \mu_1 + j\mu_2)^2 = 0,$$

and approximating the finite sums of coefficients by their infinite sums, e.g. $\sum_{j=0}^{T-1} \beta^j$ is approximated by $\frac{1}{1-\beta}$, $\sum_{j=0}^{T-1} j\beta^j$ approximated by $\frac{\beta}{(1-\beta)^2}$ and $\sum_{j=0}^{T-1} j^2\beta^j$ approximated by $\frac{\beta(1+\beta)}{(1-\beta)^3}$, the above system of two equations can now be written as:

$$\begin{aligned}(1-\beta)\sum_{j=0}^{T-1} \beta^j y_{T-j} - \mu_1 + \mu_2\frac{\beta}{1-\beta} &\cong 0\\ (1-\beta)^2\sum_{j=0}^{T-1} j\beta^j y_{T-j} - \mu_1\beta + \mu_2\frac{\beta(1+\beta)}{1-\beta} &\cong 0.\end{aligned} \qquad (5.66)$$

Denoting the smoothed series by $s_1(t)$:=(1-β)y_t+βs_1(t-1) $\cong$ (1-β) $\sum_{j=0}^{t-1} \beta^j y_{t-j}$ and the *doubly smoothed series*
$s_2(t)$:=(1-β) $s_1(t)$+βs_2(t-1) = (1-β)$^2 y_t$+β(1-β)s_1(t-1) + βs_2(t-1)
$\cong (1-\beta)\sum_{j=0}^{t-1} \beta^j s_1(t-j)$, it is immediately seen from substitution that:

$$\begin{aligned}s_2(T) &\cong (1-\beta)^2\sum_{j=0}^{T-1} \beta^j \sum_{i=0}^{T-j-1} \beta^i y_{t-j-i} = (1-\beta)^2\sum_{k=0}^{T-1}(k+1)\beta^k y_{T-k}\\ &= (1-\beta)^2\sum_{k=0}^{T-1} k\beta^k y_{T-k} + (1-\beta)\,s_1(T),\end{aligned}$$

so that (5.66) can be rewritten as:

$$\begin{aligned}s_1(T) - \mu_1 + \mu_2\frac{\beta}{1-\beta} &\cong 0\\ s_2(T) - (1-\beta)\,s_1(T) - \mu_1\beta + \mu_2\frac{\beta(1+\beta)}{1-\beta} &\cong 0.\end{aligned}$$

Solving for μ_1 and μ_2, updating formulas are obtained:
$\mu_1(T) \cong 2s_1(T)$-$s_2(T)$=2(1-β)y_T+2βs_1(T-1)-(1-β)$^2 y_T$-β(1-β)s_1(T-1)-βs_2(T-1)
$= (1-\beta)^2 y_T + \beta(\beta+1)s_1(T-1) - \beta s_2(T-1)$ and
$\mu_2(T) \cong \frac{1-\beta}{\beta}(s_1(T) - s_2(T)) = \frac{1-\beta}{\beta}\left(\beta(1-\beta)y_T + \beta^2 s_1(T-1) - \beta s_2(T-1)\right)$.
Alternatively, $s_1(T)$ and $s_2(T)$ can also be expressed as a function of $\mu_1(T)$ and

$\mu_2(T)$: $s_1(T) \cong \mu_1(T) - \frac{\beta}{1-\beta}\mu_2(T)$ and $s_2(T) \cong \mu_1(T) - 2\frac{\beta}{1-\beta}\mu_2(T)$, so that: $\mu_1(T) \cong (1-\beta)^2 y_T + \beta^2 (\mu_1(T-1) - \mu_2(T-1))$ and using $y_T(h)$ for $h = 1$ a simple updating is obtained:

$$\mu_1(T) \cong \mu_1(T-1) + \mu_2(T-1) + (1-\beta)^2 (y_T - y_{T-1}(1)),$$
$$\begin{aligned}\mu_2(T) &\cong (1\text{-}\beta)^2 y_T + \beta(1\text{-}\beta)\mu_1(T\text{-}1)\text{-}\beta^2\mu_2(T\text{-}1)\text{-}(1\text{-}\beta)\mu_1(T\text{-}1)+2\beta\mu_2(T\text{-}1)\\ &= (1\text{-}\beta)^2 y_T - (1-\beta)^2\mu_1(T-1) + \left(2\beta - \beta^2\right)\mu_2(T-1)\\ &= \mu_2(T-1) + (1-\beta)^2 (y_T - y_{T-1}(1)).\end{aligned}$$

From the updating formulas $\mu_1(T)$ and $\mu_2(T)$ it may be assumed that y_T was perfectly forecast at period $T-1$ (i.e. $y_{T-1}(1) = y_T$), so that these formulas become: $\mu_1(T) \cong \mu_1(T-1) + \mu_2(T-1)$ and $\mu_2(T) \cong \mu_2(T-1)$ and the forecast lines are the same at period $(T-1)$ and at period T.

Remark 5.3 *Double exponential smoothing can also be viewed as a special case of Holt's linear trend method. First SES is used to update the local estimate of the level. Then the same operation, with the same smoothing parameter, is applied to the sequence of estimates of the level. It is possible to show that this method arises if the underlying ARIMA $(0,2,2)$ model has equal roots. With this method the number of smoothing parameters is reduced to 1.*

Holt-Winters forecasting procedure

This method is a further generalization. Now seasonal variation of either an *additive* or *multiplicative* form is introduced into (5.64). Therefore, let SI_t denote a seasonal index in time period t, s the number of periods in one year (e.g. $s = 4$), and δ the smoothing parameter for updating the seasonal indices. In the additive case, the seasonal indices are constrained to *sum to zero* over a full year, in the multiplicative case, the y should *average to unity.* E.g. the recurrence form of the updating equations for the local level L_T in the additive case is:

$$L_T = \alpha (y_T - SI_{T-s}) + (1-\alpha)(L_{T-1} + T_{T-1}), \tag{5.67}$$

with a local trend term T_T and a seasonal index SI_T satisfying:

$$SI_T = \delta (y_T - L_T) + (1-\delta) SI_{T-s}. \tag{5.68}$$

The h-step (ahead) forecast at period T is then given by:

$$y_T(h) = L_T + hT_T + SI_{T-s+h} \tag{5.69}$$

for $h = 1, 2, ..., s$. Alternative error correction formulae are also available to update L_T, T_T and I_T, though, as for Holt's linear trend method, they make the smoothing parameters appear to be different. The seasonal form of exponential smoothing is usually called *Holt-Winters seasonal forecasting.*

Generalized exponential smoothing (GES)

Generalized exponential smoothing (GES) originates from Brown (1962). Suppose that the previous methods can be generalized to fitting an arbitrary function in the neighborhood of the forecast horizon T. Such a function can represent a time series in terms of polynomials and sinusoids, in the form

$$y_t = \sum_{i=1}^{m} \mu_i f_i(t) + \varepsilon_t = \boldsymbol{\mu}'\mathbf{f}(t) + \varepsilon_t \tag{5.70}$$

where $\{\mu_i\}$ denotes a set of constants and $f_i(t)$ is a known function of time ($\boldsymbol{\mu} := (\mu_1, \mu_2, \ldots, \mu_m)'$ and $\mathbf{f}(t) := (f_1(t), f_2(t), \ldots, f_m(t))'$). The time varying vector $\mathbf{f}(t)$ can be defined with a regular fixed transition matrix $\mathbf{A}$ as $\mathbf{f}(t) := \mathbf{A}\mathbf{f}(t-1)$. Although the components of $\mathbf{f}(t)$ are deterministic functions, the method recognizes that the coefficients $\{\mu_i\}$ can also change through time as for the double exponential smoothing model so that the model (5.70) only applies locally. Thus, estimates of $\{\mu_i\}$ can be updated through time as more data observations become available. As the model only holds locally it is natural to estimate the parameters not by least squares, but by giving more weight to recent observations. This suggests using *discounted least squares*, where the discounted sum of squared errors at time t is

$$S_t = \sum_{j=0}^{\infty} \beta^j \left(y_{t-j} - \sum_{i=1}^{m} \mu_i(t) f_i(t-j)\right)^2. \tag{5.71}$$

This sum is minimized w.r.t. $\{\mu_i(t)\}$. In practice the GES method is usually formulated so that the $\{f_i(t)\}$ are functions of the time difference from the end of the series rather than from the origin.

Example 5.8 *If $f_i(t)$ in (5.70) are just simple functions of time, as e.g. for the simple linear function $y_t = \mu_1 + \mu_2 t$,*

so that $\mathbf{f}(t) := (1, t)'$, which can be defined with a fixed transition matrix $\mathbf{A}$ as $\mathbf{f}(t) := \mathbf{A}\mathbf{f}(t-1)$ with $\mathbf{A} := \begin{bmatrix} 1 & 0 \\ 1 & 2 \end{bmatrix}$. In this case the GES method coincides with the double exponential smoothing method.

In general, GES reduces to SES if the only function fitted is a constant (then $f(t) = 1$ and A=1); when a constant and a linear term are fitted, then GES reduces to double exponential smoothing. If data is seasonal, however, then GES is very different from the Holt-Winters method. The latter fits only one seasonal index for each period, while GES fits several sine and cosine terms.

5.9.6 Forecasts with density functions

Density forecasts are forecasts of the complete probability distribution or of the complete pdf of the stochastic variables under investigation. As supplements to *point forecasts*, both *interval forecasts* and density (or pdf) forecasts provide a description of forecast uncertainty, about which no information is available if only point forecasts are available.

To date -contrary to point and even interval forecasts- little attention has been paid to forecasts of complete pdfs instead of just the first two moments. There are several reasons for this observation:

1. The analytic construction of density forecasts requires restrictive and sometimes dubious assumptions (e.g. Gaussian innovations, no parameter estimation uncertainty).

2. Until recently, there was little demand for density forecasts, whereas now there is a large need for pdf forecasts (e.g. for portfolio values).

3. The problem of density forecast evaluation appears difficult (e.g. correct simultaneous conditional calibration of prediction intervals at various confidence intervals).

Evaluating the accuracy of interval and density forecasts is receiving increasing attention nowadays.

For interval forecasts the first question is whether the coverage (e.g. 95 or 99 per cent) is correct *ex post*, i.e. whether the relative frequency with which outcomes are observed to fall in their respective forecast intervals is equal to their announced probability. Christoffersen (1998) argues that this *unconditional* hypothesis is inadequate in a time series context and defines an efficient sequence of interval forecasts with respect to a given information set as one which has correct conditional coverage. He presents an LR test statistic for conditional coverage testing, which combines a test of unconditional coverage with a test of independence.

For density forecasts, the first question again concerns forecast evaluation by goodness-of-fit statistics as LR test statistics. As also indicated by Wallis (2003) and Clements (2004), an alternative group of goodness-of-fit tests is based on the probability integral transform(ation). For density forecasts for the observed stochastic variable y_t with corresponding distribution function $P_t(y_t)$ this probability integral transform(ation) is simply defined as $z_t := P_t(y_t)$, so that z_t is the forecast probability of observing an outcome no larger than that actually realized. If $P_t(y_t)$ is correct, i.e. if it is the same as the theoretical distribution function $F_t(y_t)$, following Diebold *et al.* (1998) it will be shown later in this section that the z_t-sequence follows an *iid* uniform distribution on the closed interval $[0, 1]$. Diebold *et al.* (1998) also present histograms of z_t for visual assesment of unconditional uniformity. Diebold *et al.* (1999) use the (non-parametric) Kolmogorov-Smirnov test on the sample distribution function of z_t when evaluating the US Survey of Professional Forecasters' density forecasts of inflation. In Clements's (2004) evaluation of the UK inflation density forecasts an assessment is given of how closely the forecasts densities match the true densities. Specifically, the null that the observed sample of UK inflation rates could have been generated by the set of proposed forecasts densities is tested. The rationale for this approach to density forecast evaluation is provided by Diebold *et al.* (1998) while showing that the ranking of two density forecasts depends on the user's loss functions measuring the deviation between the density forecast and the true density. In the sequel these ideas will be formalized and somewhat extended.

Given that the problem of density forecast evaluation is intrinsically linked to the forecasts user's loss function, let $\{f_t(y_t|\Im_t)\}_{t=1}^{h} = f_t(y_t)$ be a sequence of pdfs for y_t with h the forecast horizon and with the available information set $\Im_t := \{y_{t-1}, y_{t-2}, ...\}$ and let $p_t(y_t) = \{p_t(y_t|\Im_t)\}_{t=1}^{h}$ be a corresponding sequence of 1-step (ahead) density forecasts (estimates of f_t), with $\{y_t\}_{t=1}^{h}$, a series of realizations. Each forecast user has a loss function $L(a, y)$, with a the *action choice* or forecast. If the user believes that the density forecast $p(y) = p_t(y_t)$ is a correct density, or $p_t(y_t) = f_t(y_t)$, then she chooses an action a^* such that:

$$a^*(p(y)) = \arg\min_{a \in A} \int L(a, y) p(y) dy$$

i.e. minimizing the expected loss. The action choice defines the loss $L(a^*, y)$ faced for every realization of the process $y \sim f(y)$. This implies that the loss distribution depends only on the action choice. The expected loss w.r.t. the true DGP is:

$$E(L(a^*, y)) = \int L(a^*, y) f(y) dy.$$

Hence, a density forecast translates into a loss distribution. Two different forecasts lead to 2 different action choices, and hence, to two different loss distributions. The better is the density forecast, the lower is its expected loss (w.r.t. the true DGP).

Ranking two forecasts

Suppose the user has the option of choosing between two forecasts ($p_j(y)$ and $p_k(y)$) in a given period. The user will weakly prefer forecast $p_j(y)$ to forecast $p_k(y)$ if

$$\int L(a_k^*, y) f(y) dy \geqq \int L(a_j^*, y) f(y) dy$$

with a_j^*, the action that minimizes expected loss when the user bases the action choice on forecast $p_j(y)$. Ideally, one would like to find a ranking of forecasts with which all users would agree, regardless of their loss function. Unfortunately, such a ranking does not exist.

Following Diebold *et al.* (1998), the next two propositions can now be formulated.

Proposition 5.4 *Let $f(y)$ be the density of y, a_j^* be the optimal action based on forecast p_j, a_k^* be the optimal action based on forecast p_k. Then there does not exist a ranking r (from bad to good) of arbitrary density forecasts p_j and p_k ($\neq f$), such that for all loss functions $L(a, y)$:*

$$r_j \geqq r_k \Leftrightarrow \int L(a_j^*, y) f(y) dy \geqq \int L(a_k^*, y) f(y) dy$$

However, if a forecast coincides with the true DGP, then it will be preferred by all forecast users, regardless of their loss function.

Proposition 5.5 *Suppose that $p_j(y) = f(y)$ so that a_j^* minimizes the expected loss w.r.t. the true probability density function. Then*

$$\int L(a_k^*, y) f(y) dy \geqq \int L(a_j^*, y) f(y) dy \ \forall k$$

So, a forecast evaluating procedure has to be established by assessing whether the forecast densities are correct, i.e. whether $\{p_t(y_t|\Im_t)\}_{t=1}^{h} = \{f_t(y_t|\Im_t)\}_{t=1}^{h}$. *If not, some users, depending on their loss functions, could potentially be better served by a different density forecast.*

Evaluating density forecasts

The task of determining whether $\{p_t(y_t|\Im_t)\}_{t=1}^{h} = \{f_t(y_t|\Im_t)\}_{t=1}^{h}$ appears difficult, because $\{f_t(y_t|\Im_t)\}_{t=1}^{h}$ is never observed (!), even after the fact. Moreover, and importantly, the true density $f_t(y_t|\Im_t)$ may exhibit structural changes (as indicated by its time subscript). But, given that the relationship between the sequence of pdfs, $\{f_t(y_t)\}$, and the sequence of density forecasts, $\{p_t(y_t)\}$, is given by the probability integral transform(ation), z_t:

$$z_t := \int_{-\infty}^{y_t} p_t(u)\,du = P_t(y_t)$$

the following lemma describes the pdf $q_t(z_t)$ of the probablity integral transform(ation):

Lemma 5.1 *Let* $f_t(y_t)$ *be the true density of* y_t*, let* $p_t(y_t)$ *be a density forecast of* y_t*, let* z_t *be the probablity integral transform(ation) of* y_t *w.r.t.* $p_t(y_t)$*, then, assuming that* $\frac{\partial P_t^{-1}(z_t)}{\partial z_t}$ *is continuous and* $\neq 0$ *over the support of* y_t*,* z_t *has support on the unit interval with density:*

$$q_t(z_t) = \left|\frac{\partial P_t^{-1}(z_t)}{\partial z_t}\right| f_t\left(P_t^{-1}(z_t)\right) = \frac{f_t\left(P_t^{-1}(z_t)\right)}{p_t\left(P_t^{-1}(z_t)\right)}.$$

Corollary 5.1 *If* $p_t(y_t) = f_t(y_t)$*, then* $q_t(z_t) = 1$ *for* $z_t \in (0,1)$*, i.e.* $z_t \sim U(0,1)$.

Even though the actual conditional densities may change over time, provided that the forecast densities match the actual densities at each t, $z_t \sim U(0,1)$ for each t so that the time subscript on $q_t(z_t)$ is in fact redundant. Moreover, the random variables z_t are *iid* for any previous time period, so that the time series $\{z_t\}_{t=1}^{h}$ is a random independently drawn sample from the $U(0,1)$ distribution, i.e. $z_t \overset{iid}{\sim} U(0,1)$. Formally, going beyond the 1-period characterization of the pdf of z_t when $p_t(y_t) = f_t(y_t)$, both the density and dependence structures of the entire z_t-sequence can be characterized for $p_t(y_t) = f_t(y_t)$ as:

Proposition 5.6 *Suppose* $\{y_t\}_{t=1}^{h}$ *is generated from* $\{f_t(y_t|\Im_t)\}_{t=1}^{h}$ *when* $\Im_t = \{y_{t-1}, y_{t-2}, ...\}$. *If a sequence of density forecasts* $\{p_t(y_t)\}_{t=1}^{h}$ *coincides with* $\{f_t(y_t|\Im_t)\}_{t=1}^{h}$*, then under the usual conditions of a nonzero Jacobian with continuous partial derivatives, the sequence of probablity integral transforms of* $\{y_t\}_{t-1}^{h}$ *w.r.t.* $\{p_t(y_t)\}_{t=1}^{h}$ *is iid* $U(0,1)$*; i.e.* $\{z_t\}_{t=1}^{h} \overset{iid}{\sim} U(0,1)$. *This implies that if a sequence of density forecasts is correctly conditionally calibrated, then every interval will be correctly conditionally calibrated and will generate an iid Bernouilli hit sequence.*

5.10 Cases and exercises about $ARIMA$ models

In this section, a number of cases and theoretical and empirical exercises on *ARIMA* models, including identification, estimation, testing, and forecasting will be presented. Since the reader is stimulated to follow a process of *learning by doing*, both practical cases and elaborated exercises and unsolved cases and exercises will be presented.

5.10.1 Theoretical exercises

This section begins by solving some simple theoretical exercises. Afterwards a number of unsolved theoretical exercises will be presented.

Solved theoretical exercises

Exercise 5.3 *Find the mean, variance and autocovariance function of the following* $MA(2)$ *process* $y_t = \varepsilon_t + 0.2\varepsilon_{t-1} - 0.4\varepsilon_{t-2}$, *where* ε_t *is a purely random process with zero mean and* $\sigma_\varepsilon^2 = 1$.

Applying the unconditional expectation operator, $E(\cdot)$, to the above $MA(2)$ process, the mean of the process is obtained:

$$E(y_t)\text{=}E(\varepsilon_t)\text{+}0.2E(\varepsilon_{t-1})\text{-}0,4E(\varepsilon_{t-2})\text{=}0\text{+}0.2 \times 0\text{+}0.4 \cup 0\text{=}0,\ \forall t \in \mathbb{Z}.$$

The autocovariance function of this $MA(2)$ process is then:

$cov(y_t, y_{t\text{-}k})\text{=}\gamma(k)\text{=}E\{[y_t\text{-}E(y_t)][y_{t-k}\text{-}E(y_{t\text{-}k})]\}\text{=}E(y_t y_{t-k})$

$\quad\text{=}E\{(\varepsilon_t\text{+}0.2\varepsilon_{t-1}\text{-}0.4\varepsilon_{t-2})(\varepsilon_{t-k}\text{+}0.2\varepsilon_{t-k-1}\text{-}0.4\varepsilon_{t-k-2})\}$

$\text{=}E(\varepsilon_t\varepsilon_{t-k})\text{+}0.2E(\varepsilon_{t-1}\varepsilon_{t-k})\text{+}0.2^2E(\varepsilon_{t-1}\varepsilon_{t-k-1})\text{-}0.2 \times 0.4E(\varepsilon_{t-1}\varepsilon_{t-k-2})$

$\quad\text{-}0.4E(\varepsilon_{t-2}\varepsilon_{t-k})\text{-}0.4 \times 0.2E(\varepsilon_{t-2}\varepsilon_{t-k-1})\text{+}0.4^2E(\varepsilon_{t-2}\varepsilon_{t-k-2})$,

so that for $k = 0$,

$cov(y_t, y_t) = \gamma(0) = var(y_t) = \sigma_\varepsilon^2 + 0.04\sigma_\varepsilon^2 + 0.16\sigma_\varepsilon^2 = 1.2$, for $k = 1$,

$cov(y_t, y_{t-1}) = \gamma(1) = 0.2\sigma_\varepsilon^2 - 0.08\sigma_\varepsilon^2 = 0.12$, for $k = 2$,

$cov(y_t, y_{t-2}) = \gamma(2) = (-0.4)\sigma_\varepsilon^2 = -0.4$, and for $k > 2$,

$cov(y_t, y_{t-k}) = \gamma(k) = 0.$■

Exercise 5.4 *Define an* $MA(q)$ *process by* $y_t = \sum_{i=0}^{q} a^{-i}\varepsilon_{t-i}$, *where* $a \in (0, 1)$ *is a constant,* ε_t *is a purely random process with zero mean and* $\sigma_\varepsilon^2 = 1$, *and* $t \in \mathbb{Z}$. *Compute the autocovariance function of* y_t. *Check whether your answer is correct for* $k = q$, *where* $\gamma(q) = a^{-q}$.

Since $E(y_t) = E\left(\sum_{i=0}^{q} a^{-i}\varepsilon_{t-i}\right) = 0$ and $\varepsilon_t \overset{iid}{\backsim} (0, 1)$, for all $k \in [0, q]$:

$$cov(y_t, y_{t-k})\text{=}E(y_t y_{t-k})\text{=}E\left\{\left(\sum_{i=0}^{q} a^{-i}\varepsilon_{t-i}\right)\left(\sum_{j=0}^{q} a^{-j}\varepsilon_{t-k-j}\right)\right\}$$

$$\text{=}\sum_{i=0}^{q} a^{-i}E\left\{\varepsilon_{t-i}\left(\sum_{j=0}^{q} a^{-j}\varepsilon_{t-k-j}\right)\right\}\text{=}\sum_{i=k}^{q} a^{-i}a^{k-i}\text{=}\sum_{i=k}^{q} a^{k-2i},$$

so that $var(y_t)=\sum_{i=0}^{q} a^{-2i}$, $cov(y_t, y_{t-k})=\gamma(k)=\sum_{i=k}^{q} a^{k-2i}$ for $0 < k \leqq q$ and $cov(y_t, y_{t-k})=0$ for $k > q$. Hence, for $k = q : \gamma(q) = a^{-q}$.■

Exercise 5.5 *Compute the autocorrelation function of the autoregressive process $y_t = 0.5y_{t-1} + \varepsilon_t$, with $\varepsilon_t \overset{iid}{\backsim} (0,1)$ and $t \in \mathbb{Z}$.*

Since the $AR(1)$ process is stationary it can be represented as the $MA(\infty)$ process $y_t = \frac{\varepsilon_t}{(1-0.5L)}=\sum_{i=0}^{\infty}(0.5)^i \varepsilon_{t-i}$. Hence, $E(y_t) = 0$ and

$$cov(y_t, y_{t-k})=\gamma(k)=E(y_t y_{t-k})=E\left\{\sum_{i=0}^{\infty}(0.5)^i \varepsilon_{t-i}\sum_{j=0}^{\infty}(0.5)^j \varepsilon_{t-k-j}\right\}$$
$$=\sigma_\varepsilon^2 (0.5)^k \sum_{i=0}^{\infty}(0.5)^i=\frac{(0.5)^k}{(1-(0.5)^2)}=\frac{4}{3}(0.5)^k,$$

so that $\rho(k)=\frac{\gamma(k)}{\gamma(0)} = 0.5^k$.■

Exercise 5.6 *Compute the autocorrelation function of the $AR(2)$ process $y_t = \frac{1}{3}y_{t-1} + \frac{2}{9}y_{t-2} + \varepsilon_t$, with $\varepsilon_t \overset{iid}{\backsim} (0,1)$ and $t \in \mathbb{Z}$.*

Construct the Yule-Walker equations by multiplying the second order stochastic difference equation by y_{t-k} for $k = 0, 1, 2, \ldots$ and taking expectations and take account of the property that the $AR(2)$ process is stationary (verify!):

$$\begin{array}{l}
E(y_t y_t)=\gamma(0)=\frac{1}{3}E(y_{t-1}y_t)+\frac{2}{9}E(y_{t-2}y_t)+E(\varepsilon_t y_t)=\frac{1}{3}\gamma(1)+\frac{2}{9}\gamma(2)+1\\
E(y_t y_{t-1})=\gamma(1)=\frac{1}{3}E(y_{t-1}y_{t-1})+\frac{2}{9}E(y_{t-2}y_{t-1})+E(\varepsilon_t y_{t-1})=\frac{1}{3}\gamma(0)+\frac{2}{9}\gamma(1)\\
E(y_t y_{t-2})=\gamma(2)=\frac{1}{3}E(y_{t-1}y_{t-2})+\frac{2}{9}E(y_{t-2}y_{t-2})+E(\varepsilon_t y_{t-2})=\frac{1}{3}\gamma(1)+\frac{2}{9}\gamma(0)\\
\vdots \qquad \vdots \qquad \vdots \qquad \vdots \qquad \vdots \qquad \vdots \qquad \vdots\\
E(y_t y_{t-k})=\gamma(k)=\frac{1}{3}E(y_{t-1}y_{t-k})+\frac{2}{9}E(y_{t-2}y_{t-k})+E(\varepsilon_t y_{t-k})=\frac{1}{3}\gamma(k\text{-}1)+\frac{2}{9}\gamma(k\text{-}2).
\end{array}$$

From the above covariances for $k = 0, 1, 2$ a deterministic system of three equations and three unknowns ($\gamma(0), \gamma(1)$ and $\gamma(2)$) can be defined. By the method of substitution, the solution of this system is: $\gamma(0) = (21/16), \gamma(1) = (9/16), \ldots$, so that the autocorrelations are given by $\left\{1, \frac{3}{7}, \frac{23}{63}, \ldots\right\}$.■

Exercise 5.7 *Suppose the stationary process $\{x_t\}$ has autocovariance function $\gamma_x(k)$. Define a new stationary process $\{y_t\}$ by $y_t := x_t - x_{t-1}$. Find the autocovariance function of $\{y_t\}$ in terms of $\gamma_x(k)$ and obtain $\gamma_y(k)$ when $\gamma_x(k) = \upsilon^{|k|}$.*

This exercise investigates the link between a stationary level process and its first differences for the autocovariance and autocorrelation functions. The Yule-Walker equations of the differenced process are:

$$\begin{array}{l}
E(y_t y_t)=E(x_t x_t)+E(x_{t-1}x_{t-1})\text{-}2E(x_t x_{t-1})=\gamma_x(0)+\gamma_x(0)\text{-}2\gamma_x(1)=2\gamma_x(0)\text{-}2\gamma_x(1)\\
E(y_t y_{t-1})=E(x_t x_{t-1})\text{-}E(x_t x_{t-2})\text{-}E(x_{t-1}x_{t-1})+E(x_{t-1}x_{t-2})=2\gamma_x(1)\text{-}\gamma_x(2)\text{-}\gamma_x(0)\\
E(y_t y_{t-2})=E(x_t x_{t-2})\text{-}E(x_t x_{t-3})\text{-}E(x_{t-1}x_{t-2})+E(x_{t-1}x_{t-3})=2\gamma_x(2)\text{-}\gamma_x(3)\text{-}\gamma_x(1)\\
\vdots \qquad \vdots \qquad \vdots \qquad \vdots \qquad \vdots
\end{array}$$

$E(y_t y_{t\text{-}k})=E(x_t x_{t\text{-}k})\text{-}E(x_t x_{t\text{-}k\text{-}1})\text{-}E(x_{t\text{-}1}x_{t\text{-}k})+E(x_{t\text{-}1}x_{t\text{-}k\text{-}1})=2\gamma_x(k)\text{-}\gamma_x(k+1)\text{-}\gamma_x(k\text{-}1)$, so that it follows from $\gamma_x(k) = \upsilon^{|k|}$ that $\gamma_y(k) = 2\upsilon^{|k|} - \upsilon^{|k+1|} - \upsilon^{|k-1|}$.■

Exercise 5.8 *Assume a time series consisting of* T *real numbers* $(y_1, y_2, ..., y_T)$ *and assume that one wants to forecast the out-of-sample value* y_{T+h} *by* $y_T(h)$, *where* h *is the forecast horizon. Then, consider that the value* $y_T(h)$, *obtained by the simple exponential smoothing (SES) method, can be written as* $y_T(h) = (1-\beta)\sum_{j=0}^{T-1}\beta^j y_{T-j}$, *with* β *the smoothing parameter satisfying* $0 < \beta < 1$.

(a) Supposing that y_t is a random variable with expectation zero, variance one and $cov(y_t, y_{t+h}) = \rho^{|h|}$ for all integers t and h and $|\rho| < 1$; show that the influence of an optimal choice of the smoothing parameter (obtained by minimizing the expected squared prediction error for an SES forecast based on an infinite sum; hence, for sufficiently large T) on the results is not too strong for positive values of ρ and fairly large intervals of β.

(b) Consider that in the SES model the *mean lag* between the observations and the forecast based on them is measurable by:

$$\tau = (1-\beta)\sum_{j=0}^{T-1}\beta^j(T-j).$$

Simplify this expression of τ as a function of β and T. What happens to the expression when the number of observations increases indefinitely?

(c) Suppose now that $y_t = \sum_{i=0}^{t-1}\lambda_i\varepsilon_{t-i}$, $\lambda_0 \neq 0$ for positive t, where the error process is assumed to be a WN process:

- Is the process $\{\varepsilon_t\}$ stationary? Show that one can always normalize $\lambda_0 = 1$ and that y_t can be put in the form $y_t = \sum_{j=1}^{t-1}\alpha_j y_{t-j} + \varepsilon_t$. How can α_js as a function of the λ_is be computed?

- Compute $E(y_{T+1}|y_1, y_2, ..., y_T)$. For which values of λ_i it is verified that $E(y_{T+1}|y_1, y_2, ..., y_T) = (1-\beta)\sum_{j=0}^{T-1}\beta^j y_{T-j}$?

- Compute now $E(y_{T+h}|y_1, y_2, ..., y_T)$ and $y_T - y_{T-1}$.

Regarding question (a) of this exercise, the optimal choice of β is determined by minimizing the expected squared forecast error (see (5.45)):

$$\min_{\beta} E(y_{T+h} - y_T(h))^2,\ h \in \mathbb{N}$$
$$\text{s.t. } y_T(h) = (1-\beta)\sum_{j=0}^{\infty}\beta^j y_{T-j}.$$

Redefining the objective as:

$$\varphi_T(h,\beta,\rho){:=}E(y_{T+h} - y_T(h))^2{=}E(y_{T+h}^2)\text{-}2E(y_{T+h}y_T(h)){+}E(y_T^2(h))$$

the right-hand side terms can be simplified as follows:

$E(y_{T+h}^2) = var(y_{T+h}) = 1;$

$2E(y_{T+h}\hat{y}_T(h)){=}E\left[y_{T+h}(1\text{-}\beta)\sum_{j=0}^{\infty}\beta^j y_{T-j}\right]{=}(1\text{-}\beta)\sum_{j=0}^{\infty}\beta^j E\left(y_{T+h}y_{T-j}\right)$

$\quad = (1\text{-}\beta)\sum_{j=0}^{\infty}\beta^j\rho^{|h+j|}{=}(1\text{-}\beta)\rho^{|h|}\sum_{j=0}^{\infty}\beta^j\rho^j{=}\frac{(1-\beta)\rho^{|h|}}{1-\beta\rho};$

$E(\hat{y}_T^2(h)){=}E\left((1\text{-}\beta)^2\left(\sum_{j=0}^{\infty}\beta^j y_{T-j}.\right)^2\right){=}(1\text{-}\beta)^2\sum_{j=0}^{\infty}\sum_{i=0}^{\infty}\beta^j\beta^i E(y_{T-j}y_{T-i})$

$$=(1\text{-}\beta)^2 \sum_{i=0}^{\infty}\sum_{j=0}^{\infty} \beta^{i+j}\rho^{|i-j|}=(1\text{-}\beta)^2 \sum_{i=0}^{\infty} \beta^{2i} + 2(1\text{-}\beta)^2 \sum_{i=0}^{\infty}\sum_{j=0}^{i-1} \beta^{i+j}\rho^{i-j}$$
$$=\frac{(1-\beta)^2}{1-\beta^2} + 2(1\text{-}\beta)^2 \sum_{j=0}^{\infty}\sum_{i=j+1}^{\infty} \beta^{i+j}\rho^{i-j}=\frac{(1-\beta)^2}{1-\beta^2}+2(1\text{-}\beta)^2 \sum_{j=0}^{\infty} \beta^{2j} \sum_{i=1}^{\infty} \beta^{i}\rho^{i}$$
$$=\frac{(1-\beta)^2}{1-\beta^2} + \frac{2(1-\beta)^2\beta\rho}{(1-\beta^2)(1-\beta\rho)}.$$

Substituting these three terms in the redefinition φ_T of the objective:

$$\varphi_T(h,\beta,\rho)=1\text{-}\frac{(1-\beta)\rho^{|h|}}{1-\beta\rho}+\frac{(1-\beta)^2}{1-\beta^2}+\frac{2(1-\beta)^2\beta\rho}{(1-\beta^2)(1-\beta\rho)}=\frac{2}{1+\beta}+\frac{2(1-\beta)(\beta\rho-\rho^{|h|}-\beta\rho^{|h|})}{(1+\beta)(1-\beta\rho)}.$$

This expression for the expected squared forecast error can be used to determine the optimal value of the smoothing parameter. Illustrating this for $h = 1$:

$$\varphi_T(1,\beta,\rho)=\frac{2}{1+\beta}\text{-}\frac{2(1-\beta)\rho}{(1+\beta)(1-\beta\rho)}=\frac{2(1-\rho)}{(1+\beta)(1-\beta\rho)}, \tag{5.72}$$

the optimal $\beta^*(1)$ satisfies:

$$\beta^*(1)=\arg\min_{\beta} \log \frac{2(1-\rho)}{(1+\beta)(1-\beta\rho)}=\arg\min_{\beta} \log \frac{1}{(1+\beta)(1-\beta\rho)},$$

where the logarithm is taken because it is a monotonic transformation of the objective function and the logarithm of the numerator is independent of β. Hence, the optimal β is derived from equating the derivative of the squared forecast error:
$\frac{d\log F_T(1,\beta,\rho)}{d\beta}=\frac{d[-\log(1+\beta)-\log(1-\beta\rho)]}{d\beta}=-\frac{1}{1+\beta}+\frac{\rho}{(1-\beta\rho)}=\frac{2\beta\rho+\rho-1}{(1+\beta)(1-\beta\rho)}$ to zero, or $\beta^*=\frac{1-\rho}{2\rho}$.[14]

Since $\beta \in (0,1)$, $\beta^* = \frac{1-\rho}{2\rho}$, when $\rho \in \left(\frac{1}{3},1\right)$ (notice that $|\rho| < 1$). When, on the other hand, $\rho \in (-1,\frac{1}{3}]$, then $\frac{d\log F_T(1,\beta,\rho)}{d\beta} < 0$ such that the expected squared forecast error decreases if one increases the smoothing parameter. Thus, in this case the expected quadratic forecast error is minimized for $\beta^*(1) = 1$.

Summarizing, the optimal choice of the smoothing parameter satisfies:

$$\beta^*(1) = 1.\text{if } \rho \in (-1,\tfrac{1}{3}] \text{ and } \beta^*(1) = \tfrac{1-\rho}{2\rho} \text{ if } \rho \in \left(\tfrac{1}{3},1\right).$$

Hence, for relative big intervals of $\beta^*(1)$ and a positive value of ρ the optimal choice of the smoothing parameter has little impact on the expected quadratic forecast error (5.72), which answers question (a) of this exercise.

Regarding question (b) of this exercise, the mean lag τ can be rewritten as:

$$\tau=(1\text{-}\beta)\textstyle\sum_{j=0}^{T-1}\beta^j(T\text{-}j)=(1\text{-}\beta)T\frac{1-\beta^T}{1-\beta}\text{-}(1\text{-}\beta)\sum_{j=0}^{T-1}\beta^j j.$$

the weighted sum $\sum_{j=0}^{T-1}\beta^j j$ can be derived from:

[14] This optimal value of β corresponds to a minimum of the squared forecast error because $\frac{d^2\log F_T(1,\beta,\rho)}{d\beta^2} = \frac{d\left(\frac{2(1-\rho)}{(1+\beta)(1-\beta\rho)}\right)}{d\beta} = \frac{1+2\beta^2\rho^2+2\beta\rho^2+\rho^2-2\beta\rho}{(1+\beta)^2(1-\beta\rho)^2} > 0$ if $\rho \in \left(\frac{1}{3},1\right)$.

$$(1\text{-}\beta)\sum_{j=0}^{T-1}\beta^j j{=}\sum_{j=1}^{T-1}\beta^j\text{-}(T\text{-}1)\beta^T{=}\frac{\beta-\beta^T}{1-\beta}\text{-}(T\text{-}1)\beta^T,$$

so that $\sum_{j=0}^{T-1}\beta^j j=\frac{\beta-\beta^T-(T-1)\beta^T(1-\beta)}{(1-\beta)^2}=\frac{\beta-T\beta^T+T\beta^{T+1}-\beta^{T+1}}{(1-\beta)^2}$.

Substituting this weighted sum in the above mean lag expression:

$$\tau=\frac{T(1-\beta^T)(1-\beta)-\beta+T\beta^T-T\beta^{T+1}+\beta^{T+1}}{1-\beta}=T+\frac{\beta^{T+1}-\beta}{1-\beta},$$

which is a simple expression of the mean lag τ as a function of β and T.

When the number T of observations increases indefinitely and since $\beta\in(0,1)$ for the mean lag it results $\tau=\lim\limits_{T\to\infty}\left\{T+\frac{\beta^{T+1}-\beta}{1-\beta}\right\}=\infty$, or the mean lag of the SES model goes to infinity when the number of observations increases indefinitely.

Regarding the first part of question (c) of this exercise, one may check whether the $\{y_t\}$ process is stationary:

$E\left(y_t\right)=E\left(\sum_{i=0}^{t-1}\lambda_i\varepsilon_{t-i}\right){=}\sum_{i=0}^{t-1}\lambda_i\left(E\varepsilon_{t-i}\right)=0$

$var\left(y_t\right)=E\left(\sum_{i=0}^{t-1}\lambda_i\varepsilon_{t-i}\right)^2=\sum_{i=0}^{t-1}\lambda_i^2E\left(\varepsilon_{t-i}^2\right)=\sigma_\varepsilon^2\sum_{i=0}^{t-1}\lambda_i^2.$

Hence, the variance of the $\{y_t\}$ process increases with increasing t, such that the process is nonstationary.

In the following, it will be demonstrated that a normalization to $\lambda_0=1$ is possible if there are invariant stochastic properties of the $\{y_t\}$ process.

Assume a normalization as follows, by redefining new λ_is: $\tilde{\lambda}_i=\frac{\lambda_i}{\lambda_0}$. Then, $\tilde{\sigma}_\varepsilon^2=\lambda_0^2\sigma_\varepsilon^2$ and $\tilde{\lambda}_0=1$, so that considering the normalized $\{\tilde{y}_t\}$ process $\tilde{y}_t=\sum_{i=0}^{t-1}\tilde{\lambda}_i\varepsilon_{t-i}$ we have for the stationarity conditions:

$E(\tilde{y}_t)=E\left(\sum_{i=0}^{t-1}\tilde{\lambda}_i\varepsilon_{t-i}\right)=\sum_{i=0}^{t-1}\tilde{\lambda}_iE\left(\varepsilon_{t-i}\right)=0,$

$var(\tilde{y}_t){=}E\left(\sum_{i=0}^{t-1}\tilde{\lambda}_i\varepsilon_{t-i}\right)^2{=}\sum_{i=0}^{t-1}\tilde{\lambda}_i^2E\left(\varepsilon_{t-i}^2\right){=}\ \tilde{\sigma}_\varepsilon^2\sum_{i=0}^{t-1}\tilde{\lambda}_i^2{=}\sigma_\varepsilon^2\sum_{i=0}^{t-1}\lambda_i^2,$

so that the new $\{\tilde{y}_t\}$ process is again nonstationary with the same stochastic properties as the original $\{y_t\}$ process. Hence, it is possible to make a normalization of $\lambda_0=1$.

In the sequel, assume that the $\{y_t\}$ process is normalized or,

$$y_t=\sum_{i=0}^{t-1}\lambda_i\varepsilon_{t-i} \tag{5.73}$$

with $\lambda_0=1$,i.e. $y_t=\sum_{i=1}^{t-1}\lambda_i\varepsilon_{t-i}+\varepsilon_t$. Now it will be shown that this *MA* process can be rewriten as an *AR*$(t-1)$ process $y_t=\sum_{j=1}^{t-1}\alpha_jy_{t-j}+\varepsilon_t$ by computing the α_js as a function of the λ_is. Therefore, equating both expressions and substituting (5.73) for y_{t-j}:

$$y_t{=}\sum_{i=1}^{t-1}\lambda_i\varepsilon_{t-i}{+}\varepsilon_t{=}\sum_{j=1}^{t-1}\alpha_jy_{t-j}{+}\varepsilon_t{=}\sum_{j=1}^{t-1}\alpha_j\sum_{i=0}^{t-j-1}\lambda_i\varepsilon_{t-j-i}{+}\varepsilon_t.$$

By comparing the corresponding coefficients of the ε_ks the α_js must result from the following system:

$$\lambda_1=\alpha_1\lambda_0$$

$$\lambda_2 = \alpha_1\lambda_1 + \alpha_2\lambda_0$$
$$\lambda_3 = \alpha_1\lambda_2 + \alpha_2\lambda_1 + \alpha_3\lambda_0$$
$$\vdots \qquad \vdots$$
$$\lambda_t = \textstyle\sum_{i=1}^{t} \lambda_{t-i}\alpha_i, \ \ \forall t \in \mathbb{N}.$$

This system can be solved with an indefinite number of equations in the α_js. This can be done by redefining the system in matrix notation and substituting for the normalization $\lambda_0 = 1$:

$$\begin{bmatrix} \lambda_1 \\ \lambda_2 \\ \vdots \\ \lambda_{t-1} \\ \lambda_t \end{bmatrix} = \begin{bmatrix} 1 & 0 & 0 & \dots & 0 \\ \lambda_1 & 1 & 0 & \dots & 0 \\ \vdots & \ddots & \ddots & \ddots & \vdots \\ \lambda_{t-2} & \dots & \lambda_1 & 1 & 0 \\ \lambda_{t-1} & \lambda_{t-2} & \dots & \lambda_1 & 1 \end{bmatrix} \begin{bmatrix} \alpha_1 \\ \alpha_2 \\ \vdots \\ \alpha_{t-1} \\ \alpha_t \end{bmatrix}.$$

Since the determinant of the undertriangular matrix is 1, this matrix is regular so that directly solving for the α_js as a function of the λ_is:

$$\begin{bmatrix} \alpha_1 \\ \alpha_2 \\ \vdots \\ \alpha_{t-1} \\ \alpha_t \end{bmatrix} = \begin{bmatrix} 1 & 0 & 0 & \dots & 0 \\ \lambda_1 & 1 & 0 & \dots & 0 \\ \vdots & \ddots & \vdots & \ddots & \vdots \\ \lambda_{t-2} & \dots & \lambda_1 & 1 & 0 \\ \lambda_{t-1} & \lambda_{t-2} & \dots & \lambda_1 & 1 \end{bmatrix}^{-1} \begin{bmatrix} \lambda_1 \\ \lambda_2 \\ \vdots \\ \lambda_{t-1} \\ \lambda_t \end{bmatrix}.$$

Regarding the second part of question (c) of this exercise one can start from the $MA(t-1)$ model (5.73) with $\lambda_0 \neq 0$ and $t \in \mathbb{N}$. First, computing and taking account of the properties of the error terms put forward in Section 5.9.1:

$$E(y_{T+1}|y_1, y_2, ..., y_T) = E\left(\sum_{i=0}^{T} \lambda_i \varepsilon_{T+1-i}|y_1, y_2, ..., y_T\right) = \sum_{i=1}^{T} \lambda_i \varepsilon_{T+1-i} = \sum_{i=0}^{T-1} \lambda_{i+1}\varepsilon_{T-i}.$$

In order to check for which values of λ_i the equation

$$E(y_{T+1}|y_1, y_2, ..., y_T) = (1-\beta)\textstyle\sum_{j=0}^{T-1} \beta^j y_{T-j}$$

is valid, it is necessary to equate the last two relations when taking account of (5.73):

$$\sum_{i=0}^{T-1} \lambda_{i+1}\varepsilon_{T-i} = (1-\beta)\sum_{j=0}^{T-1} \beta^j y_{T-j} = (1-\beta)\sum_{j=0}^{T-1} \beta^j \sum_{i=0}^{T-j-1} \lambda_i \varepsilon_{T-j-i},$$

so that the λ_is can be solved from the following system by equating the corresponding coefficients for similar ε_ks:

$$\lambda_1 = (1-\beta)\lambda_0$$
$$\lambda_2 = (1-\beta)(\beta\lambda_0 + \lambda_1) = (1-\beta)(\beta\lambda_0 + (1-\beta)\lambda_0) = (1-\beta)\lambda_0$$
$$\vdots \qquad \vdots$$
$$\lambda_t = (1-\beta)\lambda_0 \ \ \forall t \in \mathbb{N}.$$

Hence, for all values of $\lambda_t = (1-\beta)\lambda_0$, $\forall t \in \mathbb{N}$, the requested equation for the conditional forecast is satisfied. For the normalization $\lambda_0 = 1$, it is clear that $\lambda_t = 1$ and the conditional forecast satisfies then:

$$E(y_{T+1}|y_1, y_2, ..., y_T) = \sum_{i=0}^{T-1} \lambda_{i+1}\varepsilon_{T-i} = (1-\beta)\sum_{j=0}^{T-1} \beta^j y_{T-j}.$$

Regarding the last part of question (c) of this exercise, the h-step (ahead) conditional forecast $E(y_{T+h}|y_1, y_2, ..., y_T) := E_T(y_{T+h})$ can be computed from the previous equation, from (5.73) and from the principles put forward in Section 5.9.1:

$$E_T(y_{T+h}) = E_T(\sum_{i=0}^{T+h-1} \lambda_i\varepsilon_{T+h-i}) = \sum_{i=h}^{T+h-1} \lambda_i\varepsilon_{T+h-i} + E_T(\sum_{i=0}^{h-1} \lambda_i\varepsilon_{T+h-i})$$
$$= \sum_{i=h}^{T+h-1} \lambda_i\varepsilon_{T+h-i} = \sum_{i=0}^{T-1} \lambda_{i+h}\varepsilon_{T-i} = \sum_{i=0}^{T-1}(1-\beta)\lambda_0\varepsilon_{T-i},\ \forall h \in \mathbb{N}.$$

The h-step conditional forecast in the last equation is independent of the time horizon h, such that the conditional forecast is equal for any time horizon.

Finally, Δy_T can be computed from (5.73) and $\lambda_t = (1-\beta)\lambda_0$ as:

$$y_T - y_{T-1} = \sum_{i=0}^{T-1} \lambda_i\varepsilon_{T-i} - \sum_{i=0}^{T-2} \lambda_i\varepsilon_{T-1-i} = \lambda_0(1-\beta)\left(\sum_{i=1}^{T-1} \varepsilon_{T-i} - \sum_{i=1}^{T-2} \varepsilon_{T-1-i}\right) + \lambda_0\varepsilon_T - \lambda_0\varepsilon_{T-1}$$
$$= \lambda_0(1-\beta)\varepsilon_{T-1} + \lambda_0\varepsilon_T - \lambda_0\varepsilon_{T-1} = \lambda_0\varepsilon_T - \beta\lambda_0\varepsilon_{T-1}. \blacksquare$$

Unsolved theoretical exercises

1. Derive the autocorrelation function for the $ARMA(2,1)$ model $y_t = \alpha_1 y_{t-1} + \alpha_2 y_{t-2} + \varepsilon_t + \beta\varepsilon_{t-1}$ for $t \in \mathbb{Z}$.

2. Consider the following $AR(1)$ model: $y_t = \alpha + \beta y_{t-1} + \sigma u_t \quad (t = 1, 2, ...)$, where u_t is a sequence of independent standard normal random variables and $t \in \mathbb{Z}$. Assume that $y_0 = \mu + \delta u_0$, where u_0 is standard normal and independent of $\{u_1, u_2, ...\}$.
   a. When $\alpha = 0$ and the regression has no intercept, discuss the behavior of $\hat{\beta}$ when β approaches one. Distinguish carefully between three cases.
   b. In a regression with intercept, assume that: $\frac{\delta(1-\beta)}{\sigma} \to 0$ and $\frac{\alpha-\mu(1-\beta)}{\sigma} \to \lambda$ as $\beta \to 1$.
   Derive the behavior of $\widetilde{\beta}$ (the LS estimator for β in the model with intercept) for the case $\lambda \neq 0$.
   c. Also for the case $\lambda = 0$.

3. Consider the $AR(2)$ process $y_t = 1.1y_{t-1} - 0.18y_{t-2} + \varepsilon_t$, with $E(\varepsilon_t\varepsilon_\tau) = \begin{cases} 1 \text{ for } t = \tau \\ 0 \text{ otherwise} \end{cases}$ and $t \in \mathbb{Z}$. Is this process stationary? If so calculate its autocovariances.

4. Verify that the functions $\phi(t) = a_1 \sin\omega_1 t + a_2 \cos\omega_1 t + a_3 \sin\omega_2 t + a_4 \sin\omega_2 t$ or

$\phi(t) = (a_1 + a_3 t)\sin\omega t + (a_2 + a_4 t)\cos\omega t$
can be written in the form (5.70).

5. Optimality conditions for SES. Suppose that $y_t = \sum_{j=1}^{t-1} \lambda_i \varepsilon_{t-i}$, $\lambda_0 \neq 0, t > 0$, where the process $\{\varepsilon_t;\ t \in \mathbb{N}\}$ is an independent process with zero expectation and constant variance σ_ε^2.

   (a) Is the process $\{y_t\}$ stationary? Show that it is always possible to normalize $\lambda = 0$ and to put y_t in the form $y_t = \sum_{j=1}^{t-1} \phi_j y_{t-j} + \varepsilon_t$. How can the ϕ_js be computed as a function of the λ_js?

   (b) Compute $E\left(y_{T+1} | y_1, y_2, \ldots, y_T\right)$. For which values of λ_j is it verified that $E\left(y_{T+1} | y_1, y_2, \ldots, y_T\right) = (1-\beta)\sum_{j=0}^{T-1} \beta^j y_{T-j}$?

6. Suppose two stationary $ARMA(p,q)$ processes $\{y_t, t \in \mathbb{Z}\}$ and $\{x_t, t \in \mathbb{Z}\}$ defined as $\alpha(L)y_t = \varphi(L)\varepsilon_t$ and $\alpha(L)x_t = \varphi(L)\eta_t$, where ε_t and η_t are WN processes with unit variance. Denoting the autocovariance functions by $\gamma_y(\cdot)$ and $\gamma_x(\cdot)$, respectively.

   (a) Show that for all $t \neq 0$: $\sum_{j=-\infty}^{\infty} \gamma_y(t-j)\gamma_x(j) = 0$ and $\sum_{j=-\infty}^{\infty} \gamma_y(-j)\gamma_x(j) = 1$. Interprete this.

   (b) Assuming that now $\alpha(L) := 1$ and considering an $r \times r$ square matrix $\mathbf{\Gamma}_x$ with typical elements $\gamma_{x,ij} = \gamma_x(j-i)$, $i, j = 1, 2, \ldots, r$, show that the elements a_{ij} of the matrix $\mathbf{A} := \mathbf{\Gamma}_x^{-1}$ are given by $a_{ij} = \gamma(j-i)$ for $q+1 \leqq i \leqq r-q$ and $1 \leqq j \leqq r$. This result can be used to get an approximated form for the inverse of the variance-covariance matrix of an autorregressive process.

7. Consider the process $\{z_t\ t \in \mathbb{Z}\}:\ z_t = \frac{4}{5}z_{t-1} + u_t$ with $\{u_t,\ t \in \mathbb{Z}\} \in WN$ with $\sigma_u^2 = \frac{9}{50}$. Assume that this process is affected by a measurement error and that we observe $y_t = z_t + \eta_t$ with $\{\eta_t,\ t \in \mathbb{Z}\} \in WN$ with $\sigma_\eta^2 = \frac{1}{2}$, but uncorrelated with u_t. Show that y_t can be written as an $ARMA(1,1)$ process. Derive a representation of y_t of the type $y_t = -\sum_{i=1}^{+\infty} \pi_i y_{t-i} + \xi_t$ with $\{\xi_t,\ t \in \mathbb{Z}\} \in WN$ such that $E(\xi_t y_{t-i}) = 0$, for all $i > 0$ and all t. What is the usefulness of such a representation?

8. Consider a stationary process $\{x_t\ t \in \mathbb{Z}\}$ defined as $(1-0.5L)x_t = (1-4L)\varepsilon_t$, where $\{\eta_t,\ t\in \mathbb{Z}\} \in WN$ with zero mean and $\sigma_\eta^2 = 1$.

   (a) Compute the innovation η_t of x_t as a function of the ε_t's. Provide the standard deviation of η_t.

   (b) What is the autocovariance function of the sum of the two white noises ε_t and η_t?

   (c) Compute the linear regression x_t^* of x_t on the future values of $x_{t+1}, x_{t+2}, \ldots$. What is the standaard deviation of $x_t - x_t^*$?

9. Consider the stationary process $\{x\ _t\ t \in \mathbb{Z}\}$ defined as $(1-4L)x_t = 3 + (1-0.5L)\varepsilon_t$, where $\{\varepsilon_t,\ t \in \mathbb{Z}\} \in WN$ with zero mean and $\sigma_\varepsilon^2 = 1$.

(a) Express x_t as a function of the ε_{t-i}s. What is the covariance between x_t and ε_{t-i}?

(b) What is the one-step ahead forecast error variance?

(c) What is the autocorrelation function of the process?

5.10.2 Solved (empirical) cases

In practice, it is very rare to have a data series which exactly or even nicely matches a theoretical ACF or PACF. Therefore one should use the estimated ACFs and PACFs to identify the underlying time series model (see Section 5.2). Recall that the evaluation of such a time series model runs as follows:

1. Verify whether there is a deterministic trend of a certain degree and correct for it, if necessary.

2. Test for unit roots (stochastic trend) and, if necessary, take the appropriate number of differences of the data (procedure described at the end of Section 5.5).

3. Identify a candidate $AR(I)MA$-model from the ACF and PACF of the series under consideration in order to detect aberrant observations.

4. Choose an appropriate $AR(I)MA$-model and estimate the unknown parameters.

5. Analyze the ACF and PACF of the residuals to test whether the $AR(I)MA$-model is adjusted.

6. If not return to step 3 .

7. Detect aberrant observations and, if necessary, adjust for them (procedure explained in Subsection 5.8.2) and verify the $AR(I)MA$-model underlying to the adjusted time series.

8. Compare the output of the estimated models, the AICs, BICs, Ljung-Box Q(and Q^*)-statistics and the sum of squared residuals.

9. Select the most appropriate model.

10. Forecast with the most appropriate model.

In this section we try to construct and estimate quarterly models for Dutch GDP, for (consumer price) inflation in Belgium and the unemployment rate in The Netherlands. Q-statistics are reported to confirm the findings, but recall their 'low power' and the criticism of Maddala & Kim (1998) (cf. the Box-Jenkins approach). Note also that in practice models will also use observations on fundamentals, rather than purely focusing on historical observations of the variable to model.

Figure 5.5: Dutch GDP (Volume Index)

Case 1.The volume index of Dutch GDP

A time series analysis based on Dutch GDP levels and Dutch GDP growth rates will be presented in this case.

(a) Analysis based on GDP levels The quarterly data, ranging from 1977^I to 2004^{II}, are taken from the OECD Statistical Compendium (http://www.oecd.org). From Figure 5.5 it seems reasonable to try to fit a second order deterministic trend on the volume index of Dutch GDP for the (assumed) sample period 1977^I-2000^{IV} (see Table 5.2; the period 2001^I-2004^{IV} will be used for forecasting), where from the resulting Durbin-Watson test statistic a (very) strong positive autocorrelation of the residuals is observed.[15]

In order to determine the possible existence of a stochastic trend an ADF test on the residuals (see Figure 5.6) of this second order deterministic trend is carried out. The BIC (defined in (5.32)) indicates that only one lag is used in calculating this test (see Table 5.3). Hence, the null of a unit root hypothesis is rejected.[16] Therefore, an *ARMA* model of these deterministic-trend-adjusted level-values will be built.

Observing the graph of the adjusted level-series we see that a potential aberrant observation can occur in 1963:Q1. Following the procedure in Section 5.4,

[15] Since this volume index largely moves between zero and one it is not advisable to take logarithms.

[16] When introducing a constant term in the ADF test, the rejection rate of the null of a unit root is lower (verify!), although the mean of the OLS residuals is always zero.

Table 5.2: Second order deterministic trend estimation

Sample: 1977^{I}-2000^{IV}

Included observations: 96

Dependent Variable: GDPNL

Method: Least Squares

Variable	Coefficient	Std. Error	t-Statistic	Prob.
C	0.570414	0.003751	152.0683	0
T	0.000835	0.000182	4.576937	0
T2	3.93E-05	1.86E-06	21.14089	0
R-squared	0.991037	Mean dependent var	0.728927	
Adjusted R-squared	0.990844	S.D. dependent var	0.130707	
S.E. of regression	0.012507	Akaike info criterion	-5.89431	
Sum squared resid	0.014547	Schwarz criterion	-5.81418	
Log likelihood	285.927	F-statistic	5141.381	
Durbin-Watson stat	0.366094	Prob(F-statistic)	0	

Figure 5.6: Plot of the second order trend residuals

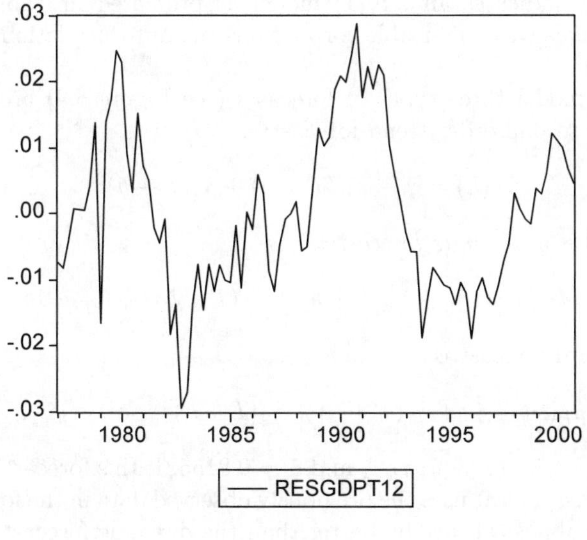

Table 5.3: Unit root test on the values adjusted for deterministic trend

Null Hypothesis: RESGDPT12 has a unit root
Exogenous: None
Lag Length: 0 (Automatic based on SIC, MAXLAG=11)

			t-Statistic	Prob.*
Augmented Dickey-Fuller test statistic			-3.10139	0.0022
Test critical values:1% level			-2.58953	
5% level			-1.94425	
10% level			-1.61451	

Augmented Dickey-Fuller Test Equation
Dependent Variable: D(RESGDPT12)
Method: Least Squares
Sample(adjusted): 1977^{II}-2000^{IV}
Included observations: 95 after adjusting endpoints

Variable	Coefficient	Std. Error	t-Statistic	Prob.
RESGDPT12(-1)	-0.18448	0.059483	-3.10139	0.0025
R-squared	0.092569	Mean dependent var	0.000126	
Adjusted R-squared	0.092569	S.D. dependent var	0.007526	
S.E. of regression	0.007169	Akaike info criterion	-7.02758	
Sum squared resid	0.004831	Schwarz criterion	-7.0007	
Log likelihood	334.8099	Durbin-Watson stat	2.355178	

an appropriate *ARMA* model is estimated after considering the correlogram in Figure 5.7.

This clearly indicates an *AR*(1) model as presented in Table 5.4, which produces residuals without sizable correlations (no autocorrelation left; see also Figure 5.8).

Using this model three types of forecasts (see Figure 5.9) are built. The first one is the second order trend forecast:

$$\hat{y}_T(h) = \hat{\gamma}_0 + \hat{\gamma}_1(T+h) + \hat{\gamma}_2(T+h)^2.$$

The second is a *dynamic forecast:*

$$\hat{y}_T(h) = \hat{\gamma}_0 + \hat{\gamma}_1(T+h) + \hat{\gamma}_2(T+h)^2 + \hat{\alpha}_T^h \hat{u}_T.$$

And the third forecast is:

$$\hat{y}_T(h) = \hat{\gamma}_0 + \hat{\gamma}_1(T+h) + \hat{\gamma}_2(T+h)^2 + \hat{\alpha}\hat{u}_{T+h-1},$$

where $\hat{u}_{T+h-1} := y_{T+h-1} - \hat{y}_{T+h-1}$ and $\hat{\alpha} = 0.815521$; this forecast is also called the *static forecast* since it uses the previously observed data in the forecast period (so that it will almost always be better than the dynamic forecast).

Assembling these three forecasts for the Dutch GDP levels in Figure 5.9, it is clear that the strong slowdown of Dutch GDP in the first years of this decade (after September 11, 2001 shock) is very difficult to forecast. Given

Figure 5.7: Correlogram of the residuals

Sample: 1977:1 2000:4
Included observations: 96

<table>
<tr><th>Autocorrelation</th><th>Partial Correlation</th><th></th><th>AC</th><th>PAC</th><th>Q-Stat</th><th>Prob</th></tr>
<tr><td>. |****** |</td><td>. |****** |</td><td>1</td><td>0.814</td><td>0.814</td><td>65.673</td><td>0.000</td></tr>
<tr><td>. |****** |</td><td>. |** |</td><td>2</td><td>0.734</td><td>0.209</td><td>119.52</td><td>0.000</td></tr>
<tr><td>. |***** |</td><td>. | . |</td><td>3</td><td>0.642</td><td>-0.003</td><td>161.16</td><td>0.000</td></tr>
<tr><td>. |**** |</td><td>. | . |</td><td>4</td><td>0.561</td><td>-0.015</td><td>193.33</td><td>0.000</td></tr>
<tr><td>. |**** |</td><td>. | . |</td><td>5</td><td>0.490</td><td>-0.004</td><td>218.18</td><td>0.000</td></tr>
<tr><td>. |*** |</td><td>.*| . |</td><td>6</td><td>0.379</td><td>-0.154</td><td>233.17</td><td>0.000</td></tr>
<tr><td>. |** |</td><td>**| . |</td><td>7</td><td>0.238</td><td>-0.231</td><td>239.15</td><td>0.000</td></tr>
<tr><td>. |*. |</td><td>.*| . |</td><td>8</td><td>0.132</td><td>-0.070</td><td>241.02</td><td>0.000</td></tr>
<tr><td>. | . |</td><td>.*| . |</td><td>9</td><td>0.015</td><td>-0.112</td><td>241.04</td><td>0.000</td></tr>
<tr><td>.*| . |</td><td>.*| . |</td><td>10</td><td>-0.080</td><td>-0.069</td><td>241.74</td><td>0.000</td></tr>
<tr><td>.*| . |</td><td>. | . |</td><td>11</td><td>-0.168</td><td>-0.043</td><td>244.87</td><td>0.000</td></tr>
<tr><td>**| . |</td><td>.*| . |</td><td>12</td><td>-0.288</td><td>-0.161</td><td>254.13</td><td>0.000</td></tr>
</table>

Table 5.4: AR(1) model of the residuals

Dependent Variable: RESGDPT12
Method: Least Squares
Sample(adjusted): 1977^{II}-2000^{IV}
Included observations: 95 after adjusting endpoints
Convergence achieved after 2 iterations

Variable	Coefficient	Std. Error	t-Statistic	Prob.
AR(1)	0.815521	0.059483	13.71018	0
R-squared	0.666617	Mean dependent var	7.80E-05	
Adjusted R-squared	0.666617	S.D. dependent var	0.012416	
S.E. of regression	0.007169	Akaike info criterion	-7.02758	
Sum squared resid	0.004831	Schwarz criterion	-7.0007	
Log likelihood	334.8099	Durbin-Watson stat	2.355178	
Inverted AR Roots	0.82			

Figure 5.8: Correlogram of the residuals of the $AR(1)$ model

Sample: 1977:1 2000:4
Included observations: 95

<table>
<tr><th>Autocorrelation</th><th>Partial Correlation</th><th></th><th>AC</th><th>PAC</th><th>Q-Stat</th><th>Prob</th></tr>
<tr><td>.*| . |</td><td>.*| . |</td><td>1</td><td>-0.179</td><td>-0.179</td><td>3.1275</td><td>0.077</td></tr>
<tr><td>. |*. |</td><td>. |*. |</td><td>2</td><td>0.105</td><td>0.076</td><td>4.2206</td><td>0.121</td></tr>
<tr><td>. | . |</td><td>. |*. |</td><td>3</td><td>0.045</td><td>0.079</td><td>4.4243</td><td>0.219</td></tr>
<tr><td>. | . |</td><td>. | . |</td><td>4</td><td>0.034</td><td>0.048</td><td>4.5385</td><td>0.338</td></tr>
<tr><td>. |*. |</td><td>. |*. |</td><td>5</td><td>0.156</td><td>0.165</td><td>7.0329</td><td>0.218</td></tr>
<tr><td>. |*. |</td><td>. |*. |</td><td>6</td><td>0.120</td><td>0.177</td><td>8.5270</td><td>0.202</td></tr>
<tr><td>. | . |</td><td>. | . |</td><td>7</td><td>-0.047</td><td>-0.027</td><td>8.7557</td><td>0.271</td></tr>
<tr><td>. | . |</td><td>.*| . |</td><td>8</td><td>0.004</td><td>-0.063</td><td>8.7578</td><td>0.363</td></tr>
<tr><td>. | . |</td><td>. | . |</td><td>9</td><td>-0.012</td><td>-0.055</td><td>8.7736</td><td>0.458</td></tr>
<tr><td>. | . |</td><td>.*| . |</td><td>10</td><td>-0.017</td><td>-0.072</td><td>8.8049</td><td>0.551</td></tr>
<tr><td>. |*. |</td><td>. | . |</td><td>11</td><td>0.076</td><td>0.023</td><td>9.4428</td><td>0.581</td></tr>
<tr><td>**| . |</td><td>**| . |</td><td>12</td><td>-0.297</td><td>-0.303</td><td>19.235</td><td>0.083</td></tr>
</table>

this strong shock effect and also the very strong positive autocorrelation of the deterministic trend residuals of the GDP levels, it may be recommendable to analyze the growth rates of Dutch GDP rather than their levels.

(b) Analysis based on GDP growth. Estimations and forecasts the GDP growth in The Netherlands (Figure 5.10) are presented below.

Examining the correlogram of the growth rates of Dutch GDP for the complete period 1997^{I}-2004^{II} in Figure 5.11, it is observed that an AR(1) is a good candidate for estimation.

Starting from the identified $AR(1)$ model for the Dutch GDP growth rates in Table 5.5 for the sample period (1977^{I}-2000^{IV}) it is observed that the large value for the JB statistic (47.9598) for the corresponding $AR(1)$ residuals reveals the possibility of one or more outliers. The plot in Figure 5.12 shows the possibility of an outlier in the first and second quarter of 1979 (at the beginning of the sample).

It is important to determine whether the possible outlier at 1979^{II} is significant or not. Calculate $\hat{\omega}_{AO}\left(1979^{II}\right)$ from (5.40) for $\tau = 1979^{II}$, $\hat{e}_t$ the estimated residuals from the $AR(1)$ model and $\hat{z}_{1979:2} = 1$, $\hat{z}_{1979:3} = 0.19724$ and zero otherwise, so that $\hat{\omega}_{AO}\left(1979^{II}\right) = 0.020912$ (see Table 5.6).

Compute now $\hat{\zeta}_{AO}(\tau)$ in (5.41) where $\hat{\sigma}_{\hat{\varepsilon}}$ is computed with the 'omit-one' method so that $SE(\hat{\omega}_{AO}(\tau)) = 0.00726$ and $\hat{\zeta}_{AO}(\tau) = 0.020912/0.00726 = 2.8804$, which is slightly significant at the 5 per cent level. Removing then its effect from the original growth series, a new value for 1979^{II} is obtained (as given in (5.42)), i.e. 0.01109.

Examining now the correlogram of the adjusted growth rate series in Figure

Figure 5.9: Different forecasts of Dutch GDP

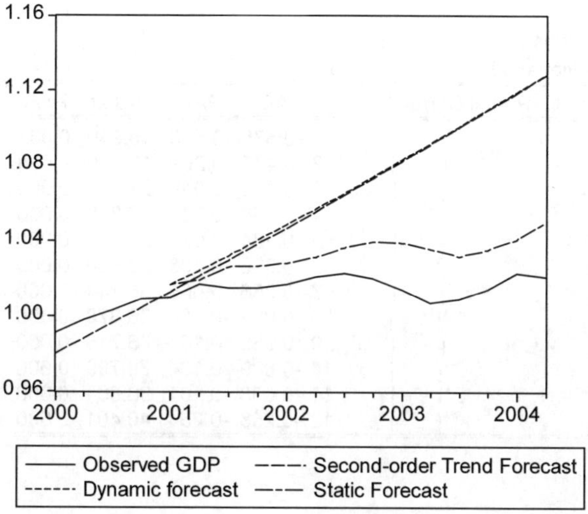

Figure 5.10: Dutch GDP growth rates

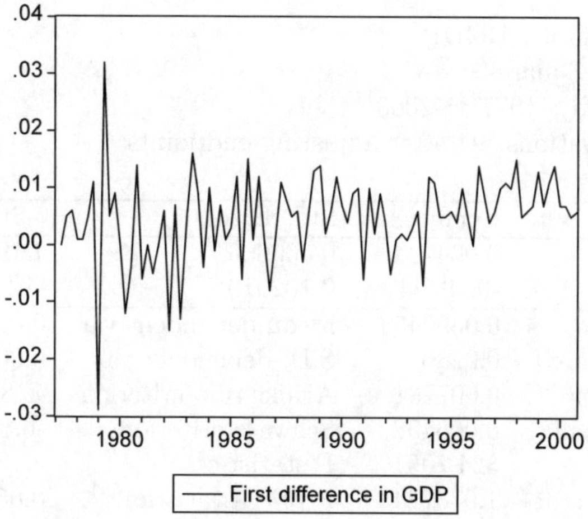

Figure 5.11: Correlogram of Dutch GDP growth rates

Sample: 1997:1 2004:2
Included observations: 30

<table>
<tr><th>Autocorrelation</th><th>Partial Correlation</th><th></th><th>AC</th><th>PAC</th><th>Q-Stat</th><th>Prob</th></tr>
<tr><td>. |**** |</td><td>. |**** |</td><td>1</td><td>0.575</td><td>0.575</td><td>10.940</td><td>0.001</td></tr>
<tr><td>. |**** |</td><td>. |**. |</td><td>2</td><td>0.470</td><td>0.209</td><td>18.520</td><td>0.000</td></tr>
<tr><td>. |**. |</td><td>. | . |</td><td>3</td><td>0.303</td><td>-0.048</td><td>21.789</td><td>0.000</td></tr>
<tr><td>. |*** |</td><td>. |**. |</td><td>4</td><td>0.350</td><td>0.201</td><td>26.310</td><td>0.000</td></tr>
<tr><td>. |**. |</td><td>. | . |</td><td>5</td><td>0.284</td><td>0.022</td><td>29.415</td><td>0.000</td></tr>
<tr><td>. |**. |</td><td>. |* . |</td><td>6</td><td>0.323</td><td>0.108</td><td>33.600</td><td>0.000</td></tr>
<tr><td>. |**. |</td><td>. | . |</td><td>7</td><td>0.256</td><td>0.005</td><td>36.344</td><td>0.000</td></tr>
<tr><td>. | . |</td><td>***| . |</td><td>8</td><td>0.028</td><td>-0.350</td><td>36.378</td><td>0.000</td></tr>
<tr><td>. |* . |</td><td>. |* . |</td><td>9</td><td>0.086</td><td>0.187</td><td>36.716</td><td>0.000</td></tr>
<tr><td>. | . |</td><td>.**| . |</td><td>10</td><td>-0.039</td><td>-0.194</td><td>36.790</td><td>0.000</td></tr>
<tr><td>. | . |</td><td>. *| . |</td><td>11</td><td>-0.038</td><td>-0.101</td><td>36.864</td><td>0.000</td></tr>
<tr><td>.**| . |</td><td>.**| . |</td><td>12</td><td>-0.258</td><td>-0.254</td><td>40.401</td><td>0.000</td></tr>
</table>

Table 5.5: AR(1) model for the Dutch GDP growth

Dependent Variable: DGDP
Method: Least Squares
Sample(adjusted): 1977^{III}-2000^{IV}
Included observations: 94 after adjusting endpoints

Variable	Coefficient	Std. Error	t-Statistic	Prob.
C	0.004734	0.000666	7.108095	0
AR(1)	-0.19724	0.102019	-1.93342	0.0563
R-squared	0.039045	Mean dependent var	0.004745	
Adjusted R-squared	0.0286	S.D. dependent var	0.007844	
S.E. of regression	0.007731	Akaike info criterion	-6.86615	
Sum squared resid	0.005498	Schwarz criterion	-6.81204	
Log likelihood	324.7091	F-statistic	3.738092	
Durbin-Watson stat	1.973724	Prob(F-statistic)	0.05626	

Figure 5.12: Residuals of the $AR(1)$ model for Dutch GDP growth

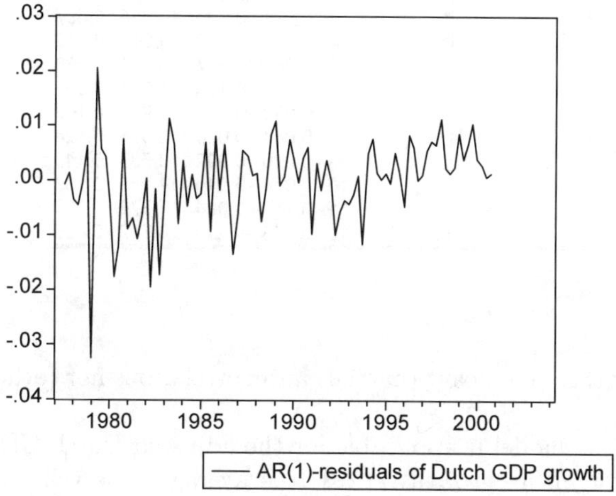

Table 5.6: Regression of the residuals against Z for outlier correction

Dependent Variable: RESDGDPAR1
Method: Least Squares
Sample(adjusted): 1977^{III}-2000^{IV}
Included observations: 94 after adjusting endpoints

Variable	Coefficient	Std. Error	t-Statistic	Prob.
Z	0.020912	0.007225	2.894163	0.0047
R-squared	0.082625	Mean dependent var	2.78E-18	
Adjusted R-squared	0.082625	S.D. dependent var	0.007689	
S.E. of regression	0.007365	Akaike info criterion	-6.97367	
Sum squared resid	0.005044	Schwarz criterion	-6.94661	
Log likelihood	328.7623	Durbin-Watson stat	1.755269	

Figure 5.13: Correlogram of the adjusted Dutch GDP growth

Sample: 1977:1 2004:2
Included observations: 95

<table>
<tr><th>Autocorrelation</th><th>Partial Correlation</th><th></th><th>AC</th><th>PAC</th><th>Q-Stat</th><th>Prob</th></tr>
<tr><td>.*| . |</td><td>.*| . |</td><td>1</td><td>-0.087</td><td>-0.087</td><td>0.7432</td><td>0.389</td></tr>
<tr><td>. |*. |</td><td>. |*. |</td><td>2</td><td>0.074</td><td>0.067</td><td>1.2906</td><td>0.525</td></tr>
<tr><td>. |*. |</td><td>. |*. |</td><td>3</td><td>0.077</td><td>0.090</td><td>1.8855</td><td>0.597</td></tr>
<tr><td>. |*. |</td><td>. |*. |</td><td>4</td><td>0.138</td><td>0.150</td><td>3.8062</td><td>0.433</td></tr>
<tr><td>. |** |</td><td>. |** |</td><td>5</td><td>0.251</td><td>0.277</td><td>10.241</td><td>0.069</td></tr>
<tr><td>. |*. |</td><td>. |*. |</td><td>6</td><td>0.123</td><td>0.179</td><td>11.810</td><td>0.066</td></tr>
<tr><td>. | . |</td><td>. | . |</td><td>7</td><td>0.032</td><td>0.030</td><td>11.920</td><td>0.103</td></tr>
<tr><td>. |*. |</td><td>. | . |</td><td>8</td><td>0.084</td><td>0.020</td><td>12.671</td><td>0.124</td></tr>
<tr><td>. |*. |</td><td>. | . |</td><td>9</td><td>0.090</td><td>0.002</td><td>13.537</td><td>0.140</td></tr>
<tr><td>. |*. |</td><td>. | . |</td><td>10</td><td>0.074</td><td>-0.040</td><td>14.138</td><td>0.167</td></tr>
<tr><td>. |*. |</td><td>. |*. |</td><td>11</td><td>0.172</td><td>0.094</td><td>17.370</td><td>0.097</td></tr>
<tr><td>.*| . |</td><td>**| . |</td><td>12</td><td>-0.174</td><td>-0.219</td><td>20.712</td><td>0.055</td></tr>
</table>

5.13 neither further significant (partial) autocorrelations nor (other) outliers are observed.

As no *ARMA* model is applicable for the adjusted Dutch GDP growth, the best forecast for future growth rates is the average observed growth obtained from the adjusted series $\hat{\mu} = 0.004475$. The implied dynamic forecast values for the Dutch GDP levels are given by:

$$\hat{y}_T(h) = y_T + h\hat{\mu}, \tag{5.74}$$

while the static forecast for Dutch GDP is specified as

$$\hat{y}_T(h) = y_{T+h-1} + \hat{\mu}. \tag{5.75}$$

Finally, assembling all the forecasts for the Dutch GDP levels in Figure 5.14, from which it is clear that compared to Figure 5.9, the forecast of the GDP level may be improved by considering the GDP growth rates rather than GDP levels.[17]

Case 2. A model for inflation based on the consumer price index in Belgium

In this section a quarterly model for the consumer price index and inflation in Belgium from the OECD Statistical Compendium (http://www.oecd.org) is constructed, estimated and used for forecasting future inflation. The original database consists of quarterly observations of the Belgian consumer price index for the period 1963^I to 2004^{III}. The inflation rate is modeled as the first

[17] Mind also the very strong positive autocorrelation in the deterministic trend residuals of the Dutch GDP levels.

Figure 5.14: Comparison of different forecasts for Dutch GDP

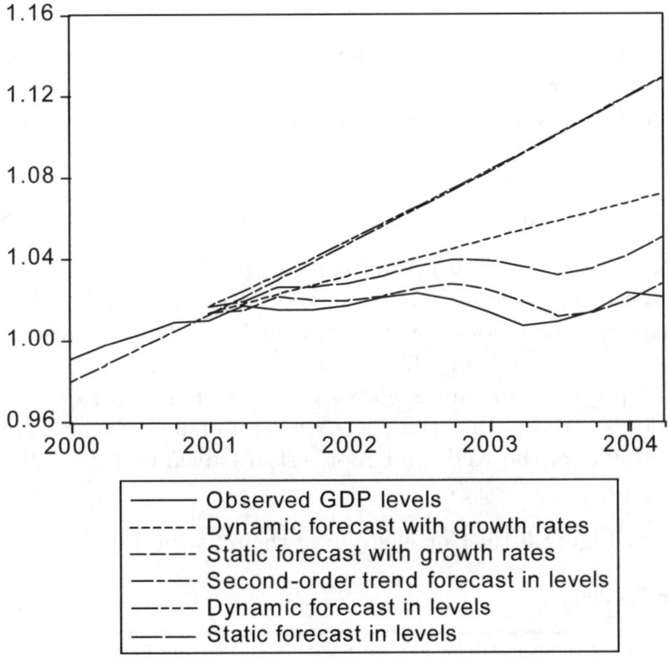

Figure 5.15: Consumer Price Inflation in Belgium

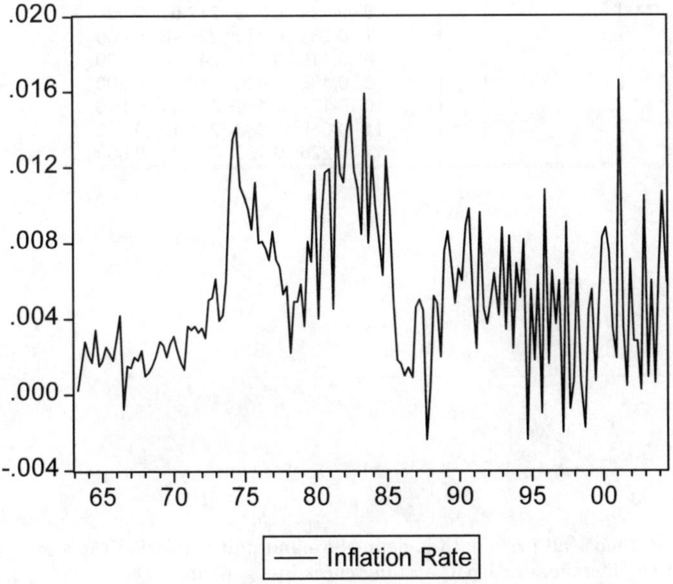

difference of this series.[18] Figure 5.15 depicts the consumer price inflation. For estimation a subsample for the period 1963^{I} to 2000^{IV} will be used.

The correlogram of the inflation rate presented in Figure 5.16 shows that an $AR(1, 2, 4)$ model could be applied to this data (see Table 5.7).

The JB test for the residuals of this estimation does not indicate the presence of aberrant observations.

Both dynamic and static forecasts are calculated for this model presented in Table 5.8 and Figure 5.17.

A model for the unemployment rate in The Netherlands

Quarterly data in this section is taken from the OECD Statistical Compendium (http://www.oecd.org). The sample range is 1970^{I} to 2004^{II}. Figure 5.18 plots the quarterly unemployment rate in The Netherlands. Estimations are conducted for the subsample of 1970^{I} to 2000^{IV}.

The first step is to examine the correlogram (see Figure 5.19):

The correlogram shows that there is a strong probability that a unit root is present. The results of the ADF unit root test are given in Table 5.9.

Figure 5.16: Correlogram of the inflation rate

Sample: 1963^{I} -2004^{III}
Observations: 166

<table>
<tr><th>Autocorrelation</th><th>Partial Correlation</th><th></th><th>AC</th><th>PAC</th><th>Q-Stat</th><th>Prob</th></tr>
<tr><td>.|*** |</td><td>.|*** |</td><td>1</td><td>0.436</td><td>0.436</td><td>32.102</td><td>0.000</td></tr>
<tr><td>.|**** |</td><td>.|*** |</td><td>2</td><td>0.515</td><td>0.402</td><td>77.250</td><td>0.000</td></tr>
<tr><td>.|*** |</td><td>.|** |</td><td>3</td><td>0.452</td><td>0.209</td><td>112.14</td><td>0.000</td></tr>
<tr><td>.|**** |</td><td>.|** |</td><td>4</td><td>0.557</td><td>0.319</td><td>165.58</td><td>0.000</td></tr>
<tr><td>.|** |</td><td>*|. |</td><td>5</td><td>0.285</td><td>-0.173</td><td>179.66</td><td>0.000</td></tr>
<tr><td>.|*** |</td><td>.|* |</td><td>6</td><td>0.434</td><td>0.096</td><td>212.51</td><td>0.000</td></tr>
<tr><td>.|** |</td><td>*|. |</td><td>7</td><td>0.261</td><td>-0.119</td><td>224.46</td><td>0.000</td></tr>
<tr><td>.|** |</td><td>.|. |</td><td>8</td><td>0.318</td><td>-0.009</td><td>242.28</td><td>0.000</td></tr>
<tr><td>.|* |</td><td>.|. |</td><td>9</td><td>0.193</td><td>-0.020</td><td>248.89</td><td>0.000</td></tr>
<tr><td>.|* |</td><td>*|. |</td><td>10</td><td>0.174</td><td>-0.166</td><td>254.31</td><td>0.000</td></tr>
<tr><td>.|* |</td><td>*|. |</td><td>11</td><td>0.068</td><td>-0.068</td><td>255.13</td><td>0.000</td></tr>
<tr><td>.|** |</td><td>.|* |</td><td>12</td><td>0.226</td><td>0.166</td><td>264.36</td><td>0.000</td></tr>
</table>

[18] Given that almost all price data is defined around unity, inflation rates were considered as absolute price differences rather than differences in logarithms.

Table 5.7: AR(1,2,4) estimation results for Belgian inflation

Dependent Variable: INF
Method: Least Squares
Sample(adjusted): 1964^{II}-2000^{IV}
Included observations: 147 after adjusting endpoints
Convergence achieved after 3 iterations

Variable	Coefficient	Std. Error	t-Statistic	Prob.
C	0.005823	0.001359	4.286208	0
AR(1)	0.198542	0.073074	2.716988	0.0074
AR(2)	0.295324	0.0797	3.705461	0.0003
AR(4)	0.333091	0.078138	4.262885	0
R-squared	0.493075	Mean dependent var	0.005463	
Adjusted R-squared	0.48244	S.D. dependent var	0.003927	
S.E. of regression	0.002825	Akaike info criterion	-8.87366	
Sum squared resid	0.001141	Schwarz criterion	-8.79229	
Log likelihood	656.2138	F-statistic	46.36436	
Durbin-Watson stat	1.879441	Prob(F-statistic)	0	

Table 5.8: Observations and forecasts for the Belgian inflation rates

	Observed Inflation	Forecast Dynamic	Static
2001^{I}	0.0019	0.011478	0.011478
2001^{II}	0.0166	0.012037	0.010136
2001^{III}	0.0035	0.014101	0.012178
2001^{IV}	0.0004	0.01331	0.012553
2002^{I}	0.0071	0.016453	0.007569
2002^{II}	0.0028	0.01703	0.01288
2002^{III}	0.0028	0.01876	0.009642
2002^{IV}	0.0002	0.01901	0.007339
2003^{I}	0.0105	0.020618	0.009055
2003^{II}	0.0009	0.021203	0.008899
2003^{III}	0.006	0.02237	0.010035
2003^{IV}	0.0006	0.022858	0.007347
2004^{I}	0.0065	0.023836	0.011212
2004^{II}	0.0107	0.024369	0.00759
2004^{III}	0.0059	0.025152	0.011866

Figure 5.17: Belgian inflation forecasts

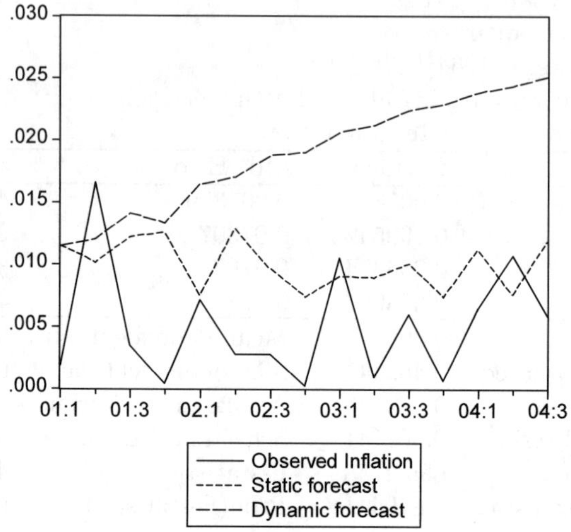

Figure 5.18: Dutch unemployment rate

Figure 5.19: Correlogram of the Dutch unemployment rate

```
Sample: 1970:1 2000:4
Included observations: 124
Autocorrelation   Partial Correlation      AC     PAC    Q-Stat  Prob
    .|*******|        .|*******|      1  0.963  0.963  117.81  0.000
    .|*******|        *|.      |      2  0.914 -0.180  224.90  0.000
    .|*******|        *|.      |      3  0.859 -0.101  320.13  0.000
    .|****** |        *|.      |      4  0.797 -0.092  402.85  0.000
    .|****** |        *|.      |      5  0.730 -0.089  472.77  0.000
    .|*****  |        .|.      |      6  0.663  0.001  531.03  0.000
    .|*****  |        .|.      |      7  0.597 -0.028  578.68  0.000
    .|****   |        .|.      |      8  0.536  0.030  617.39  0.000
    .|****   |        .|.      |      9  0.478 -0.010  648.45  0.000
    .|***    |        .|.      |     10  0.424 -0.006  673.11  0.000
    .|***    |        .|.      |     11  0.373 -0.026  692.31  0.000
    .|**     |        .|.      |     12  0.324 -0.023  706.95  0.000
```

Table 5.9: Unit root tests of Dutch unemployment

Null Hypothesis: UNNL has a unit root
Exogenous: Constant
Lag Length: 1 (Automatic based on SIC, MAXLAG=12)

	t-Statistic	Prob.*
Augmented Dickey-Fuller test statistic	-2.07488	0.2552
Test critical values:1% level	-3.48465	
5% level	-2.88525	
10% level	-2.57949	

Augmented Dickey-Fuller Test Equation
Dependent Variable: D(UNNL)
Method: Least Squares
Sample(adjusted): $1970^{III} - 2000^{IV}$
Included observations: 122 after adjusting endpoints

Variable	Coefficient	Std. Error	t-Statistic	Prob.
UNNL(-1)	-0.02897	0.013963	-2.07488	0.0402
D(UNNL(-1))	0.373257	0.083516	4.469281	0
C	0.171778	0.082397	2.084762	0.0392
R-squared	0.168554	Mean dependent var		0.016967
Adjusted R-squared	0.15458	S.D. dependent var		0.342885
S.E. of regression	0.315272	Akaike info criterion		0.553518
Sum squared resid	11.82814	Schwarz criterion		0.622469
Log likelihood	-30.7646	F-statistic		12.06204
Durbin-Watson stat	2.133223	Prob(F-statistic)		0.000017

Figure 5.20: Changes in the Dutch unemployment rate

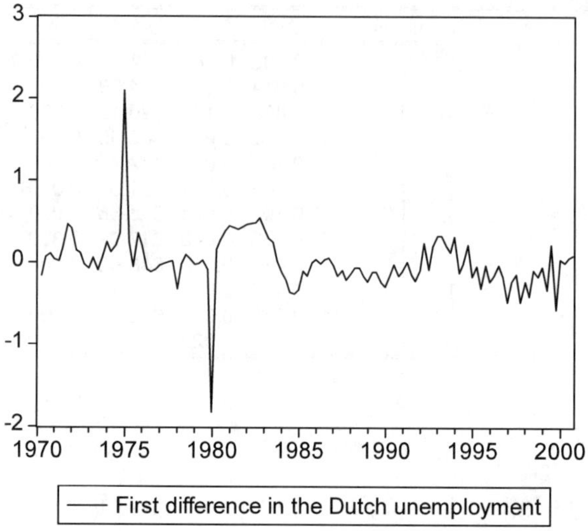

The ADF unit root test supports the existence of a unit root in the Dutch unemployment rate. Therefore one must use the change in the unemployment rate for estimation and forecasting.

The JB statistic of this series is very high (1324.349), strongly indicating the presence of abberant observations. The maximum observation is 2.1 at 1975:1 and the minimum is -1.83 at 1980:1 which are definitely larger than four times the standard deviation (0.34).

The correlogram of the changes in the Dutch unemployment rate calls for using the $AR(1)$ model (as shown in Table 5.10).

Further analysis of the residuals shows that there is no (partial) autocorrelation left in the series.

Similar to the first case on Dutch GDP correction for an outlier at 1975^I has to be applied as in (5.42), which after correction takes up new value 0.1998. Yet the JB statistic (849.6795) still points at another outlier at 1980:1. The new $AR(1)$ coefficient is 0.420880 so that the adjusted value becomes $-1.83 + 1.8586 = 0.0286$. Finally the adjusted series looks like in Figure 5.22.

The correlogram of the adjusted series is given in Figure 5.23 and indicates that an $AR(2)$ model would be appropiate.

After estimating such an $AR(2)$ model in Table 5.11 it is observed that there is no further (partial) autocorrelation detected, so that one can proceed with forecasting.

Forecasts for the changes in the Dutch unemployment rate are presented in Table 5.12 and Figure 5.24. Based on these forecasts of the changes then the

Figure 5.21: Correlogram of changes in the Dutch unemployment rate

Sample: 1970:1 2000:4
Included observations: 123

Autocorrelation	Partial Correlation		AC	PAC	Q-Stat	Prob
.|*** |	.|*** |	1	0.372	0.372	17.406	0.000
.|** |	.|* |	2	0.273	0.157	26.904	0.000
.|** |	.|* |	3	0.260	0.138	35.544	0.000
.|** |	.|* |	4	0.213	0.066	41.383	0.000
.|* |	*|. |	5	0.069	-0.092	42.008	0.000
.|. |	*|. |	6	0.007	-0.076	42.014	0.000
.|. |	.|. |	7	0.020	0.002	42.068	0.000
.|. |	.|. |	8	0.011	0.015	42.085	0.000
*|. |	*|. |	9	-0.072	-0.068	42.792	0.000
.|. |	.|. |	10	-0.045	0.005	43.068	0.000
.|. |	.|. |	11	-0.042	-0.014	43.311	0.000
*|. |	.|. |	12	-0.083	-0.053	44.262	0.000

Table 5.10: AR(1) model of the changes in the Dutch unemployment rate

Dependent Variable: DUNNL
Method: Least Squares
Sample(adjusted): $1970^{III} - 2000^{IV}$
Included observations: 122 after adjusting endpoints
Convergence achieved after 2 iterations

Variable	Coefficient	Std. Error	t-Statistic	Prob.
AR(1)	0.373244	0.084279	4.428653	0
R-squared	0.137357	Mean dependent var	0.016967	
Adjusted R-squared	0.137357	S.D. dependent var	0.342885	
S.E. of regression	0.318467	Akaike info criterion	0.557564	
Sum squared resid	12.27193	Schwarz criterion	0.580548	
Log likelihood	-33.0114	Durbin-Watson stat	2.112891	

Figure 5.22: Adjusted series

Figure 5.23: Correlogram of the corrected Dutch unemployment rate

Sample: 1970:1 2000:4
Included observations: 123

<table>
<tr><th>Autocorrelation</th><th>Partial Correlation</th><th></th><th>AC</th><th>PAC</th><th>Q-Stat</th><th>Prob</th></tr>
<tr><td>.|***** |</td><td>.|***** |</td><td>1</td><td>0.656</td><td>0.656</td><td>54.207</td><td>0.000</td></tr>
<tr><td>.|***** |</td><td>.|*** |</td><td>2</td><td>0.638</td><td>0.364</td><td>105.90</td><td>0.000</td></tr>
<tr><td>.|**** |</td><td>.|. |</td><td>3</td><td>0.532</td><td>0.055</td><td>142.15</td><td>0.000</td></tr>
<tr><td>.|*** |</td><td>.|. |</td><td>4</td><td>0.452</td><td>-0.026</td><td>168.56</td><td>0.000</td></tr>
<tr><td>.|** |</td><td>**|. |</td><td>5</td><td>0.296</td><td>-0.192</td><td>180.00</td><td>0.000</td></tr>
<tr><td>.|** |</td><td>*|. |</td><td>6</td><td>0.218</td><td>-0.082</td><td>186.27</td><td>0.000</td></tr>
<tr><td>.|* |</td><td>.|. |</td><td>7</td><td>0.169</td><td>0.056</td><td>190.07</td><td>0.000</td></tr>
<tr><td>.|* |</td><td>.|. |</td><td>8</td><td>0.096</td><td>0.001</td><td>191.31</td><td>0.000</td></tr>
<tr><td>.|. |</td><td>*|. |</td><td>9</td><td>-0.007</td><td>-0.125</td><td>191.31</td><td>0.000</td></tr>
<tr><td>.|. |</td><td>.|. |</td><td>10</td><td>0.006</td><td>0.056</td><td>191.32</td><td>0.000</td></tr>
<tr><td>.|. |</td><td>.|* |</td><td>11</td><td>0.010</td><td>0.108</td><td>191.33</td><td>0.000</td></tr>
<tr><td>*|. |</td><td>*|. |</td><td>12</td><td>-0.101</td><td>-0.182</td><td>192.76</td><td>0.000</td></tr>
</table>

Table 5.11: AR(1,2) model of the corrected changes in the Dutch unemployment rate

Dependent Variable: DUNNL2
Method: Least Squares
Sample(adjusted): $1970^{IV} - 2000^{IV}$
Included observations: 121 after adjusting endpoints
Convergence achieved after 2 iterations

Variable	Coefficient	Std. Error	t-Statistic	Prob.
AR(1)	0.423982	0.08507	4.983918	0
AR(2)	0.361539	0.084927	4.257042	0
R-squared	0.510652	Mean dependent var	0.016185	
Adjusted R-squared	0.50654	S.D. dependent var	0.232851	
S.E. of regression	0.16357	Akaike info criterion	-0.76676	
Sum squared resid	3.183871	Schwarz criterion	-0.72055	
Log likelihood	48.38885	Durbin-Watson stat	2.04217	

forecasts for levels of the Dutch unemployment rate are shown in Table 5.13 and Figure 5.25.

5.10.3 Empirical exercises

1. A researcher uses a sample of 200 quarterly observations. The sample autocorrelation function (ACF) is computed with the following results:

k	1	2	3	4	5	6	7	8	9	10
$\hat{\rho}_k$	0.83	0.71	0.60	0.45	0.44	0.35	0.29	0.20	0.11	-0.01

The sample partial autocorrelation function (PACF) is determined as:

k	1	2	3	4	5	6	7	8	9	10
$\hat{\theta}_{kk}$	0.83	0.16	-0.09	0.05	0.04	-0.05	0.01	0.10	-0.03	-0.01

- What is meant by the sample ACF? Does the pattern indicate that an autoregressive or moving average representation is more appropriate? Why?

- What is meant by the sample PACF? Why is the first partial autocorrelation coefficient equal to the first autocorrelation coefficient?

- Do the ACF and PACF patterns indicate that an autoregressive or moving average representation is more appropriate? Why?

Suppose now that the researcher estimates the following $AR(2)$ model:

$$y_t = \underset{(5.67)}{50} + \underset{(0.07)}{0.74}\, y_{t-1} + \underset{(0.07)}{0.16}\, y_{t-2} + \hat{\varepsilon}_t$$

- How would you check whether an $ARMA(2,1)$ model is more appropriate?

- Compute from the above estimates an estimate for $E(y_t)$.

Table 5.12: Forecast values and observations of the changes in the Dutch unemployment rate

	Observed	Forecast	
	Inflation	Dynamic	Static
2001^{I}	-0.320000	0.059851	0.059851
2001^{II}	-0.270000	0.057914	-0.103136
2001^{III}	0.030000	0.046193	-0.230168
2001^{IV}	-0.010000	0.040523	-0.084896
2002^{I}	0.020000	0.033882	0.006606
2002^{II}	0.210000	0.029016	0.004864
2002^{III}	0.120000	0.024552	0.096267
2002^{IV}	0.170000	0.020900	0.126801
2003^{I}	0.390000	0.017738	0.115462
2003^{II}	0.370000	0.015076	0.226815
2003^{III}	0.120000	0.012805	0.297873
2003^{IV}	0.300000	0.010880	0.184647
2004^{I}	0.380000	0.009242	0.170579
2004^{II}	0.230000	0.007852	0.269575

Figure 5.24: Forecasts of changes in the Dutch unemployment rate

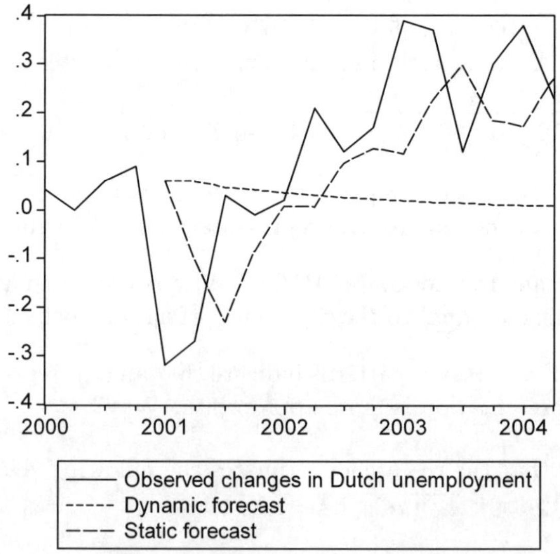

Table 5.13: Forecast values and observations of the Dutch unemployment rate

	Observed	Forecast	
	Inflation	Dynamic	Static
2001^{I}	2.69	3.06985	3.06985
2001^{II}	2.42	2.74791	2.58686
2001^{III}	2.45	2.46619	2.18983
2001^{IV}	2.44	2.49052	2.3651
2002^{I}	2.46	2.47388	2.44661
2002^{II}	2.67	2.48902	2.46486
2002^{III}	2.79	2.69455	2.76627
2002^{IV}	2.96	2.8109	2.9168
2003^{I}	3.35	2.97774	3.07546
2003^{II}	3.72	3.36508	3.57681
2003^{III}	3.84	3.7328	4.01787
2003^{IV}	4.14	3.85088	4.02465
2004^{I}	4.52	4.14924	4.31058
2004^{II}	4.75	4.52785	4.78957

Figure 5.25: Forecasts for levels of the Dutch unemployment rate

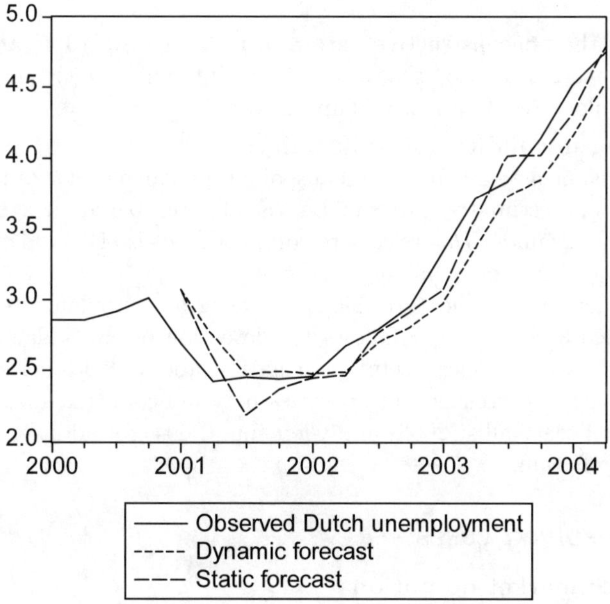

- Suppose the last two quarterly observations (e.g. the 3rd and 4th quarter of 1999) were 550 and 600 respectively, compute predictions for the next two quarters (i.e. 2000-1 and 2000-2).
- Can you say anything sensible about the predicted value for the first quarter of 2025 (and its accuracy)?

2. Density forecasts are evaluated by assessing whether z_t is *iid* $U(0,1)$ and simple tests of *iid* $U(0,1)$ behavior are Kolmogorov-Smirnov tests. But such tests are not very practical since they are not constructive i.e. when rejection occurs, the tests generally provide no guidance as to *why*. If e.g. such a statistic rejects the H_0 of *iid* $U(0,1)$, is it because of violation of unconditional uniformity, violation of *iid*, or both? Moreover even if one knows that rejection comes from a violation of uniformity, one would like to know more:

- What, precisely, is the *nature* of the violation of uniformity, and how important is it?
- Similarly, even knowing that rejection comes from a violation of *iid*, what precisely is its nature?, or
- Is z_t heterogenous but independent, or is z_t dependent? If z_t is dependent, is the dependence operational primarily through the conditional mean, or are higher-ordered conditional moments, such as the variance, relevant?.
- Is the dependence strong and important, or is *iid* an adequate approximation, even if it is strictly false?

Because of the nonconstructive nature of tests of *iid* $U(0,1)$ and the nonconstructive nature of related separate tests of *iid* and $U(0,1)$, more revealing methods of explorative data analysis are needed.

Evaluating unconditional uniformity:
Suggestion: visual assessment by means of graphical density estimates (e.g. histograms). The estimated pdf can be visually compared to a $U(0,1)$ and confidence intervals under H_0 are easy to compute (see also Diebold *et al.* (1998).

Evaluating whether z_t is *iid*:
Visual assessment using the (graphical) correlogram supplemented with the usual confidence intervals. This allows for detection of particular dependence patterns in z_t and supplies useful information about deficiencies of density forecasts (e.g. serial correlation in z_t-series may indicate that mean dynamics are ill modeled); see Wallis (2003) and Clements (2004) for additional empirical issues on UK inflation.

5.10.4 Unsolved cases

Case 1: Stock market quotations

The idea of this exercise is to estimate the best possible model for the four series in the database. The database contains daily data from 03/01/1994 until

31/12/2004 for the stock market quotations of *BASF, Danone, KBC, Telinfo (Telindus)* (Brussels Stock Exchange; Data source: http://finance.yahoo.com/). Before going through the different model selection steps, you should create four series of (logarithmic) returns. Take into account the following steps (compare with the summarizing chart in Figure 5.2 of the Box-Jenkins approach).

1. Detect aberrant observations and, if necessary, correct for them (procedure explained in Subsection 5.8.2).
2. Test for unit roots and, if necessary, correct the data for deterministic or stochastic trends (procedure described at the end of Section 5.5).
3. Identify a candidate model from the ACF and PACF of the series under consideration.
4. Choose an appropriate model and estimate the unknown parameters.
5. Analyze the ACF and PACF of the residuals to test whether the model is adjusted.
6. If not return to step 3.
7. Compare the output of the estimated models, the AICs, BICs, Ljung-Box Q-statistics and the sum of squared residuals.
8. Select the most appropriate model.

Case 2: Gross Domestic Product

Select an as recent as possible time series of quarterly data on real GDP of an OECD member country (except for The Netherlands and New Zealand), e.g. using Datastream or the CD-ROM of the OECD Statistical Compendium. The idea is to use part of the observations, e.g. until the final quarter of 2000, to estimate an ARIMA model or a seasonal ARIMA model. Starting from the most appropriate time series model, estimate 1-step, 4-step and 8-step predictions for the period after 2000 (that is from 2001, first quarter, until e.g. 2006) and compare (part of) the forecast data with the actual ones. Try to minimize the average squared forecast error could be used and compare the predicted values with the actual recent data using AIC and BIC. Do not forget to perform a graphical analysis! Schiff and Phillips (2000) can be used as a guideline.

Chapter 6

Cointegration and Transfer Functions

In this chapter, cointegration and causality tests will be analyzed. Once causality is established, a transfer function model as in (4.1) could be formulated. These three steps will be undertaken in this chapter, divided over three sections.

6.1 Cointegration

Yule (1926) suggested that regressions based on trending time series data can be spurious. This problem of *spurious regressions* was further pursued by Granger and Newbold (1974), which also led to the development of the concept of *cointegration*. The pathbreaking paper is Granger (1981), that was fully recognized only several years later.

To demonstrate the occurrence of cointegration, consider a simple two-equation model used in Engle and Granger (1987). Later, formal definitions and properties will be added.

6.1.1 A simple example

Suppose that two time series y_t and x_t are each integrated of order 1, i.e. they both have one unit root or $y_t, x_t \sim I(1)$, and the corresponding errors are possibly correlated WN errors. From these two series it is assumed that two new DGPs can be built:

$$\begin{aligned} x_t + \beta y_t &= u_t \quad u_t = u_{t-1} + \varepsilon_{1t} \qquad (RW) \\ x_t + \alpha y_t &= e_t \quad e_t = \rho e_{t-1} + \varepsilon_{2t} \; (|\rho| < 1) \end{aligned} \tag{6.1}$$

with $\begin{pmatrix} \varepsilon_{1t} \\ \varepsilon_{2t} \end{pmatrix} \overset{iid}{\sim} N(0, \Sigma)$, where $\Sigma := \begin{pmatrix} \sigma_{11} & \sigma_{12} \\ \sigma_{21} & \sigma_{22} \end{pmatrix}$. This model is internally consistent only if $\alpha \neq \beta$; the reason for this constraint is that if $\alpha = \beta$, it is impossible to find any values of y_t and x_t that satisfy both equations simultaneously in (6.1).

Note that the second equation in (6.1) implies that e_t is $I(0)$. Solving this system for $\alpha \neq \beta$:

$$x_t + \frac{\beta}{\alpha}(e_t - x_t) = u_t \Rightarrow (\alpha - \beta) x_t + \beta e_t = \alpha u_t,$$

the reduced form relations for y_t and x_t satisfy:

$$x_t = \alpha (\alpha - \beta)^{-1} u_t - \beta (\alpha - \beta)^{-1} e_t \tag{6.2}$$

and

$$y_t = -(\alpha - \beta)^{-1} u_t + (\alpha - \beta)^{-1} e_t. \tag{6.3}$$

In general, linear combinations of the integrated variables y_t and x_t also have unit roots. In the above case it is logic that $\{x_t\}$ and $\{y_t\}$ are $I(1)$, since $\{u_t\}$ is an *RW* process and $\{x_t\}$ and $\{y_t\}$ depend linearly on the common $I(1)$ process $\{u_t\}$. However, if there exists some linear combination, say $x_t - \alpha y_t$, that is stationary, x_t and y_t are said to be cointegrated, and $(1\ , -\alpha)'$ is the *cointegrating vector*; $x_t - \alpha y_t$ is then the long run (equilibrium) relationship. Note that if $\rho = 1$, then e_t is also $I(1)$ and thus there is no cointegration. Hence, the null hypothesis $\rho = 1$ is a test of the hypothesis of no cointegration so that the Engle-Granger test is based on the second equation in (6.1).

There is also an autoregressive representation for system (6.1), by rewiting it as

$$\begin{aligned} \Delta x_t &= \beta\theta x_{t-1} + \alpha\beta\theta y_{t-1} + v_{1t} \\ \Delta y_t &= -\theta x_{t-1} - \alpha\theta y_{t-1} + v_{2t}, \end{aligned}$$

where $\theta := \frac{1-\rho}{\alpha-\beta}$ and v_{1t} and v_{2t} are linear combinations of the error terms u_t and e_t. Writing $z_t := x_t + \alpha y_t$ it can be observed from the second equation in (6.1) that $z_t = \rho z_{t-1} + \varepsilon_{2t}$, or $\Delta z_t = (\rho - 1) z_{t-1} + \varepsilon_{2t}$ i.e.,

$$\Delta x_t = \Delta z_t - \alpha \Delta y_t = -\alpha \Delta y_t + (\rho - 1) z_{t-1} + \varepsilon_{2t},$$

which is in the form of an ECM model, as in (4.11), where z_{t-1} represents past disequilibrium.

Now, a formal definition will be presented, as well as the conditions for the existence of cointegration.

6.1.2 Definition and properties

Definition 6.1 *In general, if $\{x_t, t \in \mathbb{Z}\}$ and $\{y_t, t \in \mathbb{Z}\}$ are integrated processes, then, if there exists a linear combination which is integrated of a lower order, both variables are said to be cointegrated.*

The *Granger Representation Theorem* (Granger (1983) and Engle and Granger (1987)) states that if a set of variables are cointegrated, then there exists an error correction representation of the data. Thus, if y_t and x_t are both $I(1)$ and have a cointegrating vector $(1, -\beta)'$, there exists an error correction representation of the form:

$$\alpha(L)\Delta y_t = \delta + \varphi(L)\Delta x_{t-1} - \gamma(y_{t-1} - \beta x_{t-1}) + \theta(L)\varepsilon_t, \tag{6.4}$$

where ε_t is WN with constant variance σ_ε^2 and $\alpha(L)$, $\varphi(L)$ and $\theta(L)$ are polynomials in the lag operator L (with $\alpha_0 =: 1$). Considering the following special case of (6.4):

$$\Delta y_t = \delta + \varphi_1 \Delta x_{t-1} - \gamma(y_{t-1} - \beta x_{t-1}) + \varepsilon_t, \tag{6.5}$$

where the error term has no moving average part and the systematic dynamics are kept as simple as possible. Intuitively, it is clear why the Granger Representation Theorem should hold. If y_t and x_t are both $I(1)$ but have a long run relationship, there must be some force which pulls the equilibrium error back towards zero. The ECM model does exactly this: it describes how y_t and x_t behave in the short run consistently with a long run relationship. If the *cointegrating parameter* β is known, all terms in (6.5) are stationary and no inferential problems arise so that it can be estimated by OLS.

Note that the precise lag structure in (6.4) is not specified by the Granger Representation Theorem, so that some specification analysis will probably be needed. Moreover, the theory is symmetric in its treatment of y_t and x_t, so that there should also exist an ECM model representation with Δx_t as the left-hand side variable. Because at least one of the variables has to adjust to deviations from the long run equilibrium, at least one of the adjustment parameters γ in the two ECM equations has to be nonzero.

The Granger Representation Theorem also holds conversely, i.e. if y_t and x_t are both $I(1)$ and have an ECM model representation, then they are necessarily cointegrated.

Assume e.g. that both the short and the long term interest rates are integrated of order 1, i.e. $r_{s,t}, r_{l,t} \sim I(1)$. Then a simple ECM for the term structure of interest rates exists, with $\varepsilon_{s,t}$ and $\varepsilon_{l,t} \in WN$ and α_s and α_l larger than zero:

$$\begin{aligned}
\underbrace{\Delta r_{s,t}}_{I(0)} &= \alpha_s \underbrace{(r_{l,t-1} - \beta r_{s,t-1})}_{\text{previous period diseq.}=I(0)} + \varepsilon_{s,t} \qquad , \alpha_s > 0 \\
\underbrace{\Delta r_{l,t}}_{I(0)} &= -\alpha_l \underbrace{(r_{l,t-1} - \beta r_{s,t-1})}_{I(0)} + \varepsilon_{l,t} \qquad , \alpha_s < 0
\end{aligned}$$

Note that the ECM term could have been written as $\alpha_s'(\beta_1 r_{l,t-1} - \beta_2 r_{s,t-1})$; normalization w.r.t. the long term interest rate yields $\alpha_s(r_{l,t-1} - \beta r_{s,t-1})$ with $\alpha_s := \alpha_s'\beta_1$ and $\beta := \beta_2/\beta_1$. The long term equilibrium is $r_{l,t} = \beta r_{s,t}$ (with cointegrating vector $[1 \; -\beta]$.

This *basic* model can be further extended with lagged variables of each interest rate as follows:

$$\begin{aligned}
\underbrace{\Delta r_{s,t}}_{I(0)} &= a_{10} + \alpha_s \underbrace{(r_{l,t-1} - \beta r_{s,t-1})}_{prev.\,period\,diseq.=I(0)} + \sum_i a_{11}(i)\,\Delta r_{s,t-i} \\
&\quad + \sum_j a_{12}(j)\,\Delta r_{l,t-j} + \varepsilon_{s,t} \\
\underbrace{\Delta r_{l,t}}_{} &= a_{20} - \alpha_l \underbrace{(r_{l,t-1} - \beta r_{s,t-1})}_{I(0)} + \sum_i a_{21}(i)\,\Delta r_{s,t-i} \\
&\quad + \sum_j a_{22}(j)\,\Delta r_{l,t-j} + \varepsilon_{l,t}.
\end{aligned}$$

Once cointegration has been established, it is of paramount importance to identify the direction of causality of the variables. In this section, no feedback relations will be considered given the univariate analysis framework.

6.2 Causality

Describing the conditional pdf of a stochastic variable y_t given its previous value and the previous ocurrence of another variable by $f(y_t|\, y_{t-1}, x_{t-1})$, *Granger causality* is absent when $f(y_t|\, y_{t-1}, x_{t-1}) = f(y_t|\, y_{t-1})$. Granger causality states that in the conditional pdf lagged values of x_t add no information to the explanation of movements of y_t beyond that provided by lagged values of y_t itself. This concept is useful in the construction of forecasting models.

Formally, Granger's (1969) concept of causality implies that the future cannot predict the past and x causes y if, taking account of prior values of x, better forecasts of y can be obtained. More formally, following Gouriéroux and Monfort (1997):

Definition 6.2 *Defining the information sets* $\Im_y(t) := \{y_t, y_{t-1}, y_{t-2}, \ldots\}$ *and* $\Im_x(t) := \{x_t, x_{t-1}, x_{t-2}, \ldots\}$ *then i)* x *is said to cause* y *at period* t *if and only if* $E(y_t|\, \Im_y(t-1), \Im_x(t-1)) \neq E(y_t|\, \Im_y(t-1))$; *ii)* x *is said to cause* y *instantaneously at period* t *if and only if*
$E(y_t|\, \Im_y(t-1), \Im_x(t)) \neq E(y_t|\, \Im_y(t-1), \Im_x(t-1))$.

Since the variable forecast based on the most recent information is always the best one, the variance of the forecast error from (5.45) satisfies:

$$var(f_t(h)|\, \Im_y(t-1), \Im_x(t-1)) \leqq var(f_t(h)|\, \Im_y(t-1)).$$

This leads to the following theorem of noncausality:

Theorem 6.1 x *does not cause* y *at period* t *if and only if*
$var(f_t(h)|\, \Im_y(t-1), \Im_x(t-1)) = var(f_t(h)|\, \Im_y(t-1))$
and x *does not cause* y *instantaneously at period* t *if and only if*
$var(f_t(h)|\, \Im_y(t-1), \Im_x(t)) = var(f_t(h)|\, \Im_y(t-1), \Im_x(t-1))$.

Thus x causes y at period t if the past of x provides additional information for the forecast of y_t with respect to considering the past of y alone.

There are various tests to determine the (direction of) causality:

- Sims (1972): If (x_t, y_t) is a 2 dimensional time series, then Sims's test is based on the regression of y_t on previous and future values of x_t: $y_t = \sum_{i=n}^{-m} \gamma_{-i} x_{t-i} + v_t$. A test that y does not cause x is an F-test with $H_0 : \gamma_1 = \gamma_2 = ... = \gamma_m = 0$, but there may be serial correlation in the error term v_t (bias in F-test)

- Geweke *et al.* (1983): it is assumed that v_t can be approximated by an $AR\,(p)$ process $y_t = \sum_{j=1}^{p} \phi_{-j} y_{t-j} + \sum_{i=n+p}^{-m} \gamma_{-i} x_{t-i} + u_t$, i.e. a regression of y on p past values of y, m future values of x and $(n+p)$ past values of x with H_0 that y does not cause x, or $H_0 : \gamma_1 = \gamma_2 = ... = \gamma_m = 0$, which can be tested by an ordinary F-test.

6.3 Transfer function modeling

Up to now, univariate time series have been considered and trends and intercepts have been treated as explanatory variables, besides own lags. One can, of course, also include other *exogenous* variables. If these exogenous variables take the form of dummies we often talk of *intervention models*. If these exogenous variables are other variables than deterministic dummies, they are called *transfer function models*, e.g. an $ARMA(p,q)$ model can be extended to $y_t = \alpha_0 + \alpha\,(L)\, y_{t-1} + \gamma(L) z_t + \varphi\,(L)\,\varepsilon_t$, $\varepsilon_t \in WN$ and $t \in \mathbb{Z}$. The polynomial $\gamma(L) := \gamma_0 + \gamma_1 L + \gamma_2 L^2 + \ldots$ 'transfers' movement in $\{z_t\}$ into the $\{y_t\}$ process but not *vice versa*, i.e. there is no feedback or reverse effect of $\{y_t\}$ on $\{z_t\}$!

A transfer function can also have more than one exogenous variable as in model (4.1). Hence, a distinction can be made between Single Output Single Input ($SOSI$) (transfer function) models and Single Output Multiple Input ($SOMI$) (transfer function) models. $SOMI$ models are a straightforward (vector) extension of $SOSI$ models; therefore the focus in this section is on $SOSI$ models.

6.3.1 Single output single input models

As the basic model, consider the following $SOSI$ model:

$$y_t = a_0 + \alpha\,(L)\, y_{t-1} + \gamma(L) z_t + \varphi\,(L)\,\varepsilon_t, \tag{6.6}$$

where, as also for $ARIMA$ models, it is assumed that the error terms are WN, or $\varepsilon_t \in WN$, and $\gamma\,(L)$ is a transfer function polynomial $\gamma\,(L) = \gamma_0 + \gamma_1 L + \gamma_2 L^2 + ...$, where the γ_is are the transfer function weights; z_t is stochastic, but there is no feedback of $\{y_t\}$ to $\{z_t\}$ and $\{z_t\}$ is also exogenous with $E\,(z_t \varepsilon_{t-s}) = 0$, $\forall s, t$. If $\gamma_0 = 0$, there is no instantaneous effect of $\{z_t\}$ on $\{y_t\}$ and $\{z_t\}$ is then called a leading indicator. The exogenous variable $\{z_t\}$ can also be used to predict future values of $\{y_t\}$.

Definition 6.3 *Cross (auto)correlations between y_t and various z_{t-i} are defined as $\rho_{yz}(i) = \frac{cov(y_t, z_{t-i})}{\sigma_y \sigma_z}$, which give rise to a cross autocorrelation function (CACF) or a cross-correlogram $\{\rho_{yz}(i)\}$.*

Example 6.1 *Consider the following example $y_t = \alpha_1 y_{t-1} + \gamma_d z_{t-d} + \varepsilon_t$, with z_t and $\varepsilon_t \in WN$, $E(z_t \varepsilon_{t-i}) = 0$, $|\alpha_1| < 1$*

From this example, it follows that using L and I operators:

$$\begin{aligned} y_t &= \frac{\gamma_d z_{t-d}}{I - \alpha_1 L} + \frac{\varepsilon_t}{I - \alpha_1 L} \\ &= \gamma_d \left(z_{t-d} + \alpha_1 z_{t-d-1} + \alpha_1^2 z_{t-d-2} + ... \right) + \frac{\varepsilon_t}{I - \alpha_1 L}, \end{aligned}$$

so that the Yule-Walker equations can be constructed from the cross-products as:

$$\begin{aligned} y_t z_t &= \gamma_d \left(z_t z_{t-d} + \alpha_1 z_t z_{t-d-1} + \alpha_1^2 z_t z_{t-d-2} + ... \right) + \frac{z_t \varepsilon_t}{I - \alpha_1 L} \\ y_t z_{t-1} &= \gamma_d \left(z_{t-1} z_{t-d} + \alpha_1 z_{t-1} z_{t-d-1} + \alpha_1^2 z_{t-1} z_{t-d-2} + ... \right) + \frac{z_{t-1} \varepsilon_t}{I - \alpha_1 L} \\ \vdots &\quad \vdots \qquad\qquad \vdots \qquad\qquad\qquad\qquad \vdots \\ y_t z_{t-d-2} &= \gamma_d \left(z_{t-d-2} z_{t-d} + \alpha_1 z_{t-d-2} z_{t-d-1} + \alpha_1^2 z_{t-d-2} z_{t-d-2} + ... \right) + \frac{z_{t-d-2} \varepsilon_t}{I - \alpha_1 L} \end{aligned}$$

Now, taking expectations, $E(y_t z_t) = 0$, $E(y_t z_{t-1}) = 0$, ..., $E(y_t z_{t-d+1}) = 0$, $E(y_t z_{t-d}) = \gamma_d \sigma_z^2$, $E(y_t z_{t-d-1}) = \gamma_d \alpha_1 \sigma_z^2$, $E(y_t z_{t-d-2}) = \gamma_d \alpha_1^2 \sigma_z^2$, or:

$$\begin{aligned} E(y_t z_{t-i}) &= 0 \qquad\qquad (i < d) \\ &= \gamma_d \alpha_1^{i-d} \sigma_z^2 \ (i \geqq d) \end{aligned}$$

and $\rho_{yz}(i) = \frac{E[y_t z_{t-i}]}{\sigma_y \sigma_z}$.

Properties

1. $\rho_{yz}(i) = 0$ until the first nonzero element in $\gamma(L)$;
2. a_0 and $\varphi(L)\varepsilon_t$ in the basic model do not affect the cross-correlogram as $E(z_{t-i}\varepsilon_t) = 0 \; \forall t, i$;
3. a spike at lag d means that there is an impact of z_{t-d} on the endogenous variable y_t;
4. the decay pattern is dictated by $\alpha(L)$;
5. usually, however, $z_t \notin WN$, but has an $ARMA$ specification according to $\delta(L) z_t = \phi(L) \varepsilon_{zt}$, where $\varepsilon_{zt} \in WN$. Then, *prewhitening* the basic model, i.e. filtering the output variable y_t with the $ARMA$ input filter $\delta(L) z_t = \phi(L) \varepsilon_{zt}$ so that the output variable is not (necessarily) 'whitened', yields

from multiplying each side of the basic *SOSI* model (6.6) by the ratio $\frac{\delta(L)}{\phi(L)}$:

$$\underbrace{\frac{\delta(L)\,y_t}{\phi(L)}}_{y_{f,t}} = \alpha_0 \frac{\delta(L)}{\phi(L)} + \underbrace{\frac{\delta(L)\,\alpha(L)}{\phi(L)} y_{t-1}}_{y_{f,t-1}} + \underbrace{\frac{\delta(L)\,\gamma(L)}{\phi(L)} z_t}_{z_{f,t}=\varepsilon_{z,t}} + \frac{\delta(L)\,\varphi(L)}{\phi(L)} \varepsilon_t, \quad (6.7)$$

where $\frac{\delta(L)\gamma(L)}{\phi(L)}$ are the *impulse response weights* or the *transfer function weights*; note that y_t and $\gamma(L)\,z_t$ have the same correlograms as $y_{f,t}$ and $\gamma(L)\,\varepsilon_{z,t}$.

Multiplying both sides of (6.7) by $\varepsilon_{z,t-k}$ and taking expectations one obtains $\gamma_{y_f\varepsilon_z}(k) = \cdots \vartheta_k \sigma^2_{\varepsilon_z}$, or the transfer function weights are determined by:

$$\vartheta_k = \rho_{y_f\varepsilon_z}(k)\frac{\sigma_{y_f}}{\sigma_{\varepsilon_z}}. \quad (6.8)$$

Identification and estimation of the transfer function model

The following identification rules should be followed:

1. Fit *ARMA*: $\delta(L)\,z_t = \phi(L)\,\varepsilon_{z,t}$ and obtain $\widehat{\varepsilon}_{z,t}$;

2. Check the cross-correlogram $\left\{\frac{\widehat{\delta}(L)}{\widehat{\phi}(L)} y_t = \widehat{y}_{f,t}, \widehat{\varepsilon}_{z,t}\right\}$, spikes are an indicator of a nonzero $\gamma(L)$ element and a decay pattern of $\alpha(L)$ coefficients prompt to a search for the best fitting $(I - \alpha(L))\,y_t = \gamma(L)\,z_t + e_t$;

3. Use e_t to obtain a first estimate of $\phi(L)$;

4. Estimate $y_t = a_0 + \alpha(L)\,y_{t-1} + \gamma(L)\,z_t + \phi(L)\,\varepsilon_t$ by nonlinear least squares (i.e. minimizing $\sum \varepsilon_t^2$ in (6.7) with respect to the unknown parameters) or by nonlinear maximum likelihood (i.e. maximizing a loglikelihood function in the unknown parameters).

6.3.2 Intervention analysis

Interventions can affect a time series in several ways: they can change the level, either abruptly or after some delay, change the trend, or lead to other, more complicated, response patterns. Following Box and Tiao (1975) breaks in the mean or in the trend are represented as a consequence of exogenous shocks on the time series. These shocks may have a permanent or a transitory effect. Starting from the general transfer function model as in (4.1):

$$y_t = \textstyle\sum_{i=1}^{k} \gamma_i(L)\,x_{it} + \eta_t \quad (6.9)$$

with $\gamma_i(L) := \frac{\omega_i(L)}{v_i(L)}$, where η_t is an *ARIMA*(p,d,q) process and $\omega_i(L)$ and $v_i(L)$ are polynomials in the lag operator L. An *intervention variable* x_{it} can assume one of the following three characterizations:

1. $x_{it} = \begin{cases} 0 \text{ if } t < T_i \\ 1 \text{ if } t \geqq T_i \end{cases}$ so that x_{it} is a *jump (or a step) variable* used to describe the influence of a phenomenon starting at T_i (e.g. a change in regulations);

2. $x_{it} = \begin{cases} 0 \text{ if } t \neq T_i \\ 1 \text{ if } t = T_i \end{cases}$ so that x_{it} is an *(im)pulse variable* used to describe the influence on y_t of a phenomenon which happened just at period T_i (e.g. a strike);

3. $x_{it} = \begin{cases} 0 \text{ if } t < T_{1i} \text{ or } t > T_{2i} \\ 1 \text{ if } T_{1i} \leqq t \leqq T_{2i} \end{cases}$ so that x_{it} is a *step variable* used to describe the influence on y_t of a transitory phenomenon happening between T_{1i} and T_{2i} (e.g. a temporary change in regulations).

When the form of the polynomials $\omega_i(L)$ and $v_i(L)$ is chosen, i.e. when their degrees are selected, and possibly the number of unit roots in $v_i(L)$ is detected, then one can estimate the various parameters by OLS or by ML. Transforming the elementary functions in 1. and 2. by rational polynomials $\frac{\omega_i(L)}{v_i(L)}$, a sizable set of appropiate set of functions is obtained.

Various choices for intervention models for different $\gamma_i(L)$ polynomials are presented in Figure 6.1.

Figure 6.1: Various intervention models for $\gamma_i(L) := \frac{\omega_i(L)}{v_i(L)}$

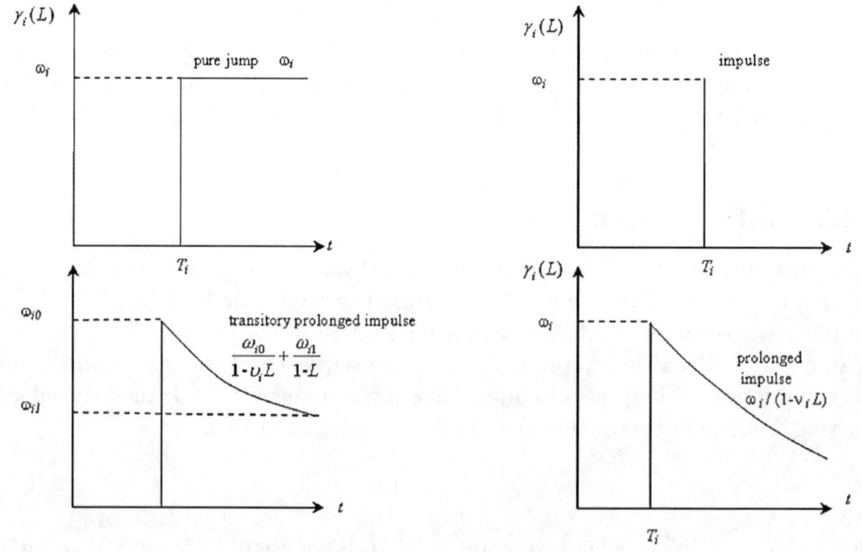

6.3.3 Theoretical and empirical examples and exercises

Example 6.2 *Consider the time series processes* $y_t = \varepsilon_t + a\eta_t + b\eta_{t-1}$ *and* $x_t = \eta_t$*, where* $\varepsilon_t, \eta_t \in WN$*, then*

$$E(y_t | \Im_y(t-1), \Im_x(t-1)) = b\eta_{t-1} = bx_{t-1}.$$

Since this expression involves x_{t-1} it is immediately observed that x causes y at period t if and only if $b \neq 0$. Moreover, $E(y_t | \Im_y(t-1), \Im_x(t) = a\eta_t + b\eta_{t-1} = ax_t + bx_{t-1}$, so that there is instantaneous causality of x to y at period t if and only if $a \neq 0$. This example involves two expressions of causality and it is concluded that these two concepts of causality have no link, since the parameters a and b are not functionally related.

Example 6.3 *As a theoretical example, consider a transfer function similar to (6.6):*

$$y_t = a_0 + a_1 y_{t-1} + c_0 z_t + \varepsilon_t$$

where z_t is now an intervention dummy and c_0 the impact effect for intervention.

According to Figure 6.1 a distinction can be made between a temporary change (pulse) and a permanent change (step), with $\varepsilon_t \in WN$, $|a_1| < 1$.

The long run effect of intervention is now $\frac{a_0+c_0}{1-a_1} - \frac{a_0}{1-a_1} = \frac{c_0}{1-a_1}$ and the transitional effects of intervention are $(I - a_1 L)\, y_t = a_0 + c_0 z_t + \varepsilon_t$, or $y_t = \frac{a_0}{1-a_1} + c_0 \sum_{i=0}^{\infty} a_1^i z_{t-i} + \sum_{i=0}^{\infty} a_1^i \varepsilon_{t-i}$.

Hence, the *impulse responses* are equal to $\frac{dy_t}{dz_t} = c_0$ and $\frac{dy_t}{dz_{t-i}} = \frac{dy_{t+i}}{dz_t}$ and $z_{t+i} = z_t = 1\ \forall i > 0$.

Similarly, $\frac{dy_{t+1}}{dz_t} = \frac{dy_t}{dz_{t-1}} = c_0 + c_0 a_1$ and $\frac{dy_{t+j}}{dz_t} = \frac{dy_t}{dz_{t-j}} = c_0\left(1 + a_1 + a_1^2 + \cdots + a_1^j\right)$ so that for $j \to \infty$, $\frac{dy_{t+j}}{dz_t} = \frac{c_0}{1-a_1}$.

Example 6.4 *Bilateral exchange rates based on Uncovered Interest rate Parity (UIP) hypothesis (Data source: OECD Statistical Compendium).*

Consider now a UIP model of bilateral exchange rates where a bilateral exchange rate is assumed to change according to the interest rate differential between two countries. The underlying model is:

$$s_t = \alpha_0 + \alpha(L) s_{t-1} + \gamma(L)(r - r^*)_t + \varphi(L)\varepsilon_t. \tag{6.10}$$

In this example the data on the USD/EURO exchange rate is analyzed and the (overnight) interest rate difference between the United States and the EMU short term interest rate.[1]

First, the stationarity behavior of the input variable will be investigated. The ADF statistics (see Section 5.5) of the interest rate differential variable indicate mixed results: a weak acceptance $(P(|\hat{\tau}| > \tau(\alpha)) = 0.07)$ of the null of

[1] Since US dollar-euro nominal exchange rate data is in the neighborhood of 1, logarithms were not taken.

nonstationarity in the (probable) case of zero mean and a strong acceptance in a (non-probable) cases of a nonzero mean or a trend.

Therefore, first a detection of an appropriate $ARMA$ model for the input is undertaken: the interest rate differential is conveniently described by an $AR(1,3)$ process (estimated by ML with standard errors between brackets):

$$\widehat{(r-r^*)}_t = \underset{(0.918)}{-0.3753} + \underset{(0.065)}{1.2653}(r-r^*)_{t-1} - \underset{(0.065)}{0.2829}(r-r^*)_{t-3}. \quad (6.11)$$

Although this is a stable third order AR process (the eigenvalues of the matrix A as in Appendix C all lie within the unit circle: $(0.955, 0.721, -0.411)$) and there is no significant residual (partial) autocorrelation, the estimated coefficients in (6.11) are not very reliable since the $AR(1)$ and $AR(3)$ coefficients are highly correlated ($\hat{\rho}_{\hat{\alpha}_1,\hat{\alpha}_3} = -0.982$). Hence, an $AR(1)$ model should be analyzed.

Since the $AR(1)$ residuals showed significant autocorrelation at lag 2, the following final input filter was selected (estimated by ML without an intercept term):

$$(I - \underset{(0.016)}{0.9830}L)\widehat{(r-r^*)}_t = (I + \underset{(0.113)}{0.3491}L^2)\hat{\varepsilon}_t. \quad (6.12)$$

There is no residual (partial) autocorrelation left and the residual variance is estimated as $\hat{\sigma}^2_{\hat{\varepsilon}} = 0.0356$. The $AR(1)$ and $MA(2)$ parameters are not correlated either. Therefore, (6.12) is selected as the input filter with which the bilateral nominal USD/EURO exchange rate has to be prewhitened.

But we find that the output variable has one unit root ($P(|\hat{\tau}| < \tau(\alpha)) = 0.93$ with zero mean and, applied on the exchange rate differences, $P(|\hat{\tau}| < \tau(\alpha)) < 0.01$), so that before estimating the final model, we must take the first differences of the nominal exchange rate.

After prewhitening we calculate the crosscorrelations between the output and the input variables and we observe that there is a (slightly) significant crosscorrelation at lag 6 so that during the analyzed sample period the interest rate differentials seem to have an impact on the nominal USD/EUR exchange rate after 6 months.

The output variabale can best be modeled by an $ARMA(3,1)$ process. Finally we obtain the following transfer function specification (following (6.10)) by ML:

$$\Delta\hat{s}_t^{USD/EUR} = \underset{(0.0028)}{0.00485} - \underset{(0.126)}{0.2718}\Delta s_{t-3}^{USD/EUR} + \underset{(0.0017)}{0.00575}(r^{EMU} - r^{US})_{t-6} + \hat{u}_t + \underset{(0.123)}{0.2597}\hat{u}_{t-1}.$$

The residual variance $\hat{\sigma}^2_{\hat{u}} = 0.00053$, there is no correlation among the estimated parameters, and the resulting residuals show no significant (partial) autocorrelation. We conclude that the UIP hypothesis is not rejected for the USD/EUR exchange rate with a 6 months lag in the observed sample period.

Exercise 6.1 *Consider the bivariate cointegrated system:*

$$\begin{aligned} x_t &= \mu + x_{t-1} + \varepsilon_t \\ y_t &= \lambda x_t + v_t \end{aligned}$$

where the stochastic disturbances are assumed to follow mutually independent WN processes with variances σ_ε^2 and σ_v^2, respectively.

- Write the one-step (ahead) and, generally, the h-step (ahead) forecasts with the corresponding forecast error variances of the implied univariate representations for the variables x and y.
- Write the one-step (ahead) and, generally, the h-step (ahead) forecasts with the corresponding forecast error variances of the above bivariate cointegrated system.
- Compare the forecast error variances based on the univariate representation, discuss this forecast accuracy comparison.

Exercise 6.2 *i) Show that x may cause y if it does not cause (y, z); ii) show that (x, z) does not cause y if and only if x does not cause y and z does not cause y; ii) consider the time series processes $y_t := \varepsilon_t + \eta_t$, $x_{t-1} := \varepsilon_t - \eta_t$, $z_{t-1} := \eta_t$, where ε_t, $\eta_t \in WN$ with the same variance; verify that it is possible that x causes (y, z) ,even if it does not cause either y or z.*

Chapter 7

Multivariate Time Series

In the previous chapters we considered the explanation of a single (economic) time series, possibly by other (economic).time series, but without any feedback from the former to the latter ones. For several reasons (estimation and forecast efficiency, the nature of the economic problem itself) it may be necessary to consider several time series simultaneously. Therefore, in this chapter the dynamics of the various (economic) time series are really interlinked, which is often the case in economic practice. In this multivariate time series analysis the following principles are taken into consideration:

- always keep parsimony
- restrictions needed to increase statistical power
- feedback should be studied (which is not present in transfer functions).

In general a multivariate time series model in inputs $\mathbf{z}_t$ and outputs $\mathbf{y}_t$ can be written as:

$$\mathbf{B}(L)\,\mathbf{y}_t + \mathbf{\Gamma}(L)\,\mathbf{z}_t + \mathbf{\Phi}(L)\,\boldsymbol{\varepsilon}_t = 0, \tag{7.1}$$

where the outputs $\{\mathbf{y}_t, t \in \mathbb{Z}\}$, which are often called *endogenous variables*, and the errors $\{\boldsymbol{\varepsilon}_t, t \in \mathbb{Z}\}$ are elements of $\mathbb{R}^n$ and the inputs $\{\mathbf{z}_t, t \in \mathbb{Z}\}$, which are often called *exogenous variables*, are elements of $\mathbb{R}^m$. Further, $\mathbf{B}(L)$, $\mathbf{\Gamma}(L)$, and $\mathbf{\Phi}(L)$ are appropiate matrix polynomials in the lag operator L:

$$\mathbf{B}(L){:=}\mathbf{B}_0 - \textstyle\sum_{i=1}^{p} \mathbf{B}_i L^i,\ \mathbf{\Gamma}(L){:=}\sum_{i=0}^{k} \mathbf{\Gamma}_i L^i, \text{ and } \mathbf{\Phi}(L){:=}\sum_{i=0}^{q} \mathbf{\Phi}_i L^i$$

with $\mathbf{\Phi}_0{:=}\mathbf{I}_n$.

The $n\times n$ matrix $\mathbf{B}_0$ is supposed to be nonsingular so that system (7.1) allows for a unique determination of the current values of the endogenous variables.

In order to illustrate the multivariate time series modeling as in (7.1), following Gouriéroux and Monfort (1997), a simplified Keynesian model will be presented.

Example 7.1 *Consider the following simple Keynesian system of three equations:*

$$\begin{aligned} GDP_t &= C_t + I_t + G_t \\ C_t &= c_1 GDP_{t-1} + \varepsilon_{1t} \\ I_t &= c_2 \left(GDP_{t-1} - GDP_{t-2}\right) + \varepsilon_{2t}, \end{aligned}$$

where the endogenous variables are gross domestic product (GDP), consumption (C), and investment (I), and the government expenditure (G_t) is the only exogenous variable. The parameters are c_1 (the marginal propensity to consume; $0 < c_1 < 1$) and c_2 (the marginal propensity to invest in a period of growth; $c_2 > 0$).

Substituting the GDP identity in the second and third equations and rewriting the resulting system in vector form:

$$\begin{bmatrix} C_t \\ I_t \end{bmatrix} = \begin{bmatrix} c_1 & c_1 \\ c_2 & c_2 \end{bmatrix} \begin{bmatrix} C_{t-1} \\ I_{t-1} \end{bmatrix} + \begin{bmatrix} 0 & 0 \\ -c_2 & -c_2 \end{bmatrix} \begin{bmatrix} C_{t-2} \\ I_{t-2} \end{bmatrix} + \begin{bmatrix} c_1 \\ c_2 \end{bmatrix} G_{t-1} + \begin{bmatrix} 0 \\ -c_2 \end{bmatrix} G_{t-2} + \begin{bmatrix} \varepsilon_{1t} \\ \varepsilon_{2t} \end{bmatrix},$$

which is a simple special case of system (7.1) with $p = 2$, $k = 2$, and $q = 0$.

If $\mathbf{\Gamma}(L) = \mathbf{0}$ and $\mathbf{\Phi}(L) = \mathbf{0}$ in model (7.1), a pure vector autoregressive (VAR) model of the order p results.

If $\mathbf{\Gamma}(L) = \mathbf{0}$, a vector autoregressive moving average ($VARMA$) model of the order (p, q) results. If there are integrated variables, a vector autoregressive integrated moving average ($VARIMA$) model results.

In this chapter, VAR models, $VAR(I)MA$ models and multivariate (structural) time series models as in (7.1) are analyzed.

7.1 Vector autoregressive models

7.1.1 A simple bivariate $VAR(1)$

A vector autoregressive (VAR) model describes the dynamic evolution of a number of variables from their own history. Following Enders (1995) VAR models will be introduced using a simple example of a first order VAR.

Example 7.2 *Consider, e.g., the following bivariate $VAR(1)$ with 10 structural parameters, where both variables affect each other's time path (e.g. the relation between advertising expenditures and sales):*

$$x_t = b_{10} - b_{12} z_t + c_{11} x_{t-1} + c_{12} z_{t-1} + \varepsilon_{xt} \quad (7.2)$$

$$z_t = b_{20} - b_{21} x_t + c_{21} x_{t-1} + c_{22} z_{t-1} + \varepsilon_{zt} \quad (7.3)$$

with $x_t, z_t \in I(0)$, $\varepsilon_{xt}, \varepsilon_{zt} \in WN$ with variances σ_x^2 and σ_z^2 and covariance $cov(\varepsilon_{xt}, \varepsilon_{zt}) = 0$.

A structural and reduced form bivariate $VAR(1)$

Rewriting this $VAR(1)$ in *structural form*:

$$\mathbf{B}\mathbf{y}_t = \mathbf{c}_0 + \mathbf{C}_1\mathbf{y}_{t-1} + \varepsilon_t, \tag{7.4}$$

where $\mathbf{B} := \begin{pmatrix} 1 & b_{12} \\ b_{21} & 1 \end{pmatrix}, \mathbf{y}_t := \begin{pmatrix} x_t \\ z_t \end{pmatrix}, \mathbf{c}_0 := \begin{pmatrix} b_{10} \\ b_{20} \end{pmatrix}, \mathbf{C}_1 := \begin{pmatrix} c_{11} & c_{12} \\ c_{21} & c_{22} \end{pmatrix}$, $\varepsilon_t := (\varepsilon_{x,t}, \varepsilon_{z,t})'$, so that the *standard or reduced form* $VAR(1)$ model of the above structural form $VAR(1)$ satisfies (provided that the structural matrix $\mathbf{B}$ is a regular matrix):

$$\mathbf{y}_t = \mathbf{a}_0 + \mathbf{A}_1\mathbf{y}_{t-1} + \mathbf{e}_t, \tag{7.5}$$

with $\mathbf{a}_0 := \mathbf{B}^{-1}\mathbf{c}_0 = (a_{10}, a_{20})'$, $\mathbf{A}_1 := \mathbf{B}^{-1}\mathbf{C}_1$, and $\mathbf{e}_t := \mathbf{B}^{-1}\varepsilon_t = (e_{1t}, e_{2t})$. Hence, (7.5) can be written as:

$$x_t = a_{10} + a_{11}x_{t-1} + a_{12}z_{t-1} + e_{1t} \tag{7.6}$$

$$z_t = a_{20} + a_{21}x_{t-1} + a_{22}z_{t-1} + e_{2t} \tag{7.7}$$

with $e_{1t} := \frac{\varepsilon_{xt} - b_{12}\varepsilon_{zt}}{1 - b_{12}b_{21}}$ and $e_{2t} := \frac{\varepsilon_{zt} - b_{21}\varepsilon_{xt}}{1 - b_{12}b_{21}}$, so that $E(e_{it}) = 0$ for $i = 1, 2$ and $var(e_{1t}) = \frac{(\sigma_x^2 + b_{12}^2\sigma_z^2)}{(1 - b_{12}b_{21})^2}$, $var(e_{2t}) = \frac{(\sigma_z^2 + b_{21}^2\sigma_x^2)}{(1 - b_{12}b_{21})^2}$ for $t = 1, 2, ..., T$.

The covariance terms are computed as:

$$E(e_{1t}e_{1,t-\tau}) = E\left\{\frac{(\varepsilon_{xt} - b_{12}\varepsilon_{zt})(\varepsilon_{x,t-\tau} - b_{12}\varepsilon_{z,t-\tau})}{(1 - b_{12}b_{21})^2}\right\} = 0 \text{ for } \tau \neq 0 \text{ and}$$

$$E(e_{1t}e_{2t}) = E\left\{\frac{(\varepsilon_{xt} - b_{12}\varepsilon_{zt})}{(1 - b_{12}b_{21})}\frac{(\varepsilon_{zt} - b_{21}\varepsilon_{xt})}{(1 - b_{12}b_{21})}\right\} = \frac{-(b_{21}\sigma_x^2 + b_{12}\sigma_z^2)}{(1 - b_{12}b_{21})^2} = cov(e_{1t}, e_{2t}),$$

so that the reduced form shocks are found to be contemporaneously correlated unless $b_{12} = b_{21} = 0$ in the (very) unlikely case that there are no contemporaneous effects of z_t on x_t *and* of x_t on z_t.

The reduced form residual variance-covariance matrix is defined as $\mathbf{\Sigma} := \begin{pmatrix} \sigma_1^2 & \sigma_{12} \\ \sigma_{21} & \sigma_2^2 \end{pmatrix}$ with $\sigma_i^2 := var(e_{it})$, $i = 1, 2$; and $\sigma_{12} = \sigma_{21} = cov(e_{1t}, e_{2t})$.

Stationarity and stability of a bivariate $VAR(1)$

Joint stationarity of the $\{x_t, t \in \mathbb{Z}\}$ and $\{z_t, t \in \mathbb{Z}\}$ processes requires that the solutions (roots) z of $|\mathbf{I}_2 - \mathbf{A}_1 z| = 0$ lie *outside* the unit circle, which means that the two roots of $(1 - a_{11}z)(1 - a_{22}z) - a_{21}a_{12}z^2 = 0$ lie outside the unit circle.[1] Another way of looking at this stationarity issue is through explicitly solving the reduced form relations (7.6) and (7.7), which can be made by rewriting these equations in operator form: $(I - a_{11}L)x_t = a_{10} + a_{12}Lz_t + e_{1t}$ and $(I - a_{22}L)z_t = a_{20} + a_{21}Lx_t + e_{2t}$ with I the identity operator ($Ix_t := x_t$) and L the lag or

[1] This stationarity condition is equivalent to the condition that the characteristic roots λ_i, $i = 1, 2$ of $|\mathbf{A}_1 - \lambda\mathbf{I}_2| = \lambda^2 - (a_{11} + a_{22})\lambda + a_{11}a_{22} - a_{12}a_{21} = 0$. are lying *within* the unit circle.

backward operator ($Lx_t := x_{t-1}$), so that $z_t = \frac{a_{20}+a_{21}Lx_t+e_{2t}}{(I-a_{22}L)}$ and replacing for Lz_t in the first equation in operator form:

$$(I - a_{11}L)\, x_t = a_{10} + a_{12}L \left[\frac{a_{20}+a_{21}Lx_t+e_{2t}}{(I-a_{22}L)}\right] + e_{1t},$$ [2]

which leads to:

$$x_t = \frac{a_{10}\,(1 - a_{22}) + a_{12}a_{20} + (I - a_{22}L)\, e_{1t} + a_{12}e_{2,t-1}}{(I - a_{11}L)\,(I - a_{22}L) - a_{12}a_{21}L^2} \tag{7.8}$$

and similarly for z_t:

$$z_t = \frac{a_{20}\,(1 - a_{11}) + a_{21}a_{10} + (I - a_{11}L)\, e_{2t} + a_{21}e_{1,t-1}}{(I - a_{11}L)\,(I - a_{22}L) - a_{12}a_{21}L^2}, \tag{7.9}$$

which are infinite lag polynomials, converging under the condition of stability (stationarity). As also argued before convergence requires that the (real or complex) roots of the second order polynomial corresponding to the denominator of (7.8) and (7.9), $(1 - a_{11}z)\,(1 - a_{22}z) - a_{21}a_{12}z^2 = 0$, lie outside the unit circle. This is called the *stability condition*, which is also verified from direct recursive substitution in (7.5):

$$\mathbf{y}_t = \mathbf{a}_0 + \mathbf{A}_1(\mathbf{a}_0 + \mathbf{A}_1\mathbf{y}_{t-2} + \mathbf{e}_{t-1}) + \mathbf{e}_t = \cdots = (\sum_{i=0}^{n} \mathbf{A}_1^i)\mathbf{a}_0 + \sum_{j=0}^{n} \mathbf{A}_1^j\mathbf{e}_{t-j} + \mathbf{A}_1^{n+1}\mathbf{y}_{t-n-1}$$

with $\lim_{n\to\infty} \mathbf{A}_1^{n+1} = 0$, or in that case

$$\mathbf{y}_t = \boldsymbol{\mu} + \sum_{j=0}^{\infty} \mathbf{A}_1^j\mathbf{e}_{t-j}, \text{ with } \boldsymbol{\mu} := (\textstyle\sum_{i=0}^{\infty} \mathbf{A}_1^i)\mathbf{a}_0 = (\mathbf{I}_2 - \mathbf{A}_1)^{-1}\mathbf{a}_0,$$

so that, if the stability condition is satisfied, the particular solution for $\mathbf{y}_t$ is $\mathbf{y}_t = \boldsymbol{\mu} + \sum_{j=0}^{\infty} \mathbf{A}_1^j\mathbf{e}_{t-j}$. Rewriting this particular solution in terms of shocks on x_t and z_t as $\mathbf{y}_t = \boldsymbol{\mu} + \sum_{j=0}^{\infty} \boldsymbol{\varphi}_{jk}(i)\mathbf{e}_{t-j}$,[3] where $\boldsymbol{\varphi}_{jk}(0)$ is the instantaneous effect of a unit change in the shock $\mathbf{e}_t$ or the *impact multiplier*, $\boldsymbol{\varphi}_{jk}(i)$ are the impulse response weights or the *delay multipliers* for period $i > 0$, $\sum_{i=0}^{\tau} \boldsymbol{\varphi}_{jk}(i)$ are the *cumulative multipliers* for τ periods and $\sum_{i=0}^{\infty} \boldsymbol{\varphi}_{jk}(i)$ is the *long run multiplier*.

Estimation and identification in a bivariate $VAR(1)$

Direct estimation of the structural bivariate $VAR(1)$ model (7.4) raises problems. Indeed, due to the feedback inherent in this system ($b_{12} \neq 0$ and $b_{21} \neq 0$), these equations cannot be estimated directly by OLS. The reason is that z_t is correlated with the error term ε_{xt} and x_t with the error term ε_{zt}.[4] Since such a

[2] Hence, we have transformed the first order VAR in the $\{x_t, t \in \mathbb{Z}\}$ and $\{z_t, t \in \mathbb{Z}\}$ sequences in (7.5) into a second order stochastic difference equation in the $\{x_t, t \in \mathbb{Z}\}$ sequence.

[3] Recall that $\mathbf{e}_t := \left(\frac{\varepsilon_{xt}-b_{12}\varepsilon_{zt}}{1-b_{12}b_{21}}, \frac{\varepsilon_{zt}-b_{21}\varepsilon_{xt}}{1-b_{12}b_{21}}\right)'$.

[4] A simultaneous equations or a GMM estimator can resolve this problem (see *infra*).

problem does not exist for the standard or reduced form $VAR(1)$ model (7.5), OLS can provide (consistent) estimates of the two unknown parameters in $\mathbf{a}_0$ and the four unknown parameters in $\mathbf{A}_1$. Moreover, by obtaining the OLS residuals from the two reduced form regressions (7.6) and (7.7), it is possible to calculate estimates of $var\,(e_{1t})$, $var\,(e_{1t})$, and $cov\,(e_{1t}, e_{2t})$. Hence, while the structural form (7.4) has 10 unknowns (unknown parameters), the standard or reduced form (7.5) has only 9 unknown parameters (6 regression coefficients and 3 parameters in $\mathbf{\Sigma}$), so that the structural form $VAR(1)$ parameters are not identifiable or estimable from the reduced form $VAR(1)$ parameters and (an) identifying restriction(s) is (are) needed to be imposed to recover the structural form parameters from the reduced form parameters. Adding one restriction, exact identification results; adding more restrictions, overidentification. Following Sims (1980) a *recursive* structure may be imposed by restricting that $b_{21} = 0$ in the original structural bivariate $VAR(1)$ equation (7.3), which implies that x_t has no contemporaneous effect on z_t, only a lagged effect![5] Imposing this restriction the structural form bivariate $VAR(1)$ in (7.4) is identified since the error terms of the reduced form $VAR(1)$ (7.5) then become $e_{1t} = \varepsilon_{xt} - b_{12}\varepsilon_{zt}$ and $e_{2t} = \varepsilon_{zt}$, so that both structural form shocks ε_{xt} and ε_{zt} affect x_t contemporaneously, but z_t is only affected by ε_{zt}.[6] Decomposing the residuals in this triangular fashion is called a *Cholesky decomposition.* Although this decomposition constrains the system such that an ε_{xt} shock has no direct effect on z_t, there is an indirect effect in that lagged values of x_t affect the contemporaneous value of z_t. The key point is that the Cholesky decomposition imposes a potentially important asymmetry on the system since an ε_{zt} shock has contemporaneous effects on both x_t and z_t, so that the Cholesky decomposition implies an *ordering* of the variables: z_t is 'prior' to or predetermined with respect to x_t. In the following, a small numerical example is considered.

Example 7.3 *The effects of unit shocks in ε_{xt} and ε_{zt} on the variables x_t and z_t can be examined.*

For example, substituting $a_{10} = a_{20} = 0$, $a_{11} = a_{22} = 0.7$, and $a_{12} = a_{21} = 0.2$ in (7.6-7.7) with mean of each series equal to zero, $var\,(e_{1t}) = var\,(e_{2t})$ and $\rho\,(e_{1t}, e_{2t}) = 0.8$ (implying that $b_{12} = -0.8$), so that the decomposed errors are represented by $e_{1t} = \varepsilon_{xt} + 0.8\varepsilon_{zt}$ and $e_{2t} = \varepsilon_{zt}$ and we can examine these effects of unit shocks in ε_{xt} and ε_{zt} on x and z variables, where the roots of $|\mathbf{I}_2 - \mathbf{A}\mathbf{z}| = 0$ are 1.11 and 2 so that the underlying $VAR(1)$ is stationary and the series have a tendency to move together, due to positive values of a_{21} and a_{12}; this is also verified from the fact that the characteristic roots, being the solution of $|\mathbf{A} - \lambda\mathbf{I}_2| = \lambda^2 - (a_{11} + a_{22})\lambda + a_{11}a_{22} - a_{12}a_{21} = 0$ yielding $\lambda_1 = 0.9$ and $\lambda_2 = 0.5$, lie within the unit circle.

[5] One could also introduce an additional restriction as e.g. $c_{21} = 0$. Such an *overidentifying* restriction can have important economic implications: if $b_{21} = 0$ *and* $c_{21} = 0$, contemporaneous ε_{xt} shocks *and* lagged values x_{t-1} do not affect z_t. Hence, the (joint) null hypothesis $b_{21} = c_{21} = 0$ is equivalent to the hypothesis that $\{z_t\}$ is exogenous to the model.

[6] Then also: $var\,(e_{1t}) = \sigma_x^2 + b_{12}^2\sigma_z^2$, $var\,(e_{2t}) = \sigma_z^2$, and $cov\,(e_{1t}, e_{2t}) = -b_{12}\sigma_z^2$.

7.1.2 $VAR(p)$ models

Consider the general $VAR(p)$ model in standard (or reduced) form:

$$\mathbf{y}_t = \mathbf{a}_0 + \mathbf{A}_1\mathbf{y}_{t-1} + \mathbf{A}_2\mathbf{y}_{t-2} + ... + \mathbf{A}_p\mathbf{y}_{t-p} + \mathbf{e}_t, \tag{7.10}$$

where the $\mathbf{A}_i$s are $n \times n$ matrices and $\mathbf{e}_t \in WN$ with $E(\mathbf{e}_t\mathbf{e}_t') = \Sigma$ and $E(\mathbf{e}_t\mathbf{e}_\theta') = \delta_{t\theta}\mathbf{\Sigma}$ (i.e. $\delta_{t\theta} = 0$ for $t \neq \theta$), or in operator form:

$$\left(\mathbf{I}_n - \mathbf{A}_1 L - \mathbf{A}_2 L^2 - ... - \mathbf{A}_p L^p\right)\mathbf{y}_t = \mathbf{A}(L)\mathbf{y}_t = \mathbf{a}_0 + \mathbf{e}_t, \tag{7.11}$$

where $\mathbf{A}(L)$ is the matrix polynomial $\mathbf{I}_n - \sum_{i=1}^{p} \mathbf{A}_i L^i$.

How can (covariance) stationarity be imposed in a $VAR(p)$?

From the standard $VAR(p)$ model (7.10) in deviations from the means gives $\mathbf{y}_t - \boldsymbol{\mu} = \sum_{i=1}^{p} \mathbf{A}_i(\mathbf{y}_{t-i} - \boldsymbol{\mu}) + \mathbf{e}_t$, with $\boldsymbol{\mu}$ defined as:
$\boldsymbol{\mu}{:=}E(\mathbf{y}_t){=}\mathbf{a}_0{+}\mathbf{A}_1\boldsymbol{\mu}{+}\mathbf{A}_2\boldsymbol{\mu}{+}...{+}\mathbf{A}_p\boldsymbol{\mu}{=}(\mathbf{I}_n\text{-}\mathbf{A}_1\text{-}\mathbf{A}_2\text{-}\dots\text{-}\mathbf{A}_p)^{-1}\mathbf{a}_0$, so that the covariance matrices are defined as:
$\mathbf{\Gamma}_s{:=}E\left\{(\mathbf{y}_t\text{-}\boldsymbol{\mu})(\mathbf{y}_{t-s}\text{-}\boldsymbol{\mu})'\right\}{=}E\left\{\sum_{i=1}^{p}[\mathbf{A}_i(\mathbf{y}_{t-i}\text{-}\boldsymbol{\mu}){+}\mathbf{e}_t][\mathbf{A}_i(\mathbf{y}_{t-i}\text{-}\boldsymbol{\mu}){+}\mathbf{e}_t]'\right\}$, so that $\mathbf{\Gamma}_s = E\left[(\mathbf{y}_{t+s} - \boldsymbol{\mu})(\mathbf{y}_t - \boldsymbol{\mu})'\right]$ and
$\mathbf{\Gamma}_{-s} := E\left[(\mathbf{y}_t - \boldsymbol{\mu})(\mathbf{y}_{t+s} - \boldsymbol{\mu})'\right]{=}E\left[(\mathbf{y}_{t-s} - \boldsymbol{\mu})(\mathbf{y}_t - \boldsymbol{\mu})'\right]$ from which $\mathbf{\Gamma}_s \neq \mathbf{\Gamma}_{-s}$ and $\mathbf{\Gamma}_s = \mathbf{\Gamma}_{-s}'$.

Stationarity requires, similar to the univariate case of an $AR(p)$ in Section 5.2, that the solutions (roots) z of $|\mathbf{I}_n - \mathbf{A}_1 z - \mathbf{A}_2 z^2 - ... - \mathbf{A}_p z^p| = 0$ lie outside the unit circle.

Estimation and identification

The standard $VAR(p)$ model (7.10) directly counts $n + pn^2$ unknown parameters to be estimated. Unquestionably, a $VAR(p)$ model will in general be *overparameterized* in that many of these unknown parameters can be properly excluded from the model. However, improperly imposing zero restrictions may waste important information. Moreover, the regressors are likely to be highly collinear, so that the t-ratio tests on individual coefficients may not be reliable guides for paring down the model. Nevertheless, the choice of the appropriate lag length p is very crucial. If p is too small, the model is misspecified; if p is too large, degrees of freedom are wasted. To check the lag length in practice, it is always better to follow the *general to specific method* (see e.g. Maddala and Kim (1998), p. 164). Therefore, we begin with the longest feasible or plausible lag length given a sufficient remaining number of degrees of freedom. Using quarterly data for example, you might start with a lag length of $p = 12$ quarters based on the *a priori* notion that 3 years is sufficiently long to capture (most of) the system's dynamics. Afterwards, you may want to verify whether a lag length considerably below 12 is (still) appropriate using an LR test statistic. Assume e.g. that you want to verify whether in the case of quarterly data a lag length of

$p = 8$ suffices. Consider the residual (reduced form) variance-covariance matrix for $p = 12$, $\mathbf{\Sigma}_{12}$, and for $p = 8$, $\mathbf{\Sigma}_8$. Since $\mathbf{\Sigma}_8$ pertains to a system of n equations and reducing the lag length from 12 to 8 implies $4n$ restrictions in each equation for a total of $4n^2$ *cross-equation restrictions*, the LR test statistic satisfies:

$$\xi_{LR}{=}T(\ln|\mathbf{\Sigma}_8| - \ln|\mathbf{\Sigma}_{12}|) \text{ or } \widetilde{\xi}_{LR}{=}(T-c)(\ln|\mathbf{\Sigma}_8| - \ln|\mathbf{\Sigma}_{12}|)$$

with Sims's (1980) correction, where T is the number of usable observations and c is the number of parameters estimated in each equation of the unrestricted system ($c = 12n{+}1$ in our example since each equation of the unrestricted model has 12 lags for each variable plus an intercept). As known from Subsection 3.1.3 in Chapter 3 ξ_{LR} and $\widetilde{\xi}_{LR}$ are asymptotically χ^2-distributed with degrees of freedom equal to the number of restrictions in the system, i.e. $4n^2$ in the example. If the calculated value of the test statistic $\widetilde{\xi}_{LR}$ were lower than the critical χ^2-value at a certain size, the null of only 8 lags would be rejected. At that point, one could seek to determine whether 4 lags were appropriate by constructing the LR test statistic $(T-c)(\ln|\mathbf{\Sigma}_4| - \ln|\mathbf{\Sigma}_8|)$. Often, this procedure will not reject the null of 8 versus 12 lags and 4 versus 8 lags, although it will reject the null of 4 versus 12 lags because you may lose a small amount of explanatory power at each stage, but, overall, the total loss in explanatory power may be significant. In such a situation, it may be safer to use the longer lag length.

This type of LR test is applicaple to any type of cross-equation restrictions. Let in general $\mathbf{\Sigma}_U$ and $\mathbf{\Sigma}_R$ be the variance-covariance matrices of the unrestricted and restricted reduced forms systems, respectively. If the equations of the unrestricted model contain different regressors (and not an equal number as usual), let c denote the maximum number of regressors contained in the longest equation. Then, asymptotically, the LR test statistic $(T - c)$ $(\ln|\mathbf{\Sigma}_R| - \ln|\mathbf{\Sigma}_U|)$ follows a χ^2-distribution with degrees of freedom equal to the number of restrictions in the system.[7]

Notice that, since LR tests are based on asymptotic theory, finite sample applications in economics can gain from using alternative criteria to determine appropriate lag lengths as the multivariate generalizations of the AIC and BIC criteria in (5.31) and (5.32):

$$AIC = T\ln\left|\hat{\mathbf{\Sigma}}\right| + 2N \quad \text{and} \quad BIC = T\ln\left|\hat{\mathbf{\Sigma}}\right| + N\ln(T)$$

where $|\mathbf{\Sigma}|$ is again the determinant of the estimated residual variance-covariance matrix and N is the total number of parameters estimated in all equations, which

[7] Another example when using quarterly data is about capturing seasonal effects by including 3 seasonal dummies in each of the n regressions of a VAR. Estimate then the unrestricted model by including the dummy variables and estimate the restricted model by excluding the dummy variables. The total number of restrictions in the system is $3n$. If the lag length is p, the equations of the unrestricted model have $c = np + 4$ unknown parameters (np lagged variables, the intercept, and the three seasonal coefficients). If the resulting LR test statistic exceeds the critical value of the asymptotic χ^2 distribution with $3n$ degrees of freedom, the restriction of no seasonal effects can be rejected.

is $n + pn^2$ in an n-variate $VAR(p)$. As in the univariate case, select the model with the lowest AIC or BIC value.

Since the right-hand side of the standard $VAR\,(p)$ model in (7.10) contains only predetermined variables and WN error terms each equation in that system may be estimated by OLS. Because of the (initial) stationarity assumption of all y_{it}s ($i = 1, 2, \ldots, n$ and $t = 1, 2, \ldots, T$), these OLS estimators are consistent and asymptotically efficient.[8]

Impulse response function(s)

Just as a stationary $AR(p)$ model has an $MA(\infty)$ representation, a stationary $VAR\,(p)$ model can be written as an infinite dimensional vector moving average ($VMA(\infty)$) model. Indeed, if the standard reduced form $VAR\,(p)$ model in (7.11) is stationary, one may directly invert the $VAR\,(p)$ specification as:

$$\mathbf{y}_t = (\mathbf{I}_n - \textstyle\sum_{i=1}^{p} \mathbf{A}_i)^{-1}\mathbf{a}_0 + (\mathbf{I}_n - \sum_{i=1}^{p} \mathbf{A}_i L^i)^{-1}\mathbf{e}_t = \boldsymbol{\mu} + \sum_{i=0}^{\infty} \boldsymbol{\Psi}_i \mathbf{e}_{t-i} = \boldsymbol{\mu} + \boldsymbol{\Psi}(L)\mathbf{e}_t, \tag{7.12}$$

which is a multivariate linear filter expression of the $\{y_t, t \in \mathbb{Z}\}$ process in the time path of the various reduced form shocks. The elements of $\boldsymbol{\Psi}(L)$ give the response on $\mathbf{y}_t$ from unit impulses to each of the elements of the reduced form residuals $\mathbf{e}_t$. Therefore, the $VMA(\infty)$ model (7.12) is called the *impulse response function* derived from the standard $VAR\,(p)$ model (7.10).

Example 7.4 *For the stationary standard bivariate $VAR\,(1)$ model (7.5) the impulse response function is derived from the bivariate $VMA(\infty)$ model $\mathbf{y}_t = \boldsymbol{\mu} + \sum_{j=0}^{\infty} \mathbf{A}_1^j \mathbf{e}_{t-j}$ with $\boldsymbol{\mu}$ defined as:*

$$\boldsymbol{\mu} := (\textstyle\sum_{i=0}^{\infty} \mathbf{A}_1^i)\mathbf{a}_0 = (\mathbf{I}_2 - \mathbf{A}_1)^{-1}\mathbf{a}_0 = \frac{1}{(1-a_{11})(1-a_{22}) - a_{12}a_{21}} \begin{pmatrix} a_{10}\,(1\text{-}a_{22})\text{-}a_{12}a_{20} \\ a_{20}\,(1\text{-}a_{11})\text{-}a_{21}a_{10} \end{pmatrix}.$$

Although this impulse response function expresses x_t and z_t in terms of the reduced form error sequences $\{e_{1t}, t \in \mathbb{Z}\}$ and $\{e_{2t}, t \in \mathbb{Z}\}$, it may be very insightful to rewrite this impulse response function in terms of the structural disturbances $\{\varepsilon_{xt}, t \in \mathbb{Z}\}$ and $\{\varepsilon_{zt}, t \in \mathbb{Z}\}$. The vector of reduced form error terms can be written as:
$\mathbf{e}_t := \begin{pmatrix} e_{1t} \\ e_{2t} \end{pmatrix} = \frac{1}{(1-b_{12}b_{21})} \begin{pmatrix} 1 & -b_{12} \\ -b_{21} & 1 \end{pmatrix} \begin{pmatrix} \varepsilon_{xt} \\ \varepsilon_{zt} \end{pmatrix}$, so that substituted in the above impulse response function this can be expressed as a function of the sequences $\{\varepsilon_{xt}, t \in \mathbb{Z}\}$ and $\{\varepsilon_{zt}, t \in \mathbb{Z}\}$ of structural disturbances:

$$\mathbf{y}_t := \begin{pmatrix} x_t \\ z_t \end{pmatrix} = \begin{pmatrix} \mu_1 \\ \mu_2 \end{pmatrix} + \frac{1}{(1\text{-}b_{12}b_{21})} \sum_{j=0}^{\infty} \begin{pmatrix} a_{11} & a_{12} \\ a_{21} & a_{22} \end{pmatrix}^j \begin{pmatrix} 1 & \text{-}b_{12} \\ \text{-}b_{21} & 1 \end{pmatrix} \begin{pmatrix} \varepsilon_{x,t\text{-}j} \\ \varepsilon_{z,t\text{-}j} \end{pmatrix}$$

[8] Even though the reduced form errors are correlated across equations, these *seemingly unrelated regressions* (cf. infra in Chapter 11) do not add to the efficiency of the estimation procedure since the regressions have identical right-hand side variables.

$$=\begin{pmatrix} \mu_1 \\ \mu_2 \end{pmatrix}+\sum_{j=0}^{\infty}\begin{pmatrix} \phi_{11}(j) & \phi_{12}(j) \\ \phi_{21}(j) & \phi_{22}(j) \end{pmatrix}\begin{pmatrix} \varepsilon_{x,t\text{-}j} \\ \varepsilon_{z,t\text{-}j} \end{pmatrix}, \text{ or more compactly, } \mathbf{y}_t =$$

$\boldsymbol{\mu}+\sum_{j=0}^{\infty}\boldsymbol{\Phi}_j\boldsymbol{\varepsilon}_{t-j}$. The elements of $\boldsymbol{\Phi}_j$ can be used to generate effects of ε_{xt} and ε_{zt} shocks on the entire time paths of the $\{x_t, t \in \mathbb{Z}\}$ and $\{z_t, t \in \mathbb{Z}\}$ sequences. The four elements of $\phi_{ik}(0)$ are *structural impact multipliers*, e.g. $\phi_{12}(0)$ is the instantaneous impact of a one-unit change in the ε_{zt} shock on x_t. Similarly, *structural cumulative multipliers* $\sum_{j=0}^{T}\phi_{ik}(j)$ and the *structural long run multiplier* $\sum_{j=0}^{\infty}\phi_{ik}(j)$ can be defined. Referring to the four sets of coefficients $\{\phi_{11}(j)\}, \{\phi_{12}(j)\}, \{\phi_{21}(j)\}$, and $\{\phi_{22}(j)\}$ as the *structural impulse response coefficients*, these coefficients can be plotted against j as a practical way to visually represent the behavior of the $\{x_t, t \in \mathbb{Z}\}$ and $\{z_t, t \in \mathbb{Z}\}$ series in response to the various structural shocks.

The structural impulse response coefficients will now be examined for (numerical) Example 7.3.

Example 7.5 *Effects of one-unit shocks in $\{\varepsilon_{xt}, t \in \mathbb{Z}\}$ and $\{\varepsilon_{zt}, t \in \mathbb{Z}\}$.*

A one-unit shock in ε_{zt} causes z_t to jump by one unit and x_t by 0.8 units. In the next period, $\varepsilon_{z,t+1}$ returns to zero but the autoregressive nature of the system is such that x_{t+1} and z_{t+1} do not immediately return to their long run values. Since $z_{t+1} = 0.2x_t+0.7z_t+\varepsilon_{z,t+1}, E_t(z_{t+1}) = 0.2\times0.8+0.7\times1 = 0.86$ and since $x_{t+1} = 0.7x_t+0.2z_t+\varepsilon_{x,t+1}+0.8\varepsilon_{z,t+1}$, $E_t(x_{t+1}) = 0.76$. Since the system is stable, the subsequent values of $\{x_t\}$ and $\{z_t\}$ will converge to their long-run levels.

On the contrary, a one-unit shock in ε_{xt} causes the value of x_t to increase by one unit; however, according to the specification in Example 7.3, there is no contemporaneous effect on the value of z_t so that $x_t = 1$ and $z_t = 0$. In the following period, the shock $\varepsilon_{x,t+1}$ returns to zero and the values for the variables are forecasted as $E_t(x_{t+1}) = 0.7x_t+0.2z_t = 0.7$ and $E_t(z_{t+1}) = 0.2x_t+0.7z_t = 0.2$.

The impulse response coefficients for the periods from $t+2$ onwards decay since the system is stationary.

Variance decomposition

Since unrestricted VAR models are in general overparameterized, these models are not particularly useful for short term forecasts. However, understanding the properties of the forecast errors is very helpful in detecting interrelationships among the variables in the system. For reasons of pedagogical exposition the bivariate standard $VAR(1)$ example in (7.5) will be continued.

Example 7.6 *Suppose the coefficients of $\mathbf{a}_0$ and $\mathbf{A}_1$ in (7.5) are known and that one is interested in conditional forecasts $E_t(\mathbf{y}_{t+h})$.*

Proceeding in a way similar as in Subsection 5.9.1 on forecasting AR processes and (5.52) in particular, the h-step (ahead) forecast error amounts to:

$$\mathbf{f}_t(h) := \mathbf{y}_{t+h} - E_t(\mathbf{y}_{t+h}) = \mathbf{e}_{t+h} + \mathbf{A}_1\mathbf{e}_{t+h-1} + \mathbf{A}_1^2\mathbf{e}_{t+h-2} + \ldots \mathbf{A}_1^{h-1}\mathbf{e}_{t+1},$$

or, equivalently, in terms of the structural disturbances in the impulse response function from Subsection 7.1.2 (see also Proposition 5.3):

$$\mathbf{f}_t(h) := \mathbf{y}_{t+h} - E_t(\mathbf{y}_{t+h}) = \boldsymbol{\varepsilon}_{t+h} + \boldsymbol{\Phi}_1\boldsymbol{\varepsilon}_{t+h-1} + \boldsymbol{\Phi}_1^2\boldsymbol{\varepsilon}_{t+h-2} + \ldots \boldsymbol{\Phi}_1^{h-1}\boldsymbol{\varepsilon}_{t+1}.$$

Focusing solely on the h-step (ahead) forecast error and the corresponding forecast variance of the $\{x_t, t \in \mathbb{Z}\}$ sequence (see (5.53)):

$$var\left[f_{xt}(h)\right]=\sigma_{\varepsilon x}^2\left[\phi_{11}^2(0)+\phi_{11}^2(1)+\ldots+\phi_{11}^2(h-1)\right]+\sigma_{\varepsilon z}^2\left[\phi_{12}^2(0)+\phi_{12}^2(1)+\ldots+\phi_{12}^2(h-1)\right].$$

Since all values of $\phi_{ik}^2(j)$ are necessarily nonnegative, the variance of the forecast error increases as the forecast horizon increases.

This h-step (ahead) forecast error variance can now be decomposed according to each of the (structural) shocks so that the proportions of $var\left[f_{xt}(h)\right]$ due to shocks in the $\{\varepsilon_{xt}, t \in \mathbb{Z}\}$ and $\{\varepsilon_{zt}, t \in \mathbb{Z}\}$ sequences are, respectively:

$$\frac{\sigma_{\varepsilon x}^2\left[\phi_{11}^2(0)+\phi_{11}^2(1)+\ldots+\phi_{11}^2(h-1)\right]}{var\left[f_{xt}(h)\right]} \quad \text{and} \quad \frac{\sigma_{\varepsilon z}^2\left[\phi_{12}^2(0)+\phi_{12}^2(1)+\ldots+\phi_{12}^2(h-1)\right]}{var\left[f_{xt}(h)\right]}.$$

Definition 7.1 *The variance decomposition measures the relative contribution to the forecast error variance of each shock as a function of the forecast horizon.*

If e.g. $\{\varepsilon_{zt}\}$ shocks do not explain anything of the forecast error variance of the $\{x_t, t \in \mathbb{Z}\}$ sequence at all forecast horizons, i.e. that this $\{x_t, t \in \mathbb{Z}\}$ sequence is exogenous. In such a situation the $\{x_t, t \in \mathbb{Z}\}$ sequence would evolve independently of the $\{\varepsilon_{zt}\}$ shocks and the $\{z_t, t \in \mathbb{Z}\}$ sequence. At the other extreme, $\{\varepsilon_{zt}\}$ shocks could explain the whole forecast error variance in the $\{x_t, t \in \mathbb{Z}\}$ sequence at all forecast horizons. Then, the $\{x_t, t \in \mathbb{Z}\}$ sequence would be entirely endogenous.

In applied economic research it is typical for a variable to explain almost all its forecast error variance at short forecast horizons and smaller proportions at longer forecast horizons. We would expect this pattern if $\{\varepsilon_{zt}\}$ shocks had little contemporaneous effect on x, but affected the $\{x_t, t \in \mathbb{Z}\}$ sequence with a lag. In practice it is useful to examine the variance decomposition at various forecast horizons. As the forecast horizon h increases the variance decomposition should converge. Moreover, if the contemporaneous correlation coefficient $\rho(e_{1t}, e_{2t})$ between the bivariate reduced form errors is significantly different from zero, it is advisable to calculate the variance decompositions under various orderings (zero parameter restrictions).

Summary 7.1 *Summarizing, impulse response analysis and variance decomposition (together called* ***innovation accounting****) can be useful tools to examine the relationships between economic variables. If the correlations among the various innovations are small, the identification problem is not so important and the alternative orderings should yield similar impulse response functions and* *variance decompositions. However, the contemporaneous movements of many economic variables are highly correlated so that alternative orderings of variables will generally matter!*

Instead of analyzing the impulse response functions (and variance decompositions) from the disturbances $\mathbf{e}_t$ or ε_t we could analyze the responses on $\mathbf{y}_t$ from new shocks in $\mathbf{v}_t$ that are linear combinations of the 'old shocks' in $\mathbf{e}_t$ (or ε_t). This issue, important for several economic applications, will be studied in the following section on *orthogonalization of shocks.*

Orthogonalization

Example 7.7 *Consider the responses of the elements of $\mathbf{y}_t$ to unit movements in new shocks that are linear combinations of the old shocks in $\mathbf{e}_t$ in the bivariate case $\mathbf{v}_t := \begin{pmatrix} e_{1t} \\ e_{2t}+0.5e_{1t} \end{pmatrix}$.*

Calling these new shocks $\mathbf{v}_t$ with $v_{1t} := e_{1t}$ and $v_{2t} := e_{2t} + 0.5e_{1t}$ or $\mathbf{v}_t = \mathbf{Q}\,\mathbf{e}_t$ with $\mathbf{Q} := \begin{pmatrix} 1 & 0 \\ 0.5 & 1 \end{pmatrix}$, the $VMA(\infty)$ representation (7.12) of a $VAR(p)$ model can be written in terms of these new shocks as:

$$\mathbf{y}_t = \boldsymbol{\mu} + \underbrace{\boldsymbol{\Psi}(L)\mathbf{Q}^{-1}}_{\boldsymbol{\Upsilon}(L)}\mathbf{Q}\mathbf{e}_t = \boldsymbol{\mu} + \boldsymbol{\Upsilon}(L)\,\mathbf{v}_t,$$

where $\boldsymbol{\Upsilon}(L)$ gives the response of $\mathbf{y}_t$ to the new shock $\mathbf{v}_t$, or $\boldsymbol{\Upsilon}(L)$ is a linear combination of the original impulse responses $\boldsymbol{\Psi}(L)$. So, which linear combinations matter? Data is of no help, so one should look after the most interesting linear combination. In order to do this *orthogonalization assumptions*, that uniquely determine the linear combinations which we find most interesting, are needed.

Orthogonal shocks We want to find orthonormal shocks, i.e. shocks with unit variance and zero covariances or the orthonormal matrix $\mathbf{Q}$ is determined as:

$$E(\mathbf{v}_t\mathbf{v}_t') = E\left(\mathbf{Q}\,\mathbf{e}_t\mathbf{e}_t'\mathbf{Q}'\right) = \mathbf{Q}\boldsymbol{\Sigma}\mathbf{Q}' = \mathbf{I}_n, \tag{7.13}$$

such that $\boldsymbol{\Sigma} = \mathbf{Q}^{-1}\left(\mathbf{Q}^{-1}\right)'$. Note, however, that there are very many matrices $\mathbf{F}$ that satisfy $\mathbf{F}\mathbf{F}' = \mathbf{Q}^{-1}\left(\mathbf{Q}^{-1}\right)'$, which becomes clear from the *Cholesky decomposition* that finds a $\mathbf{Q}^*$ such that $\mathbf{Q}^* = \mathbf{R}\mathbf{Q}$, with $\mathbf{R}$ an orthonormal matrix, i.e.
$\mathbf{R}\mathbf{R}'=\mathbf{I}$, and $\mathbf{Q}^*\mathbf{Q}^{*\prime}=\mathbf{R}\mathbf{Q}\mathbf{Q}'\mathbf{R}'=\mathbf{R}\mathbf{R}'=\mathbf{I}$ or $\boldsymbol{\Sigma}=\mathbf{Q}^{-1}\mathbf{R}\mathbf{R}'\left(\mathbf{Q}^{-1}\right)'=(\mathbf{R}\mathbf{Q})^{-1}\left[(\mathbf{R}\mathbf{Q})^{-1}\right]'$. Which $\mathbf{Q}$ should we choose? Since $\boldsymbol{\Upsilon}(L) := \boldsymbol{\Psi}(L)\,\mathbf{Q}^{-1}$, specifying a desired property on $\boldsymbol{\Upsilon}(L)$ can determine $\mathbf{Q}$. This brings us to *Sims's orthogonalization* (this orthogonalization is specific for $\boldsymbol{\Upsilon}(0)$). Sims (1980) suggests to specify properties of $\boldsymbol{\Upsilon}(0)$ (i.e. the instantaneous response of each variable to each orthogonalized shock $\mathbf{v}_t$). Since $\boldsymbol{\Psi}(0) = \mathbf{I}_n$ (each shock only affects its own

variable contemporaneously) and $\mathbf{A}(0) = \mathbf{I}_n$ (no variable appears contemporaneously in the other variable's regression), then, unless $\mathbf{\Sigma}$ is diagonal (orthogonal shocks to start with), every diagonalizing matrix $\mathbf{Q}$ will have off-diagonal elements, thus $\mathbf{\Upsilon}(0)$ cannot be equal to $\mathbf{I}$. This means that some shocks will have effects on more than one variable! Sims (1980) suggests to choose a lower triangular $\mathbf{\Upsilon}(0)$, that gives for the bivariate case:

$$\begin{pmatrix} x_t \\ z_t \end{pmatrix} = \begin{pmatrix} \tau_{0xx} & 0 \\ \tau_{0zx} & \tau_{0zz} \end{pmatrix} \begin{pmatrix} v_{1t} \\ v_{2t} \end{pmatrix} + \Upsilon_1 v_{t-1} + ...,$$

where τ_{0zx} involves the contemporaneous correlation between the original shocks $\mathbf{e}_t$, and the second shock, v_{2t}, does not affect the first variable x_t contemporaneously, but both shocks can affect z_t contemporaneously (also on the implied autoregressive representation).

In summary, one can uniquely specify $\mathbf{Q}$ and, hence, which linear combination of the original shocks $\mathbf{e}_t$ will be used to plot impulse responses by requiring that: i) the errors are orthogonal; ii) the instanteneous response of one variable to the other shock is zero? This second requirement is equivalent to iii) the VAR is estimated by OLS with contemporaneous x in the z equation, but not vice versa.

Luckily, the Cholesky decomposition produces a lower triangular $\mathbf{Q}$. Since $\mathbf{\Upsilon}(0) = \mathbf{\Psi}(0)\mathbf{Q}^{-1} = \mathbf{Q}^{-1}$, the Cholesky decomposition produces the Sims orthogonalization already. Ideally, one should use economic theory to determine the order of orthogonalization (e.g. GDP is later published, so many other variables cannot respond within the same period to a GDP shock).

Blanchard and Quah (1989) suggest that, instead of restricting immediate response, it is interesting to examine shocks such that the long term response of one variable to a shock equals zero (i.e. a restricted long run response); e.g. demand shocks have no long run impact on GDP (which is at least $I(1)$) so that they are called *transitory* shocks. The impact of permanent and transitory shocks will be analyzed in the next section, using an example of nominal and real wage shocks.

Permanent and transitory shocks

Based on Blanchard and Quah (1989), assume that:

- two uncorrelated structural shocks occur in a model, e.g. real and nominal shocks on wages;

- since the components of $\mathbf{y}_t$ are assumed to be integrated of order 1, the differenced vector is stationary or $\Delta\mathbf{y}_t \backsim I(0)$; in this example:

$$\Delta\mathbf{y}_t := \begin{pmatrix} \Delta rw_t \\ \Delta nw_t \end{pmatrix} = \begin{pmatrix} \Delta \log(\text{real wages})_t \\ \Delta \log(\text{nominal wages})_t \end{pmatrix};$$

- the linear dynamic structural model satisfies a $VAR(p)$ (see also (7.4) for the bivariate $VAR(1)$ model):

$$\mathbf{B}\Delta\mathbf{y}_t = \sum_{i=1}^{p} \mathbf{C}_i \Delta\mathbf{y}_{t-i} + \varepsilon_t \tag{7.14}$$

with $\varepsilon_t \in WN$;

- normalizing restrictions are imposed on $\mathbf{B}$ and on the structural residual covariance matrix, or $\mathbf{B} := \begin{pmatrix} 1 & b_{12} \\ b_{21} & 1 \end{pmatrix}$ and $E\left(\varepsilon_t \varepsilon_t'\right) = \mathbf{\Omega} = \begin{pmatrix} \omega_{11} & 0 \\ 0 & \omega_{22} \end{pmatrix}$ (it is assumed that there is no correlation between permanent and transitory structural shocks), with $\mathbf{C}_1, \cdots, \mathbf{C}_p$ being unrestricted parameter matrices.

The data can recover only in the (unrestricted) reduced form of the structural form model (7.14), which yields the standard $VAR(p)$ representation with two structural shocks:

$$\Delta\mathbf{y}_t = \mathbf{B}^{-1} \sum_{i=1}^{p} \mathbf{C}_i \Delta\mathbf{y}_{t-i} + \mathbf{B}^{-1}\varepsilon_t = \sum_{i=1}^{p} \mathbf{A}_i \Delta\mathbf{y}_{t-i} + \mathbf{e}_t = \mathbf{A}(L)\Delta\mathbf{y}_t + \mathbf{e}_t \tag{7.15}$$

with $E\left(\mathbf{e}_t \mathbf{e}_t'\right) = \mathbf{\Sigma} = \begin{pmatrix} \sigma_1^2 & \sigma_{12} \\ \sigma_{21} & \sigma_2^2 \end{pmatrix}$. The reduced form (7.15) can now be estimated straightforwardly to obtain estimators for $\mathbf{A}_1, \mathbf{A}_2, \cdots, \mathbf{A}_p, \mathbf{\Sigma}$ with $\mathbf{\Sigma} = E\left(\mathbf{e}_t \mathbf{e}_t'\right) = E\left(\mathbf{B}^{-1}\varepsilon_t \varepsilon_t' \mathbf{B}^{-1\prime}\right) = \mathbf{B}^{-1}\mathbf{\Omega}\mathbf{B}^{-1\prime}$ or $\mathbf{B\Sigma B}' = \mathbf{\Omega}$, i.e.

$$\mathbf{B\Sigma B}' = \begin{pmatrix} 1 & b_{12} \\ b_{21} & 1 \end{pmatrix} \begin{pmatrix} \sigma_1^2 & \sigma_{12} \\ \sigma_{21} & \sigma_2^2 \end{pmatrix} \begin{pmatrix} 1 & b_{21} \\ b_{12} & 1 \end{pmatrix} = \begin{pmatrix} \omega_{11} & 0 \\ 0 & \omega_{22} \end{pmatrix},$$

$$= \begin{pmatrix} \left(\sigma_1^2 + b_{12}\sigma_{21} + b_{12}\sigma_{12} + b_{12}^2\sigma_2^2\right) & \left(\sigma_{12} + b_{12}\sigma_2^2 + b_{21}\sigma_1^2 + b_{21}b_{12}\sigma_{12}\right) \\ \left(\sigma_{21} + b_{21}\sigma_1^2 + b_{12}\sigma_2^2 + b_{12}b_{21}\sigma_{12}\right) & \left(\sigma_2^2 + b_{21}\sigma_{12} + b_{21}\sigma_{21} + b_{21}^2\sigma_1^2\right) \end{pmatrix}$$
$$= \begin{pmatrix} \left(\sigma_1^2 + 2b_{12}\sigma_{12} + b_{12}^2\sigma_2^2\right) & \left(\sigma_{12}\left(1 + b_{21}b_{12}\right) + b_{12}\sigma_2^2 + b_{21}\sigma_1^2\right) \\ \left(\sigma_{12}\left(1 + b_{12}b_{21}\right) + b_{21}\sigma_1^2 + b_{12}\sigma_2^2\right) & \left(\sigma_2^2 + 2b_{21}\sigma_{12} + b_{21}^2\sigma_1^2\right) \end{pmatrix} = \mathbf{\Omega},$$

where the equality $\sigma_{12} = \sigma_{21}$ was used in the last line. Hence, the following system of three nonlinear equations (restrictions) is obtained:

$$\begin{aligned} \sigma_1^2 + 2b_{12}\sigma_{12} + b_{12}^2\sigma_2^2 &= \omega_{11} \\ \sigma_{12}\left(1 + b_{21}b_{12}\right) + b_{12}\sigma_2^2 + b_{21}\sigma_1^2 &= 0 \\ \sigma_2^2 + 2b_{21}\sigma_{12} + b_{21}^2\sigma_1^2 &= \omega_{22}, \end{aligned} \tag{7.16}$$

where $b_{12}, b_{21}, \sigma_1^2$,and σ_2^2 are four unknown parameters in three equations. For this system to have (a) solution(s), one more equation is needed. As also proposed by Blanchard and Quah (1989), this additional equation is a long run restriction that nominal shocks do not affect real variables for $t \longrightarrow \infty$. But even with such a restriction that we are going to discuss immediately the nonlinearity in the above three equations is not solved. In other words, the solution of the nonlinear system will be subject to the occurrence of 'local minima' (unless some 'grid search' over the relevant parameter space could be possible): for each different set of initial values for $b_{12}, b_{21}, \sigma_1^2$, and σ_2^2 a different solution will come up and each different solution will generate a different impulse response function. Such an impulse response function, derived from $VAR(p)$ model (7.15), can be written similar to (7.12):

$$\Delta \mathbf{y}_t = \begin{pmatrix} \Delta rw_t \\ \Delta nw_t \end{pmatrix} = \begin{pmatrix} \Psi_{11}(L) & \Psi_{12}(L) \\ \Psi_{21}(L) & \Psi_{22}(L) \end{pmatrix} \begin{pmatrix} e_{1t} \\ e_{2t} \end{pmatrix} = \mathbf{\Psi}(L)\mathbf{e}_t \tag{7.17}$$

where $\mathbf{\Psi}(L)$ consists of polynomials in the lag operator L. The time paths of the effects of the various shocks on the real and nominal wage rates are implied by the coefficients in $\mathbf{\Psi}(L)$.

The two residuals in (7.15) have been effectively transformed into two shocks in (7.17). Consider two types of orthogonal shocks, each of which could be the source of variation in the observed movements in real and nominal wage rates. The structural form disturbance ε_{1t} in (7.14) is a real exogenous shock which reflects changes in economic variables such as endowment, productivity and technology. The shock may affect the real and nominal exchange rates in the short and long run. The other structural form disturbance ε_{2t} in (7.14) is assumed to be a nominal exogenous shock, which is caused by e.g. nominal money supply shocks or devaluation of the exchange rate. By assumption the nominal shocks have only a temporary effect on the value of the real rate but may have permanent effect on the nominal rate. The assumption of the temporary effect on the real wage rate acts as an identifying restriction.

How can now this long run restriction be determined? The long run neutrality of nominal shocks is represented by the restriction that the sum of the coefficients in $\Psi_{12}(L)$ is zero, i.e.$\sum_{j=0}^{\infty} \Psi_{12}(j) = 0$, where $\Psi_{12}(j)$ is the jth coefficient in $\Psi_{12}(L)$, or the effect of e_2 on Δrw after j periods.; $\sum_{j=0}^{\infty} \Psi_{12}(j)$ is the cumulative effect of e_2 on Δrw over time and the long run effect of e_2 on Δrw is zero. The restriction that aggregate demand has no long run impact on the logarithmic real wages has been imposed. This restriction follows from the assumption that the natural rate of unemployment hypothesis holds in the long run.

7.1.3 Structural VAR models

Introductory remarks

The original meaning of a 'structural' model in econometrics is explained in Hurwicz (1962): "A model is structural if it allows us to predict the effect of 'interventions' - deliberate policy actions, or changes in the economy or in nature of known types". To make such a prediction, the model must tell us how the intervention correspond to changes in some elements of the model (parameters, equations, observable or unobservable random variables) and it must be true that the changed model is an accurate characterization of the behavior being modeled after the intervention. There is no sharp distinction among interventions that change equations, change parameters in equations, change disturbance terms in equations, or change the value of variables in a system.

According to Sims (2002) nowadays a model is often called 'structural' when its parameters have behavioral interpretations, regardless of whether the old (above) definition applies, and on the other hand models that are in fact structural in the old sense are thought of as *reduced form* because they contain parameters or equations that do not have unique behavioral interpretations. As seen in Subsection 7.1.2 a structural VAR or $SVAR$ model can be defined but identifying restrictions have to introduced.

Specification of an $SVAR(p)$ model

From the structural equations (7.4) of the bivariate $VAR(1)$ model the structural $VAR(p)$ or $SVAR(p)$ model can be written as $\mathbf{B}(L)\mathbf{y}_t = \mathbf{c} + \boldsymbol{\varepsilon}_t$, where $\mathbf{B}(L) := \mathbf{B}_0 - \sum_{i=1}^{p} \mathbf{B}_i L^i$ and $\boldsymbol{\varepsilon}_t$ is a WN process with covariance matrix $E(\boldsymbol{\varepsilon}_t\boldsymbol{\varepsilon}_t') = \boldsymbol{\Omega}$. Since it is assumed that $\mathbf{B}_0$ has full rank so that as in the bivariate case (7.5) the *reduced form* $\mathbf{B}_0^{-1}\mathbf{B}(L)\mathbf{y}_t = \mathbf{A}(L)\mathbf{y}_t = \mathbf{B}_0^{-1}\mathbf{c} + \mathbf{B}_0^{-1}\boldsymbol{\varepsilon}_t = \mathbf{a}_0 + \mathbf{e}_t$ exists so that the system can be solved to determine $\mathbf{y}_t$ from past values of $\mathbf{y}$ and current shocks $\boldsymbol{\varepsilon}_t$. In the $SVAR(p)$ model the conditional distribution of $\mathbf{y}_t$ given past $\mathbf{y}$s is therefore determined by the coefficients in the reduced form lag polynomial $\mathbf{A}(L) := \mathbf{B}_0^{-1}\mathbf{B}(L) = \mathbf{I}_n - \mathbf{B}_0^{-1}\sum_{i=1}^{p} \mathbf{B}_i L^i$, by $\mathbf{a}_0 := \mathbf{B}_0^{-1}\mathbf{c}$, and by the parameters of the conditional distribution of $\mathbf{e}_t := \mathbf{B}_0^{-1}\boldsymbol{\varepsilon}_t$ given past $\mathbf{y}$s.

Structural VMA specifications and $SVAR$ models Considering a 'structural' $VMA(\infty)$ model derived from the above structural form as:

$$\mathbf{y}_t = (\mathbf{B}(1))^{-1}\mathbf{c} + (\mathbf{B}(L))^{-1}\boldsymbol{\varepsilon}_t = \boldsymbol{\mu} + \mathbf{D}(L)\boldsymbol{\varepsilon}_t$$

with the infinite dimensional polynomial $\mathbf{D}(L)$ defined as:

$$\mathbf{D}(L) := (\mathbf{B}(L))^{-1} = (\mathbf{B}_0 - \textstyle\sum_{i=1}^{p} \mathbf{B}_i L^i)^{-1} = \sum_{j=0}^{p} \mathbf{D}_j L^j,$$

'a kind of a reduced form $VAR(\infty)$' can be considered by redefining the $\mathbf{D}_j$-coefficient matrices when comparing the coefficients for equal lags in:

$$\mathbf{y}_t = \boldsymbol{\mu} + \mathbf{D}_0\boldsymbol{\varepsilon}_t + \mathbf{D}_1\boldsymbol{\varepsilon}_{t-1} + \mathbf{D}_2\boldsymbol{\varepsilon}_{t-2} + \ldots = \boldsymbol{\mu} + \mathbf{w}_t + \mathbf{V}_1\mathbf{w}_{t-1} + \mathbf{V}_2\mathbf{w}_{t-2} + \ldots. \quad (7.18)$$

From the rewriting in (7.18) $\mathbf{w}_t := \mathbf{D}_0\boldsymbol{\varepsilon}_t, \mathbf{V}_1\mathbf{w}_{t-1} := \mathbf{D}_1\boldsymbol{\varepsilon}_{t-1}, \mathbf{V}_2\mathbf{w}_{t-2} := \mathbf{D}_2\boldsymbol{\varepsilon}_{t-2}$, etc. Hence, $\mathbf{V}_1\mathbf{w}_{t-1} = \mathbf{V}_1\mathbf{D}_0\boldsymbol{\varepsilon}_{t-1} = \mathbf{D}_1\boldsymbol{\varepsilon}_{t-1}$, so that by equating coefficients corresponding to the same lagged error: $\mathbf{V}_1\mathbf{D}_0 = \mathbf{D}_1$, or the new coefficient matrix $\mathbf{V}_1$ satisfies $\mathbf{V}_1 = \mathbf{D}_1\mathbf{D}_0^{-1}$ Proceeding in a similar way for $\mathbf{V}_2\mathbf{w}_{t-2}, \mathbf{V}_3\mathbf{w}_{t-3}$, the sequence of matrices $\{\mathbf{V}_i = \mathbf{D}_i\mathbf{D}_0^{-1}, i=1,2,3,\ldots\}$ is obtained.

Once one has an $SVAR(p)$ model this type of reduced form expression can uniquely be derived. However, in many practical cases the econometrician will not have the $SVAR(p)$ representation at her disposal but only the so-called 'reduced form representation' on the right-hand side of (7.18): $\boldsymbol{\mu} + \mathbf{w}_t + \mathbf{V}_1\mathbf{w}_{t-1} + \mathbf{V}_2\mathbf{w}_{t-2} + \ldots = \boldsymbol{\mu} + \mathbf{V}(L)\mathbf{w}_t$, e.g. when the effects of various types of shocks have to be studied. In such a case extra restrictions should be imposed such that a unique $SVAR$ representation can be obtained. In Section 7.1.2 Blanchard and Quah's (1989) model was introduced with long-term restrictions which, however, were not sufficient for deriving a unique representation of the $SVAR$ model, but in Lastrapes (2002) the problem of uniqueness is solved via the estimation of the (unique) *Cholesky factor* $\mathbf{D}(1)$ of the 'long run covariance matrix' $\mathbf{V}(1)\Xi\,[\mathbf{V}(1)]' = \mathbf{D}(1)\,[\mathbf{D}(1)]'$, where Ξ is the covariance matrix of the $\{\mathbf{w}_t, t \in \mathbb{Z}\}$ process, or $\Xi := E(\mathbf{w}_t\mathbf{w}_t')$ (see also Cholesky decomposition in (7.13)).

Orthogonal shocks on $SVARs$ As implied by the subsection on orthogonal shocks the covariance matrix of n variate orthonormal shocks $\mathbf{v}_t := \mathbf{Q}\,\mathbf{e}_t = \mathbf{Q}\mathbf{B}_0^{-1}\varepsilon_t$, which are shocks with unit variance and zero covariances, satisfies from (7.13) for an arbitrary $n \times n$ orthonormal matrix $\mathbf{Q}$ (or $\mathbf{Q}^{-1} = \mathbf{Q}'$) in Sims's orthogonalization procedure (see Sims (1980)):

$$E(\mathbf{v}_t\mathbf{v}_t') = \mathbf{Q}\Sigma\mathbf{Q}' = \mathbf{Q}\mathbf{B}_0^{-1}\Omega(\mathbf{B}_0^{-1})'\mathbf{Q}' = \mathbf{I}_n, \tag{7.19}$$

with the reduced form covariance matrix given by $\Sigma = E(\mathbf{e}_t\mathbf{e}_t') = \mathbf{Q}^{-1}(\mathbf{Q}^{-1})' = \mathbf{Q}'\mathbf{Q}$.

Notice that expression (7.19) is not unique at all since another orthonormal transformation such that the covariance matrix remains the same can always be found.[9] This could also be argued from the fact that the number of free elements in the orthonormal matrix $\mathbf{Q}$ is $n(n+1)/2$ while $\mathbf{B}_0$ has n^2 unknown elements. Hence, we could never derive the structural form parameters in $\mathbf{B}_0$ from the reduced form covariance matrix Σ because there are fewer equations than unknowns so that additional covariance restrictions should be introduced as in (7.16) for the bivariate $VAR(1)$ case. Also contemporaneous identifying restrictions on the matrix $\mathbf{B}_0$ will be necessary.

Identifying restrictions in $SVAR$ models: contemporaneous and covariance restrictions

Since $\mathbf{A}(0) = \mathbf{I}_n$ from the specification of the reduced form lag polynomial above and, otherwise and except for $\mathbf{B}_0$, from the construction of this reduced form polynomial $\mathbf{A}(L)$ it is clear that the number of parameters in $\mathbf{A}(L)$ (exactly) matches the number of parameters in $\mathbf{B}(L)$ ($\mathbf{A}_i := -\mathbf{B}_0^{-1}\mathbf{B}_i$ for $i = 1, 2, \ldots, p$), the reduced form $VAR(p)$ as a whole has $n^2 - n(n+1)/2 = (n^2 - n)/2 = n(n-1)/2$ parameters less than the $SVAR(p)$ model. Since the properties of the data are determined by the reduced form parameters, any attempt to determine the structural (form) parameters from properties of the data will face indeterminacy, unless we can find $(n^2 - n)/2$ identifying restrictions. Remind that this was just one restriction in the bivariate case in (7.5), but the restriction in that section is in the tradition of Blanchard and Quah (1989) a long run restriction, i.e. on the sum of coefficients in $\mathbf{B}^{-1}(L) =: \Psi(L)$ in the bivariate impulse response function (7.17) and not on the contemporaneous $\mathbf{B}_0$ matrix. Notice that these

[9] According to the *Cholesky decomposition*, one can always find a $\mathbf{Q}^*$ such that $\mathbf{Q}^* = \mathbf{R}\mathbf{Q}$, with $\mathbf{R}$ another orthonormal matrix $\mathbf{R}\mathbf{R}' = \mathbf{I}$, and $\mathbf{Q}^*\mathbf{Q}^{*\prime} = \mathbf{R}\mathbf{Q}\mathbf{Q}'\mathbf{R}' = \mathbf{R}\mathbf{R}' = \mathbf{I}$.

restrictions are derived from a more elegant economic theory than the 'delayed reaction' theory underlying most zero restrictions on $\mathbf{B}_0$ in Sims's tradition, but that the Blanchard-Quah (1989) types of restrictions generally apply on quite simple VAR models with low dimension (generally with only two orthogonal shocks, 'aggregate demand shocks' and 'aggregate supply shocks').[10]

7.2 $VARMA$ models

A $VARMA(p,q)$ model can directly be defined from (7.1) $\mathbf{B}(L)\,\mathbf{y}_t + \mathbf{\Phi}(L)\,\varepsilon_t = \mathbf{0}$, which can be written as a $VMA(\infty)$ as in (7.12).

The identification, estimation, derivation of impulse response functions, and forecasting of $VARMA(p,q)$ models can be derived from the direct extension of the univariate $ARMA(p,q)$ models in Section 5.4 and Subsection 5.9.3. Also multivariate cointegration can be extended from the univariate approach in Chapter 6, but this is not straighforward. Hence, it will be discussed in the next section.

7.2.1 Multivariate cointegration

Definition

If all the components of a vector $\mathbf{x}_t$ are integrated of order $d > 0$, or $\mathbf{x}_t \sim I(d)$, then if there exists a linear combination which is integrated of a lower order b, both variables are said to be *cointegrated of order* $d - b$ with $b < d$, or $\mathbf{x}_t \sim CI(d,b)$. In this case, there exists a nonzero vector $\boldsymbol{\alpha}$ such that $\boldsymbol{\alpha}'\mathbf{x}_t \sim I(d-b)$ and the vector $\boldsymbol{\alpha}$ is then called the *cointegrating vector* (if $\mathbf{x}_t$ has $n > 2$ components, there may be more than one cointegrating vector).

The relationship $\alpha' x_t = 0$ is called a *long term relationship.*

Proposition 7.1 *If a long term relationship exists in the bivariate case, this long term relationship is unique.*

Proof. This can easily be proven by contradiction. Suppose there exist two distinct cointegrating vectors (parameters) α and γ such that $\{x_t + ay_t\}$ and $\{x_t + cy_t\}$ are both $I(0)$. Then also $(\alpha - \gamma)\,y_t$ is $I(0)$ because subtracting one $I(d)$ process from another cannot lead to a series integrated of order $(d+1)$ or higher. Hence, $(\alpha - \gamma)\,y_t \sim I(1)$ and $\alpha x_t \sim I(0)$. ■

Lemma 7.1 *Any $VAR(p)$ model $\mathbf{x}_t = \boldsymbol{\mu} + \sum_{i=1}^{p} \mathbf{\Pi}_i \mathbf{x}_{t-i} + \mathbf{e}_t$ with $\mathbf{e}_t \in WN$ with variance-covariance matrix $\mathbf{\Sigma}$ can be rewritten in an error correction form (ECM)*

$$\Delta\mathbf{x}_t = \mathbf{\Pi}\mathbf{x}_{t-1} + \textstyle\sum_{i=1}^{p-1} \mathbf{\Theta}_i \Delta\mathbf{x}_{t-i} + \mathbf{e}_t \qquad (7.20)$$

[10] For example, it might be argued that the long run effects of a monetary policy in a three-dimensional VAR on money stock, prices, and wages should all be the same. This might then be taken to mean that, in a system where these variables appear in differenced form, the elements in the column of $\mathbf{B}^{-1}(1)$ corresponding to these three variables should all be the same.

with $\mathbf{\Pi}$ *equal to the long run matrix* $-\mathbf{\Pi}(1)$ *and* $\mathbf{\Theta}_i := -(\mathbf{\Pi}_2+\mathbf{\Pi}_3+...+\mathbf{\Pi}_i), i = 1, 2, \ldots, p-1$.

Proof By induction:

For $p = 2 : \mathbf{x}_t - \mathbf{x}_{t-1} = \boldsymbol{\mu} + (\mathbf{\Pi}_1 - \mathbf{I}_n)\mathbf{x}_{t-1} - \mathbf{\Pi}_2(-\mathbf{x}_{t-2}) + \mathbf{e}_t = \boldsymbol{\mu} + (\mathbf{\Pi}_1 - \mathbf{I}_n)\mathbf{x}_{t-1} - \mathbf{\Pi}_2\Delta\mathbf{x}_{t-1} + \mathbf{\Pi}_2\mathbf{x}_{t-1} + \mathbf{e}_t = \boldsymbol{\mu} + (\mathbf{\Pi_1} + \mathbf{\Pi_2} - \mathbf{I_n})\mathbf{x}_{t-1} - \mathbf{\Pi}_2\Delta\mathbf{x}_{t-1} + \mathbf{e}_t$.

For $p = 3 : \Delta\mathbf{x}_t = \boldsymbol{\mu} + (\mathbf{\Pi}_1 - \mathbf{I}_n)\mathbf{x}_{t-1} - \mathbf{\Pi}_2(-\mathbf{x}_{t-1}) - \mathbf{\Pi}_3(-\mathbf{x}_{t-3}) + \mathbf{e}_t = \boldsymbol{\mu} + (\mathbf{\Pi}_1 + \mathbf{\Pi}_2 - \mathbf{I}_n)\mathbf{x}_{t-1} - \mathbf{\Pi}_2\Delta\mathbf{x}_{t-1} - \mathbf{\Pi}_3(-\mathbf{x}_{t-3}) + \mathbf{e}_t = \boldsymbol{\mu} + (\mathbf{\Pi}_1 + \mathbf{\Pi}_2 - \mathbf{I}_n)\mathbf{x}_{t-1} - \mathbf{\Pi}_2\Delta\mathbf{x}_{t-1} - \mathbf{\Pi}_3\Delta\mathbf{x}_{t-2} - \mathbf{\Pi}_3(-\mathbf{x}_{t-2}) + \mathbf{e}_t = \boldsymbol{\mu} + (\mathbf{\Pi}_1 + \mathbf{\Pi}_2 - \mathbf{I}_n)\mathbf{x}_{t-1} - \mathbf{\Pi}_2\Delta\mathbf{x}_{t-1} - \mathbf{\Pi}_3\Delta\mathbf{x}_{t-2} - \mathbf{\Pi}_3(-\mathbf{x}_{t-2}) + \mathbf{e}_t = \boldsymbol{\mu} + (\mathbf{\Pi}_1 + \mathbf{\Pi}_2 + \mathbf{\Pi}_3 - \mathbf{I}_n)\mathbf{x}_{t-1} - (\mathbf{\Pi}_2 + \mathbf{\Pi}_3)\Delta\mathbf{x}_{t-1} - \mathbf{\Pi}_3\Delta\mathbf{x}_{t-2} + \mathbf{e}$.

Proceeding for a general $p > 3$, we can rewrite the $VAR(p)$ as the vector error correction ($VECM$) model (7.20).∎

If in the above multivariate cointegration there exist exactly r linearly independent cointegrating vectors with $r \leqq n-1$, then these can be gathered into an $n \times r$ matrix $\boldsymbol{\alpha}$; the rank of $\boldsymbol{\alpha}$ will be r and is called the *cointegrating rank.*

For the case where $d = b = 1$ with cointegrating rank r, the Granger representation theorem holds (see also Chapter 6 on univariate cointegration).

Theorem 7.1 *The* ***Granger Representation Theorem*** *for the multivariate case proves that if there is an* $n \times 1$ *vector* $\mathbf{x}_t \sim CI(1,1)$ *with cointegrating rank* $r < n$*, then* $\exists\ \boldsymbol{\beta}$ *(*$n \times r$ *matrix) such that* $\boldsymbol{\beta}'\mathbf{x}_t \sim I(0)$*, and* $\mathbf{x}_t$ *has a* $VARMA$ *representation, which is a direct extension of (7.10):*

$$\mathbf{\Pi}(L)\mathbf{x}_t = \boldsymbol{\mu} + \mathbf{D}(L)\mathbf{e}_t \tag{7.21}$$

with i) $\mathbf{e}_t \in WN$, ii) *rank* $(\mathbf{\Pi}(1)) = r$, *and* iii) $\mathbf{\Pi}(1) = \boldsymbol{\alpha}\boldsymbol{\beta}'$ *with* $\boldsymbol{\alpha}$ *and* $\boldsymbol{\beta}$ $n \times r$ *matrices of rank* r *(for the proof see Engle and Granger (1987)).*

Three cases can now be distinguished:

Case 1: $\mathbf{\Pi}(1)$ is of full rank ($r = n$) and any linear combination of $\mathbf{x}_{t-p}$ is stationary; in this case a usual $VAR(p)$ can be run in levels;

Case 2: $\mathbf{\Pi}(1)$ has less than full rank but at least a rank of 1 ($0 < r < n$); in this case there are *some* linear combinations of $\mathbf{x}_{t-p}$ (or of $\mathbf{x}_t$) that are stationary so that $\mathbf{x}_t$ is cointegrated; in this case the $VAR(p)$ in levels is consistent but inefficient (if you know the cointegrating vector) and the $VAR(p)$ in differences $\Delta\mathbf{x}_t$ is misspecified;

Case 3: $\mathbf{\Pi}(1)$ has rank zero ($r = 0$) so that no linear combination of $\mathbf{x}_t$ is stationary; then $\Delta\mathbf{x}_t$ is stationary with no cointegration; in this case we run a usual $VAR(p)$ in first differences $\Delta\mathbf{x}_t$.

Testing multivariate cointegration

Johansen's method (1988) tests the rank of $\mathbf{\Pi}$ (or, equivalently, the number of columns in the $n \times r$ matrix $\boldsymbol{\beta}$ of cointegrating vectors) by means of an LR test statistic (see Subsection 3.1.3).

This test about the dimension of a subspace of the cointegrating vectors can be derived as follows; consider a $VAR(p)$ process $\{\mathbf{x}_t, t \in \mathbb{Z}\}$ as in (7.11) with polynomial matrices $\mathbf{\Pi}(L)$ instead of $\mathbf{A}(L)$, the components of which are nonstationary $I(1)$ and $\mathbf{e}_t \in WN$ with variance-covariance matrix $\mathbf{\Sigma}$. Consider the null H_0: the dimension of the subspace of the cointegrating vectors is r.

Under the assumption of normality of the errors $\mathbf{e}_t$ the conditional loglikelihood can be written as:

$$\ln \ell = -n\frac{T}{2}\ln(2\pi) - \frac{T}{2}\ln|\mathbf{\Sigma}|$$
$$-\frac{1}{2}\sum_{t=1}^{T}\left(\Delta\mathbf{x}_t - \mathbf{\Pi}\mathbf{x}_{t-1} - \sum_{i=1}^{p-1}\mathbf{\Theta}_i\Delta\mathbf{x}_{t-i}\right)'\mathbf{\Sigma}^{-1}\left(\Delta\mathbf{x}_t - \mathbf{\Pi}\mathbf{x}_{t-1} - \sum_{i=1}^{p-1}\mathbf{\Theta}_i\Delta\mathbf{x}_{t-i}\right),$$ so that an ML estimator can be derived from maximizing the *concentrated loglikelihood function* with respect to the parameter matrices $\mathbf{\Pi}_1, \mathbf{\Pi}_2, \ldots, \mathbf{\Pi}_{p-1}$. From the Granger Representation Theorem 7.1 $\mathbf{\Pi} = -\boldsymbol{\alpha}\boldsymbol{\beta}'$, where $\boldsymbol{\alpha}$ is the $n \times r$ matrix of error correction terms, β' is the $r \times n$ matrix of coefficients of the cointegrating vectors, and $\boldsymbol{\beta}'\mathbf{x}_{t-1}$ are the r cointegrated variables. Since the interest is in $\boldsymbol{\alpha}$ and $\boldsymbol{\beta}$, an auxiliary OLS regression of $\Delta\mathbf{x}_t$ on $\Delta\mathbf{x}_{t-1}, \Delta\mathbf{x}_{t-2}, \ldots, \Delta\mathbf{x}_{t-p+1}$ is performed and the resulting residuals are called $\mathbf{r}_{0t}$; regressing $\mathbf{x}_{t-1}$ on these variables the resulting OLS residuals are called $\mathbf{r}_{1t}$. Referring to the $VECM$ equation (7.20), the regression equation is thus reduced to $\mathbf{r}_{0t} = -\boldsymbol{\alpha}\boldsymbol{\beta}'\mathbf{r}_{1t} + \mathbf{e}_t^*$ with the same variance-covariance matrix $\mathbf{\Sigma}$. Taking $\boldsymbol{\alpha}$ as fixed, the loglikelihood can also be concentrated with respect to $\boldsymbol{\beta}$ and $\mathbf{\Sigma}$. For solving the multivariate regression problem of $\mathbf{r}_{0t}$ on $\mathbf{r}_{1t}$, the $n \times n$ matrices of sum of squares and sums of products of $\mathbf{r}_{0t}$ and $\mathbf{r}_{1t}$ can be defined as $\begin{bmatrix}\mathbf{S}_{00} & \mathbf{S}_{01}\\ \mathbf{S}_{10} & \mathbf{S}_{11}\end{bmatrix}$. Maximizing the loglikelihood function with respect to $\boldsymbol{\alpha}$ holding $\boldsymbol{\beta}$ constant and then maximizing with respect to $\boldsymbol{\beta}$ in a second step, $\hat{\boldsymbol{\alpha}}' = \left(\boldsymbol{\beta}'\mathbf{S}_{11}\boldsymbol{\beta}\right)^{-1}\boldsymbol{\beta}'\mathbf{S}_{10}$, so that the conditional maximum of the loglikelihood function (which is the maximum of the concentrated likelihood) is:

$[L_c(\boldsymbol{\beta})]^{-2/T} = -\left|\mathbf{S}_{00} - \mathbf{S}_{01}\boldsymbol{\beta}\left(\boldsymbol{\beta}'\mathbf{S}_{11}\boldsymbol{\beta}\right)^{-1}\boldsymbol{\beta}'\mathbf{S}_{10}\right|$, which can be maximized with respect to $\boldsymbol{\beta}$ (implying the minimization of the determinant with respect to $\boldsymbol{\beta}$). Following Johansen (1991) this problem reduces to $\min\limits_{\boldsymbol{\beta}}\frac{|\boldsymbol{\beta}'\mathbf{S}_{11}\boldsymbol{\beta} - \boldsymbol{\beta}'\mathbf{S}_{10}\mathbf{S}_{00}^{-1}\mathbf{S}_{01}\boldsymbol{\beta}||\mathbf{S}_{00}|}{|\boldsymbol{\beta}'\mathbf{S}_{11}\boldsymbol{\beta}|}$, which is a problem of eigenvalues and eigenvectors because the maximum of the loglikelihood function is found by finding the eigenvalues of $\left|\mathbf{S}_{10}\mathbf{S}_{00}^{-1}\mathbf{S}_{01} - \lambda\mathbf{S}_{11}\right| = \mathbf{0}$. These eigenvalues are canonical correlations between $\mathbf{r}_{1t}$ and $\mathbf{r}_{0t}$, i.e. after conditioning on the lagged variables $\Delta\mathbf{x}_{t-1}, \Delta\mathbf{x}_{t-2}, \ldots, \Delta\mathbf{x}_{t-p+1}$ those linear combinations of $\mathbf{x}_{t-1}$ that are highly correlated with linear combinations of $\Delta\mathbf{x}_t$ are look for. If λ_i are the canonical correlations from the solution of $\left|\mathbf{S}_{11}^{-1}\mathbf{S}_{10}\mathbf{S}_{00}^{-1}\mathbf{S}_{01} - \lambda\mathbf{I}\right| = \mathbf{0}$, then $1-\lambda_i$ are the eigenvalues of $\left(\mathbf{I} - \mathbf{S}_{11}^{-1}\mathbf{S}_{10}\mathbf{S}_{00}^{-1}\mathbf{S}_{01}\right)$.

Since the value of a determinant of a matrix is equal to the product of the eigenvalues one has

$$\prod_{i=1}^{n}(1-\lambda_i) = \left|\mathbf{I} - \mathbf{S}_{11}^{-1}\mathbf{S}_{10}\mathbf{S}_{00}^{-1}\mathbf{S}_{01}\right| = \frac{\left|\mathbf{S}_{11}^{-1} - \mathbf{S}_{10}\mathbf{S}_{00}^{-1}\right|}{|\mathbf{S}_{11}|} = \frac{\left|\mathbf{S}_{00} - \mathbf{S}_{01}\mathbf{S}_{11}^{-1}\mathbf{S}_{10}\right|}{|\mathbf{S}_{00}|}.$$

Table 7.1: Trace and maximum eigenvalue tests without intercept and deterministic trend

$n-r$	1	2	3	4	5
99%	6.51	16.31	29.75	45.58	66.52
95%	3.84	12.53	24.31	39.89	59.46
90%	2.86	10.47	21.63	36.58	55.44
Test: order r versus order $r+1$					
99%	6.51	15.69	22.99	28.82	35.17
95%	3.84	11.44	17.89	23.80	30.04
90%	2.86	9.52	15.59	21.58	27.62

Theorem 7.2 *The LR test statistic of the null (the dimension of the subspace of the cointegrating vectors is r) is*

$\xi_{LR} = 2\left(\ln\hat{\ell} - \ln\ell_c\left(\hat{\boldsymbol{\beta}}\right)\right) = -T\sum_{k=r+1}^{n}\ln(1-\hat{\lambda}_k)$, where the $\hat{\lambda}_k$s are the $n-r$ smallest eigenvalues of $\left|\mathbf{S}_{10}\mathbf{S}_{00}^{-1}\mathbf{S}_{01} - \lambda\mathbf{S}_{11}\right| = \mathbf{0}$ (see Johansen (1988) for the proof).

The LR test statistics make use of the fact if there are r cointegrating vectors, then these $n-r$ smallest eigenvalues are zero. The corresponding r eigenvectors are chosen as cointegrating vectors. This way of imposing $n-r$ restrictions yields an asymptotically efficient estimator of the cointegrating vectors.

Another interesting test tests the null that the cointegration subspace is of order r versus $r+1$. The LR test statistic is then $-T\ln(1-\hat{\lambda}_{r+1})$. The latter test is called the *maximum eigenvalue test* based on $(r+1)^{th}$ largest estimated eigenvalue testing and the former test is called the *trace test* testing whether the smallest $n-r$ estimated eigenvalues are significantly different from zero.

Although these test statistics are both LR test statistics, they are not (asymptotically) distributed as the usual χ^2-distribution! Instead, the appropriate (asymptotic) distributions are multivariate extensions of the Dickey-Fuller distributions (with critical values in Figure 5.1). These asymptotic distributions depend on the assumptions about deterministic trends in the data. Johansen (1988) computes the 99%, 95%, and 90% quantiles of this multivariate extension of the DF distribution.

An excerpt of it is presented in Table 7.1, where the upper part gives the critical values of the *trace test* at the levels 1%, 5%, and 10%, and the lower part the corresponding critical values of the *maximum eigenvalue test,* both under the assumption of no deterministic trend and no intercept. As you can observe, the critical values depend on the number $(n-r)$ of nonstationary components under the null.[11]

If an intercept is present without any deterministic trend another set of critical values is used (see Table 7.2).

[11] Note that when $n-r=1$ the two test statistics are identical and thus have the same distribution.

Table 7.2: Trace and maximum eigenvalue tests with intercept and no deterministic trend

$n-r$	1	2	3	4	5
99%	6.65	20.04	35.65	54.46	76.07
95%	3.76	15.41	29.68	47.21	68.52
90%	2.69	13.33	26.79	43.95	64.84
Test: order r versus order $r+1$					
99%	6.65	18.63	25.52	32.24	38.77
95%	3.76	14.07	20.97	27.07	33.46
90%	2.69	12.07	18.60	24.60	30.90

To use these tests we denote the estimated eigenvalues of the long run matrix $\hat{\mathbf{\Pi}}$ in non-increasing order as $\hat{\lambda}_1 \geqq \hat{\lambda}_2 \geqq \ldots \geqq \hat{\lambda}_n$ where a nonzero eigenvalue correspond to a cointegrating vector and the cointegration tests boil down to testing that the corresponding $\hat{\lambda}$s are significantly different from zero. If the test is $H_0 : rank \leqq r$ versus $H_1 : \ r < rank \leqq n$, we have the trace test in Theorem 7.2, testing whether the smallest $n-r$ eigenvalues are significantly different from zero. It the test is $H_0 : rank \leqq r$ versus the more restrictive alternative $H_1 : \ rank = r+1$ we have the maximum eigenvalue test based on the $(r+1)^{th}$ largest eigenvalue.

Remark 7.1 *To see how cointegration affects the limiting impulse response of the levels of the vector $\mathbf{x}_t$ where all elements are assumed to be integrated of order one, consider the implied MA representation of the stationary differences of $\mathbf{x}_t : (I-L)\mathbf{x}_t = \mathbf{B}(L)\boldsymbol{\eta}_t$. Since under cointegration $\boldsymbol{\alpha}'\mathbf{x}_t \sim I(0)$ is an extra restriction, it must imply a restriction on $\mathbf{B}(L)$: the elements of $\mathbf{x}_t$ are said to be cointegrated with cointegrating vectors $\boldsymbol{\alpha}_{\ i}$ if and only if $\boldsymbol{\alpha}_i'\mathbf{B}(1) = \mathbf{0}$, which implies that the rank of $\mathbf{B}(1)$ is equal to the number of elements of $\mathbf{x}_t$ minus the number of cointegrating vectors $\boldsymbol{\alpha}_i$ (this is called the reduced rank property of $\mathbf{B}(1)$ in the case of cointegration). Since $\mathbf{B}(1)$ is the limiting impulse response of the levels of $\mathbf{x}_t$ so that, for example, $\mathbf{B}(1)_{yz}$ is the long run response of y to a unit z shock, here it is considered the very simple and common bivariate case $\boldsymbol{\alpha} = (1,-1)'$ to study the effect of cointegration on $\mathbf{B}(1)$. The reduced rank of $\mathbf{B}(1)$ means then: $\boldsymbol{\alpha}_i'\mathbf{B}(1) = (1,-1)\begin{bmatrix} \mathbf{B}(1)_{yy} & \mathbf{B}(1)_{yz} \\ \mathbf{B}(1)_{zy} & \mathbf{B}(1)_{zz} \end{bmatrix} = \mathbf{0}$, so that $\mathbf{B}(1)_{yy} = \mathbf{B}(1)_{zy}$ and $\mathbf{B}(1)_{yz} = \mathbf{B}(1)_{zz}$. In other words, each variable's long run response to a shock must be the same. The reason is intuitive: if y and z had different long run responses to a shock, the difference $y-z$ would not be stationary. Since this difference must be stationary from the (assumed) cointegration property, the response of this difference must vanish then.*

7.2.2 Multivariate causality

Suppose two variables similar to Section 6.2. One variable has no response to the shocks in the other variable. The shock variable fails to *Granger cause* the variable that does not respond.

Definition 7.2 *Granger causality can be defined as follows:* z_t *Granger causes* x_t *if* z_t *helps to forecast* x_t *given the past of* x_t.

Example 7.8 z_t *does not Granger cause* x_t *if* $b(L) = 0$ *in the following bivariate* $VAR(1)$ *model:*

$$\begin{aligned} x_t &= a(L)\, y_{t-1} + b(L)\, z_{t-1} + \delta_t \\ z_t &= c(L)\, y_{t-1} + d(L)\, z_{t-1} + \omega_t \end{aligned}$$

In autoregressive representation the above $VAR(1)$ model becomes:

$$\begin{pmatrix} x_t \\ z_t \end{pmatrix} = \begin{pmatrix} a(L) & b(L) \\ c(L) & d(L) \end{pmatrix} \begin{pmatrix} x_{t-1} \\ z_{t-1} \end{pmatrix} + \begin{pmatrix} \delta_t \\ \omega_t \end{pmatrix}$$

$$\begin{pmatrix} I - L\,a(L) & -L\,b(L) \\ -L\,c(L) & I - L\,d(L) \end{pmatrix} \begin{pmatrix} x_t \\ z_t \end{pmatrix} = \begin{pmatrix} \delta_t \\ \omega_t \end{pmatrix}$$

or $\begin{pmatrix} a^*(L) & b^*(L) \\ c^*(L) & d^*(L) \end{pmatrix} \begin{pmatrix} x_t \\ z_t \end{pmatrix} = \begin{pmatrix} \delta_t \\ \omega_t \end{pmatrix}$. Thus z does not Granger cause $x \Leftrightarrow b^*(L) = 0$, or if the autoregressive matrix lag polynomial is lower triangular.

The setting in this example can be directly extended to vector processes $\{\mathbf{x}_t, t \in \mathbb{Z}\}$ and $\{\mathbf{z}_t, t \in \mathbb{Z}\}$. Further details can be found in Gouriéroux and Monfort (1997), pp. 369-382.

Hence, Granger causality is defined as the presence of feedback from one variable to the another, with Granger noncausality defined as the absence of such feedback. Following Granger (1969) and Section 6.2, the following theorem can be formulated.

Theorem 7.3 *The variable* y *does not Granger causethe variable* z *if the marginal pdf* $f_z(\cdot)$ *does not depend on* $\Im_y(t-1)$, *i.e.* $f_z(z_t | \Im_x(t-1), \cdot) = f_z(z_t | \Im_z(t-1), \cdot)$.

7.2.3 Dynamic simultaneous equations models

Consider the following $VARMA(p,q)$ model $\mathbf{B}(L)\,\mathbf{y}_t = \mathbf{\Phi}(L)\,\varepsilon_t$ with the matrix (valued) polynomials $\mathbf{B}(L) := I - \sum_{i=1}^{p} \mathbf{B}_i L^i$ and $\mathbf{\Phi}(L) := I + \sum_{j=1}^{q} \mathbf{\Phi}_j L^j$, and where $\mathbf{y}_t$ can be partitioned as $\mathbf{y}_t := \binom{\mathbf{x}_t}{\mathbf{z}_t}$,where $\mathbf{x}_t \in \mathbb{R}^g$ and $\mathbf{z}_t \in \mathbb{R}^l$. Correspondingly, the $VARMA(p,q)$ model can be partitioned as:

$$\begin{pmatrix} \mathbf{B}_{11}(L) & \mathbf{B}_{12}(L) \\ \mathbf{B}_{21}(L) & \mathbf{B}_{22}(L) \end{pmatrix} \begin{pmatrix} \mathbf{x}_t \\ \mathbf{z}_t \end{pmatrix} = \begin{pmatrix} \mathbf{\Phi}_{11}(L) & \mathbf{\Phi}_{12}(L) \\ \mathbf{\Phi}_{21}(L) & \mathbf{\Phi}_{22}(L) \end{pmatrix} \begin{pmatrix} \varepsilon_{1t} \\ \varepsilon_{2t} \end{pmatrix} \tag{7.22}$$

with $\varepsilon_{1t} \overset{iid}{\sim} \varepsilon_{2t}$.

Suppose that the following restrictions are imposed $\mathbf{B}_{21}(L) = 0$, $\mathbf{\Phi}_{12}(L) = 0$ and $\mathbf{\Phi}_{21}(L) = 0$ which means that $\mathbf{x}_t$ is *endogenous* and $\mathbf{z}_t$ is *exogenous*. This leads to the set of structural equations:

$$\mathbf{B}_{11}(L)\,\mathbf{x}_t + \mathbf{B}_{12}(L)\,\mathbf{z}_t = \mathbf{\Phi}_{11}(L)\,\varepsilon_{1t}, \tag{7.23}$$

which is called a *dynamic simultaneous equations model* or *DSEM*, and to a *VARMA* process generating the exogenous variables:

$$\mathbf{B}_{22}(L)\,\mathbf{z}_t = \mathbf{\Phi}_{22}(L)\,\varepsilon_{2t}. \tag{7.24}$$

The structural equations (7.23) can be expressed as:

$$\sum_{i=0}^{r} \mathbf{B}_{11,i}\mathbf{x}_{t-i} + \sum_{j=0}^{s} \mathbf{B}_{12,j}\mathbf{z}_{t-j} = \mathbf{\Phi}_{11}(L)\,\varepsilon_{1t} = \boldsymbol{\eta}_t.$$

Note that the 'simultaneity' of the model is a consequence of the property that $\mathbf{B}_{11,0} \neq \mathbf{I}_g$.The reduced form from these structural equations satisfies

$$\mathbf{x}_t = -\mathbf{B}_{11,0}^{-1}\sum_{i=1}^{r} \mathbf{B}_{11,i}\mathbf{x}_{t-i} - \mathbf{B}_{11,0}^{-1}\sum_{j=0}^{s} \mathbf{B}_{12,j}\mathbf{z}_{t-j} + \mathbf{B}_{11,0}^{-1}\boldsymbol{\eta}_t \tag{7.25}$$

and the final form satisfies

$$\mathbf{x}_t = -\mathbf{B}_{11}^{-1}(L)\mathbf{B}_{12}(L)\,\mathbf{z}_t - \mathbf{B}_{11}^{-1}(L)\boldsymbol{\eta}_t, \tag{7.26}$$

where $-\mathbf{B}_{11}^{-1}(L)\mathbf{B}_{12}(L)$ is an ∞ dimensional matrix polynomial containing dynamic multipliers that describe the responses of $\mathbf{x}_t$ to unit shocks in the exogenous variables $\mathbf{z}_t$.

Premultiplying the final form (7.26) by $|\mathbf{B}_{11}(L)|$ and denoting the adjoint matrix associated with $\mathbf{B}_{11}(L)$ as $\mathbf{B}_{11}^{*}(L)$, yields the set of *fundamental dynamic equations*.

$$|\mathbf{B}_{11}(L)|\,\mathbf{x}_t = -\mathbf{B}_{11}^{*}(L)\mathbf{B}_{12}(L)\,\mathbf{z}_t + \mathbf{B}_{11}^{*}(L)\boldsymbol{\eta}_t, \tag{7.27}$$

in which each endogenous variable depends only on its own lagged values and on the exogenous variables.

7.2.4 *DSEMs* and *SVARs*

DSEMs and *SVARs* can both be thought of as versions of the general linear stochastic difference equation model, that do not consider the *MA* part in the above *VARMA*(p,q) model, both models can be viewed as special cases of $\mathbf{B}(L)\mathbf{y}_t = \mathbf{c} + \varepsilon_t$. The usual *DSEM* has two standard forms. In one standard form, the system satisfies a Granger causal ordering, i.e. (7.22) can be expressed as $\begin{pmatrix} \mathbf{B}_{11}(L) & \mathbf{B}_{12}(L) \\ \mathbf{0} & \mathbf{B}_{22}(L) \end{pmatrix}\begin{pmatrix} \mathbf{x}_t \\ \mathbf{z}_t \end{pmatrix} = \mathbf{c} + \begin{pmatrix} \varepsilon_{1t} \\ \varepsilon_{2t} \end{pmatrix}$, where the $\mathbf{z}_t$s are *exogenous*, while in the other standard form, only $\mathbf{B}_{11,0}$ is assumed to be block-triangular so that the $\{\mathbf{z}_s, s \leqq t\}$ are *predetermined*. For both types of models the usual claim for a structural interpretation is that an equation or block of equations can be altered or replaced to represent a particular intervention (as e.g. a change in policy behavior).

The most important difference between $DSEMs$ and $SVARs$ is that $SVARs$ usually possess a diagonal variance-covariance matrix $E(\varepsilon_t\varepsilon_t') =: \mathbf{\Omega}$ (see Subsections 7.1.1- 7.1.3) and $DSEMs$ usually not. Hence, one can also incorporate restrictions in $\mathbf{\Omega}$ into a $DSEM$ framework similar to (7.16) in Section 7.1.2 for the simple bivariate $SVAR(1)$ framework and the exogeneity and predeterminedness assumptions in an $SVAR$ framework can also be used.

7.3 Exercises and cases

In this section theoretical and empirical exercises and also three solved empirical cases will be presented.

7.3.1 Solved theoretical exercises

Exercise 7.1 *Consider the simple bivariate* $VAR(1)$ *model:* $y_t = ay_{t-1} + bz_{t-1} + v_t$ *and* $z_t = cy_{t-1} + dz_{t-1} + w_t$, *or in matrix notation:*

$$\left[I - \begin{pmatrix} a & b \\ c & d \end{pmatrix} L\right] \mathbf{x}_t = \mathbf{e}_t \text{ and } \Pi(1) = \begin{bmatrix} 1-a & -b \\ -c & 1-d \end{bmatrix}.$$

An example of a singular $\Pi(1)$ is $b = 1-a$ and $c = 1-d$ because then $\Pi(1) = \begin{bmatrix} b & -b \\ -c & c \end{bmatrix}$. The ECM representation is from (7.20) and the Granger Representation Theorem 7.1: $\Delta x_t = -\boldsymbol{\alpha\beta}'\mathbf{x}_{t-1} + \sum_{i=1}^{p-1} \mathbf{\Theta}_i \Delta\mathbf{x}_{t-i} + \mathbf{e}_t$ with $\Pi(1) = \begin{bmatrix} b \\ -c \end{bmatrix} \begin{bmatrix} 1 & -1 \end{bmatrix}$ leads to $\boldsymbol{\alpha} = \begin{bmatrix} b \\ -c \end{bmatrix}$ and $\boldsymbol{\beta} = \begin{bmatrix} 1 \\ -1 \end{bmatrix}$, or the ECM representation becomes:

$$\begin{aligned} \Delta y_t &= -b\left(y_{t-1} + z_{t-1}\right) + v_t, \\ \Delta z_t &= c\left(y_{t-1} - z_{t-1}\right) + w_t. \end{aligned}$$

Subtracting these two equations,

$$\begin{aligned} \Delta\left(y_t - z_t\right) &= \left(-b + c\right)\left(y_{t-1} - z_{t-1}\right) + \left(v_t - w_t\right), \text{ or} \\ y_t - z_t &= \left(1 - \left(b + c\right)\right)\left(y_{t-1} - z_{t-1}\right) + \left(v_t - w_t\right), \end{aligned}$$

so that $y_t - z_t$ follows a stationary AR(1) model (since $0 < (b+c) < 2)$.

It is very easy to derive the $MA(\infty)$ representation from the system in Δy_t and Δz_t, because from the latter equation.$y_t - z_t = \left(I - \left(I - (b+c)\right)L\right)^{-1}(v_t - w_t) =:$ $(I - \theta L)^{-1}(v_t - w_t)$ and substituting this into the above system: $\begin{bmatrix} \Delta y_t \\ \Delta z_t \end{bmatrix} = (I - w_t L)^{-1} \begin{bmatrix} (I-\theta L) - b & b \\ c & (I-\theta L) - c \end{bmatrix} \begin{bmatrix} v_t \\ w_t \end{bmatrix}$ with $\theta := 1 - b - c$. Evaluating the right-hand side at $L = 1$ to get: $(b+c)^{-1} \begin{bmatrix} c & b \\ c & b \end{bmatrix}$. Denoting the last matrix $\check{\mathbf{\Pi}}(L)$ the following long term restrictions are obtained: $\boldsymbol{\beta}'\check{\mathbf{\Pi}}(1) = \mathbf{0}$ and $\check{\mathbf{\Pi}}(1)\boldsymbol{\alpha} = \mathbf{0}$.■

Exercise 7.2 *Write the bivariate VAR process* $\begin{pmatrix} I-0.9L & 0.1L \\ 0.4L & I-0.6L \end{pmatrix}\begin{pmatrix} y_{1t} \\ y_{2t} \end{pmatrix} = \varepsilon_t$ *as a VECM and perform a cointegration analysis.*

The determinant of $\mathbf{B}(L)$ is given by $(I-L)(I-0.5L)$. Then, the $VECM$ satisfies

$$(I-L)\begin{pmatrix} y_{1t} \\ y_{2t} \end{pmatrix} = (I-0.5L)^{-1}\begin{pmatrix} I-0.6L & -0.1L \\ -0.4L & I-0.9L \end{pmatrix}\varepsilon_t.$$

Hence; the process $\{\mathbf{y}_t, t \in \mathbb{Z}\}$ is $I(1)$. The matrix $\mathbf{H}(L) = (I-0.5L)^{-1}\begin{pmatrix} I-0.6L & -0.1L \\ -0.4L & I-0.9L \end{pmatrix}$ is such that $\mathbf{H}(1) = 2\begin{pmatrix} 0.4 & -0.1 \\ -0.4 & 0.1 \end{pmatrix}$.

$\mathbf{H}(1)$ has rank equal to 1 and its nullspace is spanned by $\alpha = \begin{pmatrix} 1 \\ 1 \end{pmatrix}$, a cointegrating vector. Then, the ECM representation in (7.20) and in Theorem 7.1 can be written as $\begin{pmatrix} 0.1 \\ 0.4 \end{pmatrix}(y_{1,t-1} + y_{2,t-1}) + \Delta \mathbf{y}_t = \varepsilon_t$.■

7.3.2 Solved empirical exercise

Exercise 7.3 *Multilateral (effective) real exchange rate.*

The univariate long run purchasing power parity (PPP) hypothesis is given by $s_t + p_t^* - p_t \backsim I(0)$, with s_t the logarithmic nominal exchange rate, p_t is the logarithmic domestic price, and p_t^* the logarithmic foreign price for the economy in which s_t is expressed. The multivariate PPP hypothesis can now be written as the stationary representation of the effective real exchange rate:

$$\rho_t^i := \sum_{j \neq i} \omega_i^j \left(s_t^{ij} + p_t^j\right) - p_t^i \backsim I(0) \ (i, j = 1, 2, \cdots, n),$$

where s_t^{ij} is the logarithmic nominal exchange rate between currencies i and j (i.e., the unit price of currency j in terms of currency i), p_t^i is the logarithmic domestic price level in country i, and ω_i^j is the relative importance of country j for country i, measured for example by the share of country j in the imports of country i with $\sum_{j \neq i} \omega_i^j = 1$.

The effective Dollar exchange rates of the Euro (the currency unit of 12 countries-members of the European Monetary Union (EMU)), the British Pound, and the Japanese Yen can for example be tested by (principally) considering the stationarity of ρ_t^i ($i = EMU, UK, Japan$) with variables contained in the multivariate vector of 7 variables
$\mathbf{x}_t := \left(\omega_i^{EMU} \cdot s_t^{EMU}, \omega_i^{UK} \cdot s_t^{UK}, \omega_i^{JP} \cdot s_t^{JP}, \omega_i^{EMU} \cdot p_t^{EMU}, \omega_i^{UK} \cdot p_t^{UK}, \omega_i^{JP} \cdot p_t^{JP}, p_t^{US}\right)'$.

In the following, cointegration tests for ρ_t^{US} will be conducted. In this example monthly data from January 1999 to November 2004 is analyzed, so that the length of the series is relatively small (71 observations). Therefore the choice of parameter k must be made prudently in order to avoid wasting degrees of

Table 7.3: Trace statistic cointegration test

Hypotheses	Eigenvalue	Trace Statistic	5%	1%
$H_0 : rank = 0$ vs $H_1 : 0 < rank \leqq 7$	0.698803	158.2771	124.24	133.57
$H_0 : rank = 0, 1$ vs $H_1 : 1 < rank \leqq 7$	0.429797	76.67769	94.15	103.18
$H_0 : rank = 0, 1, 2$ vs $H_1 : 2 < rank \leqq 7$	0.255526	38.47779	68.52	76.07

Table 7.4: Maximum eigen value statistic cointegration test

Hypotheses	Eigenvalue	Max-Eigenvalue Statistic	5%	1%
$H_0 : rank = 0$ vs $H_1 : rank = 1$	0.698803	81.59943	45.28	51.57
$H_0 : rank = 0, 1$ vs $H_1 : rank = 2$	0.429797	38.1999	39.37	45.1
$H_0 : rank = 0, 1, 2$ vs $H_1 : rank = 3$	0.255526	20.06526	33.46	38.77

freedom. To put ourselves on the safe side we choose to limit the lag length to 1.

The trace statistic cointegration test (with an intercept and without a deterministic trend) is given in Table 7.3 and the maximum eigen value statistic is presented in Table 7.4 accompanied by the corresponding 5% and 1% critical values (see Table 7.2).

Both tests indicate the existence of one cointegrating equation with a high degree of statistical significance.

The results of the cointegration estimation produce the following cointegrating or long term relationship (being the cointegrating vector from $\hat{\beta}$) normalized with respect to p_t^{US}:

$$\begin{aligned}
\hat{\rho}_t^{US} &= \omega_i^{EMU}(\underset{(0.00055)}{-0.002705} \cdot s_t^{EMU} + \underset{(0.00604)}{0.07291} \cdot p_t^{EMU}) + \\
&\quad \omega_i^{UK}(\underset{(0.00294)}{-0.001948} \cdot s_t^{UK} + \underset{(0.01536)}{0.09629} \cdot p_t^{UK}) + \\
&\quad \omega_i^{JP}(\underset{(0.00025)}{0.00028} \cdot s_t^{JP} + \underset{(0.01151)}{0.08358} \cdot p_t^{JP}) - p_t^{US}.
\end{aligned}$$

7.3.3 Solved empirical cases

Case 1

Corsetti *et al.* (2004) show that incomplete asset markets and a low price elasticity of tradables can account cuantitatively for the volatility of real exchange rates and their negative correlation with cross-country consumption ratios. It is found that two different patterns of the international transmission of productivity improvements generates the observed degree of international risk sharing: one associated with a strengthening, the other with a deterioration of the terms of trade and real exchange rate. There is evidence of the effect of technology shocks to US manufacturing, which questions the presumption that terms of trade movements foster international risk pooling.

Case 2

In Bruneau and De Bandt (2003) $SVAR$ models and the corresponding impulse response functions of monetary and fiscal policy in France, Germany and the euro area for the period $1979^{I} - 2000^{II}$ are estimated and tested. During these two decades monetary shocks exhibited significant correlation while fiscal shocks are found to be uncorrelated between France and Germany. At the same time, euro area fiscal shocks are largely impulsed by Germany. However, it is difficult to conclude that the latter shocks reflect purely idiosyncratic shocks, as they often reveal differences in the timing of fiscal adjustment.

Case 3

Peersman and Straub (2004) provide evidence for the impact of technology, labor supply, monetary policy and aggregate spending shocks on hours worked in the euro area. The empirics are derived from an $SVAR$ model consistent with both sticky prices and real business cycle models. Positive responses of hours to technology shocks are observed and technology shocks are found to explain business cycle fluctuations.

7.3.4 Unsolved exercises

Exercise 7.4 *Consider three $I(1)$ variables y_t, x_t, and z_t. Assume that y_t and x_t are cointegrated and that x_t and z_t are also cointegrated. Does this imply that y_t and z_t are also cointegrated? Why (not)?*

Exercise 7.5 *Show that in Exercise 7.1 with cointegration the standard VAR in differences $\Delta y_t = a(L)\Delta y_{t-1} + b(L)\Delta z_{t-1} + v_t$, $\Delta z_t = c(L)\Delta y_{t-1} + d(L)\Delta z_{t-1} + w_t$ is misspecified.*

Exercise 7.6 *Consider a nonstationary time series $\mathbf{y}_t \sim I(1)$ and a noncentral moment matrix $\mathbf{M}_T := \frac{1}{T^2}\sum_{t=1}^{T}\mathbf{y}_t\mathbf{y}_t'$. Assume that the components of $\mathbf{y}_t$ are cointegraited, that the subspace of the cointegrating vectors is of order one, and that it is spanned by a vector of the form $(-1, \alpha_2, \ldots, \alpha_n)' =: \boldsymbol{\alpha}$. Defining a* *solution* $\breve{\alpha}_T$ *of the problem:* $\min_{\alpha_2,\ldots,\alpha_n} (-1, \alpha_2, \ldots, \alpha_n)' \mathbf{M}_T \begin{bmatrix} -1 \\ \alpha_2 \\ \vdots \\ \alpha_n \end{bmatrix}$ *.i) Show that this problem is equivalent to performing OLS on the static model* $y_{1t} = \alpha_2 y_{2t} + \ldots + \alpha_n y_{nt} + \varepsilon_t$*. ii) Verify that the estimator* $\breve{\alpha}_T$ *is a consistent estimator of the cointegrating vector in which the first component is equal to* -1*.*

Exercise 7.7 *Consider a static simultaneous equation model defined as* $\mathbf{B}_0 \mathbf{y}_t + \mathbf{\Gamma}_0 \mathbf{z}_t = \varepsilon_t$*, where it is assumed that the disturbance term follows a normal distribution with mean zero and variance-covariance* $\mathbf{\Omega}$*. The diagonal elements of* $\mathbf{B}_0$ *are equal to one. i) Derive the pdf of* $\mathbf{y}_t$*; ii) Verify that it is not possible to derive the elements of* $\mathbf{B}_0, \mathbf{\Gamma}_0$*, and* $\mathbf{\Omega}$ *as functions of this pdf unambigously.*

Exercise 7.8 *Consider Case 2 on Gross Domestic Product in Section 5.10.4. Check whether the point forecasts can be significantly improved by including GDP of the two or three most important trading partners of the OECD country under investigation by using a VAR model.*

Exercise 7.9 *Univariate Long-Run Purchasing Power Parity (PPP). Examine the quarterly data on the US Dollar/UK Pound exchange rate, US and UK interest rates (downloadable from the OECD's Statistical Compendium http:// www.oecd.org).*
1) Test for a unit root in these three series.
2) Consider the US/UK univariate PPP model (as in Exercise 7.3) and test the corresponding regression for cointegration.

Exercise 7.10 *French GDP, wages, and unemployment. Examine the quarterly data about French GDP, wage rate, and standardized unemployment rate (downloadable from the OECD's Statistical Compendium http://www.oecd.org).*
1) Test for a unit root in the series.
2) Consider the regression model of French GDP on the wage rate. Test for cointegration.
3) Consider the regression model of wage rate on the French GDP. Test for cointegration.
4) Compare the estimation results in 2) and 3) and provide an economic interpretation.
6) Consider also the multivariate relationship among GDP, wages, and unemployment rate. Test for the existence of multivariate cointegration. Interpret the results.
5) Test for causality in the GDP-wage-unemployment relationship.

Chapter 8

Varying Parameter Models

Two types of varying parameter models will be analyzed. On the one hand, the specification and estimation of models with regimes determined by observable variables and regimes determined by unobservable variables and on the other hand, estimation, testing and forecasting of models with varying volatility. The former type of models concentrate on the conditional mean and are called *regime switching models* while the latter type concentrate on the conditional variance and are called *volatility models.*

8.1 Regime switching models

As put forward in Priestley (1988) and Franses and van Dijk (2000), p. 69, "a natural approach to modeling economic time series with nonlinear models seems to be to define different *states of the world* or *regimes*, and to allow for the possibility that the dynamic behavior of economic variables depends on the regime that occurs at any given point in time". For example, the means and autocorrelations of returns and squared returns on stock market indexes may vary during a time period, say a week, so that we can say that each day of the week constitutes a different regime. In this example seasonal effects are represented by a *deterministic* regime process in the sense that the regime that occurs at any given point in time is known with certainty in advance. This is in contrast to situations in which the regime process is *stochastic.* The relevance for such stochastic regime switching behavior in financial time series has been stressed by various authors. For example, LeBaron (1992) demonstrates that the autocorrelations of stock returns are related to the volatility of these returns; in particular, autocorrelations tend to be larger during periods of low volatility and smaller during periods of high volatility. The periods of low and high volatility can be interpreted as different regimes. Kräger and Kugler (1993) argue that exchange rates might show regime switching behavior, in particular under a regime of managed floating such as occurred in the 1980s when it was attempted to stabilize the US dollar exchange rate. Alternatively, the level of the exchange rate rather than its change may determine the regimes as in the target zone literature of the Exchange Rate Mechanism (ERM) area.

Nowadays, there is a growing literature about the occurrence of various regimes generated by a stochastic process. These stochastic regime switching models differ in the way the regimes evolve over time: they are determined either by (an) observable variable(s) or by an unobservable stochastic process. This will be discussed in the following sections.

8.1.1 Specification and estimation of regime switching models based on observables

The most representative model, which assumes that the regime occurring at period t is determined by an observable variable q_t, is the *Threshold Autoregressive* (TAR) model (see Tong (1978) and (1990)). The TAR model assumes that the regime is determined by the value of the threshold variable q_t relative to a threshold value c. The *Self-Exciting* TAR or $SETAR$ model arises as a special case of the TAR model when q_t is a lagged value of the time series itself (or $q_t = y_{t-d}$ for any positive integer d).

Example 8.1 *An example of a 2-regime $SETAR$ model with $d = 1$ and an underlying $AR(1)$ process is:*

$$\begin{aligned} y_t &= \alpha_{01} + \alpha_{11} y_{t-1} + \varepsilon_t \quad \text{if } y_{t-1} \leqq c \\ &= \alpha_{02} + \alpha_{12} y_{t-1} + \varepsilon_t \quad \text{if } y_{t-1} > c, \end{aligned}$$

with $\varepsilon_t \in WN$ and constant variance, or alternatively expressed using an indicator function:

$$y_t = (\alpha_{01} + \alpha_{11} y_{t-1}) [1 - I(y_{t-1} > c)] + (\alpha_{02} + \alpha_{12} y_{t-1}) I(y_{t-1} > c) + \varepsilon_t,$$

where $I(y_{t-1} > c)$ is an indicator function with $I(y_{t-1} > c) = 1$ if $y_{t-1} > c$ and $I(y_{t-1} > c) = 0$ if $y_{t-1} \leqq c$. A more gradual transition between the different regimes can be obtained by replacing the indicator function $I(y_{t-1} > c)$ by a continuous function $G(y_{t-1}; \gamma, c)$, which changes smoothly from (regime) 1 to (regime) 0 as y_{t-1} increases, resulting in a *Smooth Transition* AR or $STAR$ model (see Teräsvirta (1994)):[1]

$$y_t = (\alpha_{01} + \alpha_{11} y_{t-1}) [1 - G(y_{t-1}; \gamma, c)] + (\alpha_{02} + \alpha_{12} y_{t-1}) G(y_{t-1}; \gamma, c) + \varepsilon_t.$$

[1] The logistic function is often chosen as an example of this continuous function G, which leads to the *Logistic* $STAR$ or $LSTAR$ model $G(y_{t-1}; \gamma, c) = \frac{1}{1+\exp(-\gamma(y_{t-1}-c))}$, where the parameter c can be interpreted as the threshold between the two regimes corresponding to $G(y_{t-1}; \gamma, c) = 0$ and $G(y_{t-1}; \gamma, c) = 1$, because the logistic function changes monotonically from 0 to 1 as y_{t-1} increases, while $G(c; \gamma, c) = 0.5$; the parameter γ determines the smoothness of the change in the value of the logistic function, and thus the transition from one regime to the other.

8.1.2 Specification and estimation of regime switching models based on unobservables

Disequilibrium models for one market

When only one market is considered a typical model consists of an unobservable demand, an unobservable supply and an observable transaction equation. Prices may be exogenous or endogenous. These two types of one-market disequilibrium models will be considered.

Model I: Exogenous prices Assume the following market model for one product:

$$
\begin{aligned}
D_t &= \boldsymbol{\beta}_1'\mathbf{x}_{1t} + \alpha_1 P_t + u_{1t} \qquad (8.1)\\
S_t &= \boldsymbol{\beta}_2'\mathbf{x}_{2t} + \alpha_2 P_t + u_{2t} \qquad (8.2)\\
Q_t &:= \min\{D_t, S_t\}, \qquad (8.3)
\end{aligned}
$$

where by assumption $u_{1t} \sim N(0, \sigma_1^2)$ and $u_{2t} \sim N(0, \sigma_2^2)$.

Consider then the probability that the observation Q_t belongs to the demand function:

$$
\begin{aligned}
\pi_t &:= P(D_t < S_t) = P\{\boldsymbol{\beta}_1'\mathbf{x}_{1t} + \alpha_1 P_t + u_{1t} < \boldsymbol{\beta}_2'\mathbf{x}_{2t} + \alpha_2 P_t + u_{2t}\}\\
&= P\{u_{1t} - u_{2t} < \boldsymbol{\beta}_2'\mathbf{x}_{2t} - \boldsymbol{\beta}_1'\mathbf{x}_{1t} + (\alpha_2 - \alpha_1) P_t\}\\
&= \int_{-\infty}^{\frac{\boldsymbol{\beta}_2'\mathbf{x}_{2t} - \boldsymbol{\beta}_1'\mathbf{x}_{1t} + (\alpha_2-\alpha_1)P_t}{\sigma}} \frac{1}{\sqrt{2\pi}} e^{-\frac{1}{2}u} du \qquad (8.4)
\end{aligned}
$$

with $\sigma := \sqrt{\sigma_1^2 + \sigma_2^2}$ (or $\sigma := \sqrt{\sigma_1^2 + \sigma_2^2 - 2cov(u_{1t}, u_{2t})}$).

Now, define the pdfs for a transaction on the demand function and supply function, respectively:

$$
\begin{aligned}
f_1(Q_t) &:= \tfrac{1}{\sqrt{2\pi}\sigma_1} \exp\left\{-\tfrac{1}{2\sigma_1^2}\left(Q_t - \boldsymbol{\beta}_1'\mathbf{x}_{1t} - \alpha_1 P_t\right)^2\right\}\\
f_2(Q_t) &:= \tfrac{1}{\sqrt{2\pi}\sigma_2} \exp\left\{-\tfrac{1}{2\sigma_2^2}\left(Q_t - \boldsymbol{\beta}_2'\mathbf{x}_{2t} - \alpha_2 P_t\right)^2\right\},
\end{aligned}
$$

with the associated complementary distribution functions:

$F_1(Q_t) := \frac{1}{\sqrt{2\pi}\sigma_1} \int_{Q_t}^{\infty} \exp\left\{-\frac{1}{2\sigma_1^2}\left(D_t - \boldsymbol{\beta}_1'\mathbf{x}_{1t} - \alpha_1 P_t\right)^2\right\} dD_t$ (*excess* demand)

$F_2(Q_t) := \frac{1}{\sqrt{2\pi}\sigma_2} \int_{Q_t}^{\infty} \exp\left\{-\frac{1}{2\sigma_2^2}\left(S_t - \boldsymbol{\beta}_2'\mathbf{x}_{2t} - \alpha_2 P_t\right)^2\right\} dS_t$ (*excess* supply).

Then, under the hypothesis that Q_t is an element of the demand function, the conditional pdf of Q_t is given by:

$$
h(Q_t | Q_t = D_t) = \frac{\int_{Q_t}^{\infty} g(Q_t, S_t) dS_t}{P(D_t < S_t)} \left(= \frac{f_1(Q_t) F_2(Q_t)}{\int_{-\infty}^{+\infty} f_1(Q_t) F_2(Q_t) dQ_t} \right).
$$

and under the hypothesis that Q_t is an element of the supply function, the conditional pdf of Q_t is given by:

$$h\left(Q_t|\, Q_t = S_t\right) = \frac{\int_{Q_t}^{\infty} g(D_t,Q_t)dD_t}{P(S_t<D_t)} = \left(\frac{f_2(Q_t)F_1(Q_t)}{\int_{-\infty}^{+\infty} f_2(Q_t)F_1(Q_t)dQ_t}\right)$$

Since Q_t is an element of the demand function with probability π_t and an element of the supply function with probability $1-\pi_t$, the unconditional pdf of Q_t is given by:

$$\begin{aligned} h\left(Q_t\right) &= h\left(Q_t|\, Q_t = D_t\right) P\left(D_t < S_t\right) + h\left(Q_t|\, Q_t = S_t\right) P\left(S_t < D_t\right) \\ &= \int_{Q_t}^{\infty} g\left(Q_t, S_t\right) dS_t + \int_{Q_t}^{\infty} g\left(D_t, Q_t\right) dD_t. \end{aligned}$$

If L_T is the loglikelihood function based on T observations, then:

$$L_T{:=}\sum_{t=1}^{T} \ln h_t \text{ with } h_t{:=}h\left(Q_t\right){=}f_1\left(Q_t\right) F_2\left(Q_t\right){+}f_2\left(Q_t\right) F_1(Q_t).$$

How to maximize? E.g. according to Section A.2.2 in Appendix A, the Newton (Raphson) iteration step is:

$$\boldsymbol{\theta}^{(k+1)} = \boldsymbol{\theta}^{(k)} - \left(\left.\frac{\partial^2 L_T}{\partial\boldsymbol{\theta}\partial\boldsymbol{\theta}'}\right|_{\boldsymbol{\theta}=\boldsymbol{\theta}^{(k)}}\right)^{-1} \left(\left.\frac{\partial L_T}{\partial\boldsymbol{\theta}}\right|_{\boldsymbol{\theta}=\boldsymbol{\theta}^{(k)}}\right),$$

where the estimation of the asymptotic variance-covariance matrix $\boldsymbol{\Sigma}_{\hat{\boldsymbol{\theta}}_{NLML}} = \lim_{T\to\infty}\left\{\left[-E\left(\frac{\partial^2 L_T}{\partial\boldsymbol{\theta}\partial\boldsymbol{\theta}'}\right)\right]^{-1}\Big|_{\boldsymbol{\theta}=\hat{\boldsymbol{\theta}}}\right\}$ can be given by the inverted Hessian matrix: $\left(-\left.\frac{\partial^2 L_T}{\partial\boldsymbol{\theta}\partial\boldsymbol{\theta}'}\right|_{\boldsymbol{\theta}=\hat{\boldsymbol{\theta}}}\right)^{-1}$ or by the estimate of the Cramèr-Rao lower bound (1.12).

Elaboration (with independent u_{1t} and u_{2t}) Assume $h_{1t} := \frac{Q_t-\boldsymbol{\beta}_1'\mathbf{x}_{1t}-\alpha_1 P_t}{\sigma_1}$ and $h_{2t} := \frac{Q_t-\boldsymbol{\beta}_2'\mathbf{x}_{2t}-\alpha_2 P_t}{\sigma_2}$ and from the pdfs $f_1\left(Q_t\right)$ and $f_2\left(Q_t\right)$, $f_{1t} := \frac{1}{\sqrt{2\pi}} e^{\frac{-h_{1t}^2}{2}}$ and $f_{2t} := \frac{1}{\sqrt{2\pi}} e^{\frac{-h_{2t}^2}{2}}$ and, correspondingly from the associated complementary distribution functions $F_1\left(Q_t\right)$ and $F_2\left(Q_t\right)$, $F_{1t}\left(Q_t\right) := \int_{h_{1t}}^{\infty} \frac{1}{\sqrt{2\pi}} e^{\frac{-u^2}{2}} du$ and $F_{2t}\left(Q_t\right) := \int_{h_{2t}}^{\infty} \frac{1}{\sqrt{2\pi}} e^{\frac{-u^2}{2}} du$.

Since $L_T := \sum_{t=1}^{T} \ln h_t = \sum_{t=1}^{T} \ln\left(f_{1t}F_{2t} + f_{2t}F_{1t}\right)$, we get $\frac{\partial f_{1t}}{\partial\boldsymbol{\beta}_1} = f_{1t}h_{1t}\mathbf{x}_{1t}$, $\frac{\partial f_{1t}}{\partial\alpha_1} = f_{1t}h_{1t}P_t$, $\frac{\partial F_{1t}}{\partial\boldsymbol{\beta}_1} = \frac{1}{\sqrt{2\pi}} e^{\frac{-h_1^2 t}{2}} \mathbf{x}_{1t} = f_{1t}\mathbf{x}_{1t}$ and $\frac{\partial F_{1t}}{\partial\alpha_1} = f_{1t}P_t$ (similarly for f_{2t}, F_{2t}). Then also:

$$\begin{aligned} \mathbf{a}_{1t} &:= \partial h_t/\,\partial\boldsymbol{\beta}_1 = \left[f_{1t}h_{1t}F_{2t} + f_{1t}f_{2t}\right]\mathbf{x}_{1t} \\ \mathbf{a}_{2t} &:= \partial h_t/\,\partial\boldsymbol{\beta}_2 = \left[f_{2t}h_{2t}F_{1t} + f_{1t}f_{2t}\right]\mathbf{x}_{2t} \\ a_{3t} &:= \partial h_t/\,\partial\alpha_1 = \left[f_{1t}h_{1t}F_{2t} + f_{1t}f_{2t}\right]P_t \\ a_{4t} &:= \partial h_t/\,\partial\alpha_2 = \left[f_{2t}h_{2t}F_{1t} + f_{1t}f_{2t}\right]P_t, \end{aligned}$$

so that the elements of the score vector satisfy: $\frac{\partial L_T}{\partial\boldsymbol{\beta}_1} = \sum_{t=1}^{T}\frac{\mathbf{a}_{1t}}{h_t}$, $\frac{\partial L_T}{\partial\boldsymbol{\beta}_2} = \sum_{t=1}^{T}\frac{\mathbf{a}_{2t}}{h_t}$, $\frac{\partial L_T}{\partial\alpha_1} = \sum_{t=1}^{T}\frac{a_{3t}}{h_t}$, $\frac{\partial L_T}{\partial\alpha_2} = \sum_{t=1}^{T}\frac{a_{4t}}{h_t}$.

Model II: Endogenous prices Consider now a structural model for a market of one product with endogenous prices:

$$\begin{aligned}
D_t &= \boldsymbol{\beta}_1'\mathbf{x}_{1t} + \alpha_1 P_t + u_{1t} && (8.5)\\
S_t &= \boldsymbol{\beta}_2'\mathbf{x}_{2t} + \alpha_2 P_t + u_{2t} && (8.6)\\
Q_t &:= \min\left(D_t, S_t\right) && (8.7)\\
\Delta P_t &= \eta\left(D_t - S_t\right) + \boldsymbol{\beta}_3'\mathbf{x}_{3t} + u_{3t}, && (8.8)
\end{aligned}$$

where it is assumed that $u_{1t} \sim N\left(0, \sigma_1^2\right)$, $u_{2t} \sim N\left(0, \sigma_2^2\right)$, and $u_{3t} \sim N\left(0, \sigma_3^2\right)$ so that we have three endogenous variables D_t, S_t, P_t and two observable endogenous variables Q_t and P_t. Hence, from the convolution rule:

$$\begin{aligned}
g\left(D_t, S_t, P_t \left| \mathbf{x}_t\right.\right) &= g\left(u_{1t}, u_{2t}, u_{3t} \left| \mathbf{x}_t\right.\right) \left| \left(\frac{\partial\left(D_t, S_t, P_t\right)}{\partial\left(u_{1t}, u_{2t}, u_{3t}\right)} \right)^{-1} \right| \\
&= g\left(u_{1t}, u_{2t}, u_{3t} \left| \mathbf{x}_t\right.\right) \left|1 + \eta\left(\alpha_2 - \alpha_1\right)\right|^{-1}. \qquad (8.9)
\end{aligned}$$

It follows from transaction function (8.7) that:

1. the conditional pdf of Q_t and P_t, given that Q_t is an element of the demand function, is given by:

$$g_1\left(Q_t, P_t \left| \mathbf{x}_t\right.\right) = \int_{Q_t}^{\infty} \frac{g(Q_t, S_t, P_t | \mathbf{x}_t) dS_t}{N_t},$$

with N_t the 'normalizing factor' given the integral of pairs of $g_1\left(Q_t, P_t \left| \mathbf{x}_t\right.\right)$ w.r.t. Q_t, or $N_t = P\left(D_t < S_t\right)$, and

2. the conditional pdf of Q_t and P_t given that Q_t is an element of the supply function:

$$g_2\left(Q_t, P_t \left| \mathbf{x}_t\right.\right) = \int_{Q_t}^{\infty} \frac{g(D_t, Q_t, P_t | \mathbf{x}_t) dD_t}{1 - N_t}.$$

Hence, the unconditional pdf of the (observable) endogenous variables Q_t, P_t is given by:[2]

$$\begin{aligned}
f\left(Q_t, P_t | \, \mathbf{x}_t\right) &= N_t g_1\left(Q_t, P_t | \, \mathbf{x}_t\right) + \left(1 - N_t\right) g_2\left(Q_t, P_t | \, \mathbf{x}_t\right) \\
&= \int_{Q_t}^{\infty} g\left(Q_t, S_t, P_t | \, \mathbf{x}_t\right) dS_t + \int_{Q_t}^{\infty} g\left(D_t, Q_t, P_t | \, X_t\right) dD_t \, .
\end{aligned}$$

The simultaneous loglikelihood function satisfies $L_T := \sum_{t=1}^{T} \ln f\left(Q_t, P_t | \, \mathbf{x}_t\right)$ or:

$$\max_{(\alpha_1, \alpha_2, \beta_1, \beta_2, \beta_3, \eta, \Sigma)} L_T = \max \sum_{t=1}^{T} \ln f\left(Q_t, P_t \left| \mathbf{x}_t\right.\right). \qquad (8.10)$$

The ML estimators from the solution of (8.10) are derived in Appendix D for uncorrelated and correlated structural disturbances.

[2] Note that, if there are other endogenous variables in the system besides Q_t and P_t, i.e. $\mathbf{y}_t$, then $f\left(Q_t, P_t | \, \mathbf{x}_t\right)$ becomes:

$$f\left(Q_t, P_t, \mathbf{y}_t | \, \mathbf{x}_t\right) = \int_{Q_t}^{\infty} g\left(Q_t, S_t, P_t, \mathbf{y}_t | \, \mathbf{x}_t\right) dS_t + \int_{Q_t}^{\infty} \left(D_t, Q_t, P_t, \mathbf{y}_t | \, \mathbf{x}_t\right) dD_t$$

8.1.3 Solved empirical case for regime switching models

Regime switching model of cash flow in and out of hedge investment funds. Hedge funds are 'skill-based' asset management firms that, for different reasons, do not qualify as mutual funds. While the mutual fund industry is well regulated and fairly transparent, the hedge fund industry is not. For example, reporting of historical returns and other information is not obligatory for hedge funds. On the other hand, access to hedge funds is typically restricted to investors with 'sophisticated understanding' and investing large sums of money. So-called *fund-of-funds* pool investments in different hedge funds and allow for smaller investments to attract a larger group of individual investors.

The data is given in a database, which was presented in the 2003 Econometric Game in Amsterdam (see http://www.vsae.nl/econometricgame/ and http://www.ua.ac.be/joseph.plasmans).

The key variables in the database are: REF - the fund's ID code; *Quarter* - quarter (1=94Q4); $CFlows$- cash flows (%), (inflow-outflow)/net asset value x 100%; $CF1-4$ - cash flows, lagged 1...4 quarters; *Returns* - raw return; $R1-6$ - raw return, lagged 1...6 quarters; *Rank*$-$ cross-sectional ranking between 0 and 1 (0 =best, 1 =worst); $RNK1-6$ - rank, lagged 1...6 quarters; NAV- net asset value;*Age*- age of fund (in months); $Offshore$ - Offshore/Onshore location dummy (1 = Offshore); $Incfee$ - incentive fees (%); $Mngfee$- management fees (%); $FOF-$ fund of funds dummy (1 =fund of funds); *Pcapinv* - personal capital invested dummy; and *Leverage*- dummy indicating whether the fund uses leverage.

Solution In the first part we estimate and test the regime switching behavior of the cash flows in and out of funds using the $SETAR$ equation as in Example 8.1. It is possible to have a higher order AR model in the different regimes. Assuming that we have a two-regime case, the AR orders might be set to p_1 and p_2 in the upper and lower regimes respectively.

$$\begin{aligned} y &= \alpha_{01} + \alpha_{11} y_{t-1} + \ldots + p_{11} y_{t-p_1} + \varepsilon_t \text{ if } y_{t-1} > c\ ; \\ &= \alpha_{02} + \alpha_{12} y_{t-1} + \ldots + p_{12} y_{t-p_2} + \varepsilon_t \text{ if } y_{t-1} \leqq c; \end{aligned}$$

In order to define the regimes as net inflows and net outflows we assume that if cash flows are positive or $cflows > 0$ then cflows is used in the first equation or if $cflows < 0$ then cflows is in the second equation. The stock returns are related to the level of volatility of these returns. In particular, as mentioned before, autocorrelation tends to be higher during periods of low volatility and lower during periods of higher volatility. Besides, periods of low and high volatility can be interpreted as distinct regimes. However, we cannot be certain that a particular regime has occurred at a particular point in time.

Identification of lag orders. We start by specifying a linear $AR(p)$ model for y_t and assume that the order p is based on information criterion based on AIC (see (5.31)) and specified for this regime switching model as:

$$AIC(p1,p2) = T_1 \ln \hat{\sigma}_1^2 + T_2 \ln \hat{\sigma}_2^2 + 2(p_1+1) + 2(p_2+1).$$

Running different AR specifications in various regimes the corresponding information criteria are obtained for significant coefficients of AR models (see Table 8.1).

Table 8.1: AIC(P1,P2)

$AIC(1,1) = -26080.6$	$AIC(2,3) = -26093$
$AIC(1,2) = -26090$	$AIC(2,4) = -26093$
$AIC(1,3) = -26090$	$AIC(3,1) = -26085.7$
$AIC(1,4) = -26090$	$AIC(3,2) = -26095.1$
$AIC(2,1) = -26083.6$	$AIC(3,3) = -26095.1$
$AIC(2,2) = -26093$	$AIC(3,4) = -26095.1$

The AIC criterion must be minimized over choices of AR. Since the lowest AIC value is obtained for the several $AR(3)$ combinations and we look for a parsimonious model, the $(AR(3), AR(2))$ combination is selected:

$$\begin{aligned}
\hat{y}_t &= 0.164348^{(**)} + 0.039089^{(**)} y_{t-1} + 0.029579^{(**)} y_{t-2} + 0.012436^{(**)} y_{t-3} \text{ if } y_t > 0, \\
\hat{y}_t &= -0.10456^{(**)} + 0.016175 y_{t-1} + 0.01778 y_{t-2}, \text{ if } y_t < 0.
\end{aligned}$$

8.2 Volatility modeling

Until the beginning of the eighties the modeling of macroeconomic and financial time series was based principally on time-invariant conditional expectations. More specifically, variances and covariances were usually assumed to be time-invariant. Even highly volatile processes as the pricing of options were modeled under time-invariancy of the (co)variances (see e.g. Black and Scholes (1973)). It is, however, generally known that highly frequent data as e.g. asset returns show, not only changing volatility, i.e. a time-varying conditional variance (or a variance which is conditional on time t information), but also volatility clustering. *Volatility clustering*, or in other words conditional heteroskedasticity, means that (positive and/or negative) shocks in asset returns succeed each other. As early as in 1963 Mandelbrot described this phenomenon as: "... *Large changes tend to be followed by large changes, and small changes tend to be followed by small changes* ...". This means that e.g. if the volatility of an asset price is high today, it will also be high tomorrow. A distinction can be made between periods of high volatility and periods where this volatility is generally low. In reality the volatility of highly frequent asset returns is certainly not time-invariant.[3] Two

[3] Neglecting the existence of varying conditional variances and covariances of error terms leads to under- or overestimation of the size of confidence intervals when time series fore-

Figure 8.1: Daily returns of ING assets from 01/04/1994 till 31/12/2004

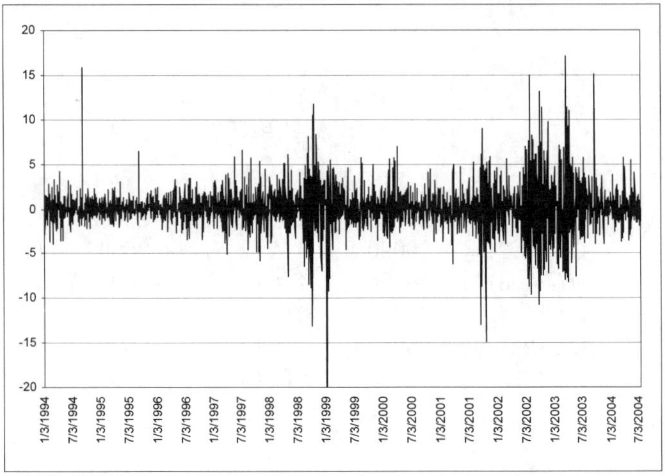

different examples of volatility clustering are observed in the time series of two Dutch asset returns, i.e. the daily ING returns and the daily Philips returns between 1/4/1994 and 31/12/2004 (see Figures 8.1 and 8.2). Note, however, by visual inspection, the occurrence of some *outliers* in both time series of asset returns.

Through the increased importance of uncertainty and risk new stochastic time series models have been developed since the beginning of the eighties which take account of time-varying (co)variances. These models offer the possibility to make forecasts w.r.t. the future volatility. Nevertheless the *efficient market hypothesis*, a forecast of volatility is sometimes possible and may be important for fast acting asset allocation, calculating the Value at Risk (VaR) of a new investment, pricing financial derivatives (e.g. pricing an option for hedging the portfolio risk,[4] etc...). Forecasts conditional on recent information are more efficient than forecasts which do not use this information. The class of *autoregressive conditional heteroskedastic* ($ARCH$) models, introduced by Engle (1982), is based on this assumption. In this kind of models it is assumed that the variance depends on the values of the process in the past (AR). Various other formulations to model conditional heteroskedasticity have been proposed and used in the literature, the principal being the class of *generalized autoregressive conditional heteroskedasticity* ($GARCH$) models. Notice that, although it is sometimes possible to predict the size of volatility, the sign of volatility is

casts are made. Hence, the risk to which the (return) variables are subject in the future is misinterpreted.

[4] For example, when hedging the downward risk of a portfolio, a put option could be bought at the value of the underlying portfolio.

Figure 8.2: Daily returns of Philips assets from 01/04/1994 till 31/12/2004

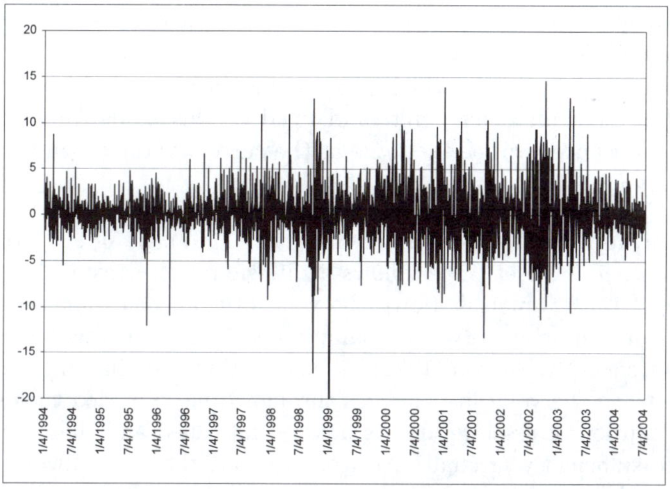

usually very difficult to predict due to asymmetric effects.[5] Moreover, although conventionally in the Markowitz tradition, where an investor has to make a trade-off between expected asset returns and current volatility, it might also be that current volatility is immediately of influence on the current asset returns.[6]

Both univariate and multivariate types of (*G*)*ARCH* models will be discussed in the following subsections.

The pre-*GARCH* literature on changing volatility discussed three concepts:

1. *Moving or rolling variances of asset returns*, or if r_t is denoted as the (logarithmic) asset return at period t, then the moving variance over a sub-sample which moves forward through time (as an approximation to the population moving variance σ_t^2) can be written as $s_t^2 = \frac{1}{T+1}\sum_{j=1}^{T} r_{t-j}^2$, which has the particular features that all observations from $t-T$ to t are given equal weight, that all observations before $t-T$ (and after t) are given no weight and that the choice of T is left to the investor (see e.g. Officer (1973) who uses the rolling standard deviation);

2. *Time spans* (or ranges), e.g. the difference between the high and low spot prices on a given day to estimate volatility for that day[7] (see Garman and Klass (1980) and Parkinson (1980)), where, as in the above case,

[5] See Nelson's (1992) 'Exponential *GARCH*' or *EGARCH* model, where positive shocks and negative shocks have a different impact.

[6] A model with such a feature will be called a '*GARCH*-in-Mean' or *GARCH*-M model.

[7] Note that it is implicitly assumed here that volatility is constant over some interval of time.

volatility estimates are extracted from asset returns data before specifying a parametric time-series model for volatility;

3. Exponential smoothing, which may be written for a known adjustment parameter λ, satisfying $0 < \lambda < 1$, as: $\sigma_t^2 = \lambda\sigma_{t-1}^2 + (1-\lambda)r_{t-1}^2 = (1-\lambda)\sum_{j=1}^{\infty}\lambda^{j-1}r_{t-j}^2$, which is a moving variance model with declining weights, no truncation point, and fixed λ; this model predicts that all future variances will be the same as the current variance (moving-average character)[8].

Moreover, as indicated by Campbell *et al.* (1997), p. 481, Merton (1980) observed that, if an asset price follows a diffusion with constant volatility (as e.g. a geometric Brownian motion), the volatility can be estimated arbitrarily accurately with an arbitrarily short sample period if asset prices are measured sufficiently frequently. Nelson (1992) has shown that a similar argument can be made even when the volatility changes through time, provided that the conditional distribution of asset returns is not too fat-tailed and that the volatility changes are sufficiently gradual. Notice, however, that it is always more efficient to directly model the time-varying character of volatility and to estimate it together with the equation for asset returns.

In general, a distinction can be made between discrete-time volatility as discussed above and continuous-time volatility; the latter concept of volatility involves three types of volatility (see e.g. Anderson *et al.* (2002)): *notional volatility* corresponding to ex-post sample return volatility (variability) over a fixed time interval, *ex-ante expected volatility* over a fixed time interval, and *instantaneous volatility* corresponding to the strength of the volatility process at a point in time. In this book we concentrate on discrete-time volatility, where also three types can be distinguished: *historical volatility* measured by the conditional variance (which in Finance can e.g. be estimated by the squared (daily) returns) or the conditional standard deviation, *realized volatility* because it was observed (by Anderson and Bollerslev (1998)) that squared daily returns provide a poor approximation of actual daily volatility so that summing squared intraday returns yields a better measure for daily volatility, and *implied volatility:* i.e. the implied volatitlity in an option price, which is the inverse function of the price (value) of an option-pricing contract according to Black-Scholes, Hull and White (1987), among others. Regarding the latter concept, it is clear that the value of an option is determined in a strong measure by the volatility of the underlying asset.

In this section, two basic models will be analyzed: univariate $ARCH$ and $GARCH$ models, and multivariate $ARCH$ and $GARCH$ models, with several variations for nonstationarity, asymmetry, among others.

[8] Recall that the SES model is in fact an $ARIMA(0,1,1)$ model.

8.2.1 Univariate *ARCH* models

Introductory remarks

As said before the (conditional) variances need not be constant over time; in many cases, certainly in Finance, periods of low and high volatility succeed each other over time.[9] Before analyzing various univariate *ARCH* models, some introductory remarks will be given.

Assume a discrete-time stochastic process $\{\varepsilon_t, t \in \mathbb{Z}\}$ with conditional moments $E_{t-1}(\varepsilon_t) = 0$ and $var_{t-1}(\varepsilon_t) = E_{t-1}(\varepsilon_t^2) =: \sigma_t^2$, $t = 1, 2, 3 \ldots$, respectively. The conditional variance depends in a nontrivial way on past observations, i.e. $\sigma_t^2 = \Psi(\varepsilon_{t-1}, \varepsilon_{t-2}, \ldots; \mathbf{x}_{t-1}, \mathbf{b})$, where $\Psi(x)$ is a nontrivial function and $\mathbf{x}_{t-1}$ a vector of lagged regressors (e.g. lags of conditional variances)[10].

In most applications ε_t is the innovation in the mean for an other stochastic process of logarithmic asset returns, say r_t, where $r_t = \zeta(\mathbf{z}_{t-1}, \mathbf{a}) + \varepsilon_t$ with $\zeta(\mathbf{z}_{t-1}, \mathbf{a})$ a function of $\mathbf{z}_{t-1}$and the parameter vector $\mathbf{a}$ ($\mathbf{z}_{t-1}$ is in the information set at time $t-1$). If the asset returns are subject to autocorrelation, these can be modeled by means of k autocorrelation parameters. In this case the function $\zeta(\mathbf{z}_{t-1}, \mathbf{a})$ models an $AR(k)$ process. The process for the asset returns together with the process for the conditional variances of these satisfy then: $r_t = \zeta(\mathbf{z}_{t-1}, \mathbf{a}) = \mu + \sum_{i=1}^{k} \rho_i r_{t-i} + \varepsilon_t$ with ε_t a noise process with zero mean and time-varying conditional variance $\sigma_t^2 = \Psi(\varepsilon_{t-1}, \varepsilon_{t-2}, \ldots; \mathbf{x}_{t-1}, \mathbf{b})$.

In fact, appropriate methods of econometric inference estimate and test the asset returns process and the volatility process simultaneously. Various ways to model the innovation process $\{\varepsilon_t, t \in \mathbb{Z}\}$ are discussed in the next (sub)sections.

The *ARCH* class of models

These models use historical data to determine the weighting and allow for the possibility of richer information sets. In particular, the importance of *ARCH* models is derived from the difference between conditional and unconditional variances. While the unconditional variances are assumed to be time-independent, the conditional variances are assumed to depend on past events which are contained in the information set at time $t-1$. As Engle (1982) proposes the conditional variance is assumed to depend linearly on the quadratic innovations from the past and on a stochastic process, $\varepsilon_t := \sigma_t \eta_t$, where $\eta_t \overset{iid}{\sim} N(0,1)$, meaning that all are identically and independently distributed as a standard normal distribution. The resulting $ARCH(q)$ model satisfies:

$$\sigma_t^2 = \omega + \textstyle\sum_{i=1}^{q} \alpha_i \varepsilon_{t-i}^2 = \omega + \alpha(L)\varepsilon_{t-1}^2 \text{ with } \omega > 0 \text{ and } \alpha_i \geqq 0,\ i = 1, \ldots, q,$$

[9] If volatility is constant over time, we often have white noise processes (if also the data are mutually independently distributed).

[10] In *ARCH* models $\mathbf{x}_{t-1}$ is empty, such that the conditional variance can be written as a linear function of past quadratic innovations.

where L is the lag operator, i.e. $L^i x_t = x_{t-i}$ and $\alpha(L)$ the corresponding lag polynomial of order q.[11]

The $ARCH(q)$ process is stationary if the sum of the parameters of the autoregressive part is smaller than one, i.e. if $\sum_{i=1}^{q} \alpha_i < 1$. The (constant) unconditional variance equals then:

$$\sigma_\varepsilon^2 = var\{\varepsilon_t\} = E\left[E_{t-1}(\varepsilon_t^2)\right] = E\{\omega + \sum_{i=1}^{q} \alpha_i \varepsilon_{t-i}^2\} = \omega + \sigma_\varepsilon^2 \sum_{i=1}^{q} \alpha_i = \frac{\omega}{1 - \sum_{i=1}^{q} \alpha_i} = \frac{\omega}{1-\alpha(1)},$$

i.e. the shock dies out in q periods (if the shock has 'to live' long, q is very high).

Note that the OLS regression with $ARCH$ errors is still valid, but more efficient nonlinear (ML) estimators exist.

Finally, how do we test for $ARCH$ effects?

$ARCH$ effects are tested by considering auxiliary regressions of the squared OLS residuals $\hat{\varepsilon}_t^2$ upon $\hat{\varepsilon}_{t-1}^2, \hat{\varepsilon}_{t-2}^2, \ldots, \hat{\varepsilon}_{t-q}^2$ and a constant:

$$\widehat{\varepsilon_t^2} = \hat{a}_0 + \hat{a}_1 \widehat{\varepsilon_{t-1}^2} + \hat{a}_2 \widehat{\varepsilon_{t-2}^2} + \ldots + \hat{a}_q \widehat{\varepsilon_{t-q}^2}, \tag{8.11}$$

and the LM test statistic $\xi_{LM} = T\widehat{R}^2 \overset{as}{\sim} \chi_q^2$ should be computed, with $\widehat{R}^2$ the multiple determination coefficient of the regression in (8.11), to test the null of no $ARCH$ effects H_0: $\alpha_1 = \alpha_2 = \ldots \alpha_q = 0$ (thus rejecting the null if $\widehat{R}^2$ is too high).

Parametric $ARCH$ models were designed to parsimoniously model the expected volatility as an explicit function of discretely observed returns. Also nonparametric methods, principally regarding the assummed continuity of prices and volatility processes are studied for ARCH effects by Anderson *et al.* (2002).

8.2.2 The symmetric *GARCH* class of models

Definition 8.1 *Empirical applications of the $ARCH(q)$ model often find a high value for q, i.e. one needs a large number of lagged innovations to describe the innovation process. Hence, a more flexible lag structure has been proposed by Bollerslev (1986), where an autoregressive term has been added. The resulting generalized $ARCH$ or $GARCH(p,q)$ model satisfies for the conditional variances*

$$\sigma_t^2 = \omega + \sum_{j=1}^{p} \beta_j \sigma_{t-j}^2 + \sum_{i=1}^{q} \alpha_i \varepsilon_{t-i}^2 = \omega + \beta(L)\sigma_{t-1}^2 + \alpha(L)\varepsilon_{t-1}^2$$

with $\omega > 0, \alpha_i \geqq 0,\ i = 1, \ldots, q$ and $\beta_j \geqq 0,\ j = 1, \ldots, p$.

[11] A variant of the $ARCH(1)$ model (the above model with $q = 1$) is the model (see Gouriéroux (1997), pp. 29-32): $r_t = \mu + \rho r_{t-1} + \varepsilon_t, \forall t, |\rho| < 1,\ E\left[\varepsilon_t | \Im_{t-1}\right] = 0$, $\varepsilon_t^2 = c + a\varepsilon_{t-1}^2 + u_t$. For this model, we find from the law of iterated expectations that $E\left[\varepsilon_t | \Im_{t-h}\right] = 0, \forall h > 0$ (since the information set $\Im_{t-h}$ is a subset of the information set $\Im_{t-1}$ and cannot contain more information), $cov\left[(\varepsilon_t, \varepsilon_{t+k}) | \Im_{t-h}\right] = 0$ (so that there is no conditional covariance between current and future innovations), $var(\varepsilon_t | \Im_{t-h}) = c\frac{1-a^h}{1-a} + a^h \varepsilon_{t-h}^2$, $var(\varepsilon_t) = \frac{c}{1-a}$, and $var(\varepsilon_t | \Im_{t-h}) - var(\varepsilon_t) = a^h \left[\varepsilon_{t-h}^2 - E(\varepsilon_{t-h}^2)\right]$.

Example 8.2 *The most commonly used model in the class of $GARCH(p,q)$ models is the simple $GARCH(1,1)$ model for the asset returns r_t, where the conditional variance of r_t is assumed to be a weighted average of three components: (i) a constant variance ω, the past forecast σ_{t-1}^2, and the past (squared) news ε_{t-1}^2. Hence, the news impact curve of the $GARCH(1,1)$ model is perfectly symmetric.*

The conditional variance in this simple $GARCH(1,1)$ model is: $\sigma_t^2=\omega+\beta\sigma_{t-1}^2+\alpha\varepsilon_{t-1}^2=\omega+(\alpha+\beta)\sigma_{t-1}^2+\alpha(\varepsilon_{t-1}^2\text{-}\sigma_{t-1}^2)=\omega+(\alpha+\beta)\sigma_{t-1}^2+\alpha\sigma_{t-1}^2(\eta_{t-1}^2\text{-}1)$, where the term $(\varepsilon_{t-1}^2-\sigma_{t-1}^2)$ has (conditional) mean zero and can be interpreted as the shock to volatility. The parameter α measures the extent to which a current volatility shock feeds through into next period's volatility, while $(\alpha+\beta)$ measures the rate at which this effect dies out over time (generally very slowly). Notice that the volatility shock is finally rewritten as $\alpha\sigma_{t-1}^2(\eta_{t-1}^2-1)$, i.e. as a demeaned χ^2- distributed random variable multiplied by the past volatility.

Unconditional variance

Taking account of the properties of the unconditional expectation operator, the unconditional expectation of the conditional variance in a $GARCH(p,q)$ model satisfies:
$E\left(\sigma_t^2\right)=E\left\{var_{t-1}\left(\varepsilon_t\right)\right\}=E_{t-1}\left\{E_{t-1}(\varepsilon_t^2)\right\}=\omega+\alpha(L)E(\varepsilon_{t-1}^2)+\beta(L)E(\sigma_{t-1}^2)\Longleftrightarrow$ $[1\text{-}\alpha(L)\text{-}\beta(L)]\,\sigma_\varepsilon^2=\omega$.

Hence, the unconditional variance of the $GARCH(p,q)$ process is given by: $\sigma_\varepsilon^2 = E\left(\varepsilon_t^2\right) = \omega \Big/ (1-\sum_{j=1}^p \beta_j - \sum_{i=1}^q \alpha_i)$, which is independent of time.

From this time-independent unconditional variance expression it is immediately clear that the $GARCH(p,q)$ model is stationary if $\sum_{i=1}^q \alpha_i+\sum_{j=1}^p \beta_j < 1$.

A $GARCH$ process rewritten as an $ARCH(\infty)$ process

The very popular $GARCH(1,1)$ model can be rewritten as:

$$\begin{aligned}\sigma_t^2 &= \omega+\alpha\varepsilon_{t-1}^2+\beta\sigma_{t-1}^2=\omega+\alpha\varepsilon_{t-1}^2+\beta\left(\omega+\alpha\varepsilon_{t-2}^2+\beta\sigma_{t-2}^2\right)=\ldots\\ &= \omega(1+\beta+\beta^2+\ldots)+\alpha(\varepsilon_{t-1}^2+\beta\varepsilon_{t-2}^2+\beta^2\varepsilon_{t-3}^2+\ldots\\ &= \tfrac{\omega}{1-\beta}+\alpha\sum_{j=1}^\infty \beta^{j-1}\varepsilon_{t-j}^2 \quad \text{if } \beta<1,\end{aligned}$$

which is an $ARCH(\infty)$ model with declining weights. Hence, the $GARCH(1,1)$ process is a very parsimonious alternative to an higher order $ARCH$ model.

In general, the conditional variance of a $GARCH(p,q)$ model can be written as an $ARCH(\infty)$ model if the sum of the coefficients belonging to the lagged conditional variances is smaller than one (which is always the case if the $GARCH(p,q)$ model is stationary; see also Bollerslev (1996)):

$$\sigma_t^2=\alpha_o\left(1-\sum_{j=1}^p\beta_j\right)^{-1}+\sum_{i=1}^\infty \delta_i\varepsilon_{t-i}^2;$$
$$\delta_i=\alpha_i+\sum_{j=1}^n\beta_j\delta_{i-j} \quad i=1,\ldots,q;\ n:=min\left\{p,i-1\right\}$$
$$\delta_i=\sum_{j=1}^n\beta_j\delta_{i-j} \quad i=q+1,\ldots$$

GARCH models rewritten as *ARMA* models

The simple $GARCH(1,1)$ model can also be written in terms of its implications for squared innovations ε_t^2:

$\varepsilon_t^2 = \omega + (\alpha+\beta)\varepsilon_{t-1}^2 + (\varepsilon_t^2 - \sigma_t^2) - \beta(\varepsilon_{t-1}^2 - \sigma_{t-1}^2)$, which makes clear that this $GARCH(1,1)$ model can be rewritten as an $ARMA(1,1)$ model for the squared innovations ε_t^2.[12]

By appropriate induction, it can be verified that the general $GARCH(p,q)$ model can be rewritten as an $ARMA(p,q)$ model in the quadratic innovations: $\varepsilon_t^2 = \omega + \sum_{i=1}^{m}(\alpha_i+\beta_i)\varepsilon_{t-i}^2 - \sum_{j=1}^{p}\beta_j v_{t-j} + v_t$, where $\alpha_i := 0$ for $i > q, \beta_i := 0$ for $i > p$, $m := max\,\{p,q\}$, $v_t := \varepsilon_t^2 - \sigma_t^2 = (\eta_t^2 - 1)\sigma_t^2$, $\eta_t \overset{iid}{\sim} D(0,1)$ with D a probability distribution function to be specified: in $GARCH$ modeling usually either a normal distribution or a Student t-distribution is chosen, such that $\eta_t \overset{iid}{\sim} N(0,1)$ or $\eta_t \sim t_\upsilon$ where υ is the number of degrees of freedom in the t-distribution. This $ARMA(m,p)$ process for the quadratic innovations can be used to determine the optimal values of p and q in the $GARCH(p,q)$ process. After rewriting the $GARCH(p,q)$ process into an $ARMA(p,q)$ process traditional Box-Jenkins analysis can be used to determine m, p and possibly q with $m > p$.[13]

Fat tails

The (unconditional) kurtosis should be equal to 3 when the error terms $\varepsilon_t := \sigma_t\eta_t$ satisfy $\eta_t \overset{iid}{\sim} N(0,1)$. $GARCH$, however, generally leads to a kurtosis larger than 3, because:

$$\begin{aligned}\kappa &= \frac{E(\varepsilon_t^4)}{[E(\varepsilon_t^2)]^2} = \frac{E\left[E_{t-1}(\varepsilon_t^4)\right]}{[E(\varepsilon_t^2)]^2} = \frac{E\left[\sigma_t^4 E_{t-1}(\eta_t^4)\right]}{[E(\varepsilon_t^2)]^2} = \frac{E\left[\sigma_t^4 E(\eta_t^4)\right]}{[E(\varepsilon_t^2)]^2} = \frac{E(\sigma_t^4)E(\eta_t^4)}{[E(\varepsilon_t^2)]^2} \\ &= 3\frac{E(\sigma_t^4)\left[E(\eta_t^2)\right]^2}{[E(\varepsilon_t^2)]^2} = 3\frac{E(\sigma_t^4)}{[E(\varepsilon_t^2)]^2} = 3\frac{E(\left[E_{t-1}(\varepsilon_t^2)\right]^2)}{[E(\varepsilon_t^2)]^2} = 3\frac{\left[E(E_{t-1}(\varepsilon_t^2))\right]^2 + var(E_{t-1}(\varepsilon_t^2))}{[E(\varepsilon_t^2)]^2} \\ &= 3\frac{\left[E(\varepsilon_t^2)\right]^2 + var(E_{t-1}(\varepsilon_t^2))}{[E(\varepsilon_t^2)]^2} = 3 + 3\frac{var(E_{t-1}(\varepsilon_t^2))}{[E(\varepsilon_t^2)]^2} = 3 + 3\frac{var(\sigma_t^2)}{[E(\varepsilon_t^2)]^2},\end{aligned} \tag{8.12}$$

where the second part of the (last) right-hand side of equality (8.12) is called the *excess kurtosis*, which is a measure of the fat-tailed characteristic of a $GARCH$ process.

[12] Notice, however, that a standard $ARMA(1,1)$ model has homoscedastic shocks, while here the shocks $(\varepsilon_t^2 - \sigma_t^2)$ are heteroscedastic.

[13] An alternative for this is to estimate the $GARCH(p,q)$ model for various combinations of p and q by a maximum likelihood method. Using an appropriate selection criterion the optimal orders can be determined then. This procedure will be followed in this book.

Volatility forecasts

Consider the simple $GARCH(1,1)$ model for asset returns again:

$$\begin{aligned}\sigma_t^2 = \omega + \alpha\varepsilon_{t-1}^2 + \beta\sigma_{t-1}^2 = & \ (1-\alpha-\beta)\omega(1-\alpha-\beta)^{-1} + \beta\sigma_{t-1}^2 + \alpha\varepsilon_{t-1}^2 \\ = & \ \sigma_\varepsilon^2 + \beta(\sigma_{t-1}^2 - \sigma_\varepsilon^2) + \alpha(\varepsilon_{t-1}^2 - \sigma_\varepsilon^2),\end{aligned}$$

where, under stationary ($\alpha + \beta < 1$), the unconditional variance satisfies from above: $\sigma_\varepsilon^2 = \omega(1-\alpha-\beta)^{-1}$ and the conditional variances and quadratic innovations are expressed as deviations from their 'long-term mean'. Rewriting the conditional variances for future periods $t+s$: $\sigma_{t+s}^2 = \sigma_\varepsilon^2 + \alpha(\varepsilon_{t+s-1}^2 - \sigma_\varepsilon^2) + \beta(\sigma_{t+s-1}^2 - \sigma_\varepsilon^2)$, the best guess as to what ε^2 will be in the future is σ_ε^2. Hence, under a stationary volatility model we get by substitution: $\sigma_{t+s}^2 = \sigma_\varepsilon^2 + (\alpha+\beta)(\sigma_{t+s-1}^2 - \sigma_\varepsilon^2)$, which can be repeated back to ε_t which is observed. More precisely, using the law of iterated expectations, the conditional expectation s periods ahead is then given by recursive substitution as $E_t\left(\sigma_{t+s}^2\right) = \sigma_\varepsilon^2 + (\alpha+\beta)^{s-1}(\sigma_{t+1}^2 - \sigma_\varepsilon^2)$, or the future conditional variance depends on the deviation of the conditional variance at period $t+1$ from the unconditional variance (being equal to the long term mean). In other words, the multiperiod volatility forecast *reverts* to its unconditional mean at rate $(\alpha+\beta)$: if the forecast horizon increases, the current information becomes less important and the forecast converges to the unconditional variance $\sigma_{t+s}^2 \to \sigma_\varepsilon^2$ if $s \to \infty$ with a decay rate $(\alpha+\beta)$.[14] Hence, the stationary $GARCH(1,1)$ model will *mean revert* to the unconditional variance $\sigma_\varepsilon^2 = \frac{\omega}{(1-\alpha-\beta)}$ and the volatility may be temporarily above or below the long run mean as shown in figure 8.3. [15] Notice that these forecasts also represent the 'term structure of the conditional variance' in the (stationary) $GARCH(1,1)$ model (volatility forecasts with varying forecast horizons).

Notice also that, instead of convergence to the constant unconditional mean σ_ε^2, the *component model* allows mean reversion to a varying variance level defined by q_t :
$\sigma_t^2 = q_t + \alpha(\varepsilon_{t-1}^2 - q_{t-1}) + \beta(\sigma_{t-1}^2 - q_{t-1})$; $q_t = \omega + \rho(q_{t-1} - \omega) + \phi(\varepsilon_{t-1}^2 - \sigma_{t-1}^2)$.

Here, σ_t^2 is still the volatility while q_t takes the place of σ_ε^2, the long run volatility. The transitory component is $\sigma_t^2 - q_t$, which will converge to zero with

[14] This relation between single-period and multiperiod forecasts is the same as in a linear $ARMA(1,1)$ model with autoregressive parameter $(\alpha+\beta)$.

[15] The long-run conditional variance σ_ε^2 is estimated for the Philips and ING asset returns in example 8.6, under an underlying normal and Student t-distribution respectively, as 4.79 and 6.08 respectively for the Philips returns, and as 3.49 and 2.67 for the ING returns. Notice also that it is frequently found that estimated residuals from $GARCH$ models still show excess kurtosis, even when one allows for conditional Student t-distributed innovations, respectively. This feature may often be due to the consideration of some extreme return observations as additive outliers. Neglecting these extreme observations, however, leads to biased parameter estimations and, hence, to biased forecasts. Franses and Ghijsels (1999) demonstrated that asset return models corrected for strong additive outliers generally yield substantial (forecast) improvement over $GARCH$ models with normally and Student t-distributed innovations for the original asset returns.

Figure 8.3: Term structure of volatility for $GARCH(1,1)$

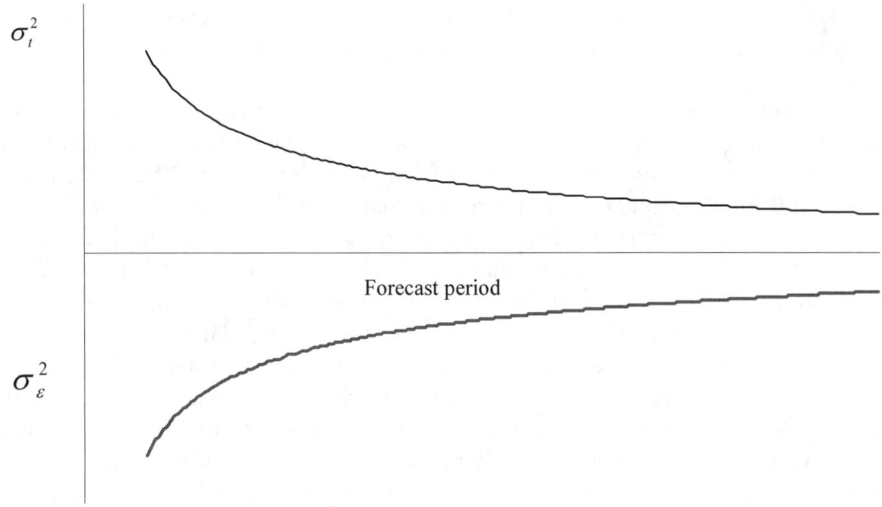

powers of $(\alpha + \beta)$. The long run component q_t will converge to ω with powers of ρ ($0 < \rho < 1$, usually being just smaller than 1; very slow convergence).

Forecasting in higher order $GARCH$ models can be analyzed in a similar fashion (see Appendix E).

Estimating and testing a $GARCH$ process

For the econometric *estimation and testing* of the unknown parameters in the $GARCH(p,q)$ model, historical sample information is used to determine the weights for such a volatility model. Intuitively, as a general rule, choose these weights so that the conditional variance is big whenever the squared returns are big and the conditional variance is small when returns are near zero. In practice, maximum likelihood (ML) estimation procedure for various underlying probability distributions of the returns (*normal, Student, Laplace*) is used.

i) Volatility clustering based on normality The appropriate test for these returns ($r_t =: y_t$) is provided by the familiar Jarque-Bera (JB) test statistic: $JB = T\left(\frac{\hat{\xi}^2}{6} + \frac{(\hat{\kappa}-3)}{24}\right) \overset{as}{\sim} \chi_2^2$, where the estimates for the skewness, $\hat{\xi}$, and the kurtosis, $\hat{\kappa}$, are defined as $\hat{\xi} := \frac{1}{T}\frac{\sum_{t=1}^{T}(y_t-\hat{\mu})^3}{\hat{\sigma}_y^3}$ and $\hat{\kappa} := \frac{1}{T}\frac{\sum_{t=1}^{T}(y_t-\hat{\mu})^4}{\hat{\sigma}_y^4}$, respectively. If the model is good, then the standardized asset residual returns

$\hat{\varepsilon}_t/\hat{\sigma}_t$ will no longer have autocorrelation in the squares,[16] which can directly be checked by the corresponding Ljung-Box statistic (see (5.33)), where the estimated squared autocorrelations are autocorrelations among the squared residuals now.[17] A similar test can be designed on the validity of the AR part by testing $ARCH(q) = GARCH(0, q)$ under the null against $GARCH(p, q)$.

If the *conditional normality assumption* for the asset returns is accepted by the above JB statistic, then the loglikelihood function for one observation t for the model $y_t = \delta' x_t + \varepsilon_t$ is given by:

$$\ln \ell_t(\boldsymbol{\theta}) = -1/2 \ln(2\pi) - 1/2 \ln(\sigma_t^2) - 1/2 \varepsilon_t^2 \sigma_t^{-2},$$

with sample loglikelihood function:

$$L(\boldsymbol{\theta}) = \sum_{t=1}^{T} \ln \ell_t(\boldsymbol{\theta}) = -1/2 \sum_{t=1}^{T} \left[\ln(2\pi) + \ln(\sigma_t^2) + \varepsilon_t^2 \sigma_t^{-2} \right])$$

The ML estimator results from the first order conditions of the maximization of this loglikelihood:

$$\partial L(\boldsymbol{\theta}) / \partial \boldsymbol{\theta} = -\frac{1}{2} \sum_{t=1}^{T} \frac{1}{\sigma_t^2} \frac{\partial \sigma_t^2}{\partial \boldsymbol{\theta}} + \frac{1}{2} \sum_{t=1}^{T} \frac{\varepsilon_t^2}{\sigma_t^2} \frac{1}{\sigma_t^2} \frac{\partial \sigma_t^2}{\partial \boldsymbol{\theta}} = \frac{1}{2} \sum_{t=1}^{T} \left[\frac{\varepsilon_t^2}{\sigma_t^2} - 1 \right] \frac{1}{\sigma_t^2} \frac{\partial \sigma_t^2}{\partial \boldsymbol{\theta}} = \mathbf{0}$$

In the above derivations $\boldsymbol{\theta}$ is defined as: $\boldsymbol{\theta} := (\ \boldsymbol{\delta}' \quad \omega \quad \boldsymbol{\alpha}' \quad \boldsymbol{\beta}'\ )' = (\boldsymbol{\delta}', \boldsymbol{\gamma}')'$ where $\boldsymbol{\gamma}$ is the vector of the unknown $GARCH$ parameters; in this formula the gradient of the loglikelihood function with respect to the $GARCH$ parameters can be evaluated as follows:

$$\frac{\partial \ln \ell_t(\boldsymbol{\theta})}{\partial \boldsymbol{\gamma}} = -\frac{1}{2} \left[\frac{1}{\sigma_t^2} - \frac{\varepsilon_t^2}{(\sigma_t^2)^2} \right] \frac{\partial \sigma_t^2}{\partial \boldsymbol{\gamma}} = \frac{1}{2} \left(\frac{1}{\sigma_t^2} \right) \frac{\partial \sigma_t^2}{\partial \boldsymbol{\gamma}} \left(\frac{\varepsilon_t^2}{\sigma_t^2} - 1 \right) = \frac{1}{2} \left(\frac{1}{\sigma_t^2} \right) g_t v_t.$$

Note that, since $E\{v_t\} = 0$, the matrix of second order partial derivatives of the loglikelihood function satisfies:

$$\frac{\partial L(\boldsymbol{\theta})}{\partial \boldsymbol{\theta} \partial \boldsymbol{\theta}'} = \frac{1}{2} \sum_{t=1}^{T} \left(\frac{\varepsilon_t^2}{\sigma_t^2} - 1 \right) \frac{1}{\sigma_t^2} \frac{\partial^2 \sigma_t^2}{\partial \boldsymbol{\theta} \partial \boldsymbol{\theta}'} - \frac{1}{2} \sum_{t=1}^{T} \left(2 \frac{\varepsilon_t^2}{\sigma_t^2} - 1 \right) \frac{1}{(\sigma_t^2)^2} \frac{\partial \sigma_t^2}{\partial \boldsymbol{\theta}} \frac{\partial \sigma_t^2}{\partial \boldsymbol{\theta}'},$$

where the part depending on the $GARCH$ parameters can be evaluated as:

$$\begin{aligned} \frac{\partial^2 \ln \ell_t(\boldsymbol{\theta})}{\partial \boldsymbol{\gamma} \partial \boldsymbol{\gamma}'} &= \frac{1}{2} v_t \left[\frac{\partial (1/\sigma_t^2) g_t}{\partial \boldsymbol{\gamma}'} \right] + \frac{1}{2} \left(\frac{1}{\sigma_t^2} \right) g_t \left[\frac{\partial (\varepsilon_t^2/\sigma_t^2) - 1}{\partial \boldsymbol{\gamma}'} \right] \\ &= \frac{1}{2} v_t \left[\frac{\partial (1/\sigma_t^2) g_t}{\partial \boldsymbol{\gamma}'} \right] - \frac{1}{2} \left(\frac{g_t}{\sigma_t^2} \right) \left(\frac{\varepsilon_t^2}{\sigma_t^2} \right) \left(\frac{g_t}{\sigma_t^2} \right)'. \end{aligned}$$

[16] Notice that, if there is volatility clustering in the asset returns, then there will be autocorrelation in the squares of these asset returns.

[17] If the null is rejected, then nonlinear dependence, such as $GARCH$, may be present. According to Diebold and Lopez (1995), pp. 17-18, the Ljung-Box test statistic for testing the existence of autocorrelation of the residual squares is adjusted for $GARCH$ innovations as follows: $Q_k^* = T(T+2) \sum_{j=1}^{k} \frac{1}{(T-j)} \frac{\hat{\sigma}^4}{\hat{\sigma}^4 + \hat{\gamma}_{r^2}(j)} \hat{\rho}_j^{*2}$, which is asymptotically distributed as χ_k^2. In this test statistic $\hat{\gamma}_{r^2}(j)$ is the estimated autocovariance of r^2 at lag j, $\hat{\sigma}^4$ the squared unconditional variance of r_t, and $\hat{\rho}_j^*$ the sample autocorrelation at lag j. If the test statistic is not adjusted, it is (much) larger and one will suspect the existence of serial correlation much quicker than necessary.

Under general regularity conditions the ML estimator is consistent, asymptotically efficient (Cramèr-Rao lower bound) and asymptotically normally distributed. Suppose that the assumption of conditional normality is weakened only to $E\left\{\varepsilon_t|\, I_t\right\} = 0$, $E\left\{\left(\varepsilon_t^2/\sigma_t^2\right)\middle|\, I_t\right\} = 1$, $E\left\{\left(\varepsilon_t^4/\sigma_t^4\right)\middle|\, I_t\right\} = \kappa$, where κ is a finite value and where the conditional variance is as defined earlier. Now the (conditionally) normal loglikelihood function is inappropriate, but the ML estimator may still be consistent! It has been shown by White (1982) that the *pseudo-MLE (or quasi-MLE or QML estimator)*, obtained by maximizing the same loglikelihood function as above, produces a consistent estimate, despite the misspecification, since the first order conditions of ML are still valid. Only the asymptotic covariance matrices for the parameter estimators must be adjusted, however (see Gouriéroux *et al.* (1984)). They demonbstrated that the QML estimators are consistent and asymptotically normal with asymptotic variance-covariance matrix: $asvar(\sqrt{T}(\hat{\boldsymbol{\theta}} - \boldsymbol{\theta}){=}J^{-1}IJ^{-1}$, where $J := \lim\limits_{T\to\infty}\left\{\left[-E\left(\frac{\partial^2 L}{\partial\boldsymbol{\theta}\partial\boldsymbol{\theta}'}\right)\right]\Big|_{\boldsymbol{\theta}=\hat{\boldsymbol{\theta}}}\right\}$ and $I := \lim\limits_{T\to\infty}\left\{-E\left(\frac{\partial L}{\partial\boldsymbol{\theta}}\frac{\partial L}{\partial\boldsymbol{\theta}'}\right)\Big|_{\boldsymbol{\theta}=\hat{\boldsymbol{\theta}}}\right\}$ and L is the used but inappropriate conditionally normal loglikelihood function. When the true conditional distribution is normal, $I = J$. Notice, however, that the QML estimator can be (asymptotically) inefficient (e.g. in the presence of asymmetries).

How to test for *GARCH*? The occurrence of $GARCH$ effects is tested in a similar way as for $ARCH$ effects. The LM test statistic for $ARCH(q)$ against $GARCH(p,q)$, $\xi_{LM} = TR^2 \overset{as}{\sim} \chi_q^2$ from the auxiliary regression (8.11), can be used for testing the $GARCH$ effects. There is, unfortunately, an indeterminacy in this test procedure. The test for $ARCH(q)$ against $GARCH(p,q)$ is exactly the same as that for $ARCH(p)$ against $ARCH(p+q)$. For carrying out this test, one can use as starting values a set of estimates that includes $\boldsymbol{\beta} = \mathbf{0}$ and any consistent estimators for the non$GARCH$ parameter $\boldsymbol{\delta}$ and for the $ARCH$ parameter $\boldsymbol{\alpha}$. Then the TR^2 for the regression at the initial iteration provides the test statistic.

Example 8.3 *An empirical example on inflation*

Bollerslev (1986) reports the results of a study of the implicit price deflator for GNP made by Engle and Kraft (1983) using an ARCH(8) model and his extension of it based on a $GARCH(1,1)$ model. Table 8.2 reports three sets of estimates for the model: $\pi_t = \beta_o + \beta_1\pi_{t-1} + \beta_2\pi_{t-2} + \beta_3\pi_{t-3} + \beta_4\pi_{t-4} + \varepsilon_t$ where $\pi_t := 100\ln\left(P_t/P_{t-1}\right)$. The data used in the regressions are quarterly observations on the implicit price deflator for GNP from 1948^{II} to 1983^{IV}.

The first set of results is obtained by OLS. The OLS residuals give no evidence of autocorrelation up to lag 10, but the squares of the residuals show significant autocorrelations at lags $1, 3, 7, 9$, and 10. The LM test statistics for $ARCH(1)$, $ARCH(4)$, and $ARCH(8)$ are all highly significant.

The declining linear lag model specified by Engle and Kraft (1983) is shown second. Note that there is a linear restriction in the model; the eight terms in the $ARCH$ part of the model are actually functions of a single parameter.

The values decline linearly from 0.179 to 0.022. The normalized residuals $\widehat{\varepsilon}_t/\widehat{\sigma}_t$ from the $ARCH(8)$ model show, again, no evidence of autocorrelation, nor do their squares. The linear restriction of the linear lag model on the unrestricted $ARCH(8)$ model appears not to be significant; the χ^2 value is 8.87, whereas $\chi^2_7(0.05) = 14.01$. The $ARCH(8)$ model therefore appears to have accounted adequately for the lagged effects in the conditional variance. But the LM test statistic for the inclusion of σ^2_{t-1} in the conditional variance is 4.57, which is significant at the 5 percent level.

Table 8.2: An example on inflation

	β_o	β_1	β_2	β_3	β_4
OLS	0.240	0.552	0.177	0.232	-0.209
	(0.080)	(0.083)	(0.089)	(0.090)	(0.080)
$\sigma^2 = 0.282(0.034)$					
$ARCH(8)$	0.138	0.423	0.222	0.377	0.175
	(0.059)	(0.081)	(0.108)	(0.078)	(0.104)
$\sigma^2_t = \underset{(0.033)}{0.058} + \underset{(0.265)}{0.808} \sum_{i=1}^{8} \frac{9-i}{36} \varepsilon^2_{t-i}$					
$GARCH(1,1)$	0.151	0.433	0.229	0.349	0.162
	(0.060)	(0.081)	(0.110)	(0.077)	(0.104)
$\sigma^2_t = \underset{(0.006)}{0.007} + \underset{(0.070)}{0.135}\varepsilon^2_{t-1} + \underset{(0.068)}{0.829}\sigma^2_{t-1}$					

Source: This example is considered in Greene (2000), pp.809-810.

Therefore, the third set of estimates is on the $GARCH(1,1)$ model. The normalized residuals and their squares once again appear to be nonautocorrelated to lags up to 10 periods. The $GARCH(1,1)$ specification appears to be adequate.

Plots of the actual inflation rate and asymptotic confidence intervals for the one-step (ahead) forecast errors are given in Bollerslev's (1986) paper for the predictions of the model estimated by OLS and by the $GARCH(1,1)$ model. The clear pattern is that in periods of very volatile inflation (the later 1940s and early to mid 1950s), the simple OLS regression is a visibly better predictor. The effect is reversed in the more stable period of the late 1950s and early 1970s.

ii) Case of a conditional Student t-distribution If the JB statistic rejects the normality assumption, a good choice of the underlying returns distribution might be the Student t-distribution. The Student t-distribution roughly corresponds to the normal distribution, except that the kurtosis is larger than that of the normal distribution. Hence, fat-tailed empirical distributions are often well described by this distribution.[18]

[18].Notice that the Student t-distribution has a higher kurtosis than the normal distribution, because the kurtosis of a conditional Student t-distribution with v degrees of freedom is equal to $3(v-2)/(v-4)$, which is larger than 3 (being the kurtosis of the underlying condi-

Consider the following model for the mean equation and suppose that the innovations follow now a conditional Student t-distribution with a conditional variance equal to σ_t^2:

$$y_t = \zeta\left(z_{t-1}; a\right) + \varepsilon_t = \mu + \varepsilon_t$$

where the conditional pdf for the innovations is $f_v\left(\varepsilon_t | \Im_{t-1}\right)$ which is given by:
$f_v\left(\varepsilon_t | \Im_{t-1}\right) = \Gamma\left((v+1)/2\right)\Gamma\left(v/2\right)^{-1}\pi^{-1/2}\left((v\text{-}2)\sigma_t^2\right)^{-1/2}\left(1+\varepsilon_t\sigma_t^{-2}\left(v\text{-}2\right)^{-1}\right)^{-(v+1)/2}$,
where v is the number of the degrees of freedom and Γ is the so called Γ-function defined as:

$$\Gamma(y) := \int_0^\infty x^{y-1}e^{-x}dx, \Gamma(1/2) = \sqrt{\pi}, \Gamma(1) = 1, \text{ and } \Gamma(n+1) = n\Gamma(n).$$

The Student t-distribution is symmetric around zero and the second and fourth (conditional) moments are given by:

$$var\left\{\varepsilon_t | \Im_{t-1}\right\} = \sigma_t^2, \quad v > 2$$
$$\left\{\varepsilon_t^4 | \Im_{t-1}\right\} = 3(v-2)(v-4)^{-1}\sigma_t^4, \quad v > 4$$

When $v = 1$, the density has the fattest tails with no expectation and when $v \to \infty$, the standard normal density is obtained.

The logarithmic likelihood function to be maximized for ML estimators of an arbitrary $GARCH(p, q)$ model is:

$$\ln \ell\left(\theta\right) = \sum_{t=1}^T \ln f_v\left(\varepsilon_t | \Im_{t-1}\right) = h(v) - 1/2\sum_{t=1}^T \ln \sigma_t^2$$
$$-\tfrac{(v+1)}{2}\sum_{t=1}^T \ln\left(1 + \varepsilon_t^2\sigma_t^{-2}(v-2)^{-1}\right),$$

where $h(v) := T\ln\left[\Gamma\left(\frac{v+1}{2}\right)\right] - T\ln\left[\Gamma\left(\frac{v}{2}\right)\right] - \frac{T}{2}\ln\pi - \frac{T}{2}\ln(v-2)$.

This function can also be maximized with restrictions on the parameters. Note however that in addition to the three parameters in the $GARCH(1,1)$ model the vector $\boldsymbol{\theta}$, a fourth parameter, i.e. the degrees of freedom parameter v, is also estimated.

iii) Case of a conditional Laplace distribution. The Laplace or double exponential distribution can also be used to model the conditional distribution of the innovations. Let u_t be a random variable that follows the Laplace distribution, its density is then given by: $f_{Laplace} = \frac{1}{2}\exp(-|u_t|)$

The expectation of u_t is zero, the variance equals 2 and the kurtosis coefficient equals 6. The innovation has pdf: $\varepsilon_t \overset{iid}{\sim} f_{Laplace}\left(\frac{\varepsilon_t}{\sigma_t\gamma}\right)\frac{1}{\sigma_t\gamma}$ with expectation zero and variance $2\sigma_t^2\gamma^2$. Restricting $\gamma = 1/\sqrt{2}$ suffices to get a variance of σ_t^2 for ε_t. The loglikelihood function associated with the Laplace pdf then reads as:

$$\ln \ell(\theta) = -\frac{T}{2}\ln 2 - \frac{1}{2}\sum_{t=1}^T \ln \sigma_t^2 - \sqrt{2}\sum_{t=1}^T \frac{|\varepsilon_t|}{\sigma_t}. \tag{8.13}$$

tional normal distribution) for $v > 4$. This property makes the Student t-distribution highly attractive for fat-tailed asset returns, as is often occurring in practice. Only if this number of the degrees of freedom increases indefinitely, both probability distributions will coincide.

Possible alternative conditional distributions of the asset returns are: the *generalized error distribution* and the *generalized Student t-distribution* (see Sections E.3 and E.4 in Appendix E).

Three examples:

Example 8.4 *If* $\eta_t \overset{iid}{\sim} N(0,1)$*, the ML method may be used to estimate the unknown parameters in the two equations of an ARMA-GARCH model. Details of this model are given in Bollerslev (*1986*) and Bera and Higgins (*1993*). Given appropriate choices for the starting values, the number of iterations can usually be kept at a moderate level.*

For an $AR(0)$-$GARCH(1,1)$ model for the logarithmic weekly returns of the Dow-Jones index for the period 1980^1-1994^{39} the following estimation results based on the standard normal distribution of η_t can be obtained.[19]

$$\hat{\varepsilon}_t = y_t - \underset{(0.72)}{0.00241} = \hat{\eta}_t\hat{\sigma}_t \text{ and } \hat{\sigma}_t = \underset{(0.0000135)}{0.0000520} + \underset{(0.017)}{0.164}\hat{\varepsilon}^2_{t-1} + \underset{(0.043)}{0.727}\hat{\sigma}^2_{t-1},$$

where estimated standard errors are given between parentheses and $\hat{\alpha}_1 + \hat{\beta}_1 = 0.891 < 1$.

Example 8.5 *Estimating by ML a normal* $GARCH(1,1)$ *for weekly US Treasury Bills on three months for the period 1962-1993:*

$$\hat{\varepsilon}_t = y_t - \underset{(0.00047)}{0.000803} = \hat{\eta}_t\hat{\sigma}_t, \text{ with } \hat{\sigma}_t = \underset{(0.00000196)}{0.0000116} + \underset{(0.0118)}{0.126}\,\hat{\varepsilon}^2_{t-1} + \underset{(0.0122)}{0.864}\,\hat{\sigma}^2_{t-1}$$

with estimated standard errors between brackets.

The JB statistic on the assumed normality of the residuals amounts 646.8022 which is (much) higher than its 1 per cent critical χ^2 value with 2 degrees of freedom. The standardized residuals in the estimated conditional variance equation show the following estimated autocorrelation coefficients up to order fifteen :
$(.027\ -.013\ -.047\ -.032\ .004\ .000\ -.009\ .014\ -.010\ -.033\ -.012\ .011\ -.002\ -.034\ -.023)'$, with corresponding LB statistic 12.6213, which is less than the 5% critical χ^2_{15}-value, so that the null hypothesis of no residual autocorrelation is not rejected.

Example 8.6 *In this example, we estimate and test the simple* $GARCH(1,1)$ *model for asset returns as in Example 8.2, applied on spot prices of the Dutch Philips and ING Group stocks.*

[19] See Franses (1998) who considers the 770 consecutive logarithmic returns on the Dow-Jones index, from week 1 in 1980 to week 39 in 1994. These returns concern the end-of-the-week returns, where a week is assumed to run from Thursday to Wednesday. Observation 408 corresponds to the week commencing Monday, October 19, 1987. Around that so-called *Black Monday*, the returns on the Dow-Jones index decreased by about 17%. Immediately after that observation, several data points that are large in absolute value can be found. Additionally, in other parts of the sample, we can observe 'bubbles', i.e. clusters of observations with large variances. As mentioned before this phenomenon is called volatility clustering or, otherwise, conditional heteroskedasticity.

Consider, first, the 1010 observations on the Philips asset returns for the period 01/04/1994 – 30/12/1997, as pictured in Figure 8.2. Some stylized facts of this time series are summarized in Table 8.3 with the figures between brackets being the p-values or excess probability values of the corresponding test statistics.[20]

Table 8.3: Philips returns

Sample mean:0.1276	Sample standard deviation: 2.0630
Sample skewness: -0.1385	Sample kurtosis:6.7509
LB-statistic of returns, based on 5 sample autocorrelations: 11.2081 (0.0474)	
LB-statistic of squared returns based on 5 autocorrelations: 76.9610 (0.0000)	
JB-test statistic on the (assumed) normality of asset returns: 583.3092 (0.0000)	

From the summary LB and JB test statistics it follows that:

1. the Philips asset returns are subject to (somewhat slight) autocorrelation at the 95 per cent confidence level, but that the squared asset returns show strong autocorrelation, so that conditional heteroskedasticity seems to be present in the Philips asset returns,

2. the sample kurtosis is higher than that of the normal distribution leading to a rejection of the null hypothesis that the sample Philips asset returns are normally distributed.

Similarly, for the (nonindustrial) Dutch ING Group, some stylized facts on the asset returns over the same sample period are presented in Table 8.4.

Table 8.4: ING returns

Sample mean : 0.0883	Sample standard deviation: 1.4500
Sample skewness: 0.2884	Sample kurtosis: 8.1435
LB-statistic, based on 5 sample autocorrelations: 3.0136 (0.6979)	
LB-statistic of squared returns based on 5 autocorrelations: 107.4423 (0.0000)	
JB-test statistic on the (assumed) normality of asset returns: 1127.4 (0.0000)	

[20] A very low p-value points to the rejection of the underlying null hypothesis. Notice that the 5% critical value for the LB-test statistic is 11.07.

This time, both skewness and kurtosis deviate from those of the normal distribution and conditional heteroskedasticity is also strongly present in the asset returns series of the ING Group.

Based on the above empirical findings the simple $GARCH(1,1)$ model of Example 8.2, based as well on an underlying normal as on a Student t-distribution of the sample returns, can conveniently be applied to the Philips and ING asset returns respectively. This leads to the empirical results in Table 8.5.

Table 8.5: Empirical results: Philips and ING-Group

Philips		*ING Group*	
Normal	Student t	Normal	Student t
$\hat{\mu} = 0.1276$ (0.0649)	$\hat{\mu} = 0.1276$ (0.0649)	$\hat{\mu} = 0.0883$ (0.0456)	$\hat{\mu} = 0.0883$ (0.0456)
$\hat{\omega} = 0.2501$ (0.0468)	$\hat{\omega} = 0.1356$ (0.0585)	$\hat{\omega} = 0.0188$ (0.0071)	$\hat{\omega} = 0.0985$ (0.0365)
$\hat{\alpha} = 0.1271$ (0.0203)	$\hat{\alpha} = 0.1142$ (0.0271)	$\hat{\alpha} = 0.0654$ (0.0063)	$\hat{\alpha} = 0.1500$ (0.0332)
$\hat{\beta} = 0.8207$ 0.0255)	$\hat{\beta} = 0.8635$ (0.0304)	$\hat{\beta} = 0.9292$ (0.0063)	$\hat{\beta} = 0.8132$ (0.0361)
$\hat{\upsilon} = 5.086$ (0.7906)		$\hat{\upsilon} = 5.6453$ (0.9261)	
$\hat{\alpha} + \hat{\beta} = 0.9478$	$\hat{\alpha} + \hat{\beta} = 0.9777$	$\hat{\alpha} + \hat{\beta} = 0.9946$	$\hat{\alpha} + \hat{\beta} = 0.9632$

All estimated $GARCH(1,1)$ parameters are significant at the 95% confidence level. Also stationarity is accepted for each model, although sometimes by the narrowest margin, as e.g. for the ING asset returns under normally distributed innovations.

Aggregation of $GARCH$ processes

Usually, in order to prevent high transaction costs and due to the availability of financial managers, financial decisions are taken weekly in many enterprises, although many $GARCH$ models are estimated on a daily basis. In general, daily expected asset returns are transformed to another time horizon, say h, by just multiplying these returns by h. Similarly, time-varying conditional standard deviations are multiplied by $\sqrt{h}$.[21]

In general, if z_t is the logarithmic spot price at time t, and if changes in this logarithmic spot price are mutually independently and normally distributed with zero mean and variance σ_ε^2, i.e. $z_t = z_{t-1} + \varepsilon_t$, with $\frac{\varepsilon_t}{\sigma_\varepsilon} \overset{iid}{\sim} N$, the h-daily return is given by $z_t - z_{t-h} = \sum_{i=0}^{h-1} \varepsilon_{t-i}$ with standard deviation $\sigma_\varepsilon\sqrt{h}$. This is the simple 'scaling rule': multiply with to transform from a 1-daily standard deviation to a h-daily standard deviation. This simple time-aggregation may

[21]Hence, when the volatility is measured in moving conditional standard deviations, the weekly volatility is obtained by multiplying the daily volatility by $\sqrt{5}$.

only be applied under strict conditions, which are generally not satisfied for high-frequent financial data. Rescaling of the data in this sense is only valid if the data is independently and identically distributed, which is contradictory to many empirical findings. Luckily, Drost and Nijman (1993) have found a method to scale volatilities to another time horizon, which are robust for three types of stationary *GARCH* processes. Defining $\{\varepsilon_t, t \in \mathbb{Z}\}$ as a stochastic process with finite fourth order moments, strong, semi-strong and weak stationary *GARCH* processes are defined, respectively. In the first case rescaled innovations are independent, in the second case rescaled innovations are uncorrelated, and, in the third case, σ_t^2 is the best linear predictor of ε_t^2 based on lagged innovations solely.

Drost and Nijman (1993) prove the following theorem: If the stochastic process $\{y_t, t = 1, 2, \ldots\}$ is a symmetrically weak $GARCH(p, q)$ process, then the stochastic process $\{y_{(h)th}, t = 1, 2, \ldots\}$ is a symmetrically weak $GARCH(r,r)$ process with $r := max(p, q)$. Some aggegation results are assembled in Section E.5 of Appendix E.

Example 8.7 *Assume that mid-of-week asset returns (based on Wednesday spot prices) are of primary interest to financial managers. Usually, there is a very limited direct GARCH effect left on a weekly basis. Therefore, it is more efficient to look at aggregated GARCH processes. The time-aggregation results of the normal $GARCH(1, 1)$ applied on the Philips asset returns of the previous example could be summarized as follows (see Section E.5):*

Conditional pdf	h	$\hat{\omega}_{(h)}$	$\hat{\alpha}_{(h)}$	$\hat{\beta}_{(h)}$
Normal pdf	1	0.21	0.13	0.83
	5	1.00	0.16	0.62
Student t-pdf	1	0.14	0.11	0.86
	5	0.65	0.18	0.71

From these time-aggregation results it seems that the conditional variances depend to a lesser extent on a weekly basis (lower $\hat{\beta}_{(h)}$s and higher $\hat{\alpha}_{(h)}$s). This is to be expected since aggregation leads to a slowly extinction of volatility fluctuations. When h grows indefinitely, both $\hat{\alpha}_{(h)}$s and $\hat{\beta}_{(h)}$s approximate zero.

8.2.3 Other functional forms of *GARCH* models

According to Subsection 8.2.2, the standard (stationary) *GARCH* model yields forecasts of future conditional variances, which are linear in current and past conditional variances and squared returns drive revisions in the forecasts. Various variants of the standard *GARCH* model can now be defined, some important ones will be discussed in this section.

Integrated *GARCH* (or *IGARCH*) models

If $\alpha + \beta$ is close to one in the simple $GARCH(1, 1)$ model there exists high persistence in the conditional variance so that its convergence to the unconditional variance is very slow. If $\alpha + \beta = 1$, the expectation of volatility in this model s periods ahead becomes from the expression in the subsection on volatility forecasts of the previous section: $E_t(\sigma_{t+s}^2) = \sigma_t^2 + s\omega$. Hence, the $GARCH(1,1)$ model with $\alpha+\beta = 1$ has a unit autoregressive root so that the current volatility affects forecasts of volatility into the indefinite future. It is therefore called *integrated* $GARCH(1,1)$ or $IGARCH(1,1)$ model. The $IGARCH(1,1)$ process for the conditional variance σ_t^2 looks like a linear random walk with drift ω. But Nelson (1990) has correctly remarked that a conventional linear random walk model is nonstationary in two senses. On the one side, it has no stationary probability distribution so that it is not strictly stationary and, on the other side, it has no unconditional first or second order moments so that it is not (covariance or second order) stationary. On the contrary, $IGARCH(1,1)$ is generally not (covariance) stationary, but is strictly stationary[22]. Arbitrary $IGARCH(p,q)$ can be studied *ceteris paribus*.

Absolute value $GARCH$ models

Such models make forecasts of the future conditional standard deviation. The latter is assumed to be linear in the current and past standard deviations and has absolute values of asset returns driving revisions in the forecasts. An example of an absolute value $GARCH(1,1)$ model is $\sigma_t = \omega + \beta\sigma_{t-1} + \alpha\sigma_{t-1}|\varepsilon_t|$.[23]

Exponential $GARCH$ or $EGARCH(p,q)$ models

Black (1976) was one of the first persons to study the phenomenon that (in several cases) current returns and future volatility are negatively correlated, this in contradiction to the Markowitz tradition, where a trade-off between future expected asset returns and current volatility is studied. As possible explanation for this phenomenon he suggested the so-called *leverage effect*. This effect stipulates a negative correlation between current return innovations and future expected conditional variances.

In this spirit Nelson (1991) described two important problems with the classic $GARCH$ models: (i) Garch models have symmetric shock effects ("signs of innovations are not important, only the sizes"), but as indicated above, this seemed to be too restrictive; (ii) $GARCH$ requires nonnegativity restrictions on the parameters (to guarantee positive conditional variances), so that each increase of ε_t results in an increase of each future σ_{t+s}^2, which excludes any form of oscillation! To take account of these disadvantages, Nelson (1991) proposed the exponential $GARCH$ ($EGARCH$) model . In the $EGARCH$ model the conditional variance depends on both the size and the sign of the innovations. This implies that the conditional variance is an asymmetric function of the lagged

[22] It can easily be shown that the $IGARCH(1,1)$ model has a stationary probability distribution when $\omega = 0$. When $\omega > 0$, Nelson (1990) has shown that there exists a nondegenerate stationary probability distribution for the volatility σ_t^2, but that this probability distribution has no finite moments.

[23] See Nelson and Foster (1994) for the econometric inference of absolute value GARCH(1,1) models.

Figure 8.4: The news Impact Curve: 'Negative news has a bigger effect than positive news'

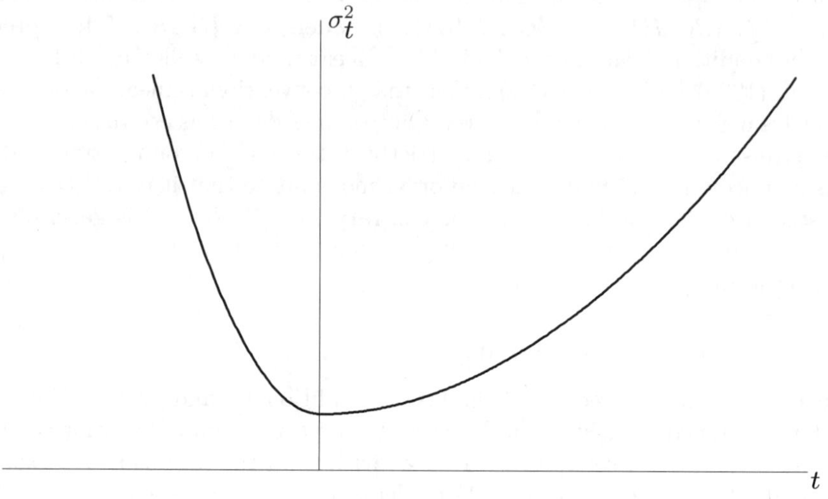

innovations or that negative and positive innovations have a different impact on volatility. Also, for equities, it is often observed that downward movements in the market are followed by higher volatilities than upward movements of the same magnitude. This is illustrated in Figure 8.4.[24]

Nelson's $EGARCH(p,q)$ model, which takes account of the asymmetric impact of news illustrated above, can be written as:[25]

$$\begin{aligned}\ln\left(\sigma_t^2\right) &= \omega + \textstyle\sum_{j=1}^{p} \beta_j \ln\left(\sigma_{t-j}^2\right) + \sum_{i=1}^{q} \alpha_i \left[\theta \eta_{t-i} + \gamma\left(\left|\eta_{t-i}\right| - E\left|\eta_t\right|\right)\right] \\ &= \omega + \textstyle\sum_{j=1}^{p} \beta_j \ln\left(\sigma_{t-j}^2\right) + \sum_{i=1}^{q} \alpha_i g\left(\eta_{t-i}\right), \qquad \eta_t \overset{iid}{\sim} N(0,1),\end{aligned}$$

where the conditional variance is always positive by construction $\sigma_t^2 = \exp\left[\ln\left(\sigma_t^2\right)\right] >$

[24] See also Acker (1998), who studied the reaction of stock return volatility to annual profit announcements in the UK and even concluded that 'good news was no news'.

[25] An alternative formulation that allows for a different treatment of positive and negative shocks is the so-called Threshold $ARCH$, or $TARCH$, model, proposed in Glosten *et al.* (1993) and Rabemananjara and Zakoian (1993). The conditional variance equation for this model is: $\sigma_t^2 = \omega + \alpha\varepsilon_{t-1}^2 + \delta I_{t-1}\left[\varepsilon_{t-1} < 0\right]\varepsilon_{t-1}^2 + \beta\sigma_{t-1}^2$, where I_{t-1} is an indicator dummy variable that takes a value of 1 when $\varepsilon_{t-1}$is negative, and a value of 0 when $\varepsilon_{t-1} \geqq 0$, and where $\delta > 0$. If δ is significantly larger than zero, then the average effect seems important. Franses (1998), p. 172, notes that when the parameters in the above equation are estimated for the weekly Dow-Jones index, the estimation results are:

$$\hat{\sigma}_t^2 = \underset{(0.000024)}{0.000123} + \underset{(0.037)}{0.042}\,\varepsilon_{t-1}^2 + \underset{(0.070)}{0.418}\,I_{t-1}\left[\varepsilon_{t-1} \prec 0\right]\varepsilon_{t-1}^2 + \underset{(0.075)}{0.492}\,\hat{\sigma}_{t-1}^2,$$

Comparing these results with those of the $GARCH$(1,1) model in Section 8.2.1, it is clear that $\hat{\beta}$ is much lower in the $EGARCH$ model and that $\hat{\alpha}$ is now insignificant. Since $\hat{\delta}$ is significantly positive in this $TARCH$ model, negative shocks to the Dow-Jones index show quite an important role.

0. The last part in the *EGARCH* formulation must catch the leverage effect. It is an asymmetric relation between the equity return and volatility changes. The sign of this leverage effect depends on the sign of θ and $g(\eta_{t-i}) := [\theta\eta_{t-i} + \gamma(|\eta_{t-i}| - E|\eta_t|)]$ has a *news impact* representing the effects of shocks in the conditional variance (recall that $\eta_{t-i} := \frac{\varepsilon_{t-i}}{\sigma_{t-i}}$ and $E|\eta_t| = \sqrt{2\pi}$) if $\eta_t \overset{iid}{\sim} N(0,1)$). If $\theta < 0$ the future conditional variances will increase more than proportionally as a consequence of a negative shock than for a positive shock of the same size. One of the most popular *EGARCH* models is the $EGARCH(1,1)$ model: $\ln(\sigma_t^2) = \omega + \beta\ln(\sigma_t^2) + \gamma\frac{|\varepsilon_{t-1}|}{\sigma_{t-1}} + \theta\frac{\varepsilon_{t-1}}{\sigma_{t-1}}$, where the coefficient of the asymmetric effect is again expected to satisfy $\theta < 0$. This $EGARCH(1,1)$ model can also be written as:

$$\begin{aligned}\ln(\sigma_t^2) &= \omega + \beta\ln(\sigma_t^2) + (\theta+\gamma)\tfrac{\varepsilon_{t-1}}{\sigma_{t-1}} \quad \text{if } \varepsilon_{t-1} > 0 \\ &= \omega + \beta\ln(\sigma_t^2) + (\theta-\gamma)\tfrac{\varepsilon_{t-1}}{\sigma_{t-1}} \quad \text{if } \varepsilon_{t-1} < 0\end{aligned}$$

Empirically it has been shown that with this *EGARCH* formulation for the conditional variance negative innovations have a significantly more increasing effect on volatility than positive shocks.

Example 8.8 *In this empirical example, an $EGARCH(1,1)$ model is applied on the sample Philips asset returns:*

$y_t = \mu + \varepsilon_t$ with $\ln(\sigma_t^2) = \omega + \beta\ln(\sigma_{t-1}^2) + \alpha\theta\eta_{t-1} + \gamma(|\eta_{t-1}| - E|\eta_{t-1}|)$ and $\eta_t \overset{iid}{\sim} N(0,1)$ the following ML estimates are found (for $\gamma = 0$):

$\underset{(0.0649)}{\hat{\mu} = 0.1276}$ $\quad \underset{(0.0299)}{\hat{\omega} = 0.0777}$ $\quad \underset{(0.0488)}{\hat{\alpha} = 0.2141}$ $\quad \underset{(0.0210)}{\hat{\beta} = 0.9504}$ $\quad \underset{(0.0845)}{\hat{\theta} = -0.1840}$

These ML estimates show that asymmetric effects are present for the Philips asset returns: the parameter θ is significantly negatively estimated so that the future conditional variances will increase more than proportionally for negative shocks (as compared to positive shocks of equal size). Again a high parameter estimate for β is found so that shocks in the conditional variance will remain present for a while in its further evolution in the future.

Note that Nelson (1991) also specified an alternative $EGARCH(1,1)$ model $\ln(\sigma_t) = \omega + \beta\ln(\sigma_{t-1}) + \alpha\left|\frac{\varepsilon_{t-1}}{\sigma_{t-1}}\right| + \gamma\frac{\varepsilon_{t-1}}{\sigma_{t-1}}$, which has asymmetry if $\gamma \neq 0$. Because of the logarithmic transformation, there is no possibility of negative variance in this *GARCH* equation. The news impact is now exponential rather than quadratic. There is some evidence that this is too strong for large shocks and that the likelihood is a little more difficult to maximize, but that this is a very good model. Nevertheless, it is more difficult to forecast.

GARCH-in-Mean (*GARCH*-M) models

Another important model in the *GARCH* literature is the *GARCH*-in-Mean, or *GARCH*-M, model. In this model the conditional variance directly influences the conditional expectation of the process in the asset returns equation:

$$y_t = \zeta\left(z_{t-1}, \sigma_t^2, a\right) + \varepsilon_t$$

An initial specification for the conditional mean equation is for example $y_t = \mu + \delta\sigma_t + \sum_{i=1}^{p} \rho_i \varepsilon_{t-i} + \varepsilon_t$.The choice for the conditional standard deviation represents the assumption that changes in the conditional variance are represented less than proportional in the mean. In this model an increase in the conditional variance is associated with an increase or decrease of the conditional returns. Hence, the idea of the $GARCH$-M model is based on the trade-off between risk and expected return (see e.g. Capital Asset Pricing Model (CAPM)). An intertemporal formulation of this CAPM suggests an approximate linear relationship between the return and the variance of the market portfolio. One of the possible flows in this model is its assumption of a constant variance of the market portfolio. In this connection, the $ARCH$-M model is a natural extension. This model states that $y_t = \gamma' x_t + \delta\sigma_t^2 + \varepsilon_t$, with $\sigma_t^2 = \omega + \sum_{i=1}^{q} \alpha_i \varepsilon_{t-i}^2$.

Among the interesting implications of this modification of the standard model is that under certain assumptions, δ is the coefficient of relative risk aversion. If we use a $GARCH(p,q)$ process for the variance equation, the following $GARCH$-M model is obtained:

$$y_t = \zeta\left(x_{t-1}, \sigma_t^2, a\right) + \varepsilon_t, \text{ with } \sigma_t^2 = \omega + \sum_{j=1}^{p} \beta_j \sigma_{t-j}^2 + \sum_{i=1}^{q} \beta_i \varepsilon_{t-i}^2.$$

Two additional examples:

Example 8.9 *A simple example of the GARCH-M(1,1) model is:* $y_t - \mu - \delta\sigma_t = \varepsilon_t = \eta_t \sigma_t$ *with* $\eta_t \overset{iid}{\sim} N(0,1)$ *distributed and* $\sigma_t^2 = \omega + \alpha_1 \varepsilon_{t-1}^2 + \beta_1 \sigma_{t-1}^2$.

When the parameters in this model are estimated for the weekly Dow-Jones index, Franses (1998) p.171, gets that $\hat{\mu} = -0.0035$ with standard error 0.004 and $\hat{\delta} = 0.317$ with standard error 0.206. Hence, the previously significant growth rate in the Dow-Jones index in the $AR(0) - GARCH(1,1)$ model (see Example 8.4) seems to be offset by the $\hat{\delta}$, which is however only significant at the 12% level. Since $\hat{\delta}$ is positive, these estimation results show that returns are positively correlated with large volatility.

Example 8.10 *Test the following GARCH-M model for the Philips asset returns:*

$$y_t = \varphi + \delta f(\sigma_t^2) + \varepsilon_t$$

with the conditional variance σ_t^2 assumed to follow a $GARCH(1,1)$ model and the function $f(\sigma_t^2)$ defined as σ_t , σ_t^2 , and $\ln \sigma_t^2$ respectively.

The following parameter estimates are obtained for these three $GARCH$-M models for the Philips asset returns, under normal as well as under Student t-distributed innovations in Table 8.6.[26]

[26] To save place the empirical results of this third model are not represented here.

Table 8.6: Two GARCH-M models for Philips returns

First Model		Second Model	
Normal	Student t	Normal	Student t
$\hat{\mu} = 0.1276$ (0.0649)	$\hat{\mu} = 0.1276$ (0.0649)	$\hat{\mu} = 0.1276$ (0.0649)	$\hat{\mu} = 0.1276$ (0.0649)
$\hat{\omega} = 0.2377$ (0.0448)	$\hat{\omega}0.1172$ (0.0524)	0.2479 (0.0464)	0.127 (0.0557)
$\hat{\alpha} = 0.1238$ (0.0198)	$\hat{\alpha} = 0.1081$ (0.0253)	$\hat{\alpha} = 0.1267$ (0.0203)	$\hat{\alpha} = 0.1114$ (0.0261)
$\hat{\beta} = 0.8267$ (0.0249)	$\hat{\beta} = 0.8734$ (0.0281)	$\hat{\beta} = 0.8217$ (0.0255)	$\hat{\beta} = 0.8677$ (0.0293)
$\hat{\delta} = 0.1508$ (0.1448)	$\hat{\delta} = 0.1410$ (0.1113)	$\hat{\delta} = 0.0256$ (0.0321)	$\hat{\delta} = 0.0231$ (0.0254)

Obviously, the parameter estimates for δ are not significant at all; this is also the case for the third model where δ is estimated as 0.1854 and 0.1719 for normally and Student t-distributed innovations respectively with standard errors 0.1480 and 0.1086 respectively. Hence, the $GARCH$-M$(1,1)$ model is rejected for the Philips asset returns, as is mostly the case for all kinds of asset returns.

The $GARCH$-M model can also be combined with the $EGARCH(p,q)$ model in order to get an $EGARCH$-in-Mean, or $EGARCH$-M(p,q), model: $y_t = \zeta\left(x_{t-1}, \sigma_t^2, a\right) + \varepsilon_t$,with $\ln(\sigma_t^2) = \omega + \sum_{j=1}^{p} \beta_j \ln(\sigma_{t-j}^2) + \sum_{i=1}^{q} \alpha_i g(\eta_{t-i})$ with $\eta_t \overset{iid}{\sim} N(0,1)$, where account has been taken of the asymmetric impact of the conditional variance on the asset returns; shocks in η_t have different effects on the volatility: if there is negative news the volatility tends to increase and the reverse with positive news.

Applying an $EGARCH$-M$(1,1)$ model on the Philips asset returns, again insignificant δ estimates are found, so that also this model is rejected for these asset returns. [27]

8.2.4 Multivariate *GARCH* processes

Modeling the volatility of n time series individually by univariate $GARCH(1,1)$ models would only be justified if these were independent of each other. However, this is often not the case in volatility applications. For example, stock returns in equity portfolios will undoubtedly exhibit some order of correlation over individuals (assets) and through time, an unexpected event in a certain sector has (often) an effect on all portfolios in that sector (e.g. 'financials', 'chemicals', etc...), certain developments, e.g. in the USA, have an effect on returns development, e.g. in Europe, etc. This so-called *volatility spillover* and *volatility*

[27] Notice that the $EGARCH$ model implies that current asset returns and future volatility are negatively correlated, while the $GARCH$-M model implies that a high volatility directly leads to periods of high expected asset returns. Generally, these two theories do economically exclude each other, so that $EGARCH-$M models for asset returns will not occur often in practice.

propagation effect are not modeled in univariate *GARCH* processes. Hence, modeling the n time series jointly should intuitively improve (co)variance predictability, thus probably improving the accuracy of estimates (a multivariate extension of volatility is also of great importance to the risk profile of a portfolio and the corresponding Value-at-Risk (VaR) values).

Assuming that $\boldsymbol{\varepsilon}_t \mid \Im_{t-1} \overset{iid}{\sim} (0, \boldsymbol{\Sigma}_t)$ for any $\boldsymbol{\varepsilon}_t \in \mathbb{R}^n$ $(t = 1, 2, \cdots, T)$, the direct multivariate extension of the univariate *GARCH* process can generally be formulated as:

$$E_{t-1}(\boldsymbol{\varepsilon}_t) = \mathbf{0} \text{ and } \boldsymbol{\Sigma}_t = \mathbf{var}_{t-1}(\boldsymbol{\varepsilon}_t) = E_{t-1}\left(\boldsymbol{\varepsilon}_t \boldsymbol{\varepsilon}_t^{'}\right) \text{ for } t = 1, 2, \ldots, T,$$

with $\boldsymbol{\Sigma}_t$ the conditional covariance matrix of size $n \times n$.

Since this straightforward generalization of univariate *GARCH* processes contains too many parameters, (e.g. symmetry) restrictions will be applied on the generating process of $\boldsymbol{\Sigma}_t$. A first restriction is involved by the *vector GARCH model* (using the so-called *vec* operator which is an operator that puts columns of a matrix under each other, deleting symmetric elements). Further restrictions lead to the diagonal *vec* operator.

Vec models

Start by focusing on $\boldsymbol{\Sigma}_t$, the $n \times n$ covariance matrix. As is clear from the reasoning above n variate *GARCH* models imply that $s = n(n+1)/2$ different equations for the conditional covariance matrices should be estimated, as all covariance terms appear twice in $\boldsymbol{\Sigma}_t$. Denote the **vec** operator, which stacks the columns of some matrix, by $\mathbf{vec}(\cdot)$, i.e. the columns are put in a vector where column $i+1$ is put directly below column i for $i = 1, ..., n-1$ and all second covariance terms are omitted (thus omiting the symmetric elements in the under-triangular part of the covariance matrix). So, only covariance terms, $\sigma_{ij,t}$, $j > i$ will be part of the resulting vector. The **vec** $GARCH(p,q)$ model looks like:

$$\mathbf{vec}(\boldsymbol{\Sigma}_t) = \mathbf{vec}(\boldsymbol{\Omega}) + \textstyle\sum_{j=1}^{p} \mathbf{B}_j \mathbf{vec}(\boldsymbol{\Sigma}_{t-j}) + \sum_{i=1}^{q} \mathbf{A}_i \mathbf{vec}(\boldsymbol{\varepsilon}_{t-i}\boldsymbol{\varepsilon}_{t-i}^{\prime}),$$

so that this **vec** $GARCH(p,q)$ is stationary if all the eigenvalues of $(\sum_{i=1}^{q} \mathbf{A}_i + \sum_{j=1}^{p} \mathbf{B}_j)$ are less than one in modulus. For the $GARCH(1,1)$ model there are $\frac{n(n+1)}{2}$ unknown parameters in $\boldsymbol{\Omega}$ and $\frac{n^2(n+1)^2}{4}$ unknown parameters in $\mathbf{B}$ and $\mathbf{A}$, respectively (still 465 unknown parameters for $n = 5$ and 21 unknown parameters for $n = 2$). Hence, the *vec GARCH* model results in too many unknown parameters, and, moreover, the positive definiteness of the conditional variance-covariance matrix $\boldsymbol{\Sigma}_t$ is not guaranteed. To visualize the *vec* model, the bivariate **vec** $GARCH(1,1)$ model is shown below:

$$\begin{bmatrix} \sigma_{1,t+1}^2 \\ \sigma_{12,t+1} \\ \sigma_{2,t+1}^2 \end{bmatrix} = \begin{bmatrix} \omega_{11} \\ \omega_{12} \\ \omega_{22} \end{bmatrix} + \begin{bmatrix} \beta_{11} & \beta_{12} & \beta_{13} \\ \beta_{21} & \beta_{22} & \beta_{23} \\ \beta_{31} & \beta_{32} & \beta_{33} \end{bmatrix} \begin{bmatrix} \sigma_{1,t}^2 \\ \sigma_{12,t} \\ \sigma_{2,t}^2 \end{bmatrix} + \begin{bmatrix} \alpha_{11} & \alpha_{12} & \alpha_{13} \\ \alpha_{21} & \alpha_{22} & \alpha_{23} \\ \alpha_{31} & \alpha_{32} & \alpha_{33} \end{bmatrix} \begin{bmatrix} \varepsilon_{1,t}^2 \\ \varepsilon_{1,t}\varepsilon_{2,t} \\ \varepsilon_{2,t}^2 \end{bmatrix}.$$

Diagonal *vec* models

To overcome the waterfall of parameters in the *vec* model, letting variances solely depend on past own squared disturbances and covariances solely on past own covariances and cross products of disturbances can restrict the model as in: $\sigma_{ij,t+1} = \omega_{ij} + \beta_{ij}\sigma_{ij,t} + \alpha_{ij}\varepsilon_{i,t}\varepsilon_{j,t}$ for $i, j = 1, 2, \cdots, n$ and $t = 1, 2, \cdots, T$, so that we can predict the (conditional) covariances in this model without worrying about the (conditional) variances. This logic implies that all off-diagonal elements in the parameter matrices equal zero, i.e. diagonal matrices remain. For the bivariate $GARCH(1,1)$ case, the diagonal *vec* representation is:

$$\begin{bmatrix}\sigma^2_{1,t+1}\\ \sigma_{12,t+1}\\ \sigma^2_{2,t+1}\end{bmatrix} = \begin{bmatrix}\omega_{11}\\ \omega_{12}\\ \omega_{22}\end{bmatrix} + \begin{bmatrix}\beta_{11} & 0 & 0\\ 0 & \beta_{22} & 0\\ 0 & 0 & \beta_{33}\end{bmatrix}\begin{bmatrix}\sigma^2_{1,t}\\ \sigma_{12,t}\\ \sigma^2_{2,t}\end{bmatrix} + \begin{bmatrix}\alpha_{11} & 0 & 0\\ 0 & \alpha_{22} & 0\\ 0 & 0 & \alpha_{33}\end{bmatrix}\begin{bmatrix}\varepsilon^2_{1,t}\\ \varepsilon_{1,t}\varepsilon_{2,t}\\ \varepsilon^2_{2,t}\end{bmatrix}.$$

Since there is no guarantee that correlations derived from this model will be between -1 and +1, we have to restrict the conditional covariance matrix e.g. to positive definiteness. The necessary conditions for the conditional covariance matrix of the diagonal *vec* model to be positive definite are: $\omega_{11} > 0$, $\omega_{22} > 0$, $\omega_{11}\omega_{22} - \omega^2_{12} > 0$, $\alpha_{11} > 0$, $\alpha_{33} > 0$, $\alpha_{11}\alpha_{33} - \alpha^2_{22} > 0$, and conditions for stationarity are $\alpha_{11} + \beta_{11} < 1$, $\alpha_{22} + \beta_{22} < 1$, and $\alpha_{33} + \beta_{33} < 1$. Now only 9 parameters remain to be estimated in this bivariate case (instead of 21 for the *vec* model). Although the diagonal *vec* model economizes on parameters, both general and diagonal *vec* models share a statistical problem. In order to have a plausible parameterization of the models, the conditional covariance matrix Σ_t should be positive definite. But the conditions herefore are often hard to justify (certainly for $n > 2$) for both models and difficult to impose during estimation. Therefore, Bollerslev, Engle, Kraft and Kroner (1993) came up with the $BEKK$ model to ensure positive definiteness of the covariance matrix.

BEKK model

The *BEKK model* is an alternative parameterization, which takes explicitly account of the positive definiteness of the conditional covariance matrix (with $\Omega = C'C$, with C an undertriangular matrix):[28]

$$\Sigma_t = C'C + \textstyle\sum_{k=1}^{K}\sum_{i=1}^{q} A'_{ik}\varepsilon_{t-i}\varepsilon'_{t-i}A_{ik} + \sum_{k=1}^{K}\sum_{j=1}^{p} G'_{jk}\Sigma_{t-j}G_{jk},$$

with $n \times n$ parameter matrices A_{ik} and G_{jk}.

As an illustration of the $BEKK$ model the simple $GARCH(1,1)$ model with $K = 1$ can be written as the following $BEKK(1,1)$ model:

$$\Sigma_t = C'C + A'\varepsilon_{t-1}\varepsilon'_{t-1}A + G'\Sigma_{t-1}G,$$

which counts $\frac{1}{2}n(n+1)$ unknown parameters in C and n^2 unknown parameters in A and G, respectively, or $\frac{5n^2+n}{2}$ unknown parameters in total (e.g. 65 unknown

[28] See Engle and Kroner (1995).

parameters in the 5 dimensional case and 11 unknown parameters for $n = 2$), which implies a serious reduction compared to the *vec* model. Hence, the $GARCH$ equations in the $BEKK$ model become for the bivariate case (for $\sigma_{i,t}^2 := \sigma_{ii,t}$, $i = 1, 2$):

$$\begin{aligned}
\sigma_{11,t} &= c_{11}^2 + c_{12}^2 + a_{11}^2\varepsilon_{1,t-1}^2 + 2a_{11}a_{21}\varepsilon_{1,t-1}\varepsilon_{2,t-1} + a_{21}^2\varepsilon_{2,t-1}^2 + g_{11}^2\sigma_{11,t-1}+\\
&\quad +2g_{11}g_{21}\sigma_{12,t-1} + g_{21}^2\sigma_{22,t-1}\\
\sigma_{12,t} &= c_{12}c_{22} + a_{11}a_{12}\varepsilon_{1,t-1}^2 + (a_{21}a_{12} + a_{11}a_{22})\varepsilon_{1,t-1}\varepsilon_{2,t-1} + a_{21}a_{22}\varepsilon_{2,t-1}^2+\\
&\quad +g_{11}g_{12}\sigma_{11,t-1} + (g_{21}g_{12} + g_{11}g_{22})\sigma_{12,t-1} + g_{21}g_{22}\sigma_{22,t-1}\\
\sigma_{22,t} &= c_{22}^2 + a_{12}^2\varepsilon_{1,t-1}^2 + 2a_{12}a_{22}\varepsilon_{1,t-1}\varepsilon_{2,t-1} + a_{22}^2\varepsilon_{2,t-1}^2 + g_{12}^2\sigma_{11,t-1} +\\
&\quad +2g_{12}g_{22}\sigma_{12,t-1} + g_{22}^2\sigma_{22,t-1},
\end{aligned}$$

which indeed results in 11 parameters (cf. 21 for the *vec* $GARCH$ model). It is also very clear that the $BEKK$ model is much more general than the diagonal *vec* model because the conditional variances and covariances do not depend only on their own past but also on the previous values of the other variances and covariances and on the other squared (and cross) innovations.

In general, for the n variate $BEKK(l,l)$ model, $s+2n^2$ parameters must be estimated (with $s = n(n-1)/2$). The certainty of positive definiteness has unfortunately increased the number of parameters when compared to the diagonal *vec* model, but the thriftiness of the $BEKK$ model is very high in comparison with the *vec* model, thereby having eliminated just a few, if any interesting, *vec* models, although it is much more general as indicated earlier. Engle and Kroner (1995) show that every *vec* model that is implied by some $BEKK$ model is unique, while the converse is not true. Furthermore, all positive definite diagonal *vec* models can be written in $BEKK$ representation. Stationarity is assured if and only if all the eigenvalues of $A \otimes A + B \otimes B$ are less than one in modulus. When the parameter matrices, except for C, in the $BEKK$ model are assumed to be diagonal, the diagonal $BEKK$ model is obtained. This model contains $s + 2n$ parameters.

Estimating multivariate $GARCH$ models

The leap from estimating univariate $GARCH$ models to estimating multivariate $GARCH$ models is theoretically as straightforward as the modeling design itself. While there are several innovation distributions for the univariate $GARCH$ models, the normal and Student t-distribution for the multivariate versions will be considered. Assume that the conditional distribution of each individual vector in the sample $\varepsilon_1, \ldots, \varepsilon_T$ is now given by the n variate conditionally normal distribution (we consider n assets). This leads to the following loglikelihood equation for the general multivariate $GARCH$ model with n variate (conditionally) normally distributed innovations:

$$L(\boldsymbol{\theta}) = -\tfrac{nT}{2}\ln(2\pi) - \tfrac{1}{2}\textstyle\sum_{t=1}^{T}\ln|\boldsymbol{\Sigma}_t| - \tfrac{1}{2}\varepsilon_t'\boldsymbol{\Sigma}_t^{-1}\varepsilon_t.$$

In the bivariate case this becomes:

$$L(\boldsymbol{\theta})=-\frac{nT}{2}\ln(2\pi)-\frac{1}{2}\sum_{t=1}^{T}\ln(\sigma_{1,t}^2\sigma_{2,t}^2-\sigma_{12,t}^2)-\frac{1}{2}\sum_{t=1}^{T}\frac{\varepsilon_{1,t}^2\sigma_{2,t}^2-2\varepsilon_{1,t}\varepsilon_{2,t}\sigma_{12,t}+\varepsilon_{2,t}^2\sigma_{1,t}^2}{\sigma_{1,t}^2\sigma_{2,t}^2-\sigma_{12,t}^2}.$$

For notational convenience the parameter space for each variance and covariance has been omitted. For example, in the bivariate diagonal *vec* model, the following parameter space would apply for $\sigma_{1,t}^2(\theta_1)$, $\sigma_{12,t}(\theta_2)$ and $\sigma_{2,t}^2(\theta_3)$:

$$\boldsymbol{\theta}:=\begin{bmatrix}\boldsymbol{\theta}_1=(\omega_{11},\alpha_{11},\beta_{11})'\\ \boldsymbol{\theta}_2=(\omega_{12},\alpha_{22},\beta_{22})'\\ \boldsymbol{\theta}_3=(\omega_{22},\alpha_{33},\beta_{33})'\end{bmatrix}=(\omega_{11},\omega_{12},\omega_{22},\alpha_{11},\alpha_{22},\alpha_{33},\beta_{11},\beta_{22,},\beta_{33})',\text{ or}$$

$$\boldsymbol{\theta}=(\boldsymbol{\omega}',\boldsymbol{\alpha}',\boldsymbol{\beta}')'\in\mathbb{R}_+^9.$$

In the case of an n variate student t-distribution, the loglikelihood for the sample $\varepsilon_1,\ldots,\varepsilon_T$ is defined as:

$$L(v,\boldsymbol{\theta})=T\ln\Gamma\left(\tfrac{v+n}{2}\right)-T\ln\Gamma\left(\tfrac{v}{2}\right)-\tfrac{Tn}{2}\ln(\pi)-\tfrac{T(n-1)}{2}\ln(v)$$
$$-\tfrac{T}{2}\ln(v-2)\text{-}\tfrac{1}{2}\sum_{t=1}^{T}\ln|\boldsymbol{\Sigma}_t|\text{-}\tfrac{v+n}{2}\sum_{t=1}^{T}\ln\left(1+\tfrac{1}{v-2}\boldsymbol{\varepsilon}_t'\boldsymbol{\Sigma}_t^{-1}\boldsymbol{\varepsilon}_t\right).$$

An important remark here is that ML estimation is quite hard when *a lot of parameters* have to be estimated as the loglikelihood function can be relatively flat in this case. Moreover, the function might not be globally concave, so that a local optimum is a possible outcome. This implies that it is important to choose a suitable maximization algorithm from Chapter 1.

Conditional correlation

The class of conditional correlation models tries to capture covariance dynamics by making assumptions on the conditional correlation between time series. This class can be split up in two models. The first model is known as the *constant conditional correlation* (CCC) model, developed by Bollerslev (1990). The second model is the *dynamic conditional correlation* (DCC), designated by Engle (2000).

Constant conditional correlation (CCC) The conditional correlation between innovations ε_i and ε_j for time t given information available up to and including time $t-1$ is defined by:

$$\rho_{ij,t}:=\frac{E_{t-1}(\varepsilon_{i,t}\varepsilon_{j,t})}{\sqrt{E_{t-1}(\varepsilon_{i,t}^2)E_{t-1}(\varepsilon_{j,t}^2)}}=\frac{\sigma_{ij,t}}{\sigma_{i,t}\sigma_{j,t}}$$

Note that this definition entails that the standardized innovations share the same correlation as the original innovations. We can readily define the $n\times n$ conditional correlation matrix $\mathbf{R}_t=\{\rho_{ij,t}\}$. Using the conditional covariance matrix Σ_t, and introducing the $n\times n$ diagonal matrix $\mathbf{D}_t:=diag(\sigma_{1,t},\ldots,\sigma_{N,t})$, i.e. a matrix with standard deviations on the diagonal and zeros on the off-diagonals, we can decompose the correlation matrix as:

$$\mathbf{R}_t=\mathbf{D}_t^{-1}\boldsymbol{\Sigma}_t\mathbf{D}_t^{-1}. \tag{8.14}$$

Pre and post multiplying the correlation matrix by $\mathbf{D}_t$ in equation (8.14), yields: $\Sigma_t = \mathbf{D}_t\mathbf{R}_t\mathbf{D}_t$. This equation is the foundation on which the class of conditional correlation models is built. The conditional variances are modeled via univariate *GARCH* models, i.e. $\sigma_{i,t}^2 = \omega_i\ \sigma_{i,t-1}^2 + \alpha_i\varepsilon_{i,t-1}^2$ for $i = 1, 2, ..., n$. The CCC model is defined by assuming that the correlation matrix is constant, i.e. $\mathbf{R}_t \equiv \mathbf{R}$, $\forall t$. Additionally, under the assumption that $\boldsymbol{\varepsilon}_t |\ \Im_{t-1} \overset{iid}{\sim} N_n(0, \boldsymbol{\Sigma}_t)$, the CCC loglikelihood function of observation t (while ignoring the constant part) becomes:

$$\begin{aligned}\ln \ell(\boldsymbol{\theta}) &= -\frac{1}{2}\ln|\boldsymbol{\Sigma}_t| - \frac{1}{2}\boldsymbol{\varepsilon}_t^{'}\boldsymbol{\Sigma}_t^{-1}\boldsymbol{\varepsilon}_t \\ &= -\frac{1}{2}\ln|\mathbf{D}_t\mathbf{R}_t\mathbf{D}_t| - \frac{1}{2}\boldsymbol{\varepsilon}_t^{'}\mathbf{D}_t^{-1}\mathbf{R}_t^{-1}\mathbf{D}_t^{-1}\boldsymbol{\varepsilon}_t \\ &= -\frac{1}{2}\ln|\mathbf{R}_t| - \frac{1}{2}\sum_{i=1}^{n}\ln\sigma_{i,t}^2 - \frac{1}{2}\boldsymbol{\varepsilon}_t^{'}\mathbf{D}_t^{-1}\mathbf{R}_t^{-1}\mathbf{D}_t^{-1}\boldsymbol{\varepsilon}_t.\end{aligned}$$

For a general n variate *CCC* model with $GARCH(1,1)$ characterizations for the conditional variances, this model requires the estimation of $s + 2n$ parameters. Bollerslev (1990) argues that, under the assumption of a constant correlation matrix, the maximum likelihood estimator of the correlation matrix equals the sample correlation matrix. The covariance matrix in the loglikelihood function will be positive definitive during optimization since the sample correlation matrix is positive definite and the conditional variances are positive. Filtering out the correlation matrix further simplifies optimization.

Dynamic Conditional Correlation (*DCC*) Tse and Tsui (2001) create the *DCC* model by introducing a recursive formula for the conditional correlation matrix:

$$\mathbf{R}_t = (1 - \theta_1 - \theta_2)\mathbf{R} + \theta_1\mathbf{R}_{t-1} + \theta_2\boldsymbol{\Psi}_{t-1}.$$

Here, $\mathbf{R}$ is some time invariant $n \times n$ matrix with $s - n$ parameters to be estimated and $\boldsymbol{\Psi}_{t-1}$, is an $n \times n$ matrix which elements are functions of lagged observations of ε_t. The length of the estimation sample used, T, should be larger than the number of variables, n, to guarantee positive definiteness of $\boldsymbol{\Psi}_{t-1}$. Tse and Tsui (2001) propose to let $\hat{\boldsymbol{\Psi}}_{t-1}$ be the correlation matrix in which the innovations are scaled by appropriate standard deviations. So, a typical element of $\hat{\boldsymbol{\Psi}}_{t-1}$, is given by:

$$\hat{\boldsymbol{\Psi}}_{ij,t-1} = \sum_{\tau=1}^{T}\hat{\varepsilon}_{i,t-\tau}^{*}\hat{\varepsilon}_{j,t-\tau}^{*} \Big/ \sqrt{\sum_{\tau=1}^{T}\hat{\varepsilon}_{i,t-\tau}^{*2}\sum_{\tau=1}^{T}\hat{\varepsilon}_{j,t-\tau}^{*2}}.$$

The restrictions $\theta_i \geqq 0$ for $i = 1, 2$ and $\theta_1 + \theta_2 \leqq 1$ are imposed so that the conditional correlation matrix is a weighted average of $\mathbf{R}$, $\mathbf{R}_{t-1}$, and $\boldsymbol{\Psi}_{t-1}$. Note that by restricting $\theta_1 = \theta_2 = 0$ the *CCC* model is obtained. This restriction can serve in an *LR* test as the null of constant conditional correlations versus the alternative of time varying correlations. Tse (2000) suggests testing the same hypothesis via an *LM* test, in which only the estimation of the *CCC* model is required. The n variate *DCC* model encompasses $s + 2n + 2$ parameters.

Estimation of the *DCC* model is straightforwardly accomplished by replacing $\mathbf{R}$ in the above loglikelihood function by $\mathbf{R}_t$. Tse and Tsui (2001) recognize that the constraints posed on θ_1 and θ_2 give rise to problems when trying to optimize the loglikelihood function as these parameters are restricted to certain values. To tackle this problem, they suggest to define $\theta_i = \lambda_i^2/(1 + \lambda_1^2 + \lambda_2^2)$ for $i = 1, 2$, where the parameters λ_1 and λ_2 are unrestricted. Optimization should be conducted with these new parameters in the loglikelihood function. After convergence to a solution, transformation to the original parameters can be executed. Engle (2000) estimates the model via QML, a two step estimation procedure that is not efficient. Since the estimation procedure suggested by Tse and Tsui (2001) is efficient, Engle's (2000) approach will not be discussed..

Asymmetric effects

Asymmetric effects may be introduced, e.g. as an alternative formulation for the multivariate Exponential *GARCH* models, since bad news is much more important than good news (see Acker (1998)). Negative news generally increases the volatility considerably, while good news has the tendency not to increase it sizeably.

According to Bollerslev *et al.* (1988), the asymmetric *vec* $GARCH(1,1)$ model can be written as:

$$\mathbf{vec}(\boldsymbol{\Sigma}_t) = \mathbf{vec}(\boldsymbol{\Omega}) + \mathbf{B}\mathbf{vec}(\boldsymbol{\Sigma}_{t-1}) + \mathbf{A}\mathbf{vec}(\boldsymbol{\varepsilon}_{t-1}\boldsymbol{\varepsilon}'_{t-1}) + \mathbf{N}\mathbf{vec}(\boldsymbol{\eta}_{t-1}\boldsymbol{\eta}'_{t-1})$$

with $\boldsymbol{\eta}_t = \boldsymbol{\varepsilon}_t$ if $\boldsymbol{\varepsilon}_t < 0$ and $\boldsymbol{\eta}_t = 0$ elsewhere, and where $\boldsymbol{\Omega}$ and $\mathbf{B}$ are symmetric $n \times n$ matrices, while the matrices $\mathbf{A}$ and $\mathbf{N}$ are of order $\frac{n(n+1)}{2} \times \frac{n(n+1)}{2}$.

Kroner (1995) has introduced asymmetric effects in the *BEKK* model, so that it can be written for a $GARCH(1,1)$ model as (*Generalized BEKK*):[29]

$$\boldsymbol{\Sigma}_t = \boldsymbol{\Omega} + \mathbf{B} \circ \boldsymbol{\Sigma}_{t-1} + \mathbf{G} \circ \mathbf{A}'\boldsymbol{\varepsilon}_{t-1}\boldsymbol{\varepsilon}'_{t-1}\mathbf{A} + \mathbf{N}'\boldsymbol{\eta}_{t-1}\boldsymbol{\eta}'_{t-1}\mathbf{N}$$

where $\boldsymbol{\Omega}, \mathbf{B}$ and $\mathbf{G}$ are symmetric $n \times n$ matrices, and A and $\mathbf{N}$ are square matrices of order n; $\circ$ stands for component wise multiplication.

Remark 8.1 *A particular type of BEKK model has a* $\mathbf{G}$ *matrix which is defined as a matrix of ones, or* $\boldsymbol{\Sigma}_t = \boldsymbol{\Omega} + \mathbf{A}'\boldsymbol{\varepsilon}_{t-1}\boldsymbol{\varepsilon}'_{t-1}\mathbf{A} + \mathbf{B} \circ \boldsymbol{\Sigma}_{t-1}$. *If this model is stationary (i.e. if all the eigenvalues of* $\mathbf{A}+\mathbf{B}$ *are smaller than one in modulus), then the unconditional covariance matrix can be written as* $\boldsymbol{\Sigma}_\varepsilon = \boldsymbol{\Omega}(\mathbf{I} - \mathbf{A} - \mathbf{B})^{-1}$ *and also the (optimal) predictor for* $\boldsymbol{\Sigma}_{t+s}$ *can be written in a similar style as in Section 8.2.2:* $E_t(\boldsymbol{\Sigma}_{t+s}) = \boldsymbol{\Sigma}_{\boldsymbol{\varepsilon}} + (\mathbf{A} + \mathbf{B})^{s-1}(\boldsymbol{\Sigma}_{t+1} - \boldsymbol{\Sigma}_{\boldsymbol{\varepsilon}})$ *and* $\boldsymbol{\Sigma}_{t+s} \rightarrow \boldsymbol{\Sigma}_{\boldsymbol{\varepsilon}}$ *if* $s \rightarrow \infty$.

[29] Kroner (1995) has programmed these multivariate *GARCH* models in Gauss; the programs can be found on http://weber.ucsd.edu/Depts/Econ/Software/ARCH.html

Empirical example: A bivariate $BEKK$ model of ABN-AMRO and ING asset returns will be considered. The $GARCH(1,1)$ results for the ING asset returns can be found in Example 8.6. Similar ML estimates, based on (conditionally) normally distributed errors, are for the ABN-AMRO asset returns and the ING asset returns, respectively (with standard errors between parentheses): for ABN-AMRO: $\hat{\mu} = 0.10591(0.037)$, $\hat{\omega} = 0.014185(0.0069)$, $\hat{\alpha} = 0.07235(0.0127)$, $\hat{\beta} = 0.924446(0.01276)$ and
for ING: $\hat{\mu} = 0.083352(0.03765)$, $\hat{\omega} = 0.029062(0.00928)$, $\hat{\alpha} = 0.077322(0.0079)$, $\hat{\beta} = 0.912118(0.008)$.

Now, the Mean Squared Forecast Error (MSFE) for univariate models is: $[\sum_{t=1}^{T}(\hat{\varepsilon}_t^2 - \hat{\sigma}_t^2)]/T$,and for multivariate models, $[\sum_{t=1}^{T}(\hat{\varepsilon}_t'\hat{\varepsilon}_t - \hat{\Sigma}_t')(\hat{\varepsilon}_t'\hat{\varepsilon}_t - \hat{\Sigma}_t)]/T$. The MSFEs for the conditional variances in the univariate models are: for the ABN-AMRO asset returns 23.28757 for (conditionally) normally distributed errors and 23.36498 for (conditionally) Student t-distributed errors, respectively and for the ING asset returns 30.49373 and 30.14669.

Applying Kroner's (1995) $BEKK$ program, a bivariately estimated model with MSFEs based on normally distributed errors is obtained: 25.598 and 34.464 for the respective conditional variances, and 14.123 for the conditional covariance $\sigma_{12,t}$ (measuring the volatility spillover). Remark that, owing to the restrictions implied by the $BEKK$ model, the within-sample MSFEs of the conditional variances did not decrease in the bivariate $GARCH(1,1)$ model, although the MSFE of the conditional covariance is pretty small.

Multivariate Component $GARCH(1,1)$ model

A special type of a multivariate $GARCH$ model is the *Multivariate Component GARCH(1,1) model*. This model can be written as:

$$\mathbf{\Sigma}_t = \mathbf{Q}_t + \mathbf{A}'(\varepsilon_{t-1}\varepsilon_{t-1}' - \mathbf{Q}_{t-1})\mathbf{A} + \mathbf{B}'(\mathbf{\Sigma}_{t-1} - \mathbf{Q}_{t-1})\mathbf{B}$$

with $\mathbf{Q}_t = \mathbf{\Omega} + \mathbf{R}'(\mathbf{Q}_{t-1} - \mathbf{\Omega})\mathbf{R} + \mathbf{F}'(\varepsilon_{t-1}\varepsilon_{t-1}' - \mathbf{\Sigma}_{t-1})\mathbf{F}$.

This model is a direct generalization of the univarate component $GARCH(1,1)$ model and the $BEKK$ $GARCH(1,1)$ model, and has the property that if $\mathbf{A}$ and $\mathbf{B}$ have all eigenvalues substantially less than one, then $\mathbf{\Sigma}_t - \mathbf{Q}_t$ will be a transitory covariance matrix and will be forecast to rapidly approach zero. If $\mathbf{R}$ has eigenvalues close to one, then $\mathbf{Q}$ will be the permanent component of volatility, which will gradually approach $\mathbf{\Omega}$ as the forecast horizon increases.

8.2.5 Exercises and cases

Solved theoretical exercise

How does the kurtosis look like for the $ARCH(1)$ model with $\sigma_t^2 = \alpha_o + \alpha_1\varepsilon_{t-1}^2$ and for the $GARCH(1,1)$ model with $\sigma_t^2 = \alpha_o + \alpha_1\varepsilon_{t-1}^2 + \beta_1\sigma_{t-1}^2$?

From the definition $var(\sigma_t^2) := E\left(\sigma_t^4\right) - \left[E\left(\sigma_t^2\right)\right]^2$, we have for the stationary $ARCH(1)$ model ($\alpha_1 < 1$) that: $E\left(\sigma_t^2\right) = E\left(E_{t-1}(\varepsilon_t^2)\right) = E(\varepsilon_t^2) = \sigma_\varepsilon^2 = \frac{\alpha_o}{1-\alpha_1}$. Then, in order to determine

$$E\left(\sigma_t^4\right),\ \sigma_t^4 = \left(\alpha_o + \alpha_1\varepsilon_{t-1}^2\right)^2 = \alpha_o^2 + 2\alpha_o\alpha_1\varepsilon_{t-1}^2 + \alpha_1^2\varepsilon_{t-1}^4,$$

or for a stationary $ARCH(1)$ model:

$$E\left(\sigma_t^4\right) = \alpha_o^2 + 2\alpha_o\alpha_1\tfrac{\alpha_o}{1-\alpha_1} + \alpha_1^2 E(\varepsilon_{t-1}^4) = \alpha_o^2 + 2\alpha_o\alpha_1\tfrac{\alpha_o}{1-\alpha_1} + \alpha_1^2 E(E_{t-1}(\varepsilon_{t-1}^4))$$
$$= \alpha_o^2 + 2\alpha_o\alpha_1\tfrac{\alpha_o}{1-\alpha_1} + 3\alpha_1^2 E(\left[E_{t-1}(\varepsilon_{t-1}^2)\right]^2) = \alpha_o^2 + 2\alpha_o\alpha_1\tfrac{\alpha_o}{1-\alpha_1} + 3\alpha_1^2 E(\sigma_{t-1}^4)$$
$$= \alpha_o^2 + 2\alpha_o\alpha_1\tfrac{\alpha_o}{1-\alpha_1} + 3\alpha_1^2 E(\sigma_t^4) = \frac{\frac{\alpha_o^2(1-\alpha_1)+2\alpha_o^2\alpha_1}{1-\alpha_1}}{1-3\alpha_1^2} = \frac{\alpha_o^2(1+\alpha_1)}{(1-3\alpha_1^2)(1-\alpha_1)} \text{ if } 3\alpha_1^2 < 1.$$

Then it can be written:

$\frac{E\left(\sigma_t^4\right)}{\left[E(\sigma_t^2)\right]^2} = \frac{\frac{\alpha_o^2(1+\alpha_1)}{(1-3\alpha_1^2)(1-\alpha_1)}}{\left(\frac{\alpha_o}{1-\alpha_1}\right)^2} = \frac{1-\alpha_1^2}{1-3\alpha_1^2}$ if $\alpha_1^2 < \frac{1}{3}$. Hence, the kurtosis for the stationary $ARCH(1)$ model is found from (8.12) as:

$$\kappa = 3 + 3\frac{var(\sigma_t^2)}{\left[E(\varepsilon_t^2)\right]^2} = 3 + 3\left[\frac{E\left(\sigma_t^4\right)}{\left[E(\sigma_t^2)\right]^2} - 1\right] = 3\frac{E\left(\sigma_t^4\right)}{\left[E(\sigma_t^2)\right]^2} = 3 + \frac{6\alpha_1^2}{1-3\alpha_1^2} \text{ if } \alpha_1^2 < \tfrac{1}{3}$$
$$= \infty \qquad \text{if } \alpha_1^2 \geqq \tfrac{1}{3}.$$

The derivation of the kurtosis for the stationary $GARCH(1,1)$ model ($\alpha_1 + \beta_1 < 1$) is similar to the previous analysis, or:

$$E\left(\sigma_t^2\right) = E\left(E_{t-1}(\varepsilon_t^2)\right) = E(\varepsilon_t^2) = \sigma_\varepsilon^2 = \frac{\alpha_o}{1-\alpha_1-\beta_1} \text{ and}$$
$$E\left(\sigma_t^4\right) = \frac{\frac{\alpha_o^2(1-\alpha_1-\beta_1)+2\alpha_o^2(\alpha_1+\beta_1)}{1-\alpha_1-\beta_1}}{(1-3\alpha_1^2-2\alpha_1\beta_1-\beta_1^2)} = \frac{\alpha_o^2(1+\alpha_1+\beta_1)}{(1-\alpha_1-\beta_1)(1-3\alpha_1^2-2\alpha_1\beta_1-\beta_1^2)} \text{ if } 3\alpha_1^2+2\alpha_1\beta_1+\beta_1^2<1.$$

Similarly as for the $ARCH(1)$ case, we have:

$\frac{E\left(\sigma_t^4\right)}{\left[E(\sigma_t^2)\right]^2} = \frac{1-\alpha_1^2-2\alpha_1\beta_1-\beta_1^2}{1-3\alpha_1^2-2\alpha_1\beta_1-\beta_1^2}$, so that the kurtosis of the stationary $GARCH(1,1)$ model satisfies:

$$\kappa = 3\frac{E\left(\sigma_t^4\right)}{\left[E(\sigma_t^2)\right]^2} = 3 + \frac{6\alpha_1^2}{1-3\alpha_1^2-2\alpha_1\beta_1-\beta_1^2} \text{ if } 3\alpha_1^2+2\alpha_1\beta_1+\beta_1^2 < 1$$
$$= \infty \qquad \text{if } 3\alpha_1^2+2\alpha_1\beta_1+\beta_1^2 \geqq 1.$$

Solved empirical cases

Case 8.1 *Four series of daily stock market quotations for the period 4/1/1994 to 23/12/2000 of the following companies: BASF, Danone, KBC and Telinfo (Telindus), obtained from the Brussels Stock Exchange, are analyzed for various types of GARCH processes.*

Figure 8.5: Logarithmic returns

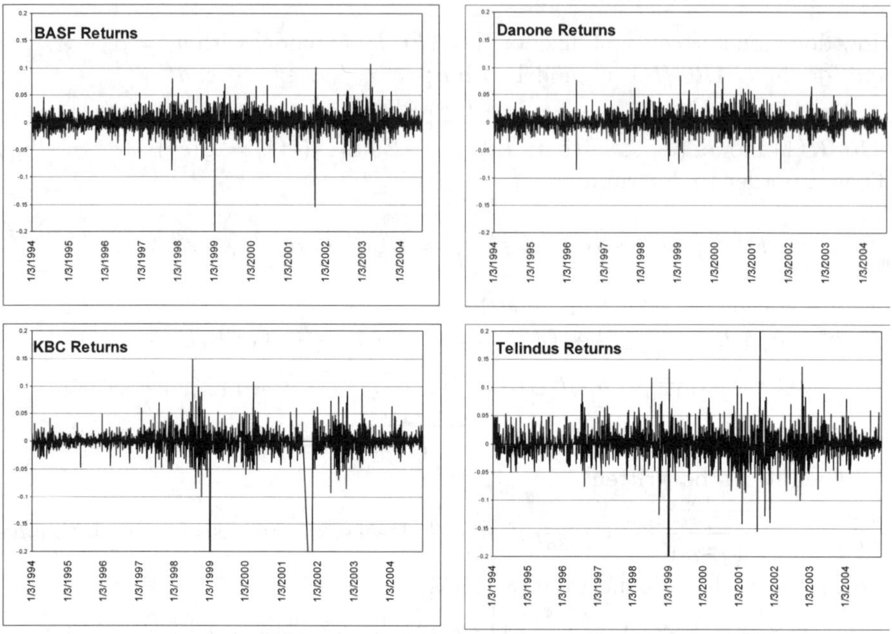

Source: http://finance.yahoo.com

The price quotations are transformed into series of logarithmic returns according to: $r_t = \ln \frac{P_t}{P_{t-1}} = \ln P_t - \ln P_{t-1}$. Daily returns up to the end of 2004 are plotted in Figure 8.5. Despite the fact that all four series means are very near to zero, visual examination suggests that the series are not WN: there is volatility clustering, because there are periods of both small and large changes.

The moments of the series are presented in Table 8.7. Examining the values of both skewness and kurtosis one may conclude that these series of returns are not normally distributed. The kurtosis of the normal distribution is 3; given that the kurtosis exceeds 3 in all four cases, the distribution presents fat tails (leptokurtic distribution) relative to the normal. Finally, to complement this analysis, we compute the JB test statistic, for testing whether the series are normally distributed or not. Since the critical value of χ^2_2 at the 1 percent level is 9.2, we conclude that the various series are not normal (see Table 8.7).

Although no significant (partial) autocorrelations could be found, there is sizable (partial) autocorrelation of squared returns that indicates volatility clustering, which is also confirmed by the LB statistics up to order 36. This calls for estimation of the returns series using $GARCH$ models.

Estimating first the most common $GARCH(1,1)$ model for the four series,

Table 8.7: Moments of the return series, and Jarque-Bera test statistic

	$BASF$	$Danone$	KBC	$Telinfo$
Mean	0.0007	0.0003	0.000	0.001
Std. Dev.	0.019	0.016	0.017	0.021
Skewness	-0.185	0.001	0.478	0.571
Kurtosis	5.339	5.063	11.173	7.542
Jarque-Bera	358.491	272.428	4334.357	1404.484

according to $r_t = \mu + \varepsilon_t$, with $\sigma_t^2 = \omega + \beta\sigma_{t-1}^2 + \alpha\varepsilon_{t-1}^2$ (see Table 8.8).

Table 8.8: GARCH $(1, 1)$ estimation

$\varepsilon_t \sim N(0, \sigma_t^2)$	*BASF*	*Danone*	*KBC*	*Telinfo*
$\hat{\mu}$	0.0008$^{(*)}$	0.0001	0.0003$^{(*)}$	0.0010$^{(*)}$
$\hat{\omega}$	1.62E-06$^{(*)}$	1.52E-06$^{(**)}$	1.21E-06$^{(**)}$	9.45E-05$^{(**)}$
$\hat{\alpha}$	0.0362$^{(**)}$	0.0379$^{(**)}$	0.0648$^{(**)}$	0.2111$^{(**)}$
$\hat{\beta}$	0.9598$^{(**)}$	0.9567$^{(**)}$	0.9344$^{(**)}$	0.5991$^{(**)}$
Ljung-Box Q_{35}	57.905$^{(**)}$	48.059	50.268$^{(*)}$	48.059
Decision	reject null	no rejection	no rejection at 1%	no rejection
$\varepsilon_t \sim t(v)$				
$\hat{\mu}$	0.0008$^{(*)}$	-0.00001	0.0005$^{(*)}$	0.00011
$\hat{\omega}$	8.03E-07	3.68E-06	3.30-07	0.00017
$\hat{\alpha}$	0.0315$^{(**)}$	0.0405$^{(**)}$	0.0459$^{(**)}$	4.1397
$\hat{\beta}$	0.9676$^{(**)}$	0.0098$^{(**)}$	0.9582$^{(**)}$	0.6907$^{(**)}$
$\hat{v}$	0.1574$^{(**)}$	0.2037$^{(**)}$	0.2529$^{(**)}$	0.4861$^{(**)}$

Statistical significance: $^{(*)}p < 0.05$, $^{(**)}p < 0.01$.

Examining the Ljung-Box statistics for the autocorrelations in the squared residuals it can be concluded that the common $GARCH(1, 1)$ fits well for all stocks except for those of BASF (and to a somewhat lesser extent (5% confidence level) for KBC stocks). Higher order models ($GARCH(2, 1)$, etc.) could not be found. The returns of BASF stocks could be described by other forms of $GARCH$ models, as e.g. asymmetric ones.

In Table 8.9, we have also computed the $GARCH$ in means ($GARCH$-M), exponential $GARCH$ ($EGARCH$), and the integrated $GARCH$ ($IGARCH$) models for normally and Student t-distributed conditional errors. The $GARCH$-M model is nowhere relevant since the estimated $\hat{\delta}$ is not significant for all series. On the contrary, the $EGARCH$ model is very relevant since $\hat{\varphi}$ for KBC stocks is significant at the 1% level and for BASF stocks at the 5% level.

Table 8.9: GARCH-M(1,1), EGARCH and IGARCHestimations

$GARCH-$M(1,1) estimation				
$\varepsilon_t \mid \Im_{t-1} \overset{iid}{\sim} N(0,\sigma_t^2)$	*BASF*	*Danone*	*KBC*	*Telinfo*
$\hat{\mu}$	0.0005	-0.0004	0.00028	-0.0008
$\hat{\omega}$	1.38E-06$^{(*)}$	1.05E-06$^{(*)}$	1.27E-06$^{(**)}$	9.86E-05$^{(**)}$
$\hat{\alpha}$	0.0332$^{(**)}$	0.0325$^{(**)}$	0.0674$^{(**)}$	0.2142$^{(**)}$
$\hat{\beta}$	0.9635$^{(**)}$	0.9635$^{(**)}$	0.9318$^{(**)}$	0.5875$^{(**)}$
$\hat{\delta}$	1.0305	03.1717	0.7357	4.9519
$\varepsilon_t \mid \Im_{t-1} \overset{iid}{\sim} t(v)$				
$\hat{\mu}$	-0.00017	-0.00131	0.0011$^{(*)}$	-0.0008
$\hat{\omega}$	8.38E-7	3.72E-7	2.82E-7$^{(*)}$	0.00011
$\hat{\alpha}$	0.0322$^{(**)}$	0.0406$^{(**)}$	0.0437$^{(**)}$	2.4827
$\hat{\beta}$	0.9668$^{(**)}$	0.9617$^{(**)}$	0.9604$^{(**)}$	0.6792$^{(**)}$
$\hat{\delta}$	0.0663	0.0986	-0.0536	0.0265
v	0.1583$^{(**)}$	0.2058$^{(**)}$	0.2539$^{(**)}$	0.4758$^{(**)}$
$EGARCH(1,1)$ estimation				
$\varepsilon_t \mid \Im_{t-1} \overset{iid}{\sim} N(0,\sigma_t^2)$	*BASF*	*Danone*	*KBC*	*Telinfo*
$\hat{\mu}$	0.0006	0.0001	0.0006$^{(*)}$	0.0014$^{(**)}$
$\hat{\omega}$	-0.1551$^{(*)}$	-0.2009$^{(*)}$	-0.1415	-2.7221$^{(**)}$
$\hat{\alpha}$	0.0958$^{(**)}$	0.0969$^{(**)}$	0.1265$^{(**)}$	0.3716$^{(**)}$
$\hat{\beta}$	0.9896$^{(**)}$	0.9844$^{(**)}$	0.9939$^{(**)}$	0.6816$^{(**)}$
$\hat{\varphi}$	-0.0322$^{(*)}$	-0.0203	0.0311$^{(**)}$	0.00799
$IGARCH(1,1)$ estimation				
$\varepsilon_t \mid \Im_{t-1} \overset{iid}{\sim} N(0,\sigma_t^2)$	*BASF*	*Danone*	*KBC*	*Telinfo*
$\hat{\mu}$	0.0008	0.0001	0.0003	0.00078
$\hat{\omega}$	5.79E-7	4.25E-7	1.22E-6$^{(**)}$	0.00006$^{(**)}$
$\hat{\alpha}$	0.0300$^{(**)}$	0.0309$^{(**)}$	0.0680$^{(**)}$	0.3908$^{(**)}$
$\hat{\beta}$	0.9700$^{(**)}$	0.9691$^{(**)}$	0.9320$^{(**)}$	0.6092$^{(**)}$
$\varepsilon_t \mid \Im_{t-1} \overset{iid}{\sim} t(v)$				
$\hat{\mu}$	0.0008$^{(*)}$	-4.02E-6	0.0005$^{(*)}$	0.00023
$\hat{\omega}$	6.25E-7	6.22E-7	5.01E-7	0.00004$^{(**)}$
$\hat{\alpha}$	0.0308$^{(**)}$	0.0390$^{(**)}$	0.0427$^{(**)}$	0.3317$^{(**)}$
$\hat{\beta}$	0.9692$^{(**)}$	0.9610$^{(**)}$	0.9573$^{(**)}$	0.6683$^{(**)}$
$\hat{v}$	0.1609$^{(**)}$	0.1954$^{(**)}$	0.2359$^{(**)}$	0.3268$^{(**)}$

Statistical significance: $^{(**)}p < 0.01$, $^{(*)}p < 0.05$.

Case 8.2 *Bala and Premaratne (2004), "Stock Market Volatility: Examining North America, Europe and Asia".*

Abstract An understanding of volatility in stock markets is important for determining the cost of capital and for assessing investment and leverage decisions as volatility is synonymous with risk. Substantial changes in volatility of financial markets are capable of having significant negative effects on risk averse investors. Using daily returns from 1992 to 2002, volatility comovement between the Singapore stock market and the markets of US, UK, Hong Kong and Japan is investigated. In order to gauge volatility comovement, econometric models of (i) Univariate $GARCH$ (ii) VAR and (iii) a multivariate and asymmetric multivariate $GARCH$ model are employed. The empirical results indicate that there is a high degree of volatility comovement between Singapore stock market and that of Hong Kong, US, Japan and UK (in that order). Results support small but significant volatility spillover from Singapore into Hong Kong, Japan and US markets despite the latter three being dominant markets. Most of the previous research concludes that spillover effects are significant only from the dominant market to the smaller market and that the volatility spillover effects are unidirectional. The study evidences that it is plausible for volatility to spill over from the smaller market to the dominant market. At a substantive level, studies on volatility comovement and spillover provide useful information for risk analysis.
[See full text at: http://repec.org/esFEAM04/up.17856.1076663719.pdf].

Case 8.3 *Balasubramanyan (2004). "Do Time-Varying Covariances, Volatility Comovement and Spillover Matter?".*

Abstract They investigate whether time varying volatility comovements and spillovers impact the true variance-covariance under a dynamic conditional correlation set up. They find that there are statistically significant asymmetric volatility spillovers and comovements between the three international markets. The use of time-varying correlations can impact the variance-covariance matrix from between 10 to 19 per cent while asymmetric volatility comovements and spillovers can impact the time-varying covariance from between 1 to 13 per cent. These results have important implications for an investor holding a portfolio of international assets attempting to minimize portfolio risk.

Case 8.4 *Koopman et al. (2004), "Forecasting daily variability of the S&P100 stock index using historical, realized and implied volatility measurements".*

Abstract The increasing availability of financial market data at intraday frequencies has not only led to the development of improved volatility measurements but has also inspired research into their potential value as an information source for volatility forecasting. In this paper it is explored that the forecasting value of historical volatility (extracted from daily return series), of implied volatility (extracted from option pricing data) and of realized volatility (computed as the sum of squared high frequency returns within a day). First they consider unobserved components and long memory models for realized volatility which is regarded as an accurate estimator of volatility. The predictive abilities of realized volatility models are compared with those of stochastic volatility models and generalized autoregressive conditional heteroskedasticity models for daily return series.

These historical volatility models are extended to include realized and implied volatility measures as explanatory variables for volatility. The main focus is on forecasting the daily variability of the Standard & Poor's 100 stock index series for which trading data (tick by tick) of almost seven years is analyzed. The forecast assessment is based on the hypothesis of whether a forecast model is outperformed by alternative models. In particular, they use superior predictive ability tests to investigate the relative forecast performances of some models. Since volatilities are not observed, realized volatility is taken as a proxy for actual volatility and is used for computing the forecast error. A stationary bootstrap procedure is required for computing the test statistic and its p-value. The empirical results show convincingly that realized volatility models produce far more accurate volatility forecasts compared to models based on daily returns. Long memory models seem to provide the most accurate forecasts.
[See full text at: http://www.tinbergen.nl/discussionpapers/04016.pdf].

Unsolved Exercises

1. Consider the $ARIMA(0,2,2)$ model $\Delta^2 y_t = 1 - \varphi_1 L - \varphi_2 L^2$.

   a. Given that the error process $\{\varepsilon_t, t \in \mathbb{Z}\}$ is a WN process (without conditional heteroskedasticity), derive the 1-step (ahead) and the h-step (ahead) forecasts for the $\{y_t, t \in \mathbb{Z}\}$ process with the corresponding forecast variances (which are the expected squared (conditional) forecast errors).

   b. Assume now that the error process $\{\varepsilon_t, t \in \mathbb{Z}\}$ is subject to conditional heteroskedasticity, compute (again) the above mentioned (conditional) forecasts and forecast variances.

   c. Compute the h-step (ahead) forecast of the conditional variance and the associated forecast error if the conditional variance of the error process $\{\varepsilon_t, t \in \mathbb{Z}\}$ is described by the $GARCH(1,1)$ process $\sigma_t^2 = \omega + \alpha_1 \varepsilon_{t-1} + \beta_1 \sigma_{t-1}^2$. How can this forecast be written if the $GARCH(1,1)$ model is stationary?

   d. Recalling that the forecast errors of the $GARCH(1,1)$ conditional variance's forecast can be written as $v_t = \sigma_t^2 \left(\eta_t^2 - 1\right)$, with $\eta_t \overset{iid}{\sim} N(0,1)$, compute the conditional variance of the forecast $GARCH(1,1)$ conditional variance.

   e. Since the restriction $E\left(\left(ln\left(\alpha_1 \eta_{t-i}^2 + \beta_1\right)\right) < 0\right.$ can be interpreted as a necessary condition for strict stationarity of the above $GARCH(1,1)$ model, show that the $IGARCH(1,1)$ model, although not (covariance) stationary, may still be strictly stationary.

2. a. Consider the Glosten *et al.* (1993) $GJR - GARCH(1,1)$ model in which, the coefficients depend on the sign of the corresponding return innovations or:
$\sigma_t^2 = \omega + \alpha_1 \varepsilon_{t-1}^2 (1 - I_{\varepsilon_{t-1}>0}) + \gamma_1 \varepsilon_{t-1}^2 I_{\varepsilon_{t-1}>0} + \beta_1 \sigma_{t-1}^2$ and assuming that the probability distribution of z_t in $v_t = \sigma_t^2 \left(z_t^2 - 1\right)$ is symmetric around zero, compute the two-step (ahead) forecasts of σ_{t+2}^2 and the k-step (ahead) forecasts of σ_{t+k}^2.

   b. In discrete $ARCH$ models or stochastic volatility models the (estimated) conditional variance depends on information till period $t-1$ (included). When *realized volatility* is used the (estimated) conditional variance depends on information within period t. If we rewrite the univariate returns as $r_t = \sigma_t \eta_t$ with η_t standardized WN error terms, the standardized returns satisfy $\eta_t = \frac{r_t}{\sigma_t}$ In the multivariate case standardized returns should satisfy $\mathbf{r}_t = \mathbf{P}_t \boldsymbol{\eta}_t$, where $\mathbf{P}_t$ is an $n \times n$ matrix satisfying $\mathbf{P}_t \mathbf{P}_t' = \boldsymbol{\Sigma}_t$ and $\boldsymbol{\Sigma}_t$ is the conditional covariance matrix; however, this root is not unique. A root can be obtained by an under-triangular Cholesky decomposition. Compute the corresponding bivariate return vector and compare to the univariate case.

   c. Consider volatility spillovers and asymmetries in volatility transmission. To that end an $EGARCH$ specification for the conditional variance equations may be considered in a possible application where the world and domestic interest rates are modeled as $R_{w,t} = \alpha_w + \beta_w \sigma_{w,t} + \sigma_{w,t} \eta_{w,t}$, and $R_{d,t} = \alpha_d + \beta_d \sigma_{d,t} + \sigma_{d,t} \eta_{d,t}$, respectively; $\eta_{w,t}$ and $\eta_{d,t}$ are standardized WN innovations. The vector $\boldsymbol{\varepsilon}_t'$ represents the vector of idiosyncratic shocks from the world and the domestic country (to be investigated). Formulate a bivariate $EGARCH$ specification and interpret, assuming that the domestic country cannot affect world volatility.

3. Consider log-returns of an asset: $r_t := y_t$, with $y_t = m_t(\mathbf{x}_t; \boldsymbol{\theta}) + \varepsilon_t$; $t = 1, 2, \ldots, T$, where $m_t(\cdot)$ is a conditional mean function, $\mathbf{x}_t$ an $m \times 1$ vector of exogenous variables which may include lagged values of the dependent variable, $\boldsymbol{\theta}$ a vector of parameters, ε_t a stochastic error term satisfying: $\varepsilon_t = \sigma_t v_t$, where σ_t is the conditional standard deviation and v_t is assumed to follow a normal WN process with unit variance. Let $\Im_{t-1}$ denote the (past) information set (available at the end of period $t-1$). Then ε_t is said to be generated by a $GARCH(p,q)$ process if the conditional distribution of ε_t is described by a normal distribution with zero mean and variance $\sigma_t^2 = \alpha_o + \sum_{i=1}^{q} \alpha_i \varepsilon_{t-i}^2 + \sum_{j=1}^{p} \beta_j \sigma_{t-j}^2$.

   a. Give and show the stationarity conditions of this $GARCH(p,q)$ process.

   b. Show/illustrate that, though the conditional error distribution is assumed to be normal, the unconditional error distribution is *non-normal* with fatter tails than in the normal distribution.

   c. Assuming that the above conditional mean function satisfies an $AR(r)$ process and that the error term ε_t is generated by an $ARCH(q)$ process as in $\varepsilon_t^2 = \alpha_o + \sum_{i=1}^q \alpha_i \varepsilon_{t-i}^2 + u_t$, with $\{u_t, t \in \mathbb{Z}\}$ a WN process with variance σ_u^2, show that the fourth moment of ε_t does not exist for all stationary $ARCH$ models; illustrate this in detail for an $ARCH(1)$ process.

d. You can find similar results for Student t-distributed WN processes. What do you suggest in order to capture asymmetry and excess kurtosis in the analysis of financial returns?

4. Perform volatility analysis with four series of daily market quotations for the period January 2000-December 2004 of the following metals: gold, silver, platinum and palladium, obtained from the London Market,[30] and for the period January 2000-September 2004 of a number of stocks as Vodafone, Shell Transport & Trading Company, Marks and Spencer, Johnston Press, Lloyds TSB, Manchester United, Investment Co., Euro Disney, HSBC Holdings, Hilton Group, GlaxoSmithKline,and British Airways from the London Stock Exchange.[31]

   (a) Obtain $GARCH$, $GARCH$-in-means, exponential $GARCH$ and Integrated $GARCH$, based on both the underlying normal as well as on the underlying t distribution.

   (b) Obtain volatility forecasts for the period January 2004- March 2005 and compare them with the actual data set.

[30] Available at http://www.kitco.com/gold.londonfix.html.

[31] See http://www.londonstockexchange.com/en-gb/ and http://www.axl.co.uk/LSE/.

Part III

Categorical and Limited Dependent Variables

This part focuses on qualitative and limited responses, duration and count data. On the one hand, models for discrete choice (or qualitative response) with binary and multiple choices, and on on the other hand, models of censored and truncated responses (including duration and count data). Hence, this part will be divided into two chapters: *discrete choice models* and *limited responses, duration and count data.*

Discrete choice models are divided into *binary choice models* and *multiple choice* or *multiple response models.* Multiple response models occur if the number of alternatives is larger than two. These models are developed to describe the probability of each of the possible outcomes as a function of specific characteristics. Within multiple response models an important distinction exists between *ordered* (ranked) response models and *unordered* response models. An ordered response model is generally more parsimonious but can only be applied if there exists a logical order of the alternatives. If alternatives cannot be ordered, we have *nominal response* or *multinomial models.*

Within this framework, it may be helpful to introduce distinctions about the classes of variables that will be considered in this part of the book.

- *Binary variables* have two categories and are used to indicate that an event has occurred or that some characteristic is present; e.g. labor force participation, with the response variable assuming a value zero if the person does not participate and a value one if she participates. The number is just a coding for the dependent variable.

- *Ordered* or *ordinal variables* have categories that can be ranked. For example, surveys ask respondents to indicate agreement or disagreement to a statement using the choices agree, neutral and disagree so that the dependent variable is given a code zero for 'agree', one for 'neutral' and two for 'disagree'. In this case, the values of the variable merely reflect an ordering or ranking.

- *Unordered or nominal variables* occur when there are multiple outcomes that cannot be ordered; a good example is to consider the occupational field chosen by an individual, or the dependent variable is given a code zero if the respondent is a 'clerk', one if she is an 'engineer', two if she is an 'economist', etc. In this latter case, the values are categories, but do not indicate a ranking.

- *Censored* or *truncated variables* occur when the value of a variable is unknown over some range of that variable.

- *Count variables* indicate the number of times that the dependent variable occurred, e.g. the number of patents, or the dependent variable takes the values zero, one, two.

How a variable is measured is not absolute; indeed, one may say that the level of measurement of a variable cannot be made in isolation from the context in which the variable is to be used. Long (1997), p.2, clarifies on this point with an example of the education variable:

> "Education can be measured as a binary variable that distinguishes those with a high school education from others. Or, it could be ordinal indicating the highest degree received: junior high, high school, college, or graduate. Or, it can be a count variable indicating the number of years of school completed. Each of these is reasonable and appropriate, depending on the substantive purpose of the analysis."

Chapter 9

Discrete Choice Models

The models in which the dependent variable assumes only discrete values are known as discrete choice models. These models have diverse applications since many economic responses are of a qualitative nature and, frequently, the dependent variable is an indication of a discrete choice, such as a 'yes or no' decision. This is an example of a binary choice model. Known types of binary choice models are the probit model, the logit model, the Weibull model, etc. When there are more alternatives, we speak about multiple choice or multiple response models. There are two types of multiple response models: ordered response models and multinomial models. For the latter type of models, two kinds of models will be analyzed: multinomial logit model and multinomial probit models.

9.1 Binary choice models

In order to consider a model as a binary choice model, the dependent variable may only take two values, i.e. zero or one.

Usually, a transformation approach for the probability that a certain event occurs is employed, which is modeled as a function of characteristics (explanatory variables) and unknown parameters:

$$\begin{aligned} P\left(\text{event } j \text{ occurs} \mid \mathbf{x}; \boldsymbol{\beta}\right) &= P\left(y = j \mid \mathbf{x}; \boldsymbol{\beta}\right) \\ &= F\left(\text{relevant variables; parameters}\right) = F\left(\mathbf{x}; \boldsymbol{\beta}\right), \end{aligned}$$

where $\mathbf{x}$ is a vector of relevant variables determining the choice for event j and $\boldsymbol{\beta}$ is the vector of unknown parameters.

We may distinguish two approaches: the *regression approach* where the distribution function and the *index function approach*, where the choice among the alternatives is made according to one single index.

9.1.1 Regression approach

Each binary decision is assumed to be represented by a *dichotomous* random variable or $y := 1$ if event E occurs and $y := 0$ if event E does not occur.

In the (above) labor force participation example, the dependent variable can be defined as $y := 1$ if the respondent decides to look for a job, and $y := 0$ if the respondent does not want to participate.The probability that the event occurs is assumed to depend on a vector of explanatory variables $\mathbf{x}$ and a vector of unknown parameters $\boldsymbol{\beta}$. An array of factors such as age, gender, marital status, level of education, work history, etc... explain the decision of participating in the labor force, and are gathered in the vector $\mathbf{x}$. The binary choice model can be written as:

$$\begin{aligned} P(y=1|\,\mathbf{x};\boldsymbol{\beta}) &= F(\mathbf{x};\boldsymbol{\beta}) \\ P(y=0|\,\mathbf{x};\boldsymbol{\beta}) &= 1-F(\mathbf{x};\boldsymbol{\beta}). \end{aligned}$$

Which are now the suitable models for the function $F(\mathbf{x};\boldsymbol{\beta})$?

- The first option is to use the *linear probability model* (as a linear regression), where it is assumed that the general nonlinear function $F(\mathbf{x};\boldsymbol{\beta})$ is linear in the variables and the parameters:

$$F(\mathbf{x};\boldsymbol{\beta}) = F(\mathbf{x}'\boldsymbol{\beta}) = \mathbf{x}'\boldsymbol{\beta} \Longrightarrow y = E(y|\,\mathbf{x};\boldsymbol{\beta}) + (y - E(y|\,\mathbf{x};\boldsymbol{\beta})) = \mathbf{x}'\boldsymbol{\beta} + \varepsilon.$$

A serious drawback of this model is that $\mathbf{x}'\boldsymbol{\beta}$ is not constrained to lie in the unit interval as a probability should. Therefore, such a model might produce nonsense probabilities.[1] For this reason, the linear probability model is less frequently used except as a basis for comparison with other models. We therefore require for a given regressor vector that $\lim_{\mathbf{x}'\boldsymbol{\beta}\to-\infty} P(y=1|\,\mathbf{x}) = 0$ and $\lim_{\mathbf{x}'\boldsymbol{\beta}\to\infty} P(y=1|\,\mathbf{x}) = 1$; in other words that the probability distribution function property for the regressor vector is required. The most widely used frameworks are the probit and the logit models. For asymmetric distributions, the Weibull model is also popular.

- *Probit model*, which uses the standard normal distribution function, or $F(\mathbf{x}'\boldsymbol{\beta})= \int_{-\infty}^{\mathbf{x}'\boldsymbol{\beta}} \frac{1}{\sqrt{2\pi}} e^{-\frac{1}{2}t^2 dt}$.
- *Logit model*, which uses the logistic distribution, or $F(\mathbf{x}'\boldsymbol{\beta})=\frac{1}{1+e^{-\mathbf{x}'\boldsymbol{\beta}}}$.
- *Weibull or complementary log-log model* (Long (1997), pp.51-52), which corresponds to $F(\mathbf{x}'\boldsymbol{\beta})=1 - e^{-e^{\mathbf{x}'\boldsymbol{\beta}}}$; this probability distribution is complementary to the *log-log model*, which is defined as $F(\mathbf{x}'\boldsymbol{\beta})= e^{-e^{-\mathbf{x}'\boldsymbol{\beta}}}$.

[1] The linear probability model has another flaw, since it might produce negative variances.

The regression approach can be studied for two basic cases: the case of repeated observations and the case in which the observations are not (sufficiently) repeated.

9.1.2 Repeated observations for discrete choice

The linear probability model

Assuming that the behavior of N individuals (units of observations) has to be investigated, the binary choice model for individual $i = 1, 2, ..., N$ can (as a starting point) be represented by the simple linear model $y_i = \mathbf{x}_i'\boldsymbol{\beta} + \varepsilon_i$ with the classic assumptions $\varepsilon_i \overset{iid}{\sim} (0, \sigma_\varepsilon^2)$ and $E(\varepsilon_i | \mathbf{x}_i; \boldsymbol{\beta}) = 0$. Then for the linear probability model in the binary choice model:
$E(\mathbf{y}_i | \mathbf{x}_i; \boldsymbol{\beta}) = 1 \times P(\mathbf{y}_i = 1 | \mathbf{x}_i; \boldsymbol{\beta}) + 0 \times P(\mathbf{y}_i = 0 | \mathbf{x}_i; \boldsymbol{\beta}) = P(\mathbf{y}_i = 1 | \mathbf{x}_i; \boldsymbol{\beta}) = \mathbf{x}_i'\boldsymbol{\beta}$, so that the linear probability model implies that $\mathbf{x}_i'\boldsymbol{\beta}$ is a probability (should always lie between zero and one). Since $P(\mathbf{y}_i = 1 | \mathbf{x}_i; \boldsymbol{\beta})$ cannot be observed in practice, a sample of $n := \sum_{i=1}^{N} n_i$ is considered, with n_i the number of repeated observations of individual i. Then, the sample proportion p_i or the ratio of the number of times that $\mathbf{y}_i = 1$ occurs in the sample with respect to n_i is a valid measure for $P(y_i = 1 | \mathbf{x}_i; \boldsymbol{\beta})$. Hence, the linear probability model can be written as:

$$p_i = \mathbf{x}_i'\boldsymbol{\beta} + \eta_i = P_i + \eta_i \quad (i = 1, 2, ..., N), \tag{9.1}$$

where $P_i := P(y_i = 1 | \mathbf{x}_i; \boldsymbol{\beta})$.

There are two possibilities: either repeated observations occur with the error term distributed as the binomial process $\eta_i \overset{iid}{\sim} B\left(0, P_i\left(1 - P_i\right)/n_i\right)$ or there are no repeated observations with the error term following a two point distribution $\eta_i \overset{iid}{\sim} B\left(0, \mathbf{x}_i'\boldsymbol{\beta}\left(1 - \mathbf{x}_i'\boldsymbol{\beta}\right)\right)$. Under repeated observations model (9.1) becomes for all individuals:

$$\mathbf{p} = F\left(\mathbf{X}'\boldsymbol{\beta}\right) + \boldsymbol{\eta} = \mathbf{X}\boldsymbol{\beta} + \boldsymbol{\eta} = \mathbf{P} + \boldsymbol{\eta} \text{ with}$$

$$E(\boldsymbol{\eta}\boldsymbol{\eta}') = \begin{bmatrix} \frac{P_1(1-P_1)}{n_1} & 0 & \cdots & 0 \\ 0 & \frac{P_2(1-P_2)}{n_2} & \cdots & 0 \\ \vdots & \vdots & \ddots & \vdots \\ 0 & 0 & \cdots & \frac{P_N(1-P_N)}{n_N} \end{bmatrix} =: \boldsymbol{\Phi}, \mathbf{X} := \begin{bmatrix} \mathbf{x}_1 \\ \mathbf{x}_2 \\ \vdots \\ \mathbf{x}_N \end{bmatrix}, \boldsymbol{\eta} := \begin{bmatrix} e_1 \\ e_2 \\ \vdots \\ e_N \end{bmatrix}, \mathbf{P} := (P_1, P_2, \ldots, P_N)'$$

and $\mathbf{p} := (p_1, p_2, \ldots, p_N)'$. The BLUE of $\boldsymbol{\beta}$ is the GLS estimator $\hat{\boldsymbol{\beta}} = \left(\mathbf{X}'\boldsymbol{\Phi}^{-1}\mathbf{X}\right)^{-1}\mathbf{X}'\boldsymbol{\Phi}^{-1}\mathbf{p}$. Since the true proportions P_i are generally not known, they can consistently be estimated by the *sample* proportions, so that the estimated GLS (EGLS) or feasible GLS (FGLS) is $\hat{\hat{\boldsymbol{\beta}}} = \left(\mathbf{X}'\hat{\boldsymbol{\Phi}}^{-1}\mathbf{X}\right)^{-1}\mathbf{X}'\hat{\boldsymbol{\Phi}}^{-1}\mathbf{p}$ with $\hat{\boldsymbol{\Phi}}$ a consistent estimator of $\boldsymbol{\Phi}$, obtained by replacing P_i by p_i $(i = 1, 2, ..., N)$.

To guarantee that the predicted proportion is in the closed interval $[0, 1]$, (E)GLS has to be used under inequality restrictions.Therefore, the relevant minimization problem is $\underset{\boldsymbol{\beta}}{Min}\ (\mathbf{p}-\mathbf{X}\boldsymbol{\beta})'\,\boldsymbol{\Phi}^{-1}(\mathbf{p}-\mathbf{X}\boldsymbol{\beta})\quad s.t.\quad 0\leq\mathbf{X}\boldsymbol{\beta}\leq\boldsymbol{\iota}$, with $\boldsymbol{\iota}$ an N vector of ones. Construct therefore the Lagrange function $\mathcal{L}=(\boldsymbol{\beta}-\mathbf{X}\boldsymbol{\beta})'\,\boldsymbol{\Phi}^{-1}(\boldsymbol{\beta}-\mathbf{X}\boldsymbol{\beta})+\boldsymbol{\lambda}_1'(\boldsymbol{\iota}-\mathbf{X}\boldsymbol{\beta})+\boldsymbol{\lambda}_2'\mathbf{X}\boldsymbol{\beta}$, so that its unconstrained maximization leads to the normal equations $2\mathbf{X}'\boldsymbol{\Phi}^{-1}\mathbf{p}=2\mathbf{X}'\boldsymbol{\Phi}^{-1}\mathbf{X}\boldsymbol{\beta}+\mathbf{X}'\boldsymbol{\lambda}_2-\mathbf{X}'\boldsymbol{\lambda}_1$, or

$$\widetilde{\boldsymbol{\beta}}=\underbrace{\left(\mathbf{X}'\boldsymbol{\Phi}^{-1}\mathbf{X}\right)^{-1}\mathbf{X}'\boldsymbol{\Phi}^{-1}\mathbf{p}}_{\hat{\boldsymbol{\beta}}}-\underbrace{\frac{1}{2}\left(\mathbf{X}'\boldsymbol{\Phi}^{-1}\mathbf{X}\right)^{-1}(\mathbf{X}'\boldsymbol{\lambda}_1-\mathbf{X}'\boldsymbol{\lambda}_2)}_{\text{if strict inequality}}.$$

The disadvantage is that there are no analytical sample properties of $\widetilde{\widetilde{\boldsymbol{\beta}}}$ (for estimated $\boldsymbol{\Phi}$); only numerical simulation properties can be obtained.

The probit model

From the linear probability model in (9.1) we have $p_i=P_i+\varepsilon_i$, where the probit model implies $P_i=F(\mathbf{x}_i'\boldsymbol{\beta})=\left(\int_{-\infty}^{\mathbf{x}_i'\boldsymbol{\beta}}\frac{1}{\sqrt{2\pi}}\exp\left(-\frac{1}{2}t^2\right)dt\right)$. Moreover, $F(p_i)=F(P_i+\varepsilon_i)$ or $F^{-1}(p_i)=F^{-1}(P_i+\varepsilon_i)$, where $F^{-1}(\cdot)$ is the inverse of a normal distribution function. Approaching $F^{-1}(P_i+\varepsilon_i)$ by a (linear) Taylor series expansion in the neighborhood of P_i, by applying the inverse function theorem (the derivative of the inverse of a function is the inverse of the derivative of a function in that point):
$F^{-1}(p_i)\simeq F^{-1}(P_i)+\left.\frac{\partial F^{-1}(p_i)}{\partial p_i}\right|_{p_i=P_i}(p_i-P_i)=F^{-1}(P_i)+\frac{\varepsilon_i}{f(F^{-1}(P_i))}$, where $f(\cdot)$ is the standard normal pdf at $F^{-1}(P_i)$, or:

$$v_i\simeq\mathbf{x}_i'\boldsymbol{\beta}+u_i, \tag{9.2}$$

with $v_i:=F^{-1}(p_i)$ the *observed probit* and $u_i:=\frac{\varepsilon_i}{f(F^{-1}(P_i))}$, where $E(u_i)=0$ and $var(u_i)=\frac{P_i(1-P_i)}{n_i(f(F^{-1}(P_i)))^2}$.

In vector-matrix notation (9.2) becomes $\mathbf{v}\simeq\mathbf{X}\boldsymbol{\beta}+\mathbf{u}$, or the EGLS (or FGLS) estimator for $\boldsymbol{\beta}$ satisfies: $\hat{\hat{\boldsymbol{\beta}}}=\left(\mathbf{X}'\hat{\boldsymbol{\Phi}}^{-1}\mathbf{X}\right)^{-1}\mathbf{X}'\hat{\boldsymbol{\Phi}}^{-1}\mathbf{v}$, where $\hat{\boldsymbol{\Phi}}$ is a diagonal matrix with the ith diagonal element $\widehat{var}(u_i)=\frac{p_i(1-p_i)}{n_i(f(F^{-1}(p_i)))^2}$.

Forecasts for the probability of an event E (and certain $\mathbf{x}_i^*$) are given by: $\hat{p}_i=F\left(\mathbf{x}_i^{*\prime}\hat{\hat{\boldsymbol{\beta}}}\right)=\int_{-\infty}^{\mathbf{x}_i^{*\prime}\hat{\hat{\boldsymbol{\beta}}}}f(t)\,dt$ with $f(\cdot)$ the standard normal pdf.

The logit model

For the ith individual $(i=1,2,...,N)$, the sample proportions are defined as the true, unknown proportion plus an error term, or $p_i:=P_i+\eta_i$, and based on the logistic distribution function: $y(z)=\frac{a}{1+be^{-cz}}$, $z:=\mathbf{x}'\boldsymbol{\beta}$, or:

$$P_i=F(\mathbf{x}_i'\boldsymbol{\beta})=\frac{1}{1+e^{-\mathbf{x}_i'\boldsymbol{\beta}}}. \tag{9.3}$$

Starting from the *odds ratio* $k_i := \frac{p_i}{(1-p_i)}$ and using a Taylor series of first order in the neighborhood of P_i:

$$\ln k_i = \ln \frac{p_i}{(1-p_i)} = \ln \frac{P_i}{1-P_i} + \left.\frac{\partial \ln\left(\frac{p_i}{(1-p_i)}\right)}{\partial p_i}\right|_{p_i=P_i} (p_i - P_i) + \ldots$$

$$\simeq \ln \frac{P_i}{1-P_i} + \frac{\eta_i}{P_i(1-P_i)} = \mathbf{x}_i'\boldsymbol{\beta} + u_i \text{ with } \eta_i \overset{iid}{\sim} B\left(0, \frac{P_i(1-P_i)}{n_i}\right).$$

Then the *observed logits* are $v_i = \ln \frac{p_i}{1-p_i} = \mathbf{x}_i'\boldsymbol{\beta} + u_i$ or $\frac{p_i}{1-p_i} = e^{v_i} = e^{\mathbf{x}_i'\boldsymbol{\beta}+u_i}$, so that $p_i = (1-p_i)\,e^{v_i} = e^{v_i} - p_i e^{v_i} = \frac{e^{v_i}}{1+e^{v_i}} = \frac{1}{1+e^{-v_i}}$, which is the logistic distribution function with $a = b = c = 1$.

The forecasts of the probability $P(\mathbf{y}_i = 1|\, \mathbf{x}_i; \boldsymbol{\beta})$ in the logit model satisfy $\widehat{p}_i = \frac{1}{1+e^{-\hat{v}_i}} = \frac{1}{1+e^{-\mathbf{x}_i'\widehat{\boldsymbol{\beta}}}}$.

Example 9.1 *Families want to buy a certain product and one wonders how this purchase decision depends on income.*

The relative frequency of families buying the product is p_i and the odds ratio satisfies $\ln \frac{p_i}{1-p_i} = v_i = \mathbf{x}_i'\boldsymbol{\beta} + u_i$ ($\mathbf{x}_i$ is a vector consisting of income levels for individual i).

Weibull or complementary log-log binary choice model

The basic Weibull or complementary log-log model can be specified as: $P(y_i = 1) = F\left(\mathbf{x}_i'\widehat{\boldsymbol{\beta}}\right) = \int_{-\infty}^{\mathbf{x}_i'\widehat{\boldsymbol{\beta}}} f(t)\,dt$, $(i = 1, 2, \ldots, N)$ with $F(t) := 1 - e^{-e^t}$ and $f_t = e^{t-e^t}$ (e.g. the dependent variable y_i is an indicator that the patent citation occurred in a particular industry); by taking natural logarithms of this distribution function twice, we find that $\ln\left[-\ln(1-P(\mathbf{y}_i{=}1|\,\mathbf{x}_i; \widehat{\boldsymbol{\beta}})\right] = \mathbf{x}_i'\widehat{\boldsymbol{\beta}} > 0$.

9.1.3 Marginal effects (or 'slopes')

a. The slope for the *probit* model is $\widehat{\frac{\partial P_i}{\partial x_{ij}}} = \frac{\partial F\left(\mathbf{x}_i'\hat{\boldsymbol{\beta}}\right)}{\partial x_{ij}} = f\left(\mathbf{x}_i'\hat{\boldsymbol{\beta}}\right)\hat{\boldsymbol{\beta}}_j$ with $f(.)$ a standard normal pdf;

b. The slope for the *logit* model is $\widehat{\frac{\partial P_i}{\partial x_{ij}}} = \frac{\partial\left(1+e^{-x_i'\hat{\beta}}\right)^{-1}}{\partial x_{ij}} = \frac{\hat{\beta}_j e^{-x_i'\hat{\beta}}}{\left(1+e^{-x_i'\hat{\beta}}\right)^2}$.

c. The slope for the *Weibull* model is $\left.\frac{\partial F(\mathbf{x}_i'\hat{\boldsymbol{\beta}})}{\partial x_{ij}}\right|_{x_i=\overline{x}_i} = f\left(\overline{x}_i'\hat{\boldsymbol{\beta}}\right)\hat{\beta}_j$, where

$f\left(\overline{x}_i'\hat{\boldsymbol{\beta}}\right) = e^{\overline{x}_i\hat{\boldsymbol{\beta}} - e^{\overline{x}_i\hat{\boldsymbol{\beta}}}}$ is the Weibull or complementary log-log pdf calculated at the mean of the estimated structural part of the model. The slopes can also be evaluated at the sample mean $\overline{\mathbf{x}}$.

For a binary independent variable d, the expression for the slope becomes:

$$P\left(y = 1|\, \overline{\mathbf{x}}^*, d = 1\right) - P\left(y = 1|\, \overline{x}^*, d = 0\right).$$

However, Greene ((2000), p. 817) indicates that *"simply taking the derivative with respect to the binary variable as if it were continuous provides an approximation that is often surprisingly accurate"*.

9.1.4 Forecasts

Forecasts can easily be made, e.g. for the probit model $\hat{\hat{p}}_i = F\left(\hat{\hat{v}}_i\right) = F\left(\mathbf{x}_i'\hat{\hat{\boldsymbol{\beta}}}\right) = \int_{-\infty}^{\mathbf{x}_i'\hat{\hat{\boldsymbol{\beta}}}} \frac{1}{\sqrt{2\pi}} e^{-\frac{1}{2}t^2} dt$; and for the logit model $\hat{\hat{p}}_i = (1 + e^{-\mathbf{x}_i'\hat{\hat{\boldsymbol{\beta}}}})^{-1}$. The decision for the best forecasts can be taken based e.g. on the minimal Theil's inequality coefficient $U = \sqrt{\frac{1}{N}\sum_{i=1}^{N}\left(p_i - \hat{\hat{p}}_i\right)^2}$, which is a goodness of fit measure (mean squared forecast error).

9.1.5 No repeated observations

Now, the case in which no repeated observations (or a very small number of maximum 5) are available will be analyzed. In general, the likelihood contribution of observation i with $y_i = 1$ is given by $P\left(y_i = 1|\, \mathbf{x}_i; \boldsymbol{\beta}\right) =: P_i$; similarly for $y_i = 0$. Hence, for N observations, the likelihood function is given as the two-points Bernoulli distribution:

$$\begin{aligned} \ell(\boldsymbol{\beta}) \quad &: = \prod_{i=1}^{N} P_i^{y_i} (1 - P_i)^{1-y_i} \\ &= \prod_{i=1}^{N} (F(\mathbf{x}_i'\boldsymbol{\beta}))^{y_i} (1 - F(\mathbf{x}_i'\boldsymbol{\beta}))^{1-y_i}, \end{aligned} \tag{9.4}$$

with $y_i = 1$ if the event occurs and $y_i = 0$ if the event does not occur, and F is, in our case, either a standard normal distribution function (probit), a logistic distribution function (logit), or a Weibull or complementary log-log distribution function.

Maximizing the logarithm of (9.4):

$$L(\boldsymbol{\beta}) := \log \ell(\boldsymbol{\beta}) = \sum_{i=1}^{N} y_i \log F(\mathbf{x}_i'\boldsymbol{\beta}) + \sum_{i=1}^{N} (1 - y_i) \log(1 - F(\mathbf{x}_i'\boldsymbol{\beta})) \tag{9.5}$$

w.r.t. $\boldsymbol{\beta}$, leads to the first order conditions for optimality:[2]
$\frac{\partial \log \ell(\boldsymbol{\beta})}{\partial \boldsymbol{\beta}} = \sum_{i=1}^{N}\left[\frac{y_i - F(\mathbf{x}_i'\boldsymbol{\beta})}{F(\mathbf{x}_i'\boldsymbol{\beta})(1-F(\mathbf{x}_i'\boldsymbol{\beta}))} f(\mathbf{x}_i'\boldsymbol{\beta})\right] \mathbf{x}_i = \mathbf{0}$; where $f(\cdot)$ is the pdf (or the derivative) of the distribution function $F(\cdot)$, the term between squared braquets is often called the *generalized residual* of the binary choice model. [3]

[2] Of course, if the objective function (9.5) is not globally concave, the algorithm yields only local maxima.

[3] This generalized residual is equal to $\frac{f(\mathbf{x}_i'\boldsymbol{\beta})}{F(\mathbf{x}_i'\boldsymbol{\beta})}$ for the positive observations ($\mathbf{y}_i = 1$) and to $\frac{-f(\mathbf{x}_i'\boldsymbol{\beta})}{1-F(\mathbf{x}_i'\boldsymbol{\beta})}$ for the zero observations ($\mathbf{y}_i = 0$). For example for the logit model, this generalized residual is equal to $\left[y_i - \frac{e^{\mathbf{x}'\boldsymbol{\beta}}}{1+e^{\mathbf{x}'\boldsymbol{\beta}}}\right]$.

The known parameter vector $\boldsymbol{\beta}$ is obtained by solving the first order conditions, e.g. by Newton-Raphson (see Subsection A.2.2 in Appendix A), obtaining the following sequence of iterations:

$$\boldsymbol{\beta}^{(k+1)} = \boldsymbol{\beta}^{(k)} - \left(\left. \frac{\partial^2 L(\boldsymbol{\beta})}{\partial \boldsymbol{\beta} \partial \boldsymbol{\beta}'} \right|_{\boldsymbol{\beta}=\boldsymbol{\beta}^{(k)}} \right)^{-1} \left. \frac{\partial L(\boldsymbol{\beta})}{\partial \boldsymbol{\beta}} \right|_{\boldsymbol{\beta}=\boldsymbol{\beta}^{(k)}} .$$

Specification tests can be carried out on the use of the correct distribution function without having to reestimate the model. The appropiate framework for such tests is the LM framework (see Section 3.1). Similar as for the conditional moment test in that section, LM tests for the binary choice model are based on the first order conditions, to verify whether they are violated or not. For example, in testing whether there are omitted variables z_i it should be evaluated whether $\sum_{i=1}^{N} \left[\frac{y_i - F(\mathbf{x}_i'\boldsymbol{\beta})}{F(\mathbf{x}_i'\boldsymbol{\beta})(1-F(\mathbf{x}_i'\boldsymbol{\beta}))} f\left(\mathbf{x}_i'\boldsymbol{\beta}\right) \right] \mathbf{z}_i$ is significantly different from zero.

9.1.6 Index function models: latent variables representation

Models with discrete dependent variables are often in the form of index function models, that is, the outcome of a discrete choice is viewed as a reflection of a *latent variable* model. For example, consider the decision of studying for a graduate degree, say a master degree. The human capital theory states that the person will analyze the expected net benefits (benefits minus costs) of undertaking such studies and will compare this net expected income, appropriately discounted, with the expected earnings (also discounted) if she does not follow these studies. Normally, it is not possible to observe the exact factors that the person included in her calculus. What is observable is the result of this calculus, that is, the decision of undertaking or not such study.

Then, these factors can be modeled as a latent, unobserved variable y_i^* such that the response function for individual i is $y_i^* = \mathbf{x}_i'\boldsymbol{\beta} + \varepsilon_i$, $i = 1, 2, \ldots, N$, where y_i^* is not observable and, hence, is called a latent variable, and represents in our case the individual i's expected benefit of studying. Only the results of y_i^*, denoted as y_i, are observable. In such case, an individual chooses to study if the utility difference exceeds a certain threshold or cutpoint τ, which can deliberately be set at zero without loss of generality in this binary choice context. Hence, for individual i results are observed $y_i = 1$ if $y_i^* > 0$ (the person undertook a degree) and $y_i = 0$ if $y_i^* \leqq 0$ (the person did not invest in human capital). Thus,
$P(y_i{=}1|\,\mathbf{x}_i;\boldsymbol{\beta}){=}\ P(y_i^* > 0|\,\mathbf{x}_i;\boldsymbol{\beta}){=}P(\mathbf{x}_i'\boldsymbol{\beta}{+}\varepsilon_i > 0){=}P(\varepsilon > -\mathbf{x}_i'\boldsymbol{\beta}|\,\mathbf{x})$
$=1 - P(\varepsilon \leqq -\mathbf{x}_i'\boldsymbol{\beta}|\,\mathbf{x}){=}1 - F(-\mathbf{x}_i'\boldsymbol{\beta})$, with $F(\cdot)$ being normal (probit), logistic (logit), Weibull, etc. distribution function. The unknown parameter vector $\boldsymbol{\beta}$ is estimated by ML.

In the next section, multiple choice or multiple response models will be analyzed.

9.2 Multiple response models

In this section, a distinction will be made between *ordered* (or ordinal) response models and *nominal* response (or multinomial) models.

9.2.1 Ordered response models

An ordinal or order variable has categories which can be ranked from low to high and the distances between categories is unknown. In this section, the focus will be on ordered response models, in particular, the ordered probit and ordered logit models.

Before considering the models with ordinal outcomes, it is noteworthy that the fact that a variable can be ordered does not mean that the variable *should* be treated as ordinal: a variable might be ordered when considered for a purpose, but be unordered or ordered differently for another purpose. An example provided by McCullagh and Nelder (1989) is illustrative. They consider the example of colors, which can be arranged according to the electromagnetic spectrum. However, this ordering is not necessarily appropriate for all purpose. Consider, e.g. customers who are about to buy a car and their preferences. There is no reason to believe that a buyer's preference of colors for a car can be ordered according to the color wheel!

Latent variable model

Assume that there are $m > 2$ alternatives (e.g. the types of jobs) and that the choice between these alternatives is based on one underlying latent variable, or for individual i the model can be written $y_i^* = \mathbf{x}_i'\boldsymbol{\beta}_i + \varepsilon_i$ and the observable variable $y_i = j$ if $\tau_{j-1} < y_i^* \leqq \tau_j$ for unknown τ_j with the property that $\tau_0 = -\infty, \tau_1 = 0$ and $\tau_m = \infty$. Hence, the probability that the alternative j is chosen, is exactly the probability that the latent variable y_i^* lies between two boundaries τ_{j-1} and τ_j. For example, if there are three alternatives, we have to check whether there exists a single index $\mathbf{x}_i'\boldsymbol{\beta}$ such that higher values of this index correspond with larger values for y_i. Then, an ordered response model can be written as:
$y_i^*{=}\mathbf{x}_i'\boldsymbol{\beta}_i + \varepsilon_i$, $y_i{=}1$ if $y_i^* \leqq 0$, $y_i{=}2$ if $0 < y_i^* \leqq \tau$ and $y_i{=}3$ if $y_i^* > \tau$.

Assuming now that $\varepsilon_i \overset{iid}{\sim} N(0,1)$ yields the *ordered probit model* and if the ε_i are iid according to a logistic distribution, an *ordered logit model* is obtained. These models can be estimated in an ML framework which will be considered now.

Identification To evaluate the consequences of the nonidentification of the mean of y_i^*, consider the univariate model $y_i^* = a + \beta x_i + \varepsilon_i$ with thresholds τ_j; with a and β being the *true* parameters (i.e. they are used to generate the data that is actually observed). Define an alternative set of parameters .

$$\alpha = a - c \text{ and } \tau_j^* = \tau_j - c, \tag{9.6}$$

with c being an arbitrary constant. The probability that $y_i = j$ is the same, whether the true or alternative parameter sets are used: $P(y = j|\, x; \alpha, \beta){=}F(\tau_j - a - \beta x) - F(\tau_{j-1} - a - \beta x){=}F(\tau_j^* - \alpha - \beta x) - F(\tau_{j-1}^* - \alpha - \beta x)$. Given that both sets of parameters generate the same probability of an observed outcome, there is no way to choose among the two sets of parameters using the observed data, since a change in the intercept in the model can always be compensated by a corresponding change in the thresholds; the ordered response model is unidentified.

There are two assumptions which are commonly used to identify the model:

1. Assume that $\tau_1 = 0$, which involves setting $c = \tau_1$ in (9.6)

2. Assume $a = 0$; which involves setting $c = a$ in (9.6).

These two assumptions identify the model imposing a constraint on one of the parameters. These different identifying assumptions lead to what are known as different parameterizations of the model. It is important to note, however, that the choice of which parameterization is to be used is arbitrary and does not affect the estimation of β (except of course for β_o), nor the tests associated with the estimation.

Estimation Let $\boldsymbol{\beta}$ be the vector with parameters of the model (including the intercept β_o) and $\boldsymbol{\tau}$ the vector containing the threshold parameters; and β_o or τ_1 is constrained to identify the model. The probability of $y_i = j$ is:

$$P(y_i = j|\, \mathbf{x}_i, \boldsymbol{\beta}, \boldsymbol{\tau}){=}F(\tau_j - \mathbf{x}_i'\boldsymbol{\beta}) - F(\tau_{j-1} - \mathbf{x}_i'\boldsymbol{\beta}).$$

The probability that any value of y_i was actually observed for the ith observation is:
$P(y_i{=}1|\, \mathbf{x}_i, \boldsymbol{\beta}, \boldsymbol{\tau})$ if $y_i{=}1,\ldots,P(y_i{=}j|\, \mathbf{x}_i, \boldsymbol{\beta}, \boldsymbol{\tau})$ if $y_i = j,\ldots,P(y_i{=}m|\, \mathbf{x}_i, \boldsymbol{\beta}, \boldsymbol{\tau})$ if $y_i{=}m$.

Provided that the observations are independent, the joint likelihood equation becomes:
$\ell(\boldsymbol{\beta}, \boldsymbol{\tau}){=} \prod_{j=1}^{m} \prod_{y_i=j} P(y_i{=}j|\, \mathbf{x}_i, \boldsymbol{\beta}, \boldsymbol{\tau}){=} \prod_{j=1}^{m} \prod_{y_i=j} F(\tau_j{-}\mathbf{x}_i'\boldsymbol{\beta}){-}F(\tau_{j-1}{-}\mathbf{x}_i'\boldsymbol{\beta})$, where $\prod_{y_i=j}$ is the multiplication over all cases where y_i is observed to be equal to j (by this operator, it is ensured that all individuals choosing a certain alternative j are considered). The logarithm of this equation can be maximized with methods mentioned in Chapter 1 to obtain ML estimators of $\boldsymbol{\beta}$ and $\boldsymbol{\tau}$. These estimators are under some regularity conditions consistent, asymptotically normal, and asymptotically efficient.

9.2.2 Multinomial models

When a variable cannot be ordered, it is called *nominal*. This section focuses on *multinomial logit* and *multinomial probit* models. The multinomial logit model is the most frequently used for nominal outcomes. The characteristics of the outcomes are used to obtain predictions of the (multiple) choice that is made by means of the conditional logit model

The multinomial logit model

This models considers all possible comparisons among the possible alternatives. In this sense, the multinomial logit is an extension of the binary logit model, which is made difficult by the large number of comparisons that are involved (e.g., for three outcomes, the multinomial logit (model) is *roughly* equivalent[4] to running three binary logit(s) (models) comparing outcomes 1 to 2, 1 to 3 and 2 to 3).

If there are e.g. m alternatives, which can be divided over individual according to utility indices, e.g. the index U_{ij} is the utility level that individual i attaches to alternative j ($i = 1, 2, \ldots, N; j = 1, 2, \ldots, m$). Now alternative j is chosen by individual i if it yields the highest utility, i.e. if $U_{ij} = \max\{U_{i1}, U_{i2}, \ldots, U_{im}\}$. Assuming that this unobservable utility index U_{ij} is equal to a constant plus a WN error term, the problem of utility selection boils down to a stochastic characterization of these error terms, with e.g. the *log-log distribution* $F(\mathbf{x}'\boldsymbol{\beta}) = e^{-e^{-\mathbf{x}'\boldsymbol{\beta}}}$ (also called an *extreme value distribution*). Under this distribution assumption of *iid* disturbances, it can be shown that $P(Y_i{=}j|\,\mathbf{x}) = \frac{e^{\boldsymbol{\beta}'\mathbf{x}_{ij}}}{1+\sum_{k=1}^{m} e^{\boldsymbol{\beta}'\mathbf{x}_{ik}}}$, so that the loglikelihood function to be maximized with respect to $\boldsymbol{\beta}$ is: $L(\boldsymbol{\beta}) = \sum_{j=1}^{m} \sum_{y_i=j} \left[\boldsymbol{\beta}'\mathbf{x}_{ij} - \ln\left(1 + \sum_{k=1}^{m} e^{\boldsymbol{\beta}'\mathbf{x}_{ik}}\right)\right]$.

Multinomial probit models

A natural alternative to the multinomial logit is the multinomial probit. This model relaxes the independence restrictions built into the multinomial logit model. Now the utilities indices can be written as $U_{ij} = \boldsymbol{\beta}'\mathbf{x}_{ij} + \varepsilon_{ij}$ with $(\varepsilon_{i1}, \varepsilon_{i2}, \ldots, \varepsilon_{im})$ mutuallyiid distributed for each individual according to a $N(\mathbf{0}, \boldsymbol{\Sigma})$, where $\boldsymbol{\Sigma}$ is the $m \times m$ covariance matrix of the error terms of all alternatives. The term in the loglikelihood that corresponds to the alternative j is:
$P(y_i{=}j|\,\mathbf{x}_i, \boldsymbol{\beta}) = P(U_{ij} > U_{ik}) = P\left(\varepsilon_{i1} - \varepsilon_{ij} < \boldsymbol{\beta}'(\mathbf{x}_{i1} - \mathbf{x}_{ij}), \ldots, \varepsilon_{im} - \varepsilon_{ij} < \boldsymbol{\beta}'(\mathbf{x}_{im} - \mathbf{x}_{ij})\right)$.
Note that this multivariate normal distribution implies that the errors can be correlated among alternatives. This is possible since it is simple to incorporate correlations among errors in a multivariate normal distribution. For example, the bivariate normal distribution is

$$F = P(X_1 < x_1, X_2 < x_2) = \int_{-\infty}^{x_2} \int_{-\infty}^{x_1} f(z_1, z_2, \rho)\, dz_1 dz_2$$

with the pdf $f(z_1, z_2, \rho) = \frac{\exp\left[-\frac{x_1^2 + x_2^2 - 2\rho x_1 x_2}{2(1-\rho^2)}\right]}{2\pi(1-\rho^2)^{1/2}}$, where ρ is the correlation between x_1 and x_2.

[4] Long (1997), pp. 149-150, provides the derivation of how the estimation of the linked set of binary logits differs from the multinomial logit given the 'no simultaneity of the samples'.

9.3 Cases and exercises

9.3.1 Solved cases

Solved case 1

We will consider the occupational decisions, and in particular, the factors that would lead a student to choose to become a teacher. Gertel *et al.* (2002) study for twenty-nine Argentinean metropolitan areas the decision of becoming a teacher, measured at the step of the educational years of university or college.

The dependent variable is a dummy variable that reflects whether the student chose a teacher training course or not. The independent variables are: a dummy variable indicating the gender of the person, *man*, the age of the individual, represented by the variable *age*, the civil status, *married*, whether the person has recently moved from another place in the same province, another province or another country or not, *migrat*, and the fact that the individual lives or not in a *bigcity* (more than a million inhabitants' city). Labor market factors are reflected by the variables *working*, that indicates whether the person is working at present or not, and *rel_wage*, the teachers' relative wage. Finally, the study incorporated per capita income level, through four levels of income: low, average, high and very high, according to whether the per capita income of the family of the student falls among the following categories or not: zero and mean level of income, mean and mean plus a standard deviation, between the mean plus a standard deviation and mean plus two standard deviations, and more than the mean plus two standard deviations- *low_y*, *avg_y*, *high_y* and *v_highy*, respectively..

Marginal effects Estimates of the linear probability, logistic, probit and log-log models were computed. [5] Table 9.1 shows the marginal effects discussed in Section 9.1.3 and also known as slopes, with all the variables valued at their means. This can be straightforwardly done in Stata.

From the slopes in Table 9.1 it is observed that females are much more likely to undertake teacher studies than men; being married and economically active strengthen the likelihood of choosing this kind of studies. By contrast, living in a *bigcity* and having finished the secondary school in a different city than the one where the student lives, decreases the probability of choosing a teacher training course. The socio-economic background of the student suggests that those preparing for teaching tend to come from less wealthy homes. Finally, there is a significant effect of the expected remuneration of the teacher (*rel_wage*) on the students' choice in Argentina.

[5] The estimates of the complementary log-log regression that Stata provides coincide with the Gompertz estimates of the SAS software. It is important to be aware of what the software you are using is doing. Unlike most programs that model the probability of a 1, SAS models the probability of a 0. For probit and logit models, given that their pdf are both symmetric, all the coefficients will have the opposite sign, given that in these cases holds: $P(y=0) = 1 - p(y=1) = 1 - F(x'\beta) = F(-x'\beta)$. This will not be the case for asymmetric models.

Table 9.1: Marginal effects

Variable	*Probit*	*Logit*	*Cloglog*	*Loglog*
Age	0.0007	0.0013	0.0010	0.0008
Married	$0.0404^{(*)}$	$0.0394^{(**)}$	$0.0376^{(**)}$	$0.0405^{(**)}$
Migrat	$-0.0665^{(*)}$	$-0.0635^{(**)}$	$-0.0615^{(**)}$	$-0.0689^{(**)}$
Man	$-0.1083^{(**)}$	$-0.1038^{(**)}$	$-0.1011^{(**)}$	$-0.1126^{(**)}$
Working	$0.0236^{(*)}$	$0.0212^{(*)}$	$0.0195^{(*)}$	$0.0260^{(*)}$
Bigcity	$-0.0502^{(**)}$	$-0.0470^{(**)}$	$-0.0452^{(**)}$	$-0.0533^{(**)}$
Low_y	$0.0454^{(**)}$	$0.0416^{(**)}$	$0.0390^{(**)}$	$0.0492^{(**)}$
High_y	$-0.0390^{(**)}$	$-0.0373^{(**)}$	$-0.0364^{(**)}$	$-0.0406^{(**)}$
V_high	$-0.0653^{(**)}$	$-0.0626^{(**)}$	$-0.0611^{(**)}$	$-0.0679^{(**)}$
Rel_wage	$0.0422^{(**)}$	$0.0402^{(**)}$	$0.0388^{(**)}$	$0.0434^{(**)}$

Statistical significance: $^{(**)}p < 0.01$, $^{(*)}p < 0.05$.

For the case of effect of a dummy variable, we compute all the variables at their means, except the binary variable we are interested in. In particular, we are interested in analyzing the influence of gender on the decision of undertaking a teacher training career, therefore, the only variable that is not valued at its mean is the gender variable, *man*. This effect can be approximated by means of the following

$$P\{teacher = 1|\,\overline{\mathbf{x}}, woman = 1\} - P\{teacher = 1|\,\overline{\mathbf{x}}, woman = 0\}$$

Considering the logit model, this marginal effect for a binary variable is computed as follows

$$\frac{\exp(\overline{\mathbf{x}}'\boldsymbol{\beta}; woman = 1)}{1 + \exp(\overline{\mathbf{x}}'\boldsymbol{\beta}; woman = 1)} - \frac{\exp(\overline{\mathbf{x}}'\boldsymbol{\beta}; woman = 0)}{1 + \exp(\overline{\mathbf{x}}'\boldsymbol{\beta}; woman = 0)}$$

Plotting probabilities over a range of a variable When we have several independent variables, examination of the effect of two variables while the remaining variables are constant can be done graphically[6].

Considering the student's decision example and the effect that relative wage and gender have on the probability of choosing to be a teacher, these effects can be computed by holding constant the remaining variables at their means, and letting wage and gender vary. To do so, first we compute the estimated probability of being a teacher, for all the variables valued at their means, except $woman = 1$, and let the adjusted salary vary. Similarly, the probability that a man chooses to be a teacher can be calculated by letting $woman = 0$, while letting the adjusted relative wage to vary.

[6] For a thorough discussion of how to plot probabilities over a range of a variable and the interpretation of marginal effects, see Long (1997, pp. 66-79).

Figure 9.1: Gender effects on the probability of choosing a teacher training career

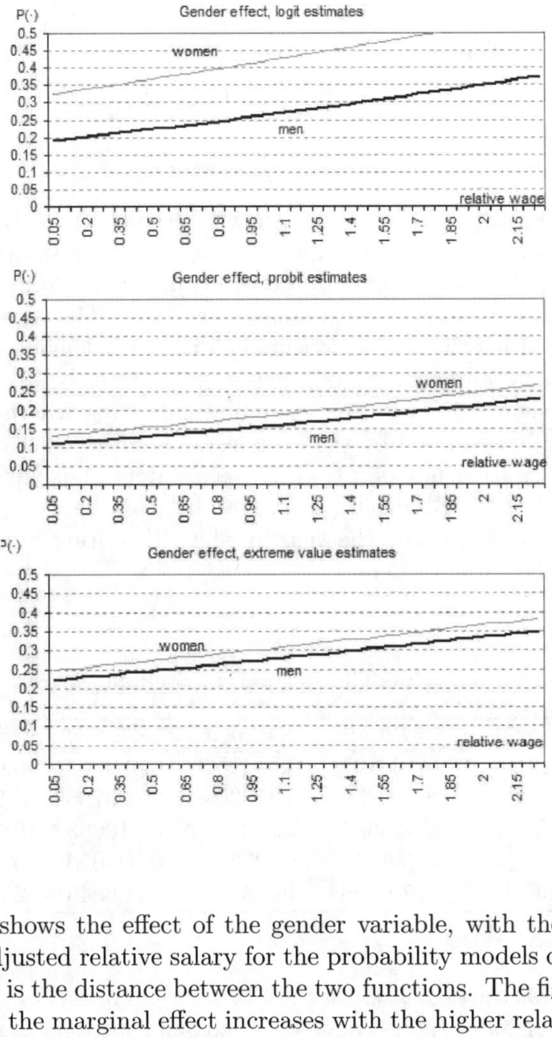

Figure 9.1 shows the effect of the gender variable, with the two functions plotted over adjusted relative salary for the probability models considered. The marginal effect is the distance between the two functions. The figure shows that for the models, the marginal effect increases with the higher relative wage. This means that the probability of choosing a teacher training career for each level of relative wage is higher for women than for men.

Forecast evaluation The first two forecast error statistics show that the smallest prediction error for the sample are both the logit model and the complementary log-log, and hence the better the forecasting ability of that model. The Theil inequality coefficient always lies between zero and one, where zero indicates a perfect fit. Therefore, according to this criterion, there is a slight preference for complementary log-log over the logit.

Table 9.2: Forecast evaluation

	Linear	*Probit*	*Logit*	*Cloglog*	*Loglog*
Root mean sq. error	0.3285	0.3277	0.3277	0.3277	0.3278
Theil ineq. coefficient	0.6332	0.6297	0.6291	0.6290	0.6311

Solved case 2

Lukach and Plasmans (2004), measure knowledge spillovers in Belgium, analyzing patent citations in patents granted to Belgian corporate applicants. They model the probability that a patent citation occurs in the citing patent belonging to a particular industry, for example, Chemistry. The dependent variable assumes a value of 1 if the citation occurs in a patent belonging to a firm of the Chemistry industry, and equals 0 otherwise.

The independent variables are: a dummy indicating whether patents in a citation pair are both owned by the same firm ($SameFirm$), an dummy indicating that the citation has occurred between patents belonging to the same industry ($SameIndustry$), and the $TimeLag$ between the citing and the cited patents. Finally, they include the year in which the citing patent was issued, represented by the variable $Year$.

Table 9.3: Estimation results

	Probit	*Logit*	*Cloglog*	*Loglog*
Timelag	0.003101	0.004957	0.004369	0.002969
Samefirm	0.270239	0.479304	0.433786	0.228997
Sameindustry	-0.419687	-0.770193	-0.710003	-0.343601
Year	-0.014244	-0.025384	-0.023058	-0.012010
Intercept	27.6766	49.43616	44.6645	23.56793

Table 9.3 shows the estimates for the Chemistry industry for the models considered. It also includes the estimates of the complementary log-log regression that Stata provides, which as commented before, coincide with the Gompertz estimates of the SAS software. Finally, we have computed the marginal effects for the models considered: probit, logit and log-log and complementary log log (see Table 9.4). It is important to note that the sign of these marginal effects do not vary when we consider different specifications.

Finally, Table 9.5 displays the root mean squared error, and the Theil inequality coefficient calculated for each one of the models. The root mean squared error indicates that in the case of the Chemistry industry, the most accurate model is the complementary log-log, since it has the smallest prediction error

Table 9.4: Marginal effects

	Probit	*Logit*	*Cloglog*	*Loglog*
Timelag	0.000841$^{(**)}$	0.000757$^{(*)}$	0.000735$^{(*)}$	0.000940$^{(**)}$
Samefirm	0.079131$^{(**)}$	0.080813$^{(**)}$	0.082242$^{(**)}$	0.075787$^{(**)}$
Sameindustry	-0.113836$^{(**)}$	-0.117629$^{(**)}$	-0.119496$^{(**)}$	-0.108845$^{(**)}$
Year	-0.003863$^{(**)}$	-0.003876$^{(**)}$	-0.003880$^{(**)}$	-0.003804$^{(**)}$

Statistical significance: $^{(**)}p < 0.01$, $^{(*)}p < 0.05$.

Table 9.5: Forecast evaluation

Variable	*Probit*	*Logit*	*Cloglog*	*Loglog*
Root mean squared error	0.391712	0.391585	0.391477	0.391896
Theil inequality coefficient	0.607218	0.606741	0.606108	0.608100

within the sample. Also according to the Theil inequality coefficient, the model that best fits the data is the complementary log-log one, since it exhibits the smallest coefficient. Hence, this model is chosen for *ex-ante* forecasting.

In addition to 'Chemistry', Lukach and Plasmans (2005) analyze in their paper seven other industries: 'Instruments', 'Pharmacy', 'Other Machinery', 'Paper, Printing and Publishing','Metal Products, excluding Machines', 'Computers and Office Machines', and 'Other Industries'.

9.3.2 Exercises

1. Show that in the logit model, e raised to the power β_i is the factor by which the odds change when the ith independent variable increases by one unit.

2. Assume a linear probability model, which is to be fit to a set of observations on a dependent variable y that assumes values zero and one, and x, a regressor varying continuously across observations. Obtain the expressions for the LS slope in the regression in terms of the mean(s) and variance of x. Interpret the result.

3. With the dataset given below, estimate a logit and a probit model and test the hypothesis that x is not influential in determining the probability that $y = 1$.

y	1	0	0	1	1	0	0	1	1	1
x	9	2	5	4	6	7	3	5	2	6

4. Construct the LM statistic for testing the hypothesis that all the slopes (but not the constant term) are equal to zero in the binomial logit model.

5. Give the loglikelihood function for the ordered probit model if $\varepsilon_i \overset{iid}{\sim} N\left(0, \sigma_\varepsilon^2\right)$ rather than $\varepsilon_i \overset{iid}{\sim} N\left(0, 1\right)$.

6. Consider a dataset that consists of 250 observations, of which the responses are

y	0	1	2	3	4
n	50	40	45	80	35

. Obtain ML estimates of the unknown parameters of the ordered probit model.

Chapter 10

Limited Responses, Duration and Count Data

10.1 Censoring and truncation

A *censored* sample is a sample for which (for a known set of independent variables) a number of observations cannot be observed, i.e., we have limited information about the dependent variable.

A *truncated* sample is a sample for which the data for independent variables are only available if the dependent variable can be observed. Therefore, truncation limits the data more severely by excluding observations based on characteristics of the dependent variable.

There are a variety of methods for censoring and truncation. We will focus on the most frequent model of censoring: the *tobit model.*

10.1.1 Problems raised by censoring

Being unable to observe certain values of the dependent variable y causes problems, and a decision should be made on how the censored observations should be handled, prior to the acceptance of the tobit model:

- Create a truncated sample by deleting the cases in which the dependent variable is censored; the model is estimated by OLS using the truncated sample.

- Create a censored dependent variable where all censored observations are assigned the value τ (the threshold or cutpoint). The models are estimated by OLS using the censored dependent variable.

10.1.2 Tobit models for censoring

The classical example of censoring is the study of Tobin (1958) for durable consumption goods; a consumer maximizes utility by purchasing durable goods subject to the budget constraint that total expenditures are not to exceed the income. However, the expenditures in durables are subject to another constraint: the expenditures in durables should be at least equal to the least expensive of the durable goods. That is, if the cheapest durable good is EUR 150, and the consumer has e.g. only EUR 149 to spend, the expenditure on durable consumption goods is null. In this example, the outcome is *censored* since one cannot know how much the consumer would have spent if a durable good could be purchased by less than EUR 150.

Taking account of unobserved heterogeneity in individual utility functions and representing this unobserved heterogeneity in the form of an error process $\left\{\varepsilon_i \overset{iid}{\sim} (0,\sigma_\varepsilon^2), i=1,2,\ldots,N\right\}$ a *tobit model* or a *censored regression* can be defined as: $y_i^* = \mathbf{x}_i'\boldsymbol{\beta} + \varepsilon_i, i = 1,2,\ldots,N$ with as observed values $y_i = y_i^*$ if $y_i^* > 0$ and equal to zero if $y_i^* \leqq 0$. Hence, the observations are censored from below at zero and the probability that $y_i = 0$ is given by:[1]
$P(y_i{=}0|\,\mathbf{x}_i;\boldsymbol{\beta}){=}P(y_i^* \leqq 0|\,\mathbf{x}_i;\boldsymbol{\beta}){=}P(\varepsilon_i \leqq -\mathbf{x}_i'\boldsymbol{\beta}|\,\mathbf{x}_i;\boldsymbol{\beta}){=}P(\frac{\varepsilon_i}{\sigma_\varepsilon} \leqq \frac{-\mathbf{x}_i'\boldsymbol{\beta}}{\sigma_\varepsilon}){=}F(-\frac{\mathbf{x}_i'\boldsymbol{\beta}}{\sigma_\varepsilon})$, which is equal to $1 - F(\cdot)$ for a symmetric distribution function.

Assume N observations with the last s y_is equal to zero and the first observable $N-s$ variables (for $i = 1,2,..,N-s$), the conditional expectation of the observable variables y_i satisfies

$$E\left(y_i|\,\mathbf{x}_i, y_i > 0;\boldsymbol{\beta}\right){=}\mathbf{x}_i'\boldsymbol{\beta}{+}E\left(\varepsilon_i|\mathbf{x}_i, y_i > 0;\boldsymbol{\beta}\right){=}\mathbf{x}_i'\boldsymbol{\beta}{+}E\left(\varepsilon_i|\,\varepsilon_i > -\mathbf{x}_i'\boldsymbol{\beta}\right),$$

with the truncated density:

$$f\left(\varepsilon_i|\,\varepsilon_i \geqq -\mathbf{x}_i'\boldsymbol{\beta}\right) = \frac{f(\varepsilon_i)}{\int_{-\mathbf{x}_i'\boldsymbol{\beta}}^{\infty} f(t)dt} \text{ for } \varepsilon_i > -\mathbf{x}_i'\boldsymbol{\beta}.$$

This leads to biased and inconsistent OLS estimates, because:

$$E\left(\varepsilon_i|\,\varepsilon_i > -\mathbf{x}_i'\boldsymbol{\beta}\right) = \sigma_\varepsilon\lambda_i, \text{ with } \lambda_i{:=}f\left(\frac{-\mathbf{x}_i'\boldsymbol{\beta}}{\sigma_\varepsilon}\right)\Big/1 - F\left(\frac{-\mathbf{x}_i'\boldsymbol{\beta}}{\sigma_\varepsilon}\right),$$

so that $E\left(y_i|\,x_i, y_i > 0\right) = x_i'\beta + \sigma_\varepsilon\lambda_i \quad (i = 1,2,...,N-s)$.

The estimation of the tobit model can be done with two different procedures: *Heckman's two-stage estimation* or using the *ML principle*. These two cases will be considered here.

Heckman's two-stage estimator

Heckman (1979) assumes that the probability $F\left(\frac{-\mathbf{x}_i'\boldsymbol{\beta}}{\sigma_\varepsilon}\right)$ is standard normal. His estimation procedure consists in two steps for

[1] For more details on truncation, see Maddala (1983), pp. 365-369, and Long (1997), pp. 212-213 for a discussion on upper and lower censoring.

1. Estimation of λ_i by a probit model and denote them by $\hat{\lambda}_i$; then we regard these estimates as 'observations of an additional explanatory variable' for the model $y_i = \mathbf{x}_i'\boldsymbol{\beta}+\sigma_\varepsilon\hat{\lambda}_i + v_i, i = 1, 2, \ldots, N$.

2. Heckman (1979) estimates the latter model by OLS (given the estimates $\hat{\lambda}_i$, which are not correlated by construction with the error term v_i), to obtain an estimate for $\boldsymbol{\beta}$ and for σ_ε.

Although the resulting estimator is consistent and asymptotically normal, it is less efficient and not easier to compute than the ML estimator. Moreover, OLS standard errors in this two step procedure are incorrect (unless the degenerate case of $\sigma_\varepsilon = 0$) since they are heteroskedastic (see Greene (2003), pp. 780-790, for more details).

MLE of the tobit model for an arbitrary threshold

The ML estimation of the tobit model involves subdividing the set of observations into two subsets: the first with uncensored observations and the second with the censored observations. For the second set of observations, one does not know the value but can proceed using the computed probability of being censored in the likelihood equation.

As indicated above, it is assumed that there are $N - s$ uncensored observations and s censored observations (threshold is assumed to be known).

For the uncensored observations, $y_i = \mathbf{x}_i'\boldsymbol{\beta} + \varepsilon_i$, $i = 1, 2, \ldots, N - s$ for the values of the latent variable $y_i^* > \tau$ (the threshold) with $\varepsilon_i \overset{iid}{\sim} (0, \sigma_\varepsilon^2)$ the loglikelihood function is $\log \ell^u(\boldsymbol{\beta},\sigma^2) = \sum^u \left[\log f\left(\frac{y_i - \mathbf{x}_i'\boldsymbol{\beta}}{\sigma_\varepsilon}\right) - \log \sigma_\varepsilon\right]$, where $-\log \sigma_\varepsilon$ is obtained from the convolution rule (when transforming from $f(\varepsilon_i)$ to $f\left(\frac{\varepsilon_i}{\sigma_\varepsilon}\right)$) and $\sum^u$ is the sum over all uncensored observations.

For censored observations, $\mathbf{x}$ is known, and about the dependent variable it is known that $y^* \leqq \tau$. Then, it can be computed $P(y_i \leqq \tau | \mathbf{x}_i; \boldsymbol{\beta}, \sigma_\varepsilon^2) = F\left(\frac{\tau - \mathbf{x}_i\boldsymbol{\beta}}{\sigma_\varepsilon}\right)$, so that the contribution of the censored observations to the likelihood function can be written as $\ell^c(\boldsymbol{\beta},\sigma_\varepsilon^2) = \prod^c F\left(\frac{\tau - \mathbf{x}_i'\boldsymbol{\beta}}{\sigma_\varepsilon}\right)$, where $\prod^c$ is the product over all censored observations.

Finally, the loglikelihood for both uncensored and censored observations satisfies:
$L(\boldsymbol{\beta},\sigma^2) = \ln \ell(\boldsymbol{\beta},\sigma^2) = \sum^u \left[\log f\left(\frac{y_i - \mathbf{x}_i'\boldsymbol{\beta}}{\sigma_\varepsilon}\right) - \log \sigma_\varepsilon\right] + \sum^c \log F\left(\frac{\tau - \mathbf{x}_i\boldsymbol{\beta}}{\sigma_\varepsilon}\right)$,
and the usual properties of the ML estimators apply if the assumptions of the tobit model are met.

10.2 Models for duration data

Duration data is the length of time until an event (e.g. failure) happens. Duration data was already used for a long time in physical and medical sciences.

Related to this is survival times, for example the length of survival after the diagnosis of a strong desease, or after an operation as a heart transplant. Social scientists have recently applied a similar methodology on strike duration, length of unemployment spells, time until business failure, etc.

This section consists of two parts: the analysis of the length of time (period) before failure and the analysis of survival times.

The crucial variable is the length of time that elapses from the beginning of some event either until its end or until the measurement is taken (which may precede termination). The observation is a cross section of *durations* $t_1, t_2, ..., t_n$ (the beginning may be different).

Censoring is a pervasive and usually unavoidable problem in the analysis of duration data, because measurement is made while the process is on-going, e.g. survival times of heart transplant patients (observations on individuals who are still alive are always censored); a sample of spells of unemployment, drawn from surveys, will probably include some individuals who are unemployed at the time the survey is taken. For these individuals the duration is at least the observed t_i but not equal to it! Therefore, estimation must take account of this censored nature.

10.2.1 Parametric models of duration

Say, the spell length (e.g. of unemployment) is assumed to be equal to a random variable T. Then, a simple approach is to apply regression analysis to a sample of observed spells. Suppose this random variable T has a continuous pdf $f(t)$, where t is a realization of T. Then the distribution function is $F(t)=\int_0^t f(s)\,ds=P(T \leqq t)$.

Generally, we are interested in the probability that the spell is of length at least t, given by the survival function:

$$S(t) = 1 - F(t) = P(T > t).$$

Given that the spell has lasted until time t, the probability that it will end in the next short interval is:

$$l(t, \Delta) = P(t < T \leqq t + \Delta | T > t).$$

Finally, if one is interested in the rate at which spells are completed, the *hazard rate* must be analyzed:

$$\begin{aligned} \lambda(t) &= \lim_{\Delta \to 0} \frac{P(t < T \leqq t + \Delta | T > t)}{\Delta} \\ &= \lim_{\Delta \to 0} \frac{F(t+\Delta) - F(t)}{\Delta S(t)} = \frac{f(t)}{S(t)} = \frac{-d \ln S(t)}{dt}. \end{aligned} \tag{10.1}$$

The hazard rate is the rate at which spells are completed after the duration t, given that they last at least until t. The hazard function is the conditional probability of a state change given the state has achieved a duration t.

The hazard function

The hazard function can be more interesting than the survival rate or the pdf. Therefore, one might consider modeling the hazard function itself. Consider e.g. the case in which the hazard rate does not vary over time, i.e. $\lambda(t) = \lambda$, a constant, which is characteristic for a process that does not have memory. If the process does not have memory, the conditional probability of failure in a given interval is the same regardless of when the observation is made. Then, from (10.1), the differential equation $\frac{-d\ln S(t)}{dt} = \lambda$ results with solution $\ln S(t) = c - \lambda t$, i.e. $S(t) = e^{c-\lambda t} = Ce^{-\lambda t}$,with C the constant of integration. The condition that $S(0) = 1$ implies that $C = 1$, and the solution is $S(t) = e^{-\lambda t}$. This solution is the exponential distribution. Since with an exponential distribution, $E(t) = 1/\lambda$, the MLE of λ would be the reciprocal of the sample mean.

A decaying hazard function is referred as a *negative duration dependence*, while an increasing hazard function is referred as a *positive duration dependence.* In this latter case, the likelihood of failure at time t, conditional upon duration up to time t, is increasing in t.

Example 10.1 *Given the conditional transition probabilities* $P_{ij}\left(D_{t-1}\right)$, *i.e. the probabilities of switching from state* i *to state* j, *given that we have been in state* i *for* d_{t-1} *periods* ($P_{ij}\left(D_{t-1}\right) = 1 - P_{ii}\left(D_{t-1}\right)$ *for* $i \neq j$), *the hazard function is* $P_{ij}\left(D_{t-1}\right) = \frac{1}{1+e^{\left(\gamma_1(i)d_{t-1}+\gamma_2(i)d_{t-2}\right)}}$ $(i = 1, 2)$ *which is strongly related to the multinomial logit in Section 9.2.2.*

Maximum likelihood estimation

The parameters of these duration models can be estimated by ML. The log-likelihood function can be formulated for observed duration data $t_1, t_2, \ldots, t_n$: $\ln L(\boldsymbol{\theta}) = \sum^u \ln f(t|\,\boldsymbol{\theta}) + \sum^c \ln S(t|\,\boldsymbol{\theta})$ with $\boldsymbol{\theta}$ the vector of parameters λ,p, $\sum^u$ the sum of the uncensored observations and $\sum^c$ the sum of the censored observations.

Formulating now the loglikelihood in terms of $f(t) := \lambda(t)S(t)$,

$$\ln \ell = \sum^u \lambda(t|\,\boldsymbol{\theta}) + \sum^{all} \ln S(t|\,\boldsymbol{\theta}),$$

where $\sum^{all}$ is the sum over all observations.

In order to estimate the asymptotic standard errors for the estimates, either the estimator of the variance-covariance matrix in (1.12), the inverse Hessian or the BHHH estimator can be used.

Accelerated failure time models

In the models studied so far, external factors do not have a role in the survival distribution. Covariates can be added to duration models in a straightforward way; however, the interpretation of the coefficients in the model is quite complex.

Consider an exponential expression for the hazard rate, $\lambda_i = e^{-\mathbf{x}_i'\beta}$, with $\mathbf{x}_i$ a constant term and a set of variables assumed not to change from time

$T = 0$ until $T = t_i$, the failure time. When expressing λ_i as a function of a set of regressors, it is as if one changed the units of measurement on the time axis. This is the reason why these models are also called *accelerated failure time models*.

Note that generally, the regressors do not bear on the question of duration dependence, which is a function of the scale parameter π.[2] The loglikelihood can be rewritten in terms of π. Therefore, let $\delta_i = 1$ if the spell is completed and $\delta_i = 0$ if the spell is censored, and let $\sigma := \frac{1}{\pi}$ and for the above hazard rate $\lambda_i = e^{-\mathbf{x}_i'\boldsymbol{\beta}}$,we can define $\pi \ln(\lambda_i t_i) := \frac{(\ln t_i - \mathbf{x}_i'\boldsymbol{\beta})}{\sigma} =: w_i$. Denoting the pdf and survival functions as $f(w_i)$ and $S(w_i)$, respectively, the (observed) random variable is $\ln t_i = \sigma w_i + \mathbf{x}_i'\boldsymbol{\beta}$.

Given that the Jacobian of the transformation from w_i to $\ln t_i$ is $dw_i/\, d\ln t_i = 1/\,\sigma$, the density function for $\ln t_i$ is $f(\ln t_i|\, \mathbf{x}_i, \boldsymbol{\beta}, \sigma) = \frac{1}{\sigma} f\left(\frac{\ln t_i - \mathbf{x}_i'\boldsymbol{\beta}}{\sigma}\right)$, the survival function is $S(\ln t_i|\, \mathbf{x}_i, \boldsymbol{\beta}, \sigma) = S\left(\frac{\ln t_i - \mathbf{x}_i'\boldsymbol{\beta}}{\sigma}\right)$, and the loglikelihood for the observed data is

$$\ln \ell(\boldsymbol{\beta}, \sigma|\text{data}) = \textstyle\sum_{i=1}^{n} [\delta_i \ln f(\ln t_i|\, \mathbf{x}_i, \boldsymbol{\beta}, \sigma) + (1 - \delta_i) \ln S(\ln t_i|\, \mathbf{x}_i, \boldsymbol{\beta}, \sigma)].$$

Heterogeneity

We will model heterogeneity in the parametric model. Assume a survival function conditioned on the v_i (individual specific) effect. The survival function is $S(t_i|\, v_i)$, and adding to that a model for the unobserved heterogeneity $f(v_i)$:

$$S(t) = E[S(t|\, v)] = \textstyle\int S(t|\, v) f(v) dv.$$

Using now the gamma distribution to incorporate heterogeneity into the above Weibull model, and assuming v has a gamma distribution with mean 1 and variance $\frac{1}{\kappa}$, we get the pdf $f(v) = \frac{\kappa^\kappa}{\boldsymbol{\Gamma}(\kappa)} e^{-\kappa v} v^{\kappa - 1}$ and the survival function $S(t|\, v) = e^{-(v\lambda t)^\pi}$.

The resulting unconditional distribution is

$$S(t) = \textstyle\int_0^\infty S(t|\, v) f(v) dv = [1 + \frac{1}{\kappa}(\lambda t)^\pi]^{-\kappa}.$$

The limiting value for $\frac{1}{\kappa} = 0$ corresponds to $var(v) = 0$, which is the Weibull survival model without heterogeneity. Finally, the hazard function for this model is $\lambda(t) = \lambda\pi(\lambda t)^{\pi - 1} [S(t)]^{\frac{1}{\kappa}}$.

10.3 Count data

In order to assess the impacts of the determinants on the number of a variable like e.g. patent applications, the discreteness of this variable has to be taken

[2]While the exponential distribution leads to a hazard function $\lambda(t) = \lambda$ and to a survival function $S(t) = e^{-\lambda t}$, the Weibull distribution leads to a hazard function $\lambda(t) = \lambda\pi(\lambda t) e^{\pi - 1}$ and to a survival function $S(t) = e^{-(\lambda t)^\pi}$.

into account. For instance, because of difficulties and uncertainty inherent to R&D activities, firms do not always apply for patents and hence a zero value is a natural outcome of this variable. Because of this property, the use of conventional linear regression models is generally inappropriate. The reasons are that some basic assumptions such as the normality of residuals or the linear adjustment of data are no longer fulfilled. There are a variety of ways of dealing explicitly with the characteristic of count data; and the usual way to deal with the discrete nonnegative nature of the patent dependent variable is to consider the simple *Poisson regression model* (Cameron and Trivedi (1998) p. 279). In this model, the probability of a count is determined by the Poisson distribution, with the mean of the distribution being a function of the independent variables. This model has the characteristic that the conditional mean of the outcome is equal to the conditional variance. However, in practice, it is often the case that the conditional variance exceeds the conditional mean. To deal with this problem, we make use of the negative binomial regression model, which allows for the variance to exceed the mean. Another problem is that the number of zeros in a sample sometimes exceeds the number predicted by both the Poisson regression model and by the negative binomial model. Zero-modified models (e.g. zero-inflated count models) explicitly model the number of predicted zeros, while allowing the variance to differ from the mean. These models are based on the Poisson distribution, which is considered in the next section.

10.3.1 Poisson model

Let y_i be the variable for observation i, where $i = 1, ..., n$, which represents a discrete variable as e.g. the number of patents (or patent applications) of firm i. The y_is are independent and assumed to have individual Poisson distributions with parameters λ_i s, which depend on a set of independent variables, that in this example are the logarithms of the firm's R&D expenditures:

$$L = P\left(Y_i = y_i | \, \mathbf{x}_i; \boldsymbol{\beta}\right) = \frac{e^{-\lambda_i}\lambda_i^{y_i}}{y_i!} \quad (y_i{=}0, 1, 2, ..),$$

where $\mathbf{x}_i$ represents the set of explanatory variables and $\boldsymbol{\beta}$ is the vector of the unknown coefficients. As a basic property of the Poisson model the (conditional) mean and variance are equal, or $E\left(y_i | \, \mathbf{x}_i; \boldsymbol{\beta}\right) = var\left(y_i | \, \mathbf{x}_i; \boldsymbol{\beta}\right) = \lambda_i$.

Assume now an exponential (conditional) mean function, i.e. $\lambda_i = e^{\boldsymbol{\beta}'\mathbf{x}_i}$,[3] which is called the *loglinear model.* Notice that then $\frac{\partial E(y_i|\mathbf{x}_i;\boldsymbol{\beta})}{\partial \mathbf{x}_{ij}} = \beta_j E\left(y_i | \, \mathbf{x}_i; \boldsymbol{\beta}\right)$, so that a one unit change in the jth regressor leads to a proportionate change in the conditional mean $E\left(y_i | \, \mathbf{x}_i; \boldsymbol{\beta}\right)$ (since $\frac{\partial E(y_i|\mathbf{x}_i;\boldsymbol{\beta})/E(y_i|\mathbf{x}_i;\boldsymbol{\beta})}{\partial \mathbf{x}_{ij}} = \beta_j$).

The unknown parameter vector $\boldsymbol{\beta}$ is estimated by the ML method through the maximization of the loglikelihood function:

$$L(\boldsymbol{\beta}) = \textstyle\sum_{i=1}^{n} [y_i \boldsymbol{\beta}'\mathbf{x}_i - e^{\boldsymbol{\beta}'\mathbf{x}_i} - \ln(y_i!)], \tag{10.2}$$

[3] An exponential mean function is taken because the nonnegativity of this (conditional) mean (and of the (conditional) variance) should be assured (e.g. a negative number of patents is not possible).

which leads to the first order conditions:

$$\frac{\partial L}{\partial \beta} = \sum_{i=1}^{n} [\mathbf{x}_i (y_i - e^{\widehat{\boldsymbol{\beta}}' \mathbf{x}_i}] = \mathbf{0},$$

from which $\widehat{\boldsymbol{\beta}}$ can be solved by using an algorithm mentioned in Chapter 1.

The ML estimation of the Poisson count model is robust to distributional misspecification as does also OLS in the linear regression model under normality. Given $\lambda_i = e^{\boldsymbol{\beta}' \mathbf{x}_i}$, the ML estimator $\widehat{\boldsymbol{\beta}}$ of the Poisson model is consistent even if the assumption of the Poisson process cannot be applied (Cameron and Trivedi (1998), Chapter 3). Since, however, the Poisson model implies the presence of heteroskedasticity in the regression model (because of the property of the Poisson model that its mean and variance are equal), the usual Poisson ML standard errors and t-statistics need to be adjusted.[4] A possible correction is $var\,(y_i | \mathbf{x}_i, \boldsymbol{\beta}) = \alpha \lambda_i = \alpha e^{\boldsymbol{\beta}' \mathbf{x}_i}$, which implies a generalization of the Poisson model if $\alpha \neq 1$. Data is called *overdispersed* if $\alpha > 1$ and *underdispersed* if $\alpha < 1$. Then the usual Poisson ML standard errors need to be multiplied by $\sqrt{\alpha}$ and t-statistics divided by $\sqrt{\alpha}$. An estimate of α is obtained after ML estimation of $\boldsymbol{\beta}$; usually $\widehat{\alpha} = (n-k)^{-1} \sum_{i=1}^{n} (y_i - \widehat{y}_i)^2 / \widehat{y}_i$ with $\widehat{y}_i = e^{\widehat{\boldsymbol{\beta}}' \mathbf{x}_i}$. It is essential that this correction be made for the Poisson regression model as it can make a much bigger difference than heteroskedasticity corrections for OLS. Often data is overdispersed and the uncorrected t-statistics are erroneously much larger than the correct(ed) t-statistics.

The Poisson model produces no equivalent of the standard R^2 goodness of fit measure for the linear regression. Greene (2003), p. 742, presents a measure based on the standardized residuals:

$$R^2 = 1 - \frac{\sum_{i=1}^{n} \left[\frac{y_i - \widehat{\lambda}_i}{\sqrt{\widehat{\lambda}_i}} \right]^2}{\sum_{i=1}^{n} \left[\frac{y_i - \bar{y}}{\sqrt{\bar{y}}} \right]^2},$$

where $\widehat{\lambda}_i = \exp(\widehat{\boldsymbol{\beta}}' \mathbf{x}_i)$ and $\bar{y} = \frac{1}{n} \sum_{i=1}^{n} y_i$. Although this measure has the advantage that it compares the fit of the count model with that provided by a model with only a constant term, it can be negative and it can fall when a variable is dropped from the model. To overcome this drawback, Cameron and Windmeijer (1993) propose a goodness of fit measure based on the *deviances* $d_i := 2 \left[y \ln \left(\frac{y_i}{\hat{\lambda}_i} \right) - (y_i - \hat{\lambda}_i) \right]$:

$$R^2 = 1 - \frac{\sum_{i=1}^{n} \left[y_i \ln \left(\frac{y_i}{\widehat{\lambda}_i} \right) - (y_i - \hat{\lambda}_i) \right]}{\sum_{i=1}^{n} \left[y_i \ln \left(\frac{y_i}{\bar{y}} \right) \right]}.$$

[4] This is similar to the OLS estimates that are consistent if the errors are heteroskedastic, but an adjustment has to be made to the standard errors then (GLS instead of OLS).

10.3.2 Negative binomial model

The *negative binomial regression* relaxes the very restrictive equality condition between conditional mean and conditional variance of the Poisson model (through which sometimes strong heteroskedasticity is present) by including individual heterogeneity, v_i, in the intensity rate $e^{\boldsymbol{\beta}'\mathbf{x}_i} = \lambda_i$ and assuming that v_i is gamma distributed. This assumption allows us to integrate out v_i from the composite pdf. The resulting negative binomial pdf is given by:

$$g\left(y_i|\mathbf{x}_i;\alpha,\boldsymbol{\beta}\right) = \frac{\Gamma(\alpha^{-1}+y_i)}{\Gamma(\alpha^{-1})\Gamma(1+y_i)}\left(\frac{\alpha^{-1}}{\alpha^{-1}+\lambda_i}\right)\left(\frac{\lambda_i}{\alpha^{-1}+\lambda_i}\right)^{y_i}$$

where $\Gamma(\cdot)$ denotes the gamma function. This model, which is particularly suited for overdispersed data, can be estimated by ML and reduces to the Poisson model if $\alpha = 0$. We can use a simple (t-)test on α to discriminate between the negative binomial and the Poisson regression model.

Although the negative binomial model is more efficient than the Poisson model, in practice the efficiency benefits are generally not so sizable. The negative binomial model should be used, however, if probabilities have to be forecast and not just the (conditional) mean has to be modeled. The negative binomial model cannot be estimated if data is underdispersed.

Another general count model is the *hurdle* model, which treats the process for zeros differently from that for the nonzero counts. In this case the (conditional) mean is no longer $e^{\boldsymbol{\beta}'\mathbf{x}_i}$, so that the Poisson count model is inconsistent in that case (and the hurdle model is consistent then; see Section 4.7 of Cameron and Trivedi (1998)).

10.3.3 Zero-inflated count models

An important property of count data is that the zero value is a natural outcome of this variable. For example, due to difficulties of the R&D process, technological and market uncertainty, competition and other factors, the firm can decide not to apply for a patent. Therefore, the zero value of the patent count has also its own informative meaning. The zero-inflated count models (such as in Lambert (1992)) take this property of many counts as patent data into account.

The *zero-inflated count model* combines two different model structures in the application to e.g. firms' patenting data. The first part is e.g. the logit model, which explains e.g. the decision to patent as opposite to the decision never to engage in such an activity:

$$P(Y_i = 0|\,\mathbf{x}_i;\boldsymbol{\beta}) = \phi_i - (1-\phi_i)e^{-\lambda_i}$$

where $\phi_i := \frac{\exp(\mathbf{z}_i'\boldsymbol{\beta})}{1+\exp(\mathbf{z}_i'\boldsymbol{\beta})}$ and $\lambda_i = e^{\boldsymbol{\beta}'\mathbf{x}_i}$, with $\mathbf{x}_i$ representing the set of explanatory variables and $\boldsymbol{\beta}$ being the vector of coefficients to be estimated, and $\mathbf{z}_i$ is the observable vector of covariances, ensuring the nonnegativity of ϕ_i.

The second part is a Poisson count data model, which explains the positive (patent) count outcomes:

$$P(Y_i = y_i | \mathbf{x}_i; \boldsymbol{\beta}) = (1 - \phi_i)\frac{\exp(-\lambda_i)\lambda_i^{y_i}}{y_i!}.$$

Now proceed with specification of the joint likelihood function for this model. Denote as $I_{Y_i=0}$ an indicator variable, which is equal to 1 if $Y_i = 0$, and zero elsewhere. The joint loglikelihood function for this model, apart from known constants, is

$$\begin{aligned} L(\boldsymbol{\beta}, \gamma) \quad \approx \quad & \textstyle\sum_{i=1}^{n} I_{Y_i=0} \ln(\mathbf{e}^{\boldsymbol{\beta}'\mathbf{z}_i} + \exp(-e^{\boldsymbol{\beta}'\mathbf{x}_i}) \\ & + \textstyle\sum_{i=1}^{n}(1 - I_{Y_i=0})(y_i\boldsymbol{\beta}'\mathbf{x}_i - e^{\boldsymbol{\beta}'\mathbf{x}_i}) - \textstyle\sum_{i=1}^{n} \ln(1 + \mathbf{e}^{\boldsymbol{\beta}'\mathbf{z}_i}). \end{aligned}$$

It is possible to consider a different specification for any valid distribution function F_i and consider a more general count model in place of the Poisson model (e.g. negative binomial or hurdle model, etc.; see Cameron and Trivedi (1998)).

10.4 Exercises

1. Compute marginal effects in the censored regression model and apply for a normal and a logistic distribution.

2. Consider the following observations from a censored normal distribution: 3.8396 7.2040 0.00000 0.00000 4.4132 8.0230 5.7971 7.0828 0.00000 0.80260 13.0670 4.3211 0.00000 8.6801 5.4571 0.00000 8.1021 0.00000 1.2526 5.6016.

   (a) Formulate the loglikelihood for a simple tobit model: $y_i^* = \mu + \varepsilon_i$, with $y_i = y_i^*$ if $\mu + \varepsilon_i > 0$ and zero otherwise, and with the error $\varepsilon_i \sim N(0, \sigma_\varepsilon^2)$.

   (b) Compute the ML estimates of μ and σ.

3. Using only the nonlimit observations, repeat Exercise 1 in the context of a censored regression model. Estimate μ and σ by using the method of moments estimator.

4. Show that the estimation of the sample selection model (Amemiya's (1984) *tobit II model*), where the dependent variable is only observed if another variable is of the binary choice type, boils down to an ML estimation of Heckman's λ_i. Comment the situation of correlated errors between the two models.

5. The Katz family of distributions is defined by the probability recursion $\frac{P(y+1)}{P(y)} = \frac{\mu+\gamma y}{1+y}$ for $y = 0, 1, 2, \ldots$ and $\mu + \gamma y \geqq 0$. Show that this yields overdispersed distributions for $0 < \gamma < 1$ and underdispersed distributions for $\gamma < 0$.

Part IV

Panel Data Analysis

Economic data can either arise in a time series or in a cross-section context. It is obvious that the behavior of cross sections (e.g. consumers, households, firms, regions) is not necessarily constant but can change over time. Moreover, it is also possible that behavior in each period of time differs across the cross sections. Databases that combine cross-sectional and time series data are called *panel data* or *longitudinal data.* Panel data econometrics is concerned with estimating and testing an econometric model taking into account differences in behavior between cross sections and differences in behavior in time for each cross section.

The analysis of panel data suggests that the individuals are heterogeneous; time series and cross-section studies that do not control for this heterogeneity run the risk of obtaining biased results.

In this introduction we will briefly summarize the main advantages and limitations of the use of panel data, but we will treat them in more detail later.

There are a number of advantages from using panel data. First, panel data are efficient when compared to random sampling of cross sections. With additional data on the same individuals, one can produce more reliable parameter estimates. Second, panel data reduces and sometimes even eliminates estimation bias. This is particularly the case for the estimation of aggregates. Consider, for example, the estimation of the income distribution that may vary over time. Such structural estimations using aggregate data can be accurately estimated by means of panel data, which allows for time varying parameters. Third, identification is easier, when compared to pure cross-section or pure time series data. Through the combination of both, time series data reflecting short run effects and cross-section databases emphasizing long run behavior, it is possible to identify the individual dynamics. Fourth, one of the characteristics of panel data is that it may alleviate the problem of multicollinearity, since when the explanatory variable varies in two dimensions it is less likely that they are highly correlated. Finally, panel data analysis can account for omitted variables due to individual and/or time effects.

According to the specification of the underlying model, this part will be subdivided in two chapters: linear panel data and nonlinear panel data.

Chapter 11

Linear Panel Data Models

The model to be studied in this chapter can be written for the ith cross section at period t ($i = 1, 2, \ldots N$ and $t = 1, 2, \ldots T$) in its most general form as:

$$\gamma_i(L)y_{it} = \mu_{it} + \boldsymbol{\beta}'_{it}\mathbf{x}_{it} + \varepsilon_{it}, \tag{11.1}$$

with y_{it} the observation of the ith individual (country, industry, town, etc.) at period t, ε_{it} the corresponding error term, $\mathbf{x}_{it}$ the corresponding vector of explanatory variables, $\boldsymbol{\beta}_{it}$ the vector of unknown parameters corresponding to this vector of explanatory variables, $\mu_{it} := \mu_i + \lambda_t + \zeta_{it}$ unknown parameters that may depend on individuals, time or may be interactive (depending both on time and individuals) and, finally, $\gamma_i(L) := I - \sum_{\tau=1}^{p} \gamma_{i\tau} L^{\tau}$ a pth order autoregressive lag polynomial with unknown parameters $\gamma_{i\tau}$ with L the lag operator and I the identity operator.

In the first instance static linear panel data will be studied with $\gamma_i(L) := I$. Dynamic linear panel data models are treated in a separate section later in this chapter.

From (11.1) it is clear that panel data characteristics can be attributed to three different types of effects:

- *Individual-specific* and time-invariant effects μ_i which are the same for a given cross section; these effects μ_i represent the (direct) impact on the endogenous variable y_i of all the time-invariant nonobservable characteristics of the individual;

- *Time-specific*, individual-invariant effects λ_t, that are the same for all cross sections at a given point in time but vary with time; this implies that each time period has a specific nonobservable characteristic;

- *Individual-time* varying effects that are deterministic (ζ_{it}) or stochastic (ε_{it}); these effects vary across cross sections and through time -i.e. they may vary with both i and t.

However, next to the advantages of panel data set out in the introduction to this part of the book, it is also important to keep in mind the limitations of panel data. First, we have to restrict the basic model (11.1) so that a sufficient number of degrees of freedom remain. Second, there are problems of data collection and design of a panel questionnaire. Among these problems, nonresponse, problems of recall, that is, when the respondent does not remember exactly, and of coverage are the most salient ones. Third, measurement errors can arise due to unclear questions, problems of incorrect recall, inappropriate informants, deliberate distortions of answers, misrecording and interviewer effects. Four, there are selectivity problems. Self selectivity, item non response and attrition problems are to be mentioned. Finally, short time span is a characteristic of most of the panels. Hence, the asymptotic properties depend crucially on the number of individuals tending to infinity. This causes some difficulties, in particular for the case of limited dependent variables (the latter models, however, are examples of nonlinear panel data models and will be discussed in the following chapter).

From (11.1), the most general static linear panel data model can be written as:

$$y_{it} = \beta_{1it} + \sum_{k=2}^{K} \beta_{kit} x_{kit} + \varepsilon_{it},$$

where $i = 1, 2, ..., N$ refers to the cross-section data points and $t = 1, 2, ..., T$ to the time periods. Further, it is mostly assumed that $E(\varepsilon_{it}) = 0 \ \forall i, t$ and $E(\varepsilon_{it}^2) = \sigma_\varepsilon^2$ (but autocorrelation and heteroskedasticity will be treated in the beginning and could be applied, *sui generis*, on each panel data model). In the most general case, the parameters $\beta_{1it}, ..., \beta_{kit}$ differ across cross sections and time. The most common special cases are the following.

1. All coefficients are assumed to be constant across cross sections and in time such that the error terms involve all differences (*unobserved heterogeneity*):

$$y_{it} = \beta_1 + \sum_{k=2}^{K} \beta_k x_{kit} + \varepsilon_{it}.$$

2. The constant term is assumed to be cross-section-specific and the other (slope) coefficients are assumed to be constant across cross sections and in time:

$$y_{it} = \beta_{1i} + \sum_{k=2}^{K} \beta_k x_{kit} + \varepsilon_{it}.$$

3. The constant term is assumed to be cross-section- and time-specific and the other (slope) coefficients are assumed to be constant across cross sections and in time:

$$y_{it} = \beta_{1it} + \sum_{k=2}^{K} \beta_k x_{kit} + \varepsilon_{it}.$$

4. All coefficients are cross-section-specific:

$$y_{it} = \beta_{1i} + \sum_{k=2}^{K} \beta_{ki} x_{kit} + \varepsilon_{it}.$$

5. All coefficients are cross-section- and time-specific:

$$y_{it} = \beta_{1it} + \sum_{k=2}^{K} \beta_{kit} x_{kit} + \varepsilon_{it}$$

In the sequel the above mentioned linear panel data models will be analyzed. Also dynamic linear panel data models will be treated and the issues of incomplete panels and attrition will be raised.

11.1 Panel data models with constant coefficients

In the linear panel data model with constant coefficients, it is assumed that all cross-sectional equations have the same intercept and the same slope coefficients. The model is $y_{it} = \beta_1 + \sum_{k=2}^{K} \beta_k x_{kit} + \varepsilon_{it}$,or $\mathbf{y}_i = \mathbf{X}_i \boldsymbol{\beta} + \boldsymbol{\varepsilon}_i$ for individual i with $i = 1, 2, ..., N$, where $\mathbf{y}_i := (y_{i1}, y_{i2}, \ldots, y_{iT})'$, $\boldsymbol{\beta} := (\beta_1, \beta_2, \ldots, \beta_K)'$, $\boldsymbol{\varepsilon}_i := (\varepsilon_{i1}, \varepsilon_{i2}, \ldots, \varepsilon_{iT})'$, and $\mathbf{X}_i := (\mathbf{x}_{i1}, \mathbf{x}_{i2}, \ldots, \mathbf{x}_{iT})'$. Note that each submatrix or vector of individual i has T observations.Since the complete variance-covariance matrix of the residuals can be represented as:

$$\boldsymbol{\Omega} = E[\boldsymbol{\varepsilon}\boldsymbol{\varepsilon}'] = \begin{pmatrix} \sigma_{11}\boldsymbol{\Omega}_{11} & \sigma_{12}\boldsymbol{\Omega}_{12} & \cdots & \sigma_{1N}\boldsymbol{\Omega}_{1N} \\ \sigma_{21}\boldsymbol{\Omega}_{21} & \sigma_{22}\boldsymbol{\Omega}_{22} & \cdots & \sigma_{2N}\boldsymbol{\Omega}_{2N} \\ & \vdots & & \\ \sigma_{N1}\boldsymbol{\Omega}_{N1} & \sigma_{N2}\boldsymbol{\Omega}_{N2} & \cdots & \sigma_{NN}\boldsymbol{\Omega}_{NN} \end{pmatrix},$$

where each element $\sigma_{ij}\boldsymbol{\Omega}_{ij}$ is a T by T matrix, the wole system for all individuals together can be written as $\mathbf{y} = \mathbf{X}\boldsymbol{\beta} + \boldsymbol{\varepsilon}$, where $\mathbf{y} := (\mathbf{y}_1', \mathbf{y}_2', \ldots, \mathbf{y}_N')'$, $\boldsymbol{\varepsilon} := (\boldsymbol{\varepsilon}_1', \boldsymbol{\varepsilon}_2', \ldots, \boldsymbol{\varepsilon}_N')'$, and $\mathbf{X} := (\mathbf{X}_1, \mathbf{X}_2, \ldots, \mathbf{X}_N)'$.

If some restrictive assumptions are made, the above model can be estimated by pooled OLS. First, the variance of the disturbance term should be constant and thus not be different across cross sections (cross-sectional heteroskedasticity is not allowed then): $\sigma_{ii} = \sigma_{jj}$ for $i \neq j$. Second, the disturbance terms may not be correlated across the cross sections (no cross-sectional or contemporaneous correlation assumed): $\sigma_{ij} = \sigma_{ji} = 0$ for $i \neq j$. Third, no autocorrelation may exist: $\boldsymbol{\Omega}_{ii} = I_T$ (no autocorrelation assumed here). As such, $\boldsymbol{\Omega}$ will have the familiar structure with diagonal elements $\sigma_{ii}\mathbf{I}_T$ and off-diagonal elements being zero. If all these three restrictive assumptions are satisfied simultaneously the panel structure is actually ignored and estimation is similar to estimation of a simple linear model without panel structure.

A more general model would incorporate the possibility of variances differing across cross sections or individuals (i.e. cross-sectional heteroskedasticity), of correlation of the disturbance terms across the cross sections (i.e. cross-sectional or contemporaneous correlation) and of autocorrelation for each cross section. In this case the variance-covariance matrix of the residuals would have the general structure. For the first order autoregressive residual process $AR(1)$: $\varepsilon_{it} = \rho_i \varepsilon_{i,t-1} + v_{it}$ with $\varepsilon_{i0} = v_{i0} / \left(1 - \rho_i^2\right)^{1/2}$, where $E(v_{it}) = 0$ and $E(v_{it}v_{js}) = \delta_{ts}\sigma_{ij}$ ($\delta_{ts} = 1$ for $t = s$ and $\delta_{ts} = 0$ for $t \neq s$) the element in the ith row and the jth

column of the residual variance-covariance matrix $\mathbf{\Omega}$ would be from subsequent substitution in this $AR(1)$ error model:

$$\sigma_{ij}\mathbf{\Omega}_{ij} = \frac{\sigma_{ij}}{1-\rho_i\rho_j}\begin{pmatrix} 1 & \rho_j & \rho_j^2 & \cdots & \rho_j^{T-1} \\ \rho_i & 1 & \rho_j & \cdots & \rho_j^{T-2} \\ \rho_i^2 & \rho_i & 1 & \cdots & \rho_j^{T-3} \\ & & \vdots & & \\ \rho_i^{T-1} & \rho_i^{T-2} & \rho_i^{T-3} & \cdots & 1 \end{pmatrix}.$$

Obviously, the best linear unbiased estimator (BLUE) of this model is the following GLS estimator $\hat{\boldsymbol{\beta}} = (\mathbf{X}'\mathbf{\Omega}^{-1}\mathbf{X})^{-1}\mathbf{X}'\mathbf{\Omega}^{-1}\mathbf{y}$ with variance-covariance matrix $(\mathbf{X}'\mathbf{\Omega}^{-1}\mathbf{X})^{-1}$.

As is well-known from linear estimation theory initial estimates of σ_{ij} and ρ_i $(i, j = 1, 2, ..., N)$ are required. These initial estimates are obtained from the LS estimators $\mathbf{b}$ of $\boldsymbol{\beta}$ and $\mathbf{e}_i$ of $\boldsymbol{\varepsilon}_i$: $\hat{\sigma}_{ij} = \mathbf{e}_i'\mathbf{e}_j/(T-K)$ (or divided by T instead of $(T-K)$ for ML) and $\hat{\rho}_i = \sum_{t=2}^{T} e_{it}e_{i,t-1}/\sum_{t=2}^{T} e_{i,t-1}^2$. The resulting estimators are in general not unbiased but consistent in T: increasing T increases the information in the sample, while increasing N increases the number of variance and covariance parameters to be estimated. To compute the *feasible GLS* (FGLS) or the *estimated GLS* (EGLS) estimator for this model, we require the full set of sample moments $\mathbf{y}_i'\mathbf{y}_j$, $\mathbf{X}_i'\mathbf{X}_j$, and $\mathbf{X}_i'\mathbf{y}_j$ for all pairs of cross-sectional units. Hence, the FGLS estimator of $\boldsymbol{\beta}$ satisfies:

$$\hat{\hat{\boldsymbol{\beta}}} = (\mathbf{X}'\hat{\mathbf{\Omega}}^{-1}\mathbf{X})^{-1}\mathbf{X}'\hat{\mathbf{\Omega}}^{-1}\mathbf{y}.$$

Remark that if there is no autocorrelation (and no time-related heteroskedasticity), $\hat{\mathbf{\Omega}}^{-1} = \hat{\mathbf{\Sigma}}^{-1} \otimes \mathbf{I}_T$ with $\hat{\mathbf{\Sigma}} = \frac{1}{(T-K)}\sum_{t=1}^{T} \mathbf{e}_t\mathbf{e}_t'$, where $\mathbf{e}_t$ is an N vector containing all N residuals for the N individuals at period t. This estimation technique can only be used when T is larger than N, i.e. when the number of time periods in the database is larger than the number of cross sections or individuals. If $T < N$ the residual variance-covariance matrix $\hat{\mathbf{\Sigma}}$ is an NxN matrix of rank T and is thus singular such that it cannot be inverted. The population matrix $\mathbf{\Omega}$, and hence Σ, is assumed to be positive definite, but, if there are not enough observations the database is too short to obtain a positive definite estimate of this covariance matrix in this (restrictive) non-autocorrelated and homoskedastic case. If residual heteroskedasticity and/or autocorrelation occurs the corresponding constant panel data model estimators can (almost) always be computed, however.

11.2 Intercepts varying over individuals

In (linear) panel data models with intercepts varying over individuals it is assumed that all slope coefficients are constant. This implies that the reaction of the dependent variable to a change in one of the independent variables is assumed to be the same across cross sections and in time. Individual heterogeneity is captured in the intercept only. The model is generally presented as follows:

$$y_{it} = \underbrace{\overline{\beta}_1 + \mu_i}_{\beta_{1i}} + \textstyle\sum_{k=2}^{K} \beta_k x_{kit} + \varepsilon_{it} \quad (i = 1, ..., N; t = 1, ..., T), \tag{11.2}$$

where $\overline{\beta}_1$ represents the mean intercept and μ_i the deviation of the intercept of the ith individual from the mean.

In what follows, a distinction is made with respect to whether the individual effects μ_i are either fixed or random.

11.2.1 μ_i fixed

An assumption of the fixed effects model is that differences across cross sections can be captured in differences in the constant term. As such, μ_i is constant. In vector notation this yields the following model for the ith unit or cross section $\mathbf{y}_i = (\overline{\beta}_1 + \mu_i)\,\boldsymbol{\iota}_T + \mathbf{X}_{si}\boldsymbol{\beta}_s + \boldsymbol{\varepsilon}_i$ with $i = 1, 2, ..., N$, where $\mathbf{y}_i := (y_{i1} \dots y_{iT})'$, $\boldsymbol{\varepsilon}_i := (\varepsilon_{i1}, \dots, \varepsilon_{iT})'$, $\boldsymbol{\iota}_T := (1, \dots, 1)'$, $\boldsymbol{\beta}_s := (\beta_2, \dots, \beta_K)'$.and $\mathbf{X}_{si} := \begin{pmatrix} x_{2i1} & \cdots & x_{Ki1} \\ \vdots & \ddots & \vdots \\ x_{2iT} & \cdots & x_{KiT} \end{pmatrix}$.

The matrix $\mathbf{X}_{si}$ is a $T \times (K-1)$ matrix of all explanatory (or right-hand-side) variables for individual i except for the constant term. It is assumed that $E(\boldsymbol{\varepsilon}_i) = 0$ and $E\left(\boldsymbol{\varepsilon}_i \boldsymbol{\varepsilon}_j'\right) = \delta_{ij}\sigma_\varepsilon^2 \mathbf{I}_T$ where $\delta_{ij} = 1$ for $i = j$ and $\delta_{ij} = 0$ for $i \neq j$. $\overline{\beta}_1$ represents the average intercept (averaged over the cross sections) such that $\overline{\beta}_1 = \frac{\sum_{i=1}^{N} \beta_{1i}}{N}$. It is further assumed that all individual-specific constant terms sum to zero such that $\sum_{i=1}^{N} \mu_i = 0$.

The appropriate estimation method of this fixed effects model is OLS, which is also the BLUE. This becomes clear when considering all NT observations:

$$\begin{pmatrix} \mathbf{y}_1 \\ \mathbf{y}_2 \\ \vdots \\ \vdots \\ \mathbf{y}_N \end{pmatrix} = \begin{pmatrix} \boldsymbol{\iota}_T & 0 & \cdots & \cdots & 0 & \mathbf{X}_{s1} \\ 0 & \boldsymbol{\iota}_T & & & \vdots & \mathbf{X}_{s2} \\ \vdots & & \ddots & & \vdots & \vdots \\ \vdots & & & \ddots & 0 & \vdots \\ 0 & \cdots & \cdots & 0 & \boldsymbol{\iota}_T & \mathbf{X}_{sN} \end{pmatrix} \begin{pmatrix} \beta_{11} \\ \beta_{12} \\ \vdots \\ \beta_{1N} \\ \boldsymbol{\beta}_s \end{pmatrix} + \begin{pmatrix} \boldsymbol{\varepsilon}_1 \\ \boldsymbol{\varepsilon}_2 \\ \vdots \\ \vdots \\ \boldsymbol{\varepsilon}_N \end{pmatrix},$$

or rewritten in compact form: $\mathbf{y} = [\mathbf{I}_N \otimes \boldsymbol{\iota}_T, \mathbf{X}_s] \binom{\boldsymbol{\beta}_1}{\boldsymbol{\beta}_s} + \boldsymbol{\varepsilon}$, with $\otimes$ the Kronecker product, so that the OLS estimator is represented by the BLUE:

$$\begin{pmatrix} \hat{\boldsymbol{\beta}}_1 \\ \hat{\boldsymbol{\beta}}_s \end{pmatrix} = \begin{pmatrix} T\mathbf{I}_N & (\mathbf{I}_N \otimes \boldsymbol{\iota}_T)' \mathbf{X}_s \\ \mathbf{X}_s' (\mathbf{I}_N \otimes \boldsymbol{\iota}_T) & \mathbf{X}_s' \mathbf{X}_s \end{pmatrix}^{-1} \begin{pmatrix} (\mathbf{I}_N \otimes \boldsymbol{\iota}_T)' \mathbf{y} \\ \mathbf{X}_s' \mathbf{y} \end{pmatrix},$$

where $T\mathbf{I}_N$ results from $(\mathbf{I}_N \otimes \boldsymbol{\iota}_T)' (\mathbf{I}_N \otimes \boldsymbol{\iota}_T) = \mathbf{I}_N \otimes \boldsymbol{\iota}_T' \boldsymbol{\iota}_T = \mathbf{I}_N \otimes T = T\mathbf{I}_N$.

Notice that the fixed effects model is often referred to as the *dummy variables model* since the above model for all NT observations can be rewritten as $\mathbf{y} = [\mathbf{d}_1, \mathbf{d}_2, ..., \mathbf{d}_N, \mathbf{X}_s] \binom{\boldsymbol{\beta}_1}{\boldsymbol{\beta}_s} + \boldsymbol{\varepsilon}$, with $\mathbf{d}_i$ $(i = 1, 2, ..., N)$ a dummy variable for the ith individual. Hence $\mathbf{y} = \mathbf{D}\boldsymbol{\beta}_1 + \mathbf{X}_s\boldsymbol{\beta}_s + \boldsymbol{\varepsilon}$ with $\mathbf{D} := (\mathbf{d}_1, \mathbf{d}_2, ..., \mathbf{d}_N)$ an $NT \times N$ matrix of dummy variables. The above mentioned OLS estimator of $\boldsymbol{\beta}_s$ can then be rewritten as $\hat{\boldsymbol{\beta}}_s = \left(\mathbf{X}_s'\mathbf{M}_D\mathbf{X}_s\right)^{-1}\left(\mathbf{X}_s'\mathbf{M}_D\mathbf{y}\right)$ with the corresponding idempotent matrix defined as

$$\mathbf{M}_D := \mathbf{I}_{NT} - \mathbf{D}(\mathbf{D}'\mathbf{D})^{-1}\mathbf{D}' = \begin{pmatrix} \mathbf{I}_T - \frac{\boldsymbol{\iota}_T\boldsymbol{\iota}_T'}{T} & 0 & \cdots & 0 \\ 0 & \mathbf{I}_T - \frac{\boldsymbol{\iota}_T\boldsymbol{\iota}_T'}{T} & \cdots & 0 \\ \vdots & & \ddots & \vdots \\ 0 & \cdots & \cdots & \mathbf{I}_T - \frac{\boldsymbol{\iota}_T\boldsymbol{\iota}_T'}{T} \end{pmatrix},$$

while the dummy variable coefficient vector $\boldsymbol{\beta}_1$ is estimated as $\hat{\boldsymbol{\beta}}_1 = (\mathbf{D}'\mathbf{D})^{-1}\mathbf{D}'(\mathbf{y} - \mathbf{X}_s\hat{\boldsymbol{\beta}}_s)$.

To calculate the above *Least Squares Dummy Variables* ($LSDV$) or *Fixed Effects* (FE) - estimator(s), the inverse of an $(N + K - 1) \times (N + K - 1)$ matrix or an $N \times N$ matrix has to be taken, which can create problems when N is very large. A more efficient method consists of formulating the model in terms of deviations from means. Consider therefore the following symmetric and idempotent matrix:

$$\begin{aligned} \mathbf{D}_T \quad : &= \mathbf{I}_T - \boldsymbol{\iota}_T \left(\boldsymbol{\iota}_T \boldsymbol{\iota}_T\right)^{-1} \boldsymbol{\iota}_T \\ &= \mathbf{I}_T - \frac{\boldsymbol{\iota}_T \boldsymbol{\iota}_T}{T} = \mathbf{I}_T - \begin{pmatrix} 1/T & \cdots & 1/T \\ \vdots & & \vdots \\ 1/T & \cdots & 1/T \end{pmatrix}, \end{aligned}$$

which is sometimes called the *annihilator* associated with the unit vector $\boldsymbol{\iota}_T$. Hence, what this matrix does is to extract deviations from group means. For example, multiplying the T dimensional vector $\mathbf{y}_i$ from the left by $\mathbf{D}_T$ results in:

$$\mathbf{D}_T\mathbf{y}_i = \begin{pmatrix} y_{i1} - \overline{y}_{i.} \\ \vdots \\ y_{iT} - \overline{y}_{i.} \end{pmatrix} = \mathbf{y}_i - \boldsymbol{\iota}_T\overline{y}_{i.},$$

where $\bar{y}_{i.} = \frac{1}{T}\boldsymbol{\iota}_T\mathbf{y}_i = \frac{1}{T}\sum_{t=1}^{T} y_{it}$ is the group mean for the dependent variable. The model in deviations from the (sample) group means becomes:

$$\underbrace{\mathbf{D}_T\mathbf{y}_i}_{=\mathbf{y}_i - \overline{y}_{i.}} = \underbrace{\mathbf{D}_T\mathbf{X}_{si}}_{=\mathbf{X}_{si} - \overline{X}_{si.}} \boldsymbol{\beta}_s + \mathbf{D}_T\boldsymbol{\varepsilon}_i$$

or per element:

$$y_{it} - \overline{y}_{i.} = \sum_{k=2}^{K} \beta_k \left(x_{kit} - \overline{x}_{ki.}\right) + \varepsilon_{it} - \overline{\varepsilon}_{i.}. \tag{11.3}$$

In matrix form this yields the following for all i:

$$(\mathbf{I}_N \otimes \mathbf{D}_T)\,\mathbf{y} = (\mathbf{I}_N \otimes \mathbf{D}_T)\,\mathbf{X}_s\boldsymbol{\beta}_s + (\mathbf{I}_N \otimes \mathbf{D}_T)\,\varepsilon.$$

The OLS estimator which is also the BLUE if the residual error terms ε_{it} are WN (*iid* across individuals and time with zero expectation and constant variance σ_ε^2), is:

$$\begin{aligned}\hat{\boldsymbol{\beta}}_{sFE} &= \left(\mathbf{X}_s^{'}\,(\mathbf{I}_N \otimes \mathbf{D}_T)\,\mathbf{X}_s\right)^{-1} \mathbf{X}_s^{'}\,(\mathbf{I}_N \otimes \mathbf{D}_T)\,\mathbf{y} \\ &= \left(\textstyle\sum_{i=1}^N \mathbf{X}_{si}^{'}\mathbf{D}_T\mathbf{X}_{si}\right)^{-1} \textstyle\sum_{i=1}^N \mathbf{X}_{si}^{'}\mathbf{D}_T\mathbf{y}_i \qquad (11.4) \\ &= \left(\textstyle\sum_{i=1}^N \sum_{t=1}^T (\mathbf{x}_{it} - \overline{\mathbf{x}}_{i.})\,(\mathbf{x}_{it} - \overline{\mathbf{x}}_{i.})'\right)^{-1} \textstyle\sum_{i=1}^N \sum_{t=1}^T (\mathbf{x}_{it} - \overline{\mathbf{x}}_{i.})\,(y_{it} - \bar{y}_{i.})\,,\end{aligned}$$

with the variance-covariance matrix being:

$$\begin{aligned} var\left(\hat{\boldsymbol{\beta}}_{sFE}\right) &= \sigma_\varepsilon^2 \left(\mathbf{X}_s^{'}\,(\mathbf{I}_N \otimes \mathbf{D}_T)\,\mathbf{X}_s\right)^{-1} \\ &= \sigma_\varepsilon^2 \left(\textstyle\sum_{i=1}^N \sum_{t=1}^T (\mathbf{x}_{it} - \overline{\mathbf{x}}_{i.})\,(\mathbf{x}_{it} - \overline{\mathbf{x}}_{i.})'\right)^{-1}. \qquad (11.5)\end{aligned}$$

Note that to calculate the above OLS estimator, an inverse has to be taken of a $(K-1)$ x $(K-1)$ matrix. As soon as $\hat{\boldsymbol{\beta}}_{sFE}$ is known the intercepts can directly be calculated as follows:

$$\begin{aligned}\hat{\beta}_{1i,FE} &= \bar{y}_{i.} - \textstyle\sum_{k=2}^K \hat{\beta}_{k,FE}\bar{x}_{ki}. \\ \hat{\mu}_{i,FE} &= \hat{\beta}_{1i,FE} - \frac{\sum_{j=1}^N \hat{\beta}_{1j,FE}}{N}.\end{aligned}$$

Because the FE estimator is based on deviations from group means it is also known as the *within estimator* or the *covariance estimator.*

Unless T is large, using the standard OLS estimate for residual variance σ_ε^2 in the variance-covariance matrix (11.5) based upon the within regression in (11.3) will underestimate the true residual variance. The reason is that in this transformed regression the error variance-covariance matrix is singular (as the T transformed errors of each individual add up to zero) and the variance of $\varepsilon_{it} - \overline{\varepsilon}_{i.}$ is $\frac{(T-1)\sigma_\varepsilon^2}{T}$ rather than σ_ε^2. A consistent estimator for σ_ε^2 is obtained as the within residual sum of squares divided by $N(T-1)$, or:

$$\hat{\sigma}_{\varepsilon FE}^2 = \frac{1}{N\,(T-1) - K + 1} \textstyle\sum_{i=1}^N \sum_{t=1}^T \left(y_{it} - \mathbf{x}_{it}'\hat{\boldsymbol{\beta}}_{FE}\right)^2, \qquad (11.6)$$

with $\hat{\boldsymbol{\beta}}_{FE} := \left(\hat{\beta}_{11,FE}, \hat{\beta}_{12,FE}, \cdots, \hat{\beta}_{1N,FE}, \hat{\boldsymbol{\beta}}_{sFE}^{'}\right)'$.[1]

[1] Note that under weak regularity conditions, the FE estimator is asymptotically normal, so that the usual statistical tests from Chapter 3 can be applied.

Essentially, the *FE* model concentrates on differences 'within' individuals, i.e. it explains to what extent y_{it} differs from $\overline{y}_{i.}$ and does not explain why $\overline{y}_{i.}$ is different from $\overline{y}_{j.}$. If, however, interest is in differences across groups, the hypothesis that all individual effects are equal can be tested with a classical F-test. Since under the null hypothesis of equality of all individual effects the efficient estimator is pooled least squares, the F-test statistic for this test about the joint significance of the individual effects ($H_0 : \beta_{11} = \beta_{12} = ... = \beta_{1N} = \overline{\beta}_1$) amounts:

$$F_{N-1,NT-N-K+1} = \frac{\left(R^2_{FE} - R^2_{Pooled}\right)/(N-1)}{\left(1 - R^2_{FE}\right)/(NT - N - K + 1)}, \tag{11.7}$$

where FE indicates the estimated fixed effects or the dummy variables, i.e. the unrestricted, model and $Pooled$ indicates the estimated pooled or restricted model with only a single overall constant term.

11.2.2 μ_i random

Contrary to the previous section, we now assume that the individual effects are not constant but random. First, it is assumed that the individual random effects are distributed with zero mean and constant variance: $E(\mu_i) = 0$ and $E\left(\mu_i^2\right) = \sigma_\mu^2$. Further, they are assumed not to be correlated across individuals (cross sections); $E\left(\mu_i\mu_j\right) = 0$ $(i \neq j)$, nor with the residuals of the model, i.e. $E\left(\mu_i\varepsilon_{jt}\right) = 0,\ \forall i, j, t$. This can generally be summarized as $\mu_i \overset{iid}{\sim} \left(0, \sigma_\mu^2\right)$ or $\mu_i \in RW$ with variance σ_μ^2. For individual i we then have the following model:

$$\mathbf{y}_i = \mathbf{X}_i\boldsymbol{\beta} + \underbrace{\mu_i\boldsymbol{\iota}_T + \boldsymbol{\varepsilon}_i}_{\text{error components } \boldsymbol{\nu}_i},$$

with $\boldsymbol{\beta} := \left(\overline{\beta}_1, \beta_2, ..., \beta_K\right)'$, where $\overline{\beta}_1$ is the mean intercept in (11.2). The expected value of the above error components term is zero, $E(\boldsymbol{\nu}_i) = E(\mu_i\boldsymbol{\iota}_T + \boldsymbol{\varepsilon}_i) = 0$, and its variance-covariance matrix is represented by:

$$\begin{aligned}\boldsymbol{\Phi}_i &= E\left(\boldsymbol{\nu}_i\boldsymbol{\nu}_i'\right) = E\left(\left(\mu_i\boldsymbol{\iota}_T + \boldsymbol{\varepsilon}_i\right)\left(\mu_i\boldsymbol{\iota}_T + \boldsymbol{\varepsilon}_i\right)'\right) \\ &= \sigma_\mu^2\boldsymbol{\iota}_T\boldsymbol{\iota}_T' + \sigma_\varepsilon^2\mathbf{I}_T \\ &= \begin{pmatrix} \sigma_\mu^2 + \sigma_\varepsilon^2 & \sigma_\mu^2 & \cdots & \sigma_\mu^2 \\ \sigma_\mu^2 & \sigma_\mu^2 + \sigma_\varepsilon^2 & \cdots & \sigma_\mu^2 \\ \vdots & \vdots & \ddots & \vdots \\ \sigma_\mu^2 & \sigma_\mu^2 & \cdots & \sigma_\mu^2 + \sigma_\varepsilon^2 \end{pmatrix}.\end{aligned} \tag{11.8}$$

In matrix notation the error components model is:

$$\mathbf{y} = \mathbf{X}\boldsymbol{\beta} + \underbrace{\boldsymbol{\mu} \otimes \boldsymbol{\iota}_T + \boldsymbol{\varepsilon}}_{\boldsymbol{\nu}}, \tag{11.9}$$

with $\mathbf{y} := \left(\mathbf{y}_1', \ldots, \mathbf{y}_N'\right)'$, $\mathbf{X} := \left(\mathbf{X}_1', \ldots, \mathbf{X}_N'\right)'$, $\boldsymbol{\mu} := \left(\mu_1, \ldots \mu_N\right)'$, $\boldsymbol{\varepsilon} := \left(\boldsymbol{\varepsilon}_1', \ldots, \boldsymbol{\varepsilon}_N'\right)$.

The error components variance-covariance matrix is the following block diagonal matrix:

$$\mathbf{\Phi} = E\left((\boldsymbol{\mu}\otimes\boldsymbol{\iota}_T+\boldsymbol{\varepsilon})(\boldsymbol{\mu}\otimes\boldsymbol{\iota}_T+\boldsymbol{\varepsilon})'\right) = \mathbf{I}_N\otimes\mathbf{\Phi}_i.$$

If σ_μ^2 and σ_ε^2 are known, the GLS estimator of $\boldsymbol{\beta}$ has the BLUE property and is given by the following expression:

$$\begin{aligned}\hat{\beta}_{RE} &= \left(\mathbf{X}'\mathbf{\Phi}^{-1}\mathbf{X}\right)^{-1}\mathbf{X}'\mathbf{\Phi}^{-1}\mathbf{y} \\ &= \left(\sum_{i=1}^N \mathbf{X}_i'\mathbf{\Phi}_i^{-1}\mathbf{X}_i\right)^{-1}\left(\sum_{i=1}^N \mathbf{X}_i'\mathbf{\Phi}_i^{-1}\mathbf{y}_i\right). \qquad (11.10)\end{aligned}$$

Partitioning this estimator as $\hat{\boldsymbol{\beta}}_{RE}' := \left(\hat{\overline{\beta}}_{1RE}, \hat{\boldsymbol{\beta}}_{sRE}'\right)$, it can be shown that $\hat{\boldsymbol{\beta}}_{sRE}$ is a matrix weighted average of the *within* (or *FE*) estimator $\hat{\boldsymbol{\beta}}_{sFE}$ and the *between* estimator. To demonstrate this, note that $\mathbf{\Phi}^{-1} = \mathbf{I}_N\otimes\mathbf{\Phi}_i^{-1} = \mathbf{I}_N\otimes\left(\frac{\boldsymbol{\iota}_T\boldsymbol{\iota}_T'}{T\sigma_1^2}+\frac{\mathbf{D}_T}{\sigma_\varepsilon^2}\right)$ with $\sigma_1^2 := T\sigma_\mu^2+\sigma_\varepsilon^2$. Substituting $\mathbf{\Phi}^{-1}$, the partitions $\mathbf{X} := (\boldsymbol{\iota}_{NT}, \mathbf{X}_s)$ and $\mathbf{X}_i := (\boldsymbol{\iota}_T, \mathbf{X}_{si})$ into (11.10):

$$\hat{\boldsymbol{\beta}}_{sRE} = \left[\frac{\mathbf{X}_s'\mathbf{Q}_1\mathbf{X}_s}{\sigma_1^2}+\frac{\sum_{i=1}^N\mathbf{X}_{si}'\mathbf{D}_T\mathbf{X}_{si}}{\sigma_\varepsilon^2}\right]^{-1}\left[\frac{\mathbf{X}_s'\mathbf{Q}_1\mathbf{y}}{\sigma_1^2}+\frac{\sum_{i=1}^N\mathbf{X}_{si}'\mathbf{D}_T\mathbf{y}_i}{\sigma_\varepsilon^2}\right], \qquad (11.11)$$

where the idempotent and symmetric matrix $\mathbf{Q}_1 := \left(I_N\otimes\frac{\boldsymbol{\iota}_T\boldsymbol{\iota}_T}{T}\right)-\frac{\boldsymbol{\iota}_{NT}\boldsymbol{\iota}_{NT}'}{NT}$, such that:

$$\mathbf{Q}_1\mathbf{X}_s = \begin{pmatrix} \overline{x}_{21.}-\overline{x}_{2..} & \cdots & \overline{x}_{K1.}-\overline{x}_{K..} \\ \overline{x}_{22.}-\overline{x}_{2..} & \cdots & \overline{x}_{K2.}-\overline{x}_{K..} \\ \vdots & & \vdots \\ \overline{x}_{2N.}-\overline{x}_{2..} & \cdots & \overline{x}_{KN.}-\overline{x}_{K..} \end{pmatrix}\otimes\boldsymbol{\iota}_T,$$

with $\overline{x}_{k..} := \sum_{i=1}^N\sum_{t=1}^T x_{kit}/NT$.[2] Then, the between estimator is given by:

$$\widetilde{\boldsymbol{\beta}}_s = \left(\mathbf{X}_s'\mathbf{Q}_1\mathbf{X}_s\right)^{-1}\mathbf{X}_s'\mathbf{Q}_1\mathbf{y}, \qquad (11.12)$$

which only utilizes the variation between individuals. Now, the random effects estimator of the slope vector $\hat{\boldsymbol{\beta}}_{sRE}$ can be rewritten as a matrix weighted average

[2] Hence, $\mathbf{Q}_1$, when multiplied by $\mathbf{X}$ or $\mathbf{y}$, has the effect of calculating the means over time for each individual ($\overline{x}_{ki.}$), expressing the individual means in terms of deviations from the overall mean, and repeating each of these N observations T times. Consequently, the between estimator implies an OLS regression on the basic model (11.2) averaged over time:

$$\overline{y}_{i.} = \overline{\beta}_1+\mu_i+\sum_{k=2}^K\beta_k\overline{x}_{ki.}+\overline{\varepsilon}_{i.} \quad (i=1,...,N).$$

of the within estimator $\hat{\boldsymbol{\beta}}_{sFE}$ and the between estimator $\widetilde{\boldsymbol{\beta}}_s$:

$$\hat{\boldsymbol{\beta}}_{sRE}=\left[\frac{\mathbf{X}_s'\mathbf{Q}_1\mathbf{X}_s}{\sigma_1^2}+\frac{\sum_{i=1}^{N}\mathbf{X}_{si}'\mathbf{D}_T\mathbf{X}_{si}}{\sigma_\varepsilon^2}\right]^{-1}\left[\left(\frac{\mathbf{X}_s'\mathbf{Q}_1\mathbf{X}_s}{\sigma_1^2}\right)\widetilde{\boldsymbol{\beta}}_s+\left(\frac{\sum_{i=1}^{N}\mathbf{X}_{si}'\mathbf{D}_T\mathbf{X}_{si}}{\sigma_\varepsilon^2}\right)\hat{\boldsymbol{\beta}}_{sFE}\right]. \tag{11.13}$$

Under certain assumptions, the weights are the inverses of the variance-covariance matrices of the respective estimators.

Next to the estimation of $\boldsymbol{\beta}$ or $\boldsymbol{\beta}_s$, one might also want to estimate the individual stochastic components μ_i.[3] According to Taub (1979), predictions of the individual random effects μ_i are:

$$\hat{\mu}_{i,RE}=\frac{\sigma_\mu^2}{\sigma_1^2}\boldsymbol{\iota}_T\left(\mathbf{y}_i-\mathbf{X}_i\hat{\boldsymbol{\beta}}_{RE}\right), \tag{11.14}$$

which can be considered as the share of the GLS residual that is allocated to $\hat{\mu}_{i,RE}$; the precise proportion depends on the known variances σ_μ^2 and σ_ε^2.[4]

Maddala (1971) has proposed the following (consistent) estimators for the unknown variances σ_μ^2 and σ_ε^2. Since the (consistent) FE residual variance estimator $\widehat{\sigma}_{\varepsilon FE}^2$ is given by (11.6) and the residual variance resulting from the between estimator (11.12) can consistently be estimated as:

$$\widetilde{\sigma}_1^2=\frac{\widetilde{\boldsymbol{\varepsilon}}'\widetilde{\boldsymbol{\varepsilon}}}{NT-K+1}, \tag{11.15}$$

with the between residual vector $\widetilde{\boldsymbol{\varepsilon}}$ defined as $\widetilde{\boldsymbol{\varepsilon}}:=\mathbf{Q}_1\mathbf{y}-\mathbf{Q}_1\mathbf{X}_s\widetilde{\beta}_s$, a consistent estimator for the variance of the individual random effects μ_i is given by:[5]

$$\begin{aligned}\widehat{\sigma}_\mu^2 &= \frac{\widetilde{\sigma}_1^2-\widehat{\sigma}_{\varepsilon FE}^2}{T} && \Leftrightarrow \text{ if } \widehat{\sigma}_\mu^2>0\\ &= 0 && \text{elsewhere.}\end{aligned} \tag{11.16}$$

[3] This may be of interest because it describes how the behavior of different individuals varies and it provides a basis for more efficient prediction of future observations about a given individual.

[4] This predictor is best linear unbiased since $E(\hat{\mu}_{i,RE})=0$ and its variance $E(\hat{\mu}_{i,RE}^2)$ is smaller than the prediction error variance of each other linear unbiased predictor. Remark also that $\hat{\mu}_{i,RE}$ is that value of μ_i that minimises the following quadratic function with respect to β and μ_i $(i=1,2,...,N)$:

$$\frac{1}{\sigma_\varepsilon^2}\sum_{i=1}^{N}(\mathbf{y}_i-\mathbf{X}_i\boldsymbol{\beta}-\mu_i\boldsymbol{\iota}_T)'(\mathbf{y}_i-\mathbf{X}_i\boldsymbol{\beta}-\mu_i\boldsymbol{\iota}_T)+\frac{\sum_{i=1}^{N}\mu_i^2}{\sigma_\mu^2}.$$

[5] It should be noted that the individual random effects (11.14) are directly computed in Eviews as:

$$\hat{\hat{\mu}}_{i,RE}=\frac{\widehat{\sigma}_\mu^2}{T\widehat{\sigma}_\mu^2+\widehat{\sigma}_\varepsilon^2}\sum_{t=1}^{T}\left(y_{it}-\mathbf{x}_{it}'\hat{\hat{\boldsymbol{\beta}}}_{RE}\right),$$

with $\hat{\hat{\boldsymbol{\beta}}}_{RE}$ a consistent FGLS estimator derived from (11.10) with estimated variance-covariance matrices $\hat{\boldsymbol{\Phi}}_i$ (although Eviews seems to commit an error in the computation of $\widehat{\sigma}_\mu^2$).

This procedure is called *estimation of variance components*.[6] A number of other (consistent) variance estimators has been suggested in the literature. Two of them will be presented here, i.e. two consistent FGLS estimators discussed in Wooldridge (2002) and Greene (2003), respectively.[7]

Wooldridge (2002), p. 261, derives a consistent estimator of σ_μ^2 from the definition of the variance-covariance matrix of the error components in (11.8), i.e. $\sigma_\mu^2 = E(\nu_{it}\nu_{is})$ for $\nu_{it} := \mu_i + \varepsilon_{it}$ all $t \neq s$. Wooldridge argues, therefore, that for each group i, there are $T(T-1)/2$ nonredundant error products that can be used to estimate σ_μ^2. Summing all these combinations and taking expectations, the following expression is obtained:

$$\begin{aligned} E\left(\sum_{t=1}^{T-1}\sum_{s=t+1}^{T}\nu_{it}\nu_{is}\right) &= \sum_{t=1}^{T-1}\sum_{s=t+1}^{T}E(\nu_{it}\nu_{is}) = \sum_{t=1}^{T-1}\sum_{s=t+1}^{T}\sigma_\mu^2 = \sigma_\mu^2\sum_{t=1}^{T-1}(T-t) \\ &= \sigma_\mu^2\left((T-1)+(T-2)+\cdots+2+1\right) = \frac{\sigma_\mu^2 T(T-1)}{2}. \end{aligned}$$

A consistent estimator of σ_μ^2 is now obtained by replacing the expectation with an average (across groups) and replacing ν_{it} with its pooled OLS residual so that, after making a degrees-of-freedom adjustment for small-sample correction, we obtain:

$$\widehat{\sigma}_\mu^2 = \frac{\sum_{i=1}^{N}\sum_{t=1}^{T-1}\sum_{s=t+1}^{T}\hat{\nu}_{it}\hat{\nu}_{is}}{\frac{1}{2}NT(T-1)-K}, \tag{11.17}$$

where $\hat{\nu}_{it}$ denote the pooled OLS residuals. Also considering a consistent estimator of the variance of these pooled OLS residuals:

$$\widehat{\sigma}_v^2 = \frac{\sum_{i=1}^{N}\sum_{t=1}^{T}\hat{\nu}_{it}^2}{NT-K}, \tag{11.18}$$

so that, given $\widehat{\sigma}_\mu^2$ and $\widehat{\sigma}_v^2$, a consistent estimator for σ_ε^2 can be directly derived from the (assumed) mutual independency of ε_{it} and μ_i in $\nu_{it} := \mu_i + \varepsilon_{it}$, or $\sigma_\varepsilon^2 = E\left(\varepsilon_{it}^2\right) = E\left(\nu_{it}^2\right) - E\left(\mu_i^2\right) = \sigma_\nu^2 - \sigma_\mu^2$, i.e. $\widehat{\sigma}_\varepsilon^2 = \widehat{\sigma}_v^2 - \widehat{\sigma}_\mu^2$.

Usually, all observations are given equal weight ('no weighting' option), but cross-section weights may be considered if the residuals are cross-sectionally heteroskedastic and contemporaneously uncorrelated, i.e. if the error-covariance matrices $\mathbf{\Phi}_i$ in (11.8) can be expressed for $i = 1, 2, ..., N$ as:

$\mathbf{\Phi}_i = E\left((\mu_i\boldsymbol{\iota}_T+\boldsymbol{\varepsilon}_i)(\mu_i\boldsymbol{\iota}_T+\boldsymbol{\varepsilon}_i)'\right) = \sigma_\mu^2\boldsymbol{\iota}_T\boldsymbol{\iota}_T' + \sigma_{\varepsilon,i}^2\mathbf{I}_T$

$= \begin{pmatrix} \sigma_\mu^2+\sigma_{\varepsilon,i}^2 & \sigma_\mu^2 & \cdots & \sigma_\mu^2 \\ \sigma_\mu^2 & \sigma_\mu^2+\sigma_{\varepsilon,i}^2 & \cdots & \sigma_\mu^2 \\ \vdots & \vdots & \ddots & \vdots \\ \sigma_\mu^2 & \sigma_\mu^2 & \cdots & \sigma_\mu^2+\sigma_{\varepsilon,i}^2 \end{pmatrix}$, then a cross-sectionally weighted

[6] Notice that if the individual variance estimator $\widehat{\sigma}_\mu^2$ is found to be negative, this may be an indication that time effects have been incorrectly omitted. Assuming that $\sigma_\mu^2 = 0$ leads to the OLS or dummy variables estimator.

[7] Some of these alternative consistent individual variance estimators may sometimes be more (asymptotically) efficient than the above Maddala (1971) individual variance estimator.

FGLS regression is appropriate[8].

Following Greene (2003), pp. 296-298, we may reformulate this type of FGLS procedure in a somewhat different and heuristic way (see also the FGLS procedure for the constant panel data model in the previous section). Consider therefore the error (or variance) components model in scalar form $y_{it} = \mathbf{x}'_{it}\boldsymbol{\beta}+\mu_i+\varepsilon_{it}$, which can be averaged over time as $\overline{y}_{i.} = \overline{\mathbf{x}}'_{i.}\boldsymbol{\beta}+\mu_i+\overline{\varepsilon}_{i.}$. Taking deviations from the group means removes the heterogeneity $y_{it}-\overline{y}_{i.} = (\mathbf{x}_{it}-\overline{\mathbf{x}}_{i.})'\boldsymbol{\beta}+(\varepsilon_{it}-\overline{\varepsilon}_{i.})$. Since $E\left[\sum_{t=1}^{T}(\varepsilon_{it}-\overline{\varepsilon}_{i.})^2\right] = (T-1)\,\sigma_\varepsilon^2$, and if $\boldsymbol{\beta}$ were observed, an unbiased estimator of σ_ε^2 based on T observations in group i would be $\tilde{\sigma}_\varepsilon^2(i) = \frac{\sum_{t=1}^{T}(\varepsilon_{it}-\overline{\varepsilon}_{i.})^2}{T-1}$.In fact, however, $\boldsymbol{\beta}$ must be estimated, which could be made from the FE model with an unbiased estimator of the residual variance σ_ε^2 for each group i as $\hat{\sigma}_{\varepsilon FE}^2(i) = \frac{\sum_{t=1}^{T}(\hat{\varepsilon}_{it}-\overline{\hat{\varepsilon}}_{i.})^2}{T-K-1}$,with $\hat{\varepsilon}_{it} := y_{it}-\mathbf{x}'_{it}\hat{\boldsymbol{\beta}}_{FE}$ the residuals from the FE (OLS) model. Averaging over all individuals we get:

$$\frac{1}{N}\sum_{i=1}^{N}\hat{\sigma}_{\varepsilon FE}^2(i) = \frac{1}{N}\sum_{i=1}^{N}\left[\frac{\sum_{t=1}^{T}(\hat{\varepsilon}_{it}-\overline{\hat{\varepsilon}}_{i.})^2}{T-K-1}\right] = \frac{\sum_{i=1}^{N}\sum_{t=1}^{T}(\hat{\varepsilon}_{it}-\overline{\hat{\varepsilon}}_{i.})^2}{NT-NK-N}.$$

The degrees of freedom correction in this average residual variance is excessive because it assumes that the constant term and $\boldsymbol{\beta}_s$ are reestimated for each individual i. The estimated parameters are the N means $\overline{y}_{i.}$ and the $(K-1)$ slopes so that the denominator in the (average) residual variance becomes $(NT-N-K+1)$. An estimate of the variance of the individual effects, σ_μ^2, can then be obtained from the pooled variance of the error components plim$\frac{\mathbf{e}'\mathbf{e}}{NT-K-1} = \sigma_\varepsilon^2+\sigma_\mu^2$. Hence, an appropriate estimator for σ_μ^2 is:

$$\widehat{\sigma}_\mu^2 = \frac{\mathbf{e}'\mathbf{e}}{NT-K-1} - \frac{1}{N}\sum_{i=1}^{N}\hat{\sigma}_{\varepsilon FE}^2(i).$$

11.2.3 Testing for random effects

When $\mu_i = 0$ or $\sigma_\mu^2 = 0$ individual error components do not occur and the OLS estimator is the BLUE. This can be tested with a Lagrange Multiplier (LM) test since the null hypothesis (classical regression model with a single constant term and estimated with OLS) is (much) simpler than the alternative (GLS). Breusch and Pagan (1980) have proposed such an LM test for the random effects model based on the OLS residuals of the classical regression model $\mathbf{y} = \mathbf{X}\boldsymbol{\beta}+\boldsymbol{\varepsilon}$ versus the error components model (11.9) (where $\mathbf{X}, \mathbf{y}, \boldsymbol{\beta}$, and $\boldsymbol{\varepsilon}$ are the matrix and vectors mentioned there, which are stacked by cross section) and showed that

[8] Eviews performs FGLS with $\sigma_{\epsilon,i}^2$ estimated from a first stage pooled OLS regression:

$$\widehat{\sigma}_{\varepsilon_i}^2 = \frac{\sum_{t=1}^{T}(y_{it}-\hat{y}_{it})^2}{T},$$

where $\hat{y}_{it}$ are the OLS fitted values.

under the null hypothesis that $\sigma_\mu^2 = 0$ the resulting test statistic:

$$g = \frac{NT}{2(T-1)} \left[\frac{\sum_{i=1}^{N} \left(\sum_{t=1}^{T} e_{it} \right)^2}{\sum_{i=1}^{N} \sum_{t=1}^{T} e_{it}^2} - 1 \right]^2 \tag{11.19}$$

is asymptotically χ^2-distributed with 1 degree of freedom (the e_{it}s are the pooled OLS residuals from a regression of $\mathbf{y}$ on $\mathbf{X}$, which are equal to the residuals $\hat{\nu}_{it}$ in (11.18)). If the test statistic exceeds the critical χ_α^2 value for 1 degree of freedom (3.84 for $\alpha = 0.05$ and 6.63 for $\alpha = 0.01$) the result of the test is to reject the null in favor of the *RE* model. But, as is noticed by Greene (2003), p. 299, "it is best to reserve judgment on that, because there is another competing specification that might induce these same results, the fixed effects model".

11.2.4 Random effects versus fixed effects

When being confronted with both options of modeling panel data, *FE* or *RE* panel data models, the question as to which model should be used inevitably arises. In most cases it is a priori not obvious how to model the individual effects, i.e. as being fixed or random. Some have argued that the individual effects should always be treated as random, since the fixed effects model is analyzed conditionally on the effects that are actually present in the sample (see Greene (2003), pp. 301-303). Furthermore, the dummy variables approach greatly reduces the degrees of freedom. On the other hand, the assumption in the random effects model that the correlation between the individual (random) effects and the regressors is zero ($\mu_i \overset{iid}{\sim} (0, \sigma_\mu^2)$ and not correlated with $\mathbf{X}_i$), might not be suited. If this assumption is not valid, the *RE* estimation yields a specification error due to omitted variables yielding a biased GLS estimator because, then, $E(\mu_i)$ will not be constant -and thus certainly not equal to zero- but will depend on $\mathbf{x}_{it}$ (see Mundlak (1978), who assumes an auxiliary linear regression between the individual effects and the explanatory variables). There are reasons to believe that in many situations μ_i and $\mathbf{x}_{it}$ are correlated. For example, consider the estimation of a production function using firm data. The output of each firm, y_{it}, may be affected by unobervable managerial ability μ_i. Firms with more efficient management tend to produce more and use more inputs $\mathbf{X}_i$. Less efficient firms tend to produce less and use fewer inputs. In this situation, μ_i and $\mathbf{X}_i$ cannot be independent so that ignoring this correlation leads to biased estimation as indicated above. The *FE* model, however, essentially eliminates the individual effects from the direct model estimation such that only the shortened parameter vector is estimated in first instance by (11.4) and thus eliminates problems that the individual effects may cause.

Whether the error components (or random effects) model is correct, in the sense that no correlation exists between the random individual effects and the regressors can be tested by the Hausman test, which is an example of the Wald test statistic as indicated in (3.16). The Hausman test statistic is based on the property that under regularity conditions $\boldsymbol{\eta} := \sqrt{n}(\hat{\boldsymbol{\beta}}_{sFE} - \hat{\boldsymbol{\beta}}_{sRE}) \overset{as}{\sim} N\left(0, n\, \mathbf{avar}(\hat{\boldsymbol{\beta}}_{sFE} - \hat{\boldsymbol{\beta}}_{sRE})\right)$, where $\hat{\boldsymbol{\beta}}_{sFE}$ is the OLS dummy variables (FE) estimator (11.4) of β_s and $\hat{\boldsymbol{\beta}}_{sRE}$ the GLS error components (RE) estimator (11.11) of β_s and $\mathbf{avar}(\hat{\boldsymbol{\beta}}_{sFE} - \hat{\boldsymbol{\beta}}_{sRE})$ the asymptotic variance-covariance matrix of the difference vector between these (reduced) FE and the RE estimators with n the total number of observations used in the sample. Hence, the Wald test statistic satisfies $\boldsymbol{\eta}'\left(n\widehat{\mathbf{avar}}(\hat{\boldsymbol{\beta}}_{sFE}\text{-}\hat{\boldsymbol{\beta}}_{sRE})\right)^{-1}\boldsymbol{\eta} \overset{as}{\sim} \chi^2_{K-1}$ with $(K-1)$ the number of different elements in $\boldsymbol{\beta}_s$.

According to Lemma 3.2 for the null that the RE model is the asymptotically efficient estimator (although both estimators are consistent, the RE estimator is asymptotically efficient because of the more general stochastic formulation of the individual effects and the Cramèr-Rao lower bound property) we obtain:[9]

$$\mathbf{avar}(\hat{\boldsymbol{\beta}}_{sFE} - \hat{\boldsymbol{\beta}}_{sRE}) = \mathbf{avar}(\hat{\boldsymbol{\beta}}_{sFE}) - \mathbf{avar}(\hat{\boldsymbol{\beta}}_{sRE}) = \mathbf{M}_1 - \mathbf{M}_0,$$

where M_1 and M_0 represent the asymptotic covariance matrices of $\hat{\boldsymbol{\beta}}_{sFE}$ and $\hat{\boldsymbol{\beta}}_{sRE}$, respectively, or $\mathbf{M}_1 := \sigma^2_\varepsilon\left(\mathbf{X}_s'\left(\mathbf{I}_N \otimes \mathbf{D}_T\right)\mathbf{X}_s\right)^{-1}$ and $\mathbf{M}_0 := \left(\frac{1}{\sigma_1^2}\mathbf{X}_s'\mathbf{Q}_1\mathbf{X}_s + \frac{1}{\sigma_\varepsilon^2}\left(\mathbf{X}_s'\left(\mathbf{I}_N \otimes \mathbf{D}_T\right)\mathbf{X}_s\right)\right)$, again with $\mathbf{D}_T := \mathbf{I}_T - \frac{\boldsymbol{\iota}_T\boldsymbol{\iota}_T'}{T}$ and $\mathbf{Q}_1 := \left(\mathbf{I}_N \otimes \frac{\boldsymbol{\iota}_T\boldsymbol{\iota}_T'}{T}\right) - \frac{\boldsymbol{\iota}_{NT}\boldsymbol{\iota}_{NT}'}{NT}$. Hence, the Hausman test statistic satisfies under the null of the error components model that:

$$\xi_H = \left(\hat{\boldsymbol{\beta}}_{sFE} - \hat{\boldsymbol{\beta}}_{sRE}\right)'(\mathbf{M}_1 - \mathbf{M}_0)^{-1}\left(\hat{\boldsymbol{\beta}}_{sFE} - \hat{\boldsymbol{\beta}}_{sRE}\right) \overset{as}{\sim} \chi^2_{K-1}. \qquad (11.20)$$

.When the null is rejected, the dummy variables or FE estimator $\hat{\boldsymbol{\beta}}_{sFE}$ is the best estimator and the error components or RE estimator $\hat{\boldsymbol{\beta}}_{sRE}$ is biased since the individual effects are found to be correlated with (the) other regressors in the model then.

Remark that FE or RE estimation can also be used for panel databases where T is relatively small compared to N.

11.3 Intercepts varying over individuals and time

In this section the previous analysis is generalized to a linear model where next to the cross-section-specific effects, also time-specific effects are included, as in the following representation:

$$y_{it} = \underbrace{\overline{\beta}_1 + \mu_i + \lambda_t}_{\beta_{1it}} + \textstyle\sum_{k=2}^{K}\beta_k x_{kit} + \varepsilon_{it} \quad (i = 1, ..., N; t = 1, ..., T). \qquad (11.21)$$

[9] Provided that no correlation exists between the stochastic individual effects and the explanatory variables as assumed for this Hausman (1978) result!

As in the previous section, a distinction can be made between FE and RE estimation.

11.3.1 μ_i and λ_t fixed: dummy variables model

If μ_i and λ_t are both assumed to be fixed, the following two normalizing restrictions are imposed without loss of generality:[10]

$$\sum_{i=1}^{N} \mu_i = 0 \text{ and } \sum_{t=1}^{T} \lambda_t = 0.$$

If we want to estimate the model by OLS and implement these restrictions, a reparameterization of the model becomes necessary. A reparameterization can be implemented in various ways, one way is to delete one of the time-specific dummies by assuming that:

$$\begin{aligned} \beta_{1i} &: = \bar{\beta}_1 + \mu_i \quad (i = 1, ..., N) \\ \lambda_t^* &: = \lambda_t - \lambda_T \quad (t = 1, ..., T-1) \Leftrightarrow \lambda_T := -\sum_{t=1}^{T-1} \lambda_t. \end{aligned}$$

Then the model becomes for individual i:

$$\mathbf{y}_i = \beta_{1i}\boldsymbol{\iota}_T + \begin{pmatrix} \mathbf{I}_{T-1} \\ \mathbf{0}' \end{pmatrix} \boldsymbol{\lambda}^* + \mathbf{X}_{si}\boldsymbol{\beta}_s + \boldsymbol{\varepsilon}_i$$

with $\boldsymbol{\lambda}^* = \left(\lambda_1^*, \lambda_2^*, ..., \lambda_{T-1}^*\right)'$.

For all individuals together it becomes:

$$\mathbf{y} = \left(\mathbf{I}_N \otimes \boldsymbol{\iota}_T, \mathbf{j}_N \otimes \begin{pmatrix} \mathbf{I}_{T-1} \\ \mathbf{0}' \end{pmatrix}, \mathbf{X}_s\right) \begin{pmatrix} \boldsymbol{\beta}_1 \\ \boldsymbol{\lambda}^* \\ \boldsymbol{\beta}_s \end{pmatrix} + \boldsymbol{\varepsilon}, \tag{11.22}$$

with $E(\boldsymbol{\varepsilon}) = 0$ and $E(\boldsymbol{\varepsilon}\boldsymbol{\varepsilon}') = \sigma_\varepsilon^2 \mathbf{I}_{NT}$ and $\mathbf{y} := (\mathbf{y}_1', \ldots, \mathbf{y}_N')'$, $\boldsymbol{\varepsilon} := (\boldsymbol{\varepsilon}_1', \ldots, \boldsymbol{\varepsilon}_N')'$, $\mathbf{X}_{si} := (\mathbf{x}_{s1}', \ldots, \mathbf{x}_{sN}')'$, $\boldsymbol{\beta}_1 := (\beta_{11}, \ldots, \beta_{1N})'$.

Thus, the OLS estimator of the $(N + T - 1 + K - 1)$ vector $(\boldsymbol{\beta}_1', \boldsymbol{\lambda}^{*\prime}, \boldsymbol{\beta}_s')'$ is the BLUE. Note that if $(N + T - 1 + K - 1)$ is too large, it might be more appropriate to formulate the model in terms of deviations from means as in the previous section, which results in the following estimator for the slope vector:

$$\hat{\boldsymbol{\beta}}_{sFE} = \left(\mathbf{X}_s'\mathbf{Q}\mathbf{X}_s\right)^{-1} \mathbf{X}_s'\mathbf{Q}\mathbf{y}, \tag{11.23}$$

but now with $\mathbf{Q} := \mathbf{I}_{NT} - \mathbf{I}_N \otimes \frac{\boldsymbol{\iota}_T\boldsymbol{\iota}_T'}{T} - \frac{\boldsymbol{\iota}_N\boldsymbol{\iota}_N'}{N} \otimes \mathbf{I}_T + \frac{\boldsymbol{\iota}_{NT}\boldsymbol{\iota}_{NT}'}{NT}$. This matrix $\mathbf{Q}$ is idempotent and transforms the values from $\mathbf{X}_s$ and $\mathbf{y}$ so that $\mathbf{Q}\mathbf{X}_s$ becomes an $NT \times NT$ matrix with elements in the form $x_{kit} - \bar{x}_{ki.} - \bar{x}_{k.t} + \bar{x}_{k..}$, where $\bar{x}_{ki.} := \frac{\sum_t x_{kit}}{T}$, $\bar{x}_{k.t} := \frac{\sum_i x_{kit}}{N}$ and $\bar{x}_{k..} := \frac{\sum_i \sum_t x_{kit}}{NT}$.

[10]Note that the normalization rule $\sum_{i=1}^{N} \mu_i = 0$ is also implied in the panel data model with intercepts varying over individuals in the previous section.

Taking the averages in equation (11.21) over t, i, and t and i, respectively, we find:

$$\begin{aligned}\bar{y}_{i.} &= \bar{\beta}_1 + \mu_i + \sum\nolimits_{k=2}^{K} \beta_k \bar{x}_{ki.} + \frac{\sum_t \varepsilon_{it}}{T} \\ \bar{y}_{.t} &= \bar{\beta}_1 + \lambda_t + \sum\nolimits_{k=2}^{K} \beta_k \bar{x}_{k.t} + \frac{\sum_i \varepsilon_{it}}{N} \\ \bar{y}_{..} &= \bar{\beta}_1 + \sum\nolimits_{k=2}^{K} \beta_k \bar{x}_{k..} + \frac{\sum_i \sum_t \varepsilon_{it}}{NT}.\end{aligned} \quad (11.24)$$

Subtracting the first two equations in (11.24) from the equations in (11.21) and from the last equation in (11.24), we get:

$$y_{it} - \bar{y}_{i.} - \bar{y}_{.t} + \bar{y}_{..} = \sum\nolimits_{k=2}^{K} \beta_k (x_{kit} - \bar{x}_{ki.} - \bar{x}_{k.t} + \bar{x}_{k..}) + v_{it} \quad (11.25)$$

with $v_{it} := \varepsilon_{it} - \frac{\sum_t \varepsilon_{it}}{T} - \frac{\sum_i \varepsilon_{it}}{N} + \frac{\sum_i \sum_t \varepsilon_{it}}{NT}$.

Thus, the estimator $\hat{\boldsymbol{\beta}}_{sFE}$ in (11.23) can be considered as the OLS estimator of equation (11.25). Furthermore, we can interpret $\hat{\boldsymbol{\beta}}_{sFE}$ as the GLS estimator of equation (11.22) where $\mathbf{QQ} = \mathbf{Q}$ and $\mathbf{Q}$ is the (generalized) inverse of the covariance matrix of the disturbances. Once $\hat{\boldsymbol{\beta}}_{sFE}$ is estimated as specified in (11.23), we can then estimate the other model parameters as:

$$\begin{aligned}\hat{\mu}_{i,FE} &= (\bar{y}_{i.} - \bar{y}_{..}) - \sum\nolimits_{k=2}^{K} \hat{\beta}_{k,FE} (\bar{x}_{ki.} - \bar{x}_{k..}), \\ \hat{\lambda}_{t,FE} &= (\bar{y}_{.t} - \bar{y}_{..}) - \sum\nolimits_{k=2}^{K} \hat{\beta}_{k,FE} (\bar{x}_{k.t} - \bar{x}_{k..}), \\ \hat{\bar{\beta}}_{1,FE} &= \bar{y}_{..} - \sum\nolimits_{k=2}^{K} \hat{\beta}_{k,FE} \bar{x}_{k..},\end{aligned} \quad (11.26)$$

with $\hat{\beta}_{k,FE}$ being an element of $\hat{\boldsymbol{\beta}}_{sFE}$ and $\hat{\bar{\beta}}_{1,FE}$ being an (FE) estimator of $\bar{\beta}_1$.

As in the previous dummy variables model, we can use a regular F-test again to check whether the parameters β_{1it} $(i = 1, ..., N; t = 1, ..., T)$ are equal or not.

11.3.2 μ_i and λ_t random: error components model

In this subsection the individual- and time-specific effects are not assumed to be constant but random. Assume that $\mu_i \overset{iid}{\sim} (0, \sigma_\mu^2)$, $\lambda_t \overset{iid}{\sim} (0, \sigma_\lambda^2)$ and $E(\mu_i \lambda_t) = 0 \; \forall i, t$; $E(\mu_i \varepsilon_{jt}) = 0 \; \forall i, j, t$, and that $E(\lambda_t \varepsilon_{i\theta}) = 0 \; \forall i, t, \theta$.

For individual i we then have the following model:

$$\mathbf{y}_i = \mathbf{X}_i \boldsymbol{\beta} + \underbrace{\mu_i \boldsymbol{\iota}_T + \mathbf{I}_T \boldsymbol{\lambda} + \boldsymbol{\varepsilon}_i}_{\text{error component terms}},$$

where $\boldsymbol{\lambda} := (\lambda_1, \lambda_2, ..., \lambda_T)'$ and $\mathbf{X}_i$ contains a constant term. The covariance matrix of the error component terms consists of the following elements:

$$\begin{aligned}\boldsymbol{\Phi}_{ii} &= E\left((\mu_i \boldsymbol{\iota}_T + \mathbf{I}_T \boldsymbol{\lambda} + \boldsymbol{\varepsilon}_i)(\mu_i \boldsymbol{\iota}_T + \mathbf{I}_T \boldsymbol{\lambda} + \boldsymbol{\varepsilon}_i)'\right) \\ &= \sigma_\mu^2 \boldsymbol{\iota}_T \boldsymbol{\iota}_T' + \sigma_\lambda^2 \mathbf{I}_T + \sigma_\varepsilon^2 \mathbf{I}_T \\ \boldsymbol{\Phi}_{ij} &= E\left((\mu_i \boldsymbol{\iota}_T + \mathbf{I}_T \boldsymbol{\lambda} + \boldsymbol{\varepsilon}_i)(\mu_j \boldsymbol{\iota}_T + \mathbf{I}_T \boldsymbol{\lambda} + \boldsymbol{\varepsilon}_j)'\right) = \sigma_\lambda^2 \mathbf{I}_T.\end{aligned}$$

When all NT observations are considered, the following model results:

$$\mathbf{y} = \mathbf{X}\boldsymbol{\beta} + \boldsymbol{\mu} \otimes \boldsymbol{\iota}_T + (\boldsymbol{\iota}_N \otimes \mathbf{I}_T)\,\boldsymbol{\lambda} + \boldsymbol{\varepsilon},$$

with $\boldsymbol{\mu} := (\mu_1, \mu_2, ..., \mu_N)'$ .The complete covariance matrix is:

$$\boldsymbol{\Phi} = \sigma_\mu^2 \left(\mathbf{I}_N \otimes \boldsymbol{\iota}_T \boldsymbol{\iota}_T'\right) + \sigma_\lambda^2 \left(\boldsymbol{\iota}_N \boldsymbol{\iota}_N' \otimes \mathbf{I}_T\right) + \sigma_\varepsilon^2 \mathbf{I}_{NT}.$$

As in the error components model without time-specific effects, the GLS estimator is the BLUE and $\boldsymbol{\beta}$ is thus estimated as follows:

$$\hat{\boldsymbol{\beta}}_{RE} = \left(\mathbf{X}'\boldsymbol{\Phi}^{-1}\mathbf{X}\right)^{-1}\mathbf{X}'\boldsymbol{\Phi}^{-1}\mathbf{y}. \tag{11.27}$$

As in the previous section, the RE estimator $\hat{\boldsymbol{\beta}}_{sRE}$ can again pe partitioned as $\hat{\boldsymbol{\beta}}_{RE}' := \left(\hat{\overline{\beta}}_{1RE}, \hat{\boldsymbol{\beta}}_{sRE}'\right)$, so that it can similarly be shown that $\hat{\boldsymbol{\beta}}_{sRE}$ can be expressed as a matrix weighted average of three other estimators. To demonstrate this, note that:

$$\boldsymbol{\Phi}^{-1} = \frac{\mathbf{Q}}{\sigma_\varepsilon^2} + \frac{\mathbf{Q}_1}{\sigma_1^2} + \frac{\mathbf{Q}_2}{\sigma_2^2} + \frac{\mathbf{Q}_3}{\sigma_3^2}, \tag{11.28}$$

with $\mathbf{Q}$ and $\mathbf{Q}_1$ defined as before and the symmetric matrices $\mathbf{Q}_2$ and $\mathbf{Q}_3$ given, respectively by $\mathbf{Q}_2 := \left(\frac{\boldsymbol{\iota}_N \boldsymbol{\iota}_N'}{N} \otimes I_T\right) - \frac{\boldsymbol{\iota}_{NT}\boldsymbol{\iota}_{NT}'}{NT}$ and $\mathbf{Q}_3 := \frac{\boldsymbol{\iota}_{NT}\boldsymbol{\iota}_{NT}'}{NT}$, while the variances are given by $\sigma_1^2 := T\sigma_\mu^2 + \sigma_\varepsilon^2$, $\sigma_2^2 := N\sigma_\lambda^2 + \sigma_\varepsilon^2$, and $\sigma_3^2 := T\sigma_\mu^2 + N\sigma_\lambda^2 + \sigma_\varepsilon^2$. Substituting (11.28) and the partitioning of $\mathbf{X} := (\mathbf{j}_{NT}, \mathbf{X}_s)$ into (11.27), we get:

$$\begin{aligned}\hat{\boldsymbol{\beta}}_{sRE} &= \left[\frac{\mathbf{X}_s'\mathbf{Q}_1\mathbf{X}_s}{\sigma_1^2} + \frac{\mathbf{X}_s'\mathbf{Q}_2\mathbf{X}_s}{\sigma_2^2} + \frac{\mathbf{X}_s'\mathbf{Q}\mathbf{X}_s}{\sigma_\varepsilon^2}\right]^{-1} \\ &\cdot \left[\left(\frac{\mathbf{X}_s'\mathbf{Q}_1\mathbf{X}_s}{\sigma_1^2}\right)\widetilde{\boldsymbol{\beta}}_s + \left(\frac{\mathbf{X}_s'\mathbf{Q}_2\mathbf{X}_s}{\sigma_2^2}\right)\breve{\boldsymbol{\beta}}_s + \left(\frac{\mathbf{X}_s'\mathbf{Q}\mathbf{X}_s}{\sigma_\varepsilon^2}\right)\hat{\boldsymbol{\beta}}_{sFE}\right]\end{aligned} \tag{11.29}$$

where $\widetilde{\boldsymbol{\beta}}_s$ is the between estimator (11.12) of the slope vector, $\breve{\boldsymbol{\beta}}_s = (\mathbf{X}_s'\mathbf{Q}_2\mathbf{X}_s)^{-1}\mathbf{X}_s'\mathbf{Q}_2\mathbf{y}$, and $\hat{\boldsymbol{\beta}}_{sFE}$ is the FE estimator (11.4). The constant $\overline{\beta}_1$ is directly estimated as $\hat{\overline{\beta}}_{1,RE} = \overline{y}_{..} - \sum_{k=2}^K \hat{\beta}_{k,RE}\overline{x}_{k..}$.

The predictions for μ_i and λ_t are:

$$\hat{\mu}_{i,RE} = \left(\frac{T\sigma_\mu^2}{\sigma_1^2}\right)\left(\overline{y}_{i.} - \hat{\overline{\boldsymbol{\beta}}}_{1,RE} - \textstyle\sum_{k=2}^K \hat{\beta}_{k,RE}\overline{x}_{k\,i.}\right), \tag{11.30}$$

$$\hat{\lambda}_{t,RE} = \left(\frac{N\sigma_\lambda^2}{\sigma_2^2}\right)\left(\overline{y}_{.t} - \hat{\overline{\boldsymbol{\beta}}}_{1,RE} - \textstyle\sum_{k=2}^K \hat{\beta}_{k,RE}\overline{x}_{k\,.t}\right). \tag{11.31}$$

Prediction of the individual effects μ_i gives information on the future behavior of individuals and is likely to be of more interest than the prediction of λ_t, which gives information on past realizations.

Again, the RE estimators (11.29), (11.30), and (11.31) depend on the unknown variances σ_ε^2, σ_μ^2, and σ_λ^2. To replace these unknowns with estimates, any of the variance estimators mentioned in the previous section, suitably modified for the time effects, can be employed. Swamy and Arora (1972) suggest as (asymptotically) unbiased estimators the between residual variance $\widetilde{\sigma_1}^2$ in (11.15), the residual variance resulting from the residuals $\breve{\varepsilon} := \mathbf{Q}_2\mathbf{y} - \mathbf{Q}_2\mathbf{X}_s\breve{\boldsymbol{\beta}}_s$, or $\breve{\sigma}_2^2 = \frac{\breve{\varepsilon}'\breve{\varepsilon}}{T-K}$, and the fixed effects residual variance resulting from the fixed effects estimator (11.23) for the slope vector $\boldsymbol{\beta}_s$. Substitution of a set of variance estimators into $\hat{\boldsymbol{\beta}}_{RE}$ leads to an FGLS estimator in the way explained in the previous section.

11.3.3 Specification tests

To check whether $\boldsymbol{\mu} = 0$, $\boldsymbol{\lambda} = 0$, or $\boldsymbol{\mu} = \boldsymbol{\lambda} = 0$, we can use the dummy variables estimator and an F-test, which compares the sums of the restricted and unrestricted squared residuals. The test for $\sigma_\mu^2 = \sigma_\lambda^2 = 0$ is given in Breusch and Pagan (1980). It is shown that under the null hypothesis $H_0 : \sigma_\mu^2 = \sigma_\lambda^2 = 0$ the LM test statistic:

$$
\begin{aligned}
g &= \frac{NT}{2}\left\{\frac{1}{T-1}\left[\frac{\sum_{i=1}^{N}\left(\sum_{t=1}^{T} e_{it}\right)^2}{\sum_i\sum_t e_{it}} - 1\right]^2\right. \\
&\quad \left. + \frac{1}{N-1}\left[\frac{\sum_{t=1}^{T}\left(\sum_{i=1}^{N} e_{it}\right)^2}{\sum_i\sum_t e_{it}} - 1\right]^2\right\}
\end{aligned}
$$

is asymptotically distributed as χ_2^2; e_{it} are, as before, the residuals of the OLS regression of $\mathbf{y}$ on $\mathbf{X}$ (see (11.19)).

11.3.4 Fixed or random effects?

Precisely as for the linear panel data models with intercepts varying over individuals alone, we can present a test which checks whether λ_t and μ_i are or are not correlated with x_{kit}. If such correlation exists we must choose the dummy variables model since the error components GLS estimator suffers from omitted variable bias then.

Hausman (1978) shows that under the null hypothesis of the RE model (no correlation between λ_t and μ_i and (the) regressors) the test statistic:

$$
m = (\hat{\boldsymbol{\beta}}_{sFE} - \hat{\boldsymbol{\beta}}_{sRE})'(\mathbf{M}_1 - \mathbf{M}_0)^{-1}(\hat{\boldsymbol{\beta}}_{sFE} - \hat{\boldsymbol{\beta}}_{sRE}) \qquad (11.32)
$$

is asymptotically distributed as χ_{K-1}^2 (see (11.20), where $\mathbf{M}_0$ is the (asymptotic) variance-covariance matrix of the GLS estimator $\hat{\boldsymbol{\beta}}_{sRE}$ and $\mathbf{M}_1$ is the (asymptotic) variance-covariance matrix of the dummy variables estimator $\hat{\boldsymbol{\beta}}_{sFE}$, respectively:

$$\begin{aligned}\mathbf{M}_1 &= \sigma_\varepsilon^2 \left(\mathbf{X}_s'\mathbf{Q}\mathbf{X}_s\right)^{-1}, \\ \mathbf{M}_0 &= \left[\frac{\mathbf{X}_s'\mathbf{Q}_1\mathbf{X}_s}{\sigma_1^2} + \frac{\mathbf{X}_s'\mathbf{Q}_2\mathbf{X}_s}{\sigma_2^2} + \frac{\mathbf{X}_s'\mathbf{Q}\mathbf{X}_s}{\sigma_\varepsilon^2}\right]^{-1}.\end{aligned}$$

An alternative procedure is a mixed approach where, say, the λ_ts are treated as fixed and the μ_is are treated as random variables. Such an approach seems to be useful if T is small and if the researcher wishes to follow some of the more general (random) specifications for μ_i and ε_{it} as mentioned in the previous section(s). A Hausman test statistic, appropriately adapted from (11.32) can be employed then.

11.4 All coefficients varying over individuals

In all previous panel data models it was assumed that all individual heterogeneity could be captured in the intercept, either in a fixed or random way, while the slope parameters were assumed to be constant across cross sections or individuals. It is possible, though, that also the slope parameters are individual-specific. This would imply that different individuals react in a different way to changes in the regressors. In this case, we have the following model:

$$y_{it} = \textstyle\sum_{k=1}^{K} \beta_{ki} x_{kit} + \varepsilon_{it} \quad (i = 1, ..., N; t = 1, ..., T),$$

where x_{1it} can be equal to 1 for all $t = 1, 2, ..., T$. A distinction can also be made between fixed and random parameters. Seemingly Unrelated Regression (SUR) is the appropriate technique for fixed parameters and the Swamy's random coefficient model for random parameters.

11.4.1 β_{ki} fixed: SUR

The SUR model is generally represented as:

$$\mathbf{y} = \mathbf{Z}\boldsymbol{\gamma} + \boldsymbol{\varepsilon} \tag{11.33}$$

or

$$\begin{pmatrix} \mathbf{y}_1 \\ \mathbf{y}_2 \\ \vdots \\ \vdots \\ \mathbf{y}_N \end{pmatrix} = \begin{pmatrix} \mathbf{X}_1 & 0 & \cdots & \cdots & 0 \\ 0 & \mathbf{X}_2 & & & \vdots \\ \vdots & & \ddots & & \vdots \\ \vdots & & & \ddots & 0 \\ 0 & \cdots & \cdots & 0 & \mathbf{X}_N \end{pmatrix} \begin{pmatrix} \boldsymbol{\beta}_1 \\ \boldsymbol{\beta}_2 \\ \vdots \\ \vdots \\ \boldsymbol{\beta}_N \end{pmatrix} + \begin{pmatrix} \boldsymbol{\varepsilon}_1 \\ \boldsymbol{\varepsilon}_2 \\ \vdots \\ \vdots \\ \boldsymbol{\varepsilon}_N \end{pmatrix}.$$

Since each individual equation has its own parameter vector $\boldsymbol{\beta}_i$, it appears that the equations are unrelated. Nevertheless, correlation across the error terms in different equations can provide links that can be exploited in the estimation process: if $E\left(\varepsilon_i\varepsilon_j^{'}\right) = \delta_{ij}\sigma_{ij}\mathbf{I}_T$, the OLS estimator has the BLUE property, but if $E\left(\varepsilon_i\varepsilon_j^{'}\right) = \sigma_{ij}\mathbf{I}_T$ with $\exists_{ij}\ \sigma_{ij} \neq 0$, the GLS estimator is the BLUE. This GLS estimator for a known contemporaneous covariance matrix $\mathbf{\Sigma}$ results from minimizing $\boldsymbol{\varepsilon}'(\mathbf{\Sigma}^{-1} \otimes \mathbf{I}_t)\boldsymbol{\varepsilon}$ with respect to $\boldsymbol{\gamma}$. In practice, $\mathbf{\Sigma}$ is unknown so that the FGLS or SUR estimator is given by $\hat{\boldsymbol{\gamma}} = \left(\mathbf{Z}'(\hat{\mathbf{\Sigma}}^{-1} \otimes \mathbf{I}_T)\mathbf{Z}\right)^{-1} \mathbf{Z}'\hat{\mathbf{\Sigma}}^{-1} \otimes \mathbf{I}_T\mathbf{y}$, where the estimator $\hat{\mathbf{\Sigma}}$ satisfies:

$$\hat{\mathbf{\Sigma}} = \frac{\hat{\varepsilon}_i'\hat{\varepsilon}_j'}{T-K}. \tag{11.34}$$

and $\mathbf{var}(\hat{\gamma}) = \left(\mathbf{Z}'(\hat{\mathbf{\Sigma}}^{-1} \otimes \mathbf{I}_T)\mathbf{Z}\right)^{-1}$.

Remark that the SUR technique can only be used when T is larger than N, i.e. when the number of time periods in the database is larger than the number of cross sections or individuals. If $T < N$ the residual covariance matrix $\hat{\mathbf{\Sigma}}$ is an $N \times N$ matrix with rank T and is thus singular such that it cannot be inverted. The number of cross-section-specific coefficients then gets very large relatively to the number of available observations in time. For this purpose, the fixed and random effects methods are more appropriate when N is large compared to T.

A real-world case with singular error covariance matrix

The above error covariance matrix $\mathbf{\Sigma}$ may also be singular through the definition of the problem. Consider e.g. the known application of *interrelated factor demands*, based on the *translog cost function* (but we could equally consider other applications as e.g. consumer demand). This cost function is a direct generalization of the familiar Cobb-Douglas (loglinear) cost function, since it is a second order approximation to the logarithm of an arbitrary cost function,[11] so that an attractive feature of this translog specification is that the cost minimizing input demand equations, if transformed into cost share equations, are linear in the logarithms of output and factor prices, with the coefficients inheriting (a subset of) the cost function parameters characterizing the technology. Those technology parameters can be estimated from a system of equations for cost shares. To be more specific, we consider the translog or loglinear cost function in three factor prices (for the inputs capital, labour, and materials, respectively) for each firm $i = 1, 2, ..., N$ and for each time period $t = 1, 2, ..., T$ (for the time being these subscripts are dropped for the clarity of the theoretical argumentation):[12]

$$\begin{aligned}\log(C) = \alpha_0 + \textstyle\sum_{j=1}^{3} \alpha_j \log(p_j) + \frac{1}{2}\sum_{j=1}^{3}\sum_{k=1}^{3} \gamma_{jk} \log(p_j)\log(p_k) \\ +\alpha_Q \log(Q) + \tfrac{1}{2}\gamma_{QQ}\left(\log(Q)\right)^2 + \textstyle\sum_{j=1}^{3} \gamma_{jQ} \log(p_j)\log(Q) + \varepsilon.\end{aligned}$$

[11] The Cobb-Douglas cost function is a first order approximation.

[12] The reasons for different input prices over individuals/states are regional variation, differential tax systems and concessions, etc..

Without loss of generality we can assume that the 3 x 3 matrix of quadratic form coefficients, $\{\gamma_{jk}\}$, is symmetric: $\gamma_{jk} = \gamma_{kj}$ $(j, k = 1, 2, 3)$. The *degree of returns to scale* η can be computed as the reciprocal of the elasticity of costs with respect to output:

$$\eta = \frac{1}{\partial \log(C)/\partial \log(Q)} = \frac{1}{\alpha_Q + \gamma_{QQ} \log(Q) + \sum_{j=1}^{3} \gamma_{jQ} \log(p_j)}.$$

Now, the link between the cost function parameters and factor demands is given by *Shephard's lemma.* To illustrate this, let x_j be the cost minimizing demand for factor input j given factor prices (p_1, p_2, p_3) and output Q. Hence, $\sum_{j=1}^{3} p_j x_j = C$, so that Shephard's lemma implies that: $\frac{\partial C}{\partial p_j} = x_j$. Since the price elasticity of total costs satisfies: $\frac{\partial \log(C)}{\partial \log(p_j)} = \frac{p_j x_j}{C}$, this price elasticity for the translog cost function amounts to:

$$\frac{\partial \log(C)}{\partial \log(p_j)} = \alpha_j + \sum_{k=1}^{3} \gamma_{jk} \log(p_k) + \gamma_{jQ} \log(Q) \quad (j = 1, 2, 3).$$

Defining cost shares as $s_j := p_j x_j / C$, we obtain the following system of cost share equations associated with the translog cost function:

$$s_j = \alpha_j + \sum_{k=1}^{3} \gamma_{jk} \log(p_k) + \gamma_{jQ} \log(Q) \quad (j = 1, 2, 3). \tag{11.35}$$

This system of cost share equations for capital, labor, and materials, respectively, is subject to cross-equation restrictions, saying e.g. that the coefficient of $\log(p_k)$ in the s_j equation equals the coefficient of $\log(p_j)$ in the s_k equation for $j \neq k$. These cross-equation restrictions are also called symmetry restrictions.[13]

Although the translog cost function has an (additive) error term ε, the cost share equations (11.35) derived from it have none. However, since, in general, the share equations do not hold exactly for each firm (unless the long run equilibrium at the cost-minimizing input combination of that firm would have exactly been reached), a random disturbance term is assumed to be added to each share equation (11.35).[14]

[13]Note that these symmetry restrictions are not a consequence of the above mentioned symmetry assumption, but are a consequence of calculus: if symmetry were not assumed, the coefficient of e.g. $\log(p_2)$ in the s_1 equation would be $(\gamma_{12} + \gamma_{21})/2$, which would equal the $\log(p_1)$ coefficient in the s_2 equation.

Moreover, the *subsitution elasticity* between inputs j and k, denoted as ϵ_{jk}, is related to the (translog) cost function as:

$$\begin{aligned} \epsilon_{jk} &= \frac{C \frac{\partial^2 C}{\partial p_j \partial p_k}}{\frac{\partial C}{\partial p_j} \frac{\partial C}{\partial p_k}} = \frac{\gamma_{jk} + s_j s_k}{s_j s_k} \quad \text{for } j \neq k, \\ &= \frac{\gamma_{jj} + s_j^2 - s_j}{s_j^2} \quad \text{for } j = k. \end{aligned}$$

[14]Since these cost share equations, measuring (relative) factor demands, do not contain the parameters α_Q and γ_{QQ}, and since these parameters occur in the expression of the degree η of returns to scale, this degree cannot be estimated from the cost share equations.

The resulting system of share equations contains fifteen parameters. The adding-up restrictions (because the cost shares should add to unity):

$$\begin{aligned}\alpha_1 + \alpha_2 + \alpha_3 &= 1\\ \gamma_{11} + \gamma_{21} + \gamma_{31} &= 0\\ \gamma_{12} + \gamma_{22} + \gamma_{32} &= 0\\ \gamma_{13} + \gamma_{23} + \gamma_{33} &= 0\\ \gamma_{1Q} + \gamma_{2Q} + \gamma_{3Q} &= 0,\end{aligned} \tag{11.36}$$

the homogeneity restrictions (requiring that the translog cost function is homogeneous of degree one in factor prices):

$$\begin{aligned}\gamma_{11} + \gamma_{12} + \gamma_{13} &= 0\\ \gamma_{21} + \gamma_{22} + \gamma_{23} &= 0\\ \gamma_{31} + \gamma_{32} + \gamma_{33} &= 0,\end{aligned} \tag{11.37}$$

and symmetry restrictions:

$$\begin{aligned}\gamma_{12} &= \gamma_{21}\\ \gamma_{13} &= \gamma_{31}\\ \gamma_{23} &= \gamma_{32}\end{aligned} \tag{11.38}$$

imply that there are in total eleven restrictions, but that only eight of them are really restrictive: given the adding-up restrictions, one of the three homogeneity restrictions is implied by the two others, and given the adding-up and homogeneity restrictions, two of the three symmetry restrictions are implied by the other. Since the sum of the dependent variables (s_{1t}, s_{2t}, s_{3t}) adds up to unity for all observations, the sum of the error terms $(\varepsilon_1, \varepsilon_2, \varepsilon_3)$ is zero for all observations, so that the 3 x 3 error covariance matrix $var(\varepsilon_1, \varepsilon_2, \varepsilon_3) =: \mathbf{\Sigma}$ is singular. The common practice to deal with this singularity is to drop one of the three equations and estimate the system composed of the remaining equations. The coefficients in the equation that was dropped can be calculated from the parameter estimates for the included two equations by using the adding-up restrictions. To incorporate the homogeneity restrictions (11.37), we have to take account of the property that these restrictions are not cross-equation restrictions, so that we can directly eliminate three parameters, say $(\gamma_{13}, \gamma_{23}, \gamma_{33})$, from the system to obtain the following set of cost share equations:

$$\begin{aligned}s_1 &= \alpha_1 + \gamma_{11}\log(p_1/p_3) + \gamma_{12}\log(p_2/p_3) + \gamma_{1Q}\log(Q) + \varepsilon_1\\ s_2 &= \alpha_2 + \gamma_{21}\log(p_1/p_3) + \gamma_{22}\log(p_2/p_3) + \gamma_{2Q}\log(Q) + \varepsilon_2\\ s_3 &= \alpha_3 + \gamma_{31}\log(p_1/p_3) + \gamma_{32}\log(p_2/p_3) + \gamma_{3Q}\log(Q) + \varepsilon_3.\end{aligned} \tag{11.39}$$

To take account of the adding-up restriction and, hence, of the singularity of the error covariance matrix, we may drop the last equation for s_3.[15] Hence,

[15] It would not matter which equation is to be dropped from the system if there were no cross-equation restrictions. To verify whether such a numerical invariance holds under cross-

only one of the three symmetry restrictions (11.38) remains, i.e. $\gamma_{12} = \gamma_{21}$. If this cross-equation restriction is imposed, system (11.39) becomes:

$$\begin{aligned} s_1 &= \alpha_1 + \gamma_{11} \log(p_1/p_3) + \gamma_{12} \log(p_2/p_3) + \gamma_{1Q} \log(Q) + \varepsilon_1 \\ s_2 &= \alpha_2 + \gamma_{12} \log(p_1/p_3) + \gamma_{22} \log(p_2/p_3) + \gamma_{2Q} \log(Q) + \varepsilon_2. \quad (11.40) \end{aligned}$$

Since the regressors are predetermined and since the equations have common regressors, this system can be estimated by SUR subject to the cross-equation restriction of symmetry. This constrained system can be cast in the unconstrained common coefficient format of $\mathbf{y}_i = \mathbf{Z}_i\boldsymbol{\delta} + \boldsymbol{\varepsilon}_i$ if $\mathbf{y}_i := \begin{pmatrix} s_{1i} \\ s_{2i} \end{pmatrix}$, $\boldsymbol{\delta}' := \begin{bmatrix} \alpha_1 & \alpha_2 & \gamma_{11} & \gamma_{12} & \gamma_{22} & \gamma_{1Q} & \gamma_{2Q} \end{bmatrix}$,

$$\mathbf{Z}_i := \begin{bmatrix} 1 & 0 & \log(p_{1i}/p_{3i}) & \log(p_{2i}/p_{3i}) & 0 & \log(Q_i) & 0 \\ 0 & 1 & 0 & \log(p_{1i}/p_{3i}) & \log(p_{2i}/p_{3i}) & 0 & \log(Q_i) \end{bmatrix}.$$

In this way restricted SUR estimates of the seven cost share equation parameters may be obtained. Notice that the rest of the fifteen cost share equation parameters $(\alpha_3, \gamma_{13}, \gamma_{21}, \gamma_{23}, \gamma_{31}, \gamma_{32}, \gamma_{33}, \gamma_{3Q})$ can be calculated using the adding-up restrictions (11.36), the homogeneity restrictions (11.37), and the symmetry restrictions (11.38). For example, γ_{33} can be computed as $\gamma_{33} = -\gamma_{31} - \gamma_{32} = -\gamma_{13} - \gamma_{23} = (\gamma_{11} + \gamma_{12}) + (\gamma_{12} + \gamma_{22}) = \gamma_{11} + 2\gamma_{12} + \gamma_{22}$, so that the point estimate of γ_{33} is given by $\hat{\gamma}_{33} = \hat{\gamma}_{11} + 2\hat{\gamma}_{12} + \hat{\gamma}_{22}$,where $(\hat{\gamma}_{11}, \hat{\gamma}_{12}, \hat{\gamma}_{22})$ are the constrained SUR estimates.

11.4.2 β_{ki} random: Swamy's random coefficient model

In the Swamy (1970) model it is assumed that for each cross section or individual i the coefficients β_{ki} are randomly distributed around a mean and have a constant variance $\beta_{ki} \overset{iid}{\sim} (\overline{\beta}_{ki}, \Delta_i) \Rightarrow \boldsymbol{\beta}_i \overset{iid}{\sim} (\overline{\boldsymbol{\beta}}, \boldsymbol{\Delta})$. The model for individual i is the following $\mathbf{y}_i = \mathbf{X}_i(\overline{\boldsymbol{\beta}} + \boldsymbol{\mu}_i) + \boldsymbol{\varepsilon}_i$ with $\boldsymbol{\mu}_i \overset{iid}{\sim} (0, \boldsymbol{\Delta})$ and $\boldsymbol{\varepsilon}_i \overset{iid}{\sim} (0, \sigma_{ii}\mathbf{I}_T)$.

In general, the model can be written as follows for all individuals $\mathbf{y}_i = \mathbf{X}\overline{\boldsymbol{\beta}} +$

equation restrictions, let Σ^* be the 2×2 matrix of error covariances for the two-equation system obtained from dropping one equation from the three-equation system (11.39). It is the appropriate submatrix of the 3×3 error variance-covariance matrix Σ. There are two ways to obtain an estimate $\hat{\boldsymbol{\Sigma}}^*$ of Σ^*:

- estimate the two equations separately, thus ignoring the cross-equation restriction, and then use the residuals to compute $\hat{\boldsymbol{\Sigma}}^*$; equivalently, estimate the three equations separately by OLS, use the three-equation residuals to calculate the full $\hat{\boldsymbol{\Sigma}}$, and then extract the appropriate submatrix from $\hat{\boldsymbol{\Sigma}}$; in this way the numerical invariance is guaranteed end it does not matter which equation to drop;

- obtain $\hat{\boldsymbol{\Sigma}}^*$ from some technique that exploits the cross-equation restriction (such as the pooled OLS) but, then, the numerical invariance is not guaranteed.

$\mathbf{Z}\boldsymbol{\mu}+\boldsymbol{\varepsilon}$, with $\mathbf{X} := (\mathbf{X}_1', \mathbf{X}_2', \ldots, \mathbf{X}_N')'$ and $\mathbf{Z} := \begin{pmatrix} \mathbf{X}_1 & 0 & \cdots & \cdots & 0 \\ 0 & \mathbf{X}_2 & & & \vdots \\ \vdots & & \ddots & & \vdots \\ \vdots & & & \ddots & 0 \\ 0 & \cdots & \cdots & 0 & \mathbf{X}_N. \end{pmatrix}$.

The covariance matrix of the compound error term $\mathbf{Z}\boldsymbol{\mu} + \boldsymbol{\varepsilon}$ is equal to the following expression $\boldsymbol{\Phi} = E\left((\mathbf{Z}\boldsymbol{\mu} + \boldsymbol{\varepsilon})(\mathbf{Z}\boldsymbol{\mu} + \boldsymbol{\varepsilon})'\right)$,or:

$$\boldsymbol{\Phi} = \begin{pmatrix} \boldsymbol{\Phi}_{11} & 0 & \cdots & \cdots & 0 \\ 0 & \boldsymbol{\Phi}_{22} & & & \vdots \\ \vdots & \vdots & \ddots & & \vdots \\ \vdots & \vdots & & \ddots & 0 \\ 0 & 0 & \cdots & 0 & \boldsymbol{\Phi}_{NN} \end{pmatrix}$$

with $\boldsymbol{\Phi}_{ii} = \mathbf{X}_i \boldsymbol{\Delta} \mathbf{X}_i' + \sigma_{ii}\mathbf{I}_T$. The GLS estimator for $\overline{\boldsymbol{\beta}}$ is:

$$\begin{aligned} \hat{\overline{\boldsymbol{\beta}}} &= \left(\mathbf{X}'\boldsymbol{\Phi}^{-1}\mathbf{X}\right)^{-1}\mathbf{X}'\boldsymbol{\Phi}^{-1}\mathbf{y} \\ &= \left(\textstyle\sum_{j=1}^N \mathbf{X}_j'\boldsymbol{\Phi}_{jj}^{-1}\mathbf{X}_j\right)^{-1} \textstyle\sum_{i=1}^N \mathbf{X}_i'\boldsymbol{\Phi}_{ii}^{-1}\mathbf{y}_i \\ &= \textstyle\sum_{i=1}^N \mathbf{W}_i \mathbf{b}_i \end{aligned} \tag{11.41}$$

with $\mathbf{W}_i := \left\{\sum_{j=1}^N \left[\boldsymbol{\Delta} + \sigma_{jj}(\mathbf{X}_j'\mathbf{X}_j)^{-1}\right]^{-1}\right\}^{-1} \left[\boldsymbol{\Delta} + \sigma_{ii}(\mathbf{X}_i'\mathbf{X}_i)^{-1}\right]^{-1}$ and

$$\mathbf{b}_i = \left(\mathbf{X}_i'\mathbf{X}_i\right)^{-1}\mathbf{X}_i'\mathbf{y}_i. \tag{11.42}$$

The covariance matrix of $\hat{\overline{\boldsymbol{\beta}}}$ is:

$$\begin{aligned} \operatorname{cov}(\hat{\overline{\boldsymbol{\beta}}}) &= \left(\mathbf{X}'\boldsymbol{\Phi}^{-1}\mathbf{X}\right)^{-1} \\ &= \left(\textstyle\sum_{j=1}^N \mathbf{X}_j'\boldsymbol{\Phi}_{jj}^{-1}\mathbf{X}_j\right)^{-1} \\ &= \left\{\textstyle\sum_{j=1}^N \left[\boldsymbol{\Delta} + \sigma_{jj}(\mathbf{X}_j'\mathbf{X}_j)^{-1}\right]^{-1}\right\}^{-1}. \end{aligned}$$

The BLUE for an individual $\boldsymbol{\beta}_i$ is (see Rao (1965, p.29):

$$\begin{aligned} \hat{\boldsymbol{\beta}}_i &= \hat{\overline{\boldsymbol{\beta}}} + \boldsymbol{\Delta}\mathbf{X}_i'\left(\mathbf{X}_i\boldsymbol{\Delta}\mathbf{X}_i' + \sigma_{ii}\mathbf{I}_T\right)^{-1}\left(\mathbf{y}_i - \mathbf{X}_i\hat{\overline{\boldsymbol{\beta}}}\right) \\ &= \left(\boldsymbol{\Delta}^{-1} + \sigma_{ii}^{-1}\mathbf{X}_i'\mathbf{X}_i\right)^{-1}\left(\sigma_{ii}^{-1}\mathbf{X}_i'\mathbf{X}_i\mathbf{b}_i + \boldsymbol{\Delta}^{-1}\hat{\overline{\boldsymbol{\beta}}}\right), \end{aligned} \tag{11.43}$$

which results from the following minimization problem:

$$\min_{\boldsymbol{\beta}_i, \overline{\boldsymbol{\beta}}} \left\{\textstyle\sum_{i=1}^N \frac{(\mathbf{y}_i - \mathbf{X}_i\boldsymbol{\beta}_i)'(\mathbf{y}_i - \mathbf{X}_i\boldsymbol{\beta}_i)}{\sigma_{ii}} + \textstyle\sum_{i=1}^N \left(\boldsymbol{\beta}_i - \overline{\boldsymbol{\beta}}\right)'\boldsymbol{\Delta}^{-1}\left(\boldsymbol{\beta}_i - \overline{\boldsymbol{\beta}}\right)\right\},$$

where variances σ_{ii} and the covariance matrix $\mathbf{\Delta}$ are estimated as in Swamy (1970).

Both the GLS estimator for $\overline{\boldsymbol{\beta}}$ in (11.41) and the BLUE for $\boldsymbol{\beta}_i$ in (11.43) depend on the unknown variances $\mathbf{\Delta}$ and σ_{ii}. Swamy (1970) shows how the OLS estimator (11.42) can be used to obtain the (asymptotically) unbiased estimators:

$$\begin{aligned} \hat{\sigma}_{ii} &= \frac{(\mathbf{y}_i - \mathbf{X}_i\mathbf{b}_i)'(\mathbf{y}_i - \mathbf{X}_i\mathbf{b}_i)}{T-K}, \\ \hat{\mathbf{\Delta}} &= \frac{\mathbf{S_b}}{N-1} - \frac{1}{N}\sum_{i=1}^{N}\hat{\sigma}_{ii}\left(\mathbf{X}_i'\mathbf{X}_i\right)^{-1}, \end{aligned}$$

with the sample inner product matrix $\mathbf{S_b} = \sum_{i=1}^{N}\mathbf{b}_i\mathbf{b}_i' - \frac{1}{N}\sum_{i=1}^{N}\mathbf{b}_i\sum_{i=1}^{N}\mathbf{b}_i'$. Under certain regularity conditions the FGLS (or estimated generalized least squares - EGLS) estimator for $\overline{\boldsymbol{\beta}}$ that uses these variance estimates is under certain regularity conditions consistent and asymptotically efficient.[16]

All kinds of tests can be formulated in a similar way as before. If e.g. homogeneity is tested under the null: $H_0 : \boldsymbol{\beta}_1 = \boldsymbol{\beta}_2 = ...\boldsymbol{\beta}_N = \overline{\boldsymbol{\beta}}$, the (Wald) test statistic:

$$g = \sum_{i=1}^{N}\frac{\left(\mathbf{b}_i - \hat{\hat{\boldsymbol{\beta}}}\right)'\mathbf{X}_i'\mathbf{X}_i\left(\mathbf{b}_i - \hat{\hat{\boldsymbol{\beta}}}\right)}{\hat{\sigma}_{ii}}, \tag{11.44}$$

with $\hat{\hat{\boldsymbol{\beta}}} := \left(\sum_{i=1}^{N}\hat{\sigma}_{ii}^{-1}\mathbf{X}_i'\mathbf{X}_i\right)^{-1}\sum_{i=1}^{N}\hat{\sigma}_{ii}^{-1}\mathbf{X}_i'\mathbf{X}_i\mathbf{b}_i$, is according to Swamy (1970) asymptotically $\chi^2_{K(N-1)}$ distributed. Whether the coefficients should be fixed or random depends on whether the coefficients are correlated with the dependent variables. If they are, SUR should be used. If not, the GLS estimator can be used. The test statistic (11.44) can also be regarded as a test for aggregation bias, but the null hypothesis in this case is a more restrictive one. Acceptance of the null suggests that aggregation bias does not exist because the coefficients are fixed and identical.

11.5 All coefficients vary over individuals and time

In this case we have the general model:

$$y_{it} = \beta_{1it} + \sum_{k=2}^{K}\beta_{kit}x_{kit} + \varepsilon_{it} \quad (i = 1, ..., N; t = 1, ..., T) \tag{11.45}$$

with $\beta_{kit} := \overline{\beta}_k + \mu_{ki} + \lambda_{kt}$.This model can be written for individual i as:

$$\mathbf{y}_i = \mathbf{X}_i\overline{\boldsymbol{\beta}} + \mathbf{X}_i\boldsymbol{\mu}_i + \overline{\mathbf{Z}}_i\boldsymbol{\lambda} + \boldsymbol{\varepsilon}_i$$

[16] A difficulty with the estimator $\hat{\mathbf{\Delta}}$ is that it may not be nonnegative definite. Then the unbiasedness property is obviously destroyed, but an appropriate solution could be to use the natural sample quantity $\widetilde{\mathbf{\Delta}} = \frac{\mathbf{S_b}}{N-1}$ in that case. This estimator will be nonnegative definite and consistent.

with $\mathbf{y}_i := (y_{i1}, \ldots, y_{iT})'$, $\overline{\boldsymbol{\beta}} := \left(\overline{\boldsymbol{\beta}}_1', \ldots, \overline{\boldsymbol{\beta}}_K'\right)'$, $\boldsymbol{\mu}_i := (\mu_{1i}, \ldots \mu_{Ki})'$,

$$\mathbf{X}_i := \begin{pmatrix} x_{1i1} & \cdots & x_{Ki1} \\ \vdots & & \vdots \\ x_{1iT} & \cdots & x_{KiT} \end{pmatrix}, \overline{\mathbf{Z}}_i := \begin{pmatrix} \mathbf{x}_{i1}' & 0 & \cdots & 0 \\ 0 & \mathbf{x}_{i2}' & \cdots & 0 \\ \vdots & & \ddots & \vdots \\ 0 & \cdots & \cdots & \mathbf{x}_{iT}' \end{pmatrix}, \boldsymbol{\lambda} := (\boldsymbol{\lambda}_1', \ldots, \boldsymbol{\lambda}_T')',$$

$\boldsymbol{\lambda}_t := (\lambda_{1t}, \ldots, \lambda_{Kt})'$, and $\mathbf{x}_{it} := (x_{1it}, x_{2it}, \ldots, x_{Kit})$.

When all NT observations are considered the model can be rewritten as:

$$\mathbf{y} = \mathbf{X}\overline{\boldsymbol{\beta}} + \mathbf{Z}\boldsymbol{\mu} + \overline{\mathbf{Z}}\boldsymbol{\lambda} + \boldsymbol{\varepsilon}$$

with $\mathbf{y} := (\mathbf{y}_1', \ldots, \mathbf{y}_N')'$, $\mathbf{X} := (\mathbf{X}_1', \ldots, \mathbf{X}_N')'$, $\boldsymbol{\mu} := (\boldsymbol{\mu}_1', \ldots, \boldsymbol{\mu}_N')'$, $\overline{\mathbf{Z}} := (\bar{\mathbf{Z}}_1, \ldots, \bar{\mathbf{Z}}_N)'$,

$$\boldsymbol{\varepsilon} := (\boldsymbol{\varepsilon}_1', \ldots, \boldsymbol{\varepsilon}_N')', \text{ and } \mathbf{Z} := \begin{pmatrix} \mathbf{X}_1 & 0 & \cdots & 0 \\ 0 & \mathbf{X}_2 & \cdots & 0 \\ \vdots & & \ddots & \vdots \\ 0 & \cdots & \cdots & \mathbf{X}_N \end{pmatrix}.$$

11.5.1 Fixed coefficients

As before the parameters $\overline{\boldsymbol{\beta}}$, $\boldsymbol{\mu}$, and $\boldsymbol{\lambda}$ can be estimated with OLS. Notice, however, that the matrix $\left[\mathbf{X}, \mathbf{Z}, \overline{\mathbf{Z}}\right]$ has dimension $(NT \text{ X } (T+N+1)K)$ and rank $(T+N+1)K$ so that $2K$ parameters are redundant (K parameters μ_{ki} and K parameters λ_{kt}). A simple procedure is to scrap $(\mu_{1N}, ..., \mu_{KN}; \lambda_{1T}, ..., \lambda_{KT})$ by redefining the corresponding columns in $\mathbf{Z}$ and $\overline{\mathbf{Z}}$ and redefining the other parameters:

$$\begin{aligned} \lambda_{kt}^* &: = \lambda_{kt} - \lambda_{kT} \quad (t = 1, 2, ..., T-1) \\ \mu_{ki}^* &: = \mu_{ki} - \mu_{kN} \quad (i = 1, 2, ..., N-1) \end{aligned}$$

11.5.2 Stochastic coefficients: Hsiao's model

According to Hsiao (1974, 1975) the following assumptions are made regarding the stochastic coefficients: $\boldsymbol{\mu}_i \overset{iid}{\sim} (0, \boldsymbol{\Delta})$, $\boldsymbol{\lambda}_t \overset{iid}{\sim} (0, \mathbf{A})$, and $\varepsilon_i \overset{iid}{\sim} (0, \boldsymbol{\sigma}_\varepsilon^2 \mathbf{I}_T)$ with $E(\boldsymbol{\mu}_i \boldsymbol{\lambda}_t) = 0 \;\; \forall i, t; E(\boldsymbol{\mu}_i \varepsilon_{jt}) = 0 \;\; \forall i, j, t; E(\boldsymbol{\lambda}_t \varepsilon_{i\theta}) = 0 \;\; \forall i, t, \theta$ and the variance-covariance matrices $\mathbf{A}$ and $\boldsymbol{\Delta}$ are assumed to be diagonal matrices with diagonal elements α_k and δ_k, respectively. Assume that all the variances are known. Then, the variance-covariance matrix of the compound error term can be written as:

$$\begin{aligned} \boldsymbol{\Phi} &= E\left[\left(\mathbf{Z}\boldsymbol{\mu} + \overline{\mathbf{Z}}\boldsymbol{\lambda} + \boldsymbol{\varepsilon}\right)\left(\mathbf{Z}\boldsymbol{\mu} + \overline{\mathbf{Z}}\boldsymbol{\lambda} + \boldsymbol{\varepsilon}\right)'\right] \\ &= \mathbf{Z}(\mathbf{I}_N \otimes \boldsymbol{\Delta})\overline{\mathbf{Z}}' + \overline{\mathbf{Z}}(\mathbf{I}_T \otimes \mathbf{A})\overline{\mathbf{Z}}' + \boldsymbol{\sigma}_\varepsilon^2 \mathbf{I}_{NT}, \end{aligned}$$

so that the GLS estimator for $\overline{\boldsymbol{\beta}}$ is $\hat{\overline{\boldsymbol{\beta}}} = \left(\mathbf{X}'\boldsymbol{\Phi}^{-1}\mathbf{X}\right)^{-1} \mathbf{X}'\boldsymbol{\Phi}^{-1}\mathbf{y}$, being the BLUE of it, with variance-covariance matrix $\mathbf{var}(\hat{\overline{\boldsymbol{\beta}}}) = \left(\mathbf{X}'\boldsymbol{\Phi}^{-1}\mathbf{X}\right)^{-1}$. If one wants a prediction of the stochastic components for each individual, one can demonstrate that the predictor $\hat{\mu} = (\mathbf{I}_N \otimes \boldsymbol{\Delta})\overline{\mathbf{Z}}'\boldsymbol{\Phi}^{-1}(\mathbf{y} - \mathbf{X}\hat{\overline{\boldsymbol{\beta}}})$ is the BLU predictor.

The variances α_k, δ_k, and $\boldsymbol{\sigma}_\varepsilon^2$, however, are generally unknown. Various methods exist to estimate these variances. According to Hildreth and Houck (1968) the general equation (11.45) can be rewritten as:

$$\begin{aligned} y_{it} &= \sum_{k=1}^K \beta_{ki} x_{kit} + v_{it} \quad (i = 1, ..., N; t = 1, ..., T) \\ \text{with } \beta_{ki} &:= \overline{\beta}_k + \mu_{ki} \text{ and } v_{it} := \sum_{k=1}^K \lambda_{ki} x_{kit} + \varepsilon_{it} \,, \end{aligned} \quad (11.46)$$

where $E(v_{it}) = 0$ and $\text{var}(v_{it}) = E(v_{it}^2) = \sum_{k=1}^K \alpha_k x_{kit}^2 + \boldsymbol{\sigma}_\varepsilon^2 =: \phi_{it}$. Consider now a T vector $\boldsymbol{\phi}_i$ containing all T variances ϕ_{it}, then $\boldsymbol{\phi}_i = \dot{\mathbf{X}}_i \boldsymbol{\alpha}$ with

$$\dot{\mathbf{X}}_i := \begin{pmatrix} x_{1i1}^2 & \cdots & \cdots & x_{Ki1}^2 \\ x_{1i2}^2 & \cdots & \cdots & x_{Ki2}^2 \\ \vdots & & & \vdots \\ x_{1iT}^2 & \cdots & \cdots & x_{KiT}^2 \end{pmatrix} \text{ and } \boldsymbol{\alpha} := \begin{pmatrix} \alpha_1 + \boldsymbol{\sigma}_\varepsilon^2 \\ \alpha_2 \\ \vdots \\ \alpha_K \end{pmatrix}.$$

Assume now that $\widetilde{\mathbf{v}}_i := \mathbf{y}_i - \mathbf{X}_i \mathbf{b}_i$ is the vector of the OLS residuals of the ith cross section ($\mathbf{b}_i = \left(\mathbf{X}_i'\mathbf{X}_i\right)^{-1} \mathbf{X}_i'\mathbf{y}_i$); the vector of the squares of these OLS residuals is represented by $\dot{\widetilde{\mathbf{v}}}_i$ so that $E(\dot{\widetilde{\mathbf{v}}}_i) = \dot{\mathbf{M}}_i \boldsymbol{\phi}_i = \mathbf{F}_i \boldsymbol{\alpha}$, where $\dot{\mathbf{M}}_i$consists of the squares of the elements of the idempotent matrix $\mathbf{M}_i := \mathbf{I}_T - \mathbf{X}_i \left(\mathbf{X}_i'\mathbf{X}_i\right)^{-1} \mathbf{X}_i'$ and $\mathbf{F}_i := \dot{\mathbf{M}}_i \dot{\mathbf{X}}_i$. Repeating this for all cross sections we get $E(\dot{\widetilde{\mathbf{v}}}) = \mathbf{F}\boldsymbol{\alpha}$ with $\dot{\widetilde{\mathbf{v}}} := (\dot{\widetilde{\mathbf{v}}}_1', \dot{\widetilde{\mathbf{v}}}_2', \ldots, \dot{\widetilde{\mathbf{v}}}_N')'$ and $\mathbf{F} := (\mathbf{F}_1', \mathbf{F}_2', \ldots, \mathbf{F}_N')'$. When OLS is applied on this equation we find the unbiased estimator $\hat{\boldsymbol{\alpha}} = \left(\mathbf{F}'\mathbf{F}\right)^{-1} \mathbf{F}'\dot{\widetilde{\mathbf{v}}}$.

According to the same procedure, but now for each time period, we find the unbiased estimator for $\boldsymbol{\delta}$. Under certain regularity conditions $\hat{\boldsymbol{\alpha}}$ and $\hat{\boldsymbol{\delta}}$ and a consistent estimator of the residual variance $\boldsymbol{\sigma}_\varepsilon^2$ is then found as $\hat{\sigma}_v^2 = \sum_{i=1}^N \sum_{t=1}^T \frac{\widetilde{v}_{it} v_{it}^*}{NT}$ and $\hat{\sigma}_\varepsilon^2 = \hat{\sigma}_v^2 - \sum_{k=1}^K \hat{\alpha}_k x_{kit}^2$,where $\widetilde{v}_{it}$ are the residuals from an OLS estimation for each individual and v_{it}^* are the residuals from an OLS estimation for each time period.

From $\hat{\boldsymbol{\alpha}}$, $\hat{\boldsymbol{\delta}}$, and $\hat{\sigma}_\varepsilon^2$ we can directly derive consistent estimators for α_1 and δ_1. Notice also that if the panel is *unbalanced*, i.e. if there are missing observations, we have to divide the ith sum of squared residuals by T_i instead of by T.

11.6 Advantages of panel data reconsidered

Conclusively, we may tentatively take the following conclusions:

- panel data is efficient (w.r.t. random sampling)
- panel data is better for aggregation (e.g. income distribution may vary over time)

- identification is easier
- less multicollinearity .

After a general discussion in the introduction to this part of the book, the first three advantages will now be analyzed in greater detail.

1. Panel data is efficient (w.r.t. random sampling)

When the panel data consists e.g. of a yearly reinterview of the same sample of cross-sectional units, one could consider to draw a random sample of individuals each year.as an alternative. Additional sample variance is introduced in this way, however, so that data based on random sampling is less efficient. This is demonstrated by the following model in mixed form:

$$y_{it} = \bar{\beta}_1 + \lambda_t + \mu_i + \varepsilon_{it} \qquad (i = 1, ..., N; t = 1, ..., T) \tag{11.47}$$

with $\bar{\beta}_1$ and λ_t fixed but unknown parameters ($\sum_{t=1}^{T} \lambda_t = 0$), μ_i and ε_{it} are noncorrelated random variables with expectation zero and variances σ_μ^2 and σ_ε^2.

For individual $i(i = 1, ..., N)$ the model becomes:

$$\mathbf{y}_i = \bar{\beta}_1 \boldsymbol{\iota}_T + \mathbf{I}_T \boldsymbol{\lambda} + \mu_i \boldsymbol{\iota}_T + \boldsymbol{\varepsilon}_i \tag{11.48}$$

with $\boldsymbol{\lambda} := (\lambda_1, \lambda_2, ..., \lambda_T)'$ and $\boldsymbol{\iota}_T := (1, 1, ..., 1)'$. The covariance matrix of the compound error term then consists of the following blocks $\boldsymbol{\Phi}_{ij} = E\{(\mu_i \boldsymbol{\iota}_T + \boldsymbol{\varepsilon}_i)(\mu_j \boldsymbol{\iota}_T + \boldsymbol{\varepsilon}_j)'\} = \sigma_\mu \boldsymbol{\iota}_T \boldsymbol{\iota}_T' + \sigma_\varepsilon^2 \mathbf{I}_T$ for $i = j$ and $\boldsymbol{\Phi}_{ij} = 0$ for $i \neq j$, so that the covariance matrix $\boldsymbol{\Phi}$ is block-diagonal:

$$\boldsymbol{\Phi} = \begin{pmatrix} \boldsymbol{\Phi}_{11} & \cdots & \cdots & 0 \\ 0 & \boldsymbol{\Phi}_{22} & \cdots & 0 \\ \vdots & \vdots & \ddots & \vdots \\ 0 & 0 & \cdots & \boldsymbol{\Phi}_{NN} \end{pmatrix} \text{ with } \boldsymbol{\Phi}^{-1} = \begin{pmatrix} \boldsymbol{\Phi}_{11}^{-1} & \cdots & \cdots & 0 \\ 0 & \boldsymbol{\Phi}_{22}^{-1} & \cdots & 0 \\ \vdots & \vdots & \ddots & \vdots \\ 0 & 0 & \cdots & \boldsymbol{\Phi}_{NN}^{-1} \end{pmatrix}$$

If we represent model (11.48) for all individuals as:

$$\mathbf{y} = \bar{\beta}_1 \boldsymbol{\iota}_{NT} + \boldsymbol{\iota}_N \otimes \mathbf{I}_T \boldsymbol{\lambda} + \boldsymbol{\mu} \otimes \boldsymbol{\iota}_T + \boldsymbol{\varepsilon} \tag{11.49}$$

with $\boldsymbol{\mu} := (\boldsymbol{\mu}_1, \boldsymbol{\mu}_2, ..., \boldsymbol{\mu}_N)'$. The unknown parameter $\bar{\beta}_1$ in (11.49) can be estimated by GLS which is the BLUE $\hat{\bar{\beta}}_1 = (\boldsymbol{\iota}_{NT}' \boldsymbol{\Phi}^{-1} \boldsymbol{\iota}_{NT})^{-1} \boldsymbol{\iota}_{NT}' \boldsymbol{\Phi}^{-1} \mathbf{y} = (\sum_i \sum_i \boldsymbol{\Phi}^{ii})^{-1} \sum_i \boldsymbol{\Phi}^{ii} y_i$, so that the cyclical component can be estimated as $\hat{\lambda}_t = (\bar{y}_{.t} - \bar{y}_{..})$, with $\sum_{t=1}^{T} \lambda_t = 0$.

The BLUE of μ_i is $\hat{\mu}_i = \frac{T\sigma_\mu^2}{T\sigma_\mu^2 + \sigma_\varepsilon^2}((\bar{y}_{i.} - \hat{\bar{\beta}}_1)$.

If we are interested in the change of the yearly means: $\lambda_t - \lambda_{t-1}$, an unbiased estimator of this change can be found from (11.47), $\bar{y}_{.t} = \bar{\beta}_1 + \lambda_t + \frac{\sum_i \mu_i}{N} + \frac{\sum_i \varepsilon_{it}}{N}$ and $\bar{y}_{..} = \bar{\beta}_1 + \frac{\sum_i \mu_i}{N} + \frac{\sum_i \sum_t \varepsilon_{it}}{NT}$ or $\bar{y}_{.t} - \bar{y}_{..} = \lambda_t + \frac{\sum_i \varepsilon_{it}}{N} - \frac{\sum_i \sum_t \varepsilon_{it}}{NT}$, or the unbiased estimator of $\lambda_t - \lambda_{t-1}$ is:

$$\hat{\lambda}_t - \hat{\lambda}_{t-1} = (\bar{y}_{.t} - \bar{y}_{..}) - (\bar{y}_{.t-1} - \bar{y}_{..}) = \bar{y}_{.t} - \bar{y}_{.t-1}. \tag{11.50}$$

With a yearly reinterview of the same sample the variance of the estimator (11.50) satisfies:

$$
\begin{aligned}
var(\hat{\lambda}_t - \hat{\lambda}_{t-1}) &= var(\bar{y}_{.t} - \bar{y}_{.t-1}) \\
&= var(\bar{y}_{.t}) + var(\bar{y}_{.t-1}) - 2cov(\bar{y}_{.t}, \bar{y}_{.t-1}) \\
&= E\{(\frac{\sum_i \mu_i}{N} + \frac{\sum_i \varepsilon_{it}}{N})^2\} + E\{(\frac{\sum_i \mu_i}{N} + \frac{\sum_i \varepsilon_{i,t-1}}{N})^2\} \\
&\quad -2E\{(\bar{\beta}_1 + \lambda_t + \frac{\sum_i \mu_i}{N} + \frac{\sum_i \varepsilon_{it}}{N})(\bar{\beta}_1 + \lambda_{t-1} + \frac{\sum_i \mu_i}{N} + \frac{\sum_i \varepsilon_{i,t-1}}{N})\} \\
&\quad -(\bar{\beta}_1 + \lambda_t)(\bar{\beta}_1 + \lambda_{t-1})
\end{aligned}
$$

$$
\begin{aligned}
&= 2\frac{\sigma_\mu^2}{N} + 2\frac{\sigma_\varepsilon^2}{N} - 2E\left\{\frac{\sum_i \mu_i}{N}\right\}^2 - 2E\left\{\left(\frac{\sum_i \varepsilon_{it}}{N}\right)\left(\frac{\sum_i \varepsilon_{i,t-1}}{N}\right)\right\} \\
&= 2\frac{\sigma_\mu^2 + \sigma_\varepsilon^2}{N} - 2\frac{\sigma_\mu^2}{N} = 2\frac{\sigma_\varepsilon^2}{N}. \qquad (11.51)
\end{aligned}
$$

There is no covariance between $\bar{y}_{.t}$ and $\bar{y}_{.t-1}$ in the case of two random samples, or (11.51) becomes in this case:

$$var(\hat{\lambda}_t - \hat{\lambda}_{t-1}) = var(\bar{y}_{.t} - \bar{y}_{.t-1}) = 2\frac{\sigma_\mu^2 + \sigma_\varepsilon^2}{N}. \qquad (11.52)$$

Hence, the estimator for $(\lambda_t - \lambda_{t-1})$, based on panel data, is more efficient than its estimator based on a completely new sample (both are unbiased estimators, but the latter estimator is not efficient). This is an illustration of the efficiency gain in the case of setting up a panel.

2. No aggregation bias in the case of panel data.

Assume a society that consists of two individuals and assume that their consumption level at period t is given by the following permanent income relationship:

$$C_{it} = \underbrace{\alpha_i Y_{it} + \beta_i Y_{i,t-1}}_{\eta_i Y_i^p} + \delta_i + \varepsilon_{it} \quad (i = 1, 2; \ \ t = 1, 2, ..., T) \qquad (11.53)$$

with C_{it} the (nominal) consumption level of individual i during period t, y_{it} the disposable income of individual i during period t, and α_i, β_i, δ_i $(i = 1, 2)$ being 6 unknown parameters.

Define now the aggregate consumption level $C_t := C_{1t} + C_{2t}$ and the aggregate income level $Y_t := Y_{1t} + Y_{2t}$ with income shares $w_{it} = \frac{y_{it}}{y_t}$ $(i = 1, 2)$. Then, the aggregate consumption function can be rewritten as:

$$
\begin{aligned}
C_t &= \alpha_1 Y_{1t} + \alpha_2 Y_{2t} + \beta_1 Y_{1,t-1} + \beta_2 Y_{2,t-1} + \delta_1 + \delta_2 + \varepsilon_{1t} + \varepsilon_{2t} \qquad (11.54) \\
&= (\alpha_1 w_{1t} + \alpha_2 w_{2t}) Y_t + (\beta_1 w_{1,t-1} + \beta_2 w_{2,t-1}) Y_{t-1} + \delta_1 + \delta_2 + \varepsilon_{1t} + \varepsilon_{2t}.
\end{aligned}
$$

It follows from the aggregate consumption function (11.54) that the coefficients of the contemporary and the previous aggregate available income depend on the income distribution during the various periods. This implies that estimation of (11.54) based on aggregate data and a prediction on the base of this estimation is advisable only if the income distribution is constant over time. But in such a case prediction is mostly not very interesting. Why? Because macroeconomic models are usually constructed to evaluate a certain economic policy and such economic policy wants to change the structure of the economy, as e.g. the income distribution. Therefore, many interesting cases of structural estimations using aggregate data could deliver misleading information. This could be prevented when estimating the unknown parameters by panel data. In the latter case no aggregation bias is committed and due analysis is attributed to often occurring heterogeneity.

3. Ability to solve the identification problem.

A third advantage of the availability of panel data is that it reduces indentification problems. Although this advantage may enter for different issues, in most cases it involves identification in the presence of endogenous regressors or measurement error, robustness to omitted variables and the identification of individual dynamics. We focus here on only one aspect (see Verbeek (2000), pp. 311 – 313, for some other issues).

Assume e.g. the estimation of a production function free of managerial bias. When specifying production functions of small enterprises (as agricultural enterprises) it is generally advisable to consider managerial capacity as an input. Since this quantity is not observable the researcher, who has the availability of only time series data or cross-sectional data, has to decide either to delete this variable or to use an approximating variable. But since managerial capacity is expected to be correlated with the other exogenous variable both approaches will lead to biased estimations. A possible solution is to use panel data and to introduce in this way a latent variable for each firm the value of which could represent managerial ability. There are also many other cases of identification improvement (e.g. through disaggregation of strongly correlated aggregate time series).

11.7 Incomplete panels and selection bias

Empirical panel databases are often incomplete because people may refuse cooperation after various waves of the panel, households may not be located again or have split up, firms may have finished business or have merged with another firm, investments may be closed down, firms may enter business at a later stage, refreshment samples may have been drawn to compensate attrition, the panel may be a rotating panel (where each period a fixed proportion of the units is replaced), etc. A consequence of all these events with *missing data* is that the resulting panel database has a total number of observations which is substantially smaller than NT. If the number of time observations differs from individual to individual, we have T_i with $N \sum_{j=1}^{N} T_j < NT$ and the formulae in the previous sections are changed accordingly, e.g. the contemporaneous variance-covariance matrix (11.34) of the SUR model can be estimated in the first stage as: $\hat{\Sigma} = \left\{ \frac{\hat{\varepsilon}'_i \hat{\varepsilon}'_j}{(T_i - K)^{1/2}(T_j - K)^{1/2}} \right\}$. We call such a panel with missing data an *unbalanced panel.*

Missing observations may be created deliberately, as in the case of *split* and *rotating panels.* Given that in a lot of cases panel data is not available, but the researcher can build up a pseudo panel from surveys, we will first treat this case.

Finally, one undesirable consequence of using incomplete panels is that of *selection bias*: if the individuals are incompletely observed because of an endogenous reason, the use of this data may lead to biased estimators and misleading tests.

11.7.1 Incomplete panels

Rotating or rolling panels

The *rotating panels* are designed to keep the same number of individuals in the sample. In each period, a fixed proportion of individuals is dropped from the sample, and is replaced by an equal number of new individuals that are freshly surveyed. The individuals are rotated mainly to replace voluntary drop outs, and to avoid interviewing again and again the same household, reducing the degree of nonresponse. Finally, the individuals are rotated in order to maintain representativeness of the sample. If one would use the same sample for example over ten years without incorporation of new individuals, the sample would loose representativeness, especially over the younger cohort of individuals, because of ageing of the ones included in the sample.

Split panels

The *split panel* design is partly a panel and partly a series of nonoverlapping cross sections. Nijman and Verbeek (1990) correctly state that "Under certain conditions, the split panel can yield more efficient estimates than the use of a panel or a series of independent cross sections".

Pseudo panels

For some countries, panel data is not available. As an alternative, there may be large random samples over the population that are repeated regularly (e.g. annually). For these repeated cross-section surveys, it might be possible to track the same households over time to obtain a panel.

It has been argued that these *pseudo panels* may be available over a longer period of time compared to a genuine panel database, and that they don't suffer from the attrition problems. Moreover, pseudo panel data can provide a relatively long time series of observations compared to panel data.

11.7.2 Attrition and selection bias

Attrition occurs when individuals having participated in one or more waves (i.e. the cross sections of the panel) leave the panel, and do not return in the panel (if they leave the panel only temporarily, we have the issue of missing observations). This can be caused by removal, emigration or decease, but also by the fact that the individual 'gets tired' of answering the questionnaire each time period. *Ignorable attrition* can also occur, when the conditional moments of the panel data results are the same as without attrition. If a balanced panel is used, this leads to an efficiency loss because the information of the incompletely observed observations is not used for the estimation. A potential consequence of using unbalanced panels is the danger of *selection* or *selectivity bias*: if individuals are incompletely observed because of an endogenous reason, the use of either the balanced subpanel (where each individual with incomplete information is discarded from the panel so that the balanced subpanel is only defined for the completely observed individuals only) or the unbalanced panel may lead to biased estimators and misleading tests. To be more specific, consider the (static) panel data model $y_{it} = \mathbf{x}_{it}'\boldsymbol{\beta} + \mu_i + \varepsilon_{it}$ and define the indicator or *response* variable $r_{it} := 1$ if (y_{it}, x_{it}) is observed and $r_{it} := 0$ if (y_{it}, x_{it}) is not observed so that $T_i = \sum_{t=1}^{T} r_{it}$. We say that the observations on (y_{it}, x_{it}) are *missing at random* if r_{it} is independent of μ_i and ε_{it}. This means that conditioning upon the outcome of the selection process does not affect the conditional distribution of y_{it} given x_{it}.[17] To specify estimators with randomly missing data, the 'available means' can be defined as $\overline{y}_{i.} := \sum_{t=1}^{T} r_{it}y_{it} / \sum_{t=1}^{T} r_{it}$ and $\overline{\mathbf{x}}_{i.} := \sum_{t=1}^{T} r_{it}\mathbf{x}_{it} / \sum_{t=1}^{T} r_{it}$, so that e.g. the FE estimator (11.4) of the slope coefficients can be rewritten as:

$$\hat{\boldsymbol{\beta}}_{sFE} = \left(\sum_{i=1}^{N} \sum_{t=1}^{T} r_{it}(\mathbf{x}_{it} - \overline{\mathbf{x}}_{i.})(\mathbf{x}_{it} - \overline{\mathbf{x}}_{i.})' \right)^{-1} \sum_{i=1}^{N} \sum_{t=1}^{T} r_{it}(\mathbf{x}_{it} - \overline{\mathbf{x}}_{i.})(y_{it} - \overline{y}_{i.}).$$

Randomly missing data, however, may be unrealistic so that in many cases the response variable r_{it} will be correlated with one or two unobservables in the model, e.g. estimating the impact of the unemployment rate on individual wages may be disturbed by the possibility that people with relatively high wages could be more likely to leave the labor market in case of increasing unemployment. If r_{it} depends upon μ_i or ε_{it}, *selection bias* may arise in the FE and RE estimators. Selection bias will be present in the estimation of the individual effects μ_i only if r_{it} depends on μ_i and not on ε_{it}. On the contrary, if r_{it} depends upon ε_{it} all the estimators ($\hat{\boldsymbol{\beta}}_s$ and $\hat{\mu}_i$) will be subject to selection bias (inconsistency), except when this dependency is purely time-invariant in the case of the FE estimator (see Verbeek and Nijman (1992 and 1996)). Hence, the FE estimator is more robust against selection bias than the RE estimator.

[17] If we want to concentrate upon the balanced sub-panel, the conditioning is upon $r_{i1} = ... = r_{iT} = 1$ and we require that r_{it} is independent of μ_i and $\varepsilon_{i1}, ..., \varepsilon_{iT}$. Then, the consistency properties of the estimators are not affected.

Concluding, we may observe that the four different estimators, FE and RE, using either the balanced subpanel or an unbalanced panel, usually all suffer differently from selection bias. As the estimators using an unbalanced panel are (asymptotically) efficient within a particular class of estimators (more informative since using more relevant observations), test statistics of the Hausman type (11.20) and (11.32) can be used to test the differences between estimators of balanced sub-panels and unbalanced panels (H_0 : plim $\hat{\boldsymbol{\beta}}^B_{FE,RE}$ = plim $\hat{\boldsymbol{\beta}}^U_{FE,RE}$, where the superscripts B and U refer to the balanced and unbalanced sample, respectively):

$$\widehat{\xi}_H = \left(\hat{\boldsymbol{\beta}}^B_{FE,RE} - \hat{\boldsymbol{\beta}}^U_{FE,RE}\right)' \left(\widehat{\mathbf{avar}}^B_{FE,RE} - \widehat{\mathbf{avar}}^U_{FE,RE}\right)^{-1} \left(\hat{\boldsymbol{\beta}}^B_{FE,RE} - \hat{\boldsymbol{\beta}}^U_{FE,RE}\right) \overset{as}{\sim} \chi^2_K,$$

where $\widehat{\mathbf{avar}}^B_{FE,RE}$ and $\widehat{\mathbf{avar}}^U_{FE,RE}$ represent (consistent) estimates of the asymptotic variance-covariance matrices of $\hat{\boldsymbol{\beta}}^B_{FE,RE}$ and $\hat{\boldsymbol{\beta}}^U_{FE,RE}$ in the case of an FE and RE estimation, respectively.

11.8 Dynamic linear panel data models

Up to now we have assumed that the vector of explanatory variables $\mathbf{x}_{it}$ is fixed and does not contain random variables (strict exogeneity assumption), but we know from daily practice that a potential feedback from y_{it} to $\mathbf{x}_{i\vartheta}$ for $\vartheta > t$ very often occurs. With an unobserved effect as in the case of panel data, the most revealing form of the strict exogeneity assumption in a linear panel data model is:

$$E(y_{it} \mid \mathbf{x}_{i1}, \mathbf{x}_{i2}, \cdots, \mathbf{x}_{iT}, \mu_i) = E(y_{it} \mid \mathbf{x}_{it}, \mu_i) = \mathbf{x}'_{it}\boldsymbol{\beta} + E(\mu_i \mid \mathbf{x}_{it}), \qquad (11.55)$$

for $t = 1, 2, \cdots, T$ and where μ_i is the unobserved individual effect in a panel data model, be it fixed or random.

Nevertheless, in any panel data application we should initially focus on two questions:

i) Is the unobserved effect ξ_i uncorrelated with $\mathbf{x}_{it}$ *for all t*?

ii) Is the strict exogeneity assumption reasonable?

As already indicated we should start from real-world applications to provide answers to these basic questions. Numerous examples can be put forward, but let us only consider three:

a) *Distributed lag model*: in many applications cases (patent awards, consumption expenditures,...) we have a model with (distributed) lags of (some) explanatory variable(s), say e.g. the Hausman, Hall, and Griliches (1984) linear model of patents awarded to a firm and current and past levels (up to five lags) of R&D expenditures:

$$patents_{it} = \theta_t + \mathbf{z}'_{it}\boldsymbol{\delta} + \zeta_0 RD_{it} + \zeta_1 RD_{i,t-1} + \cdots + \zeta_5 RD_{i,t-5} + \mu_i + \varepsilon_{it},$$

where θ_t denotes a time-varying intercept, RD_{it} is spending on R&D for firm i at period t and $\mathbf{z}_{it}$ contains exogenous variables such as firm size; the unobservable variable μ_i represents firm heterogeneity, that might be correlated with current, past, and future R&D expenditures;

b) *Lagged dependent variables*: of course, this situation where the strict exogeneity assumption is necessarily false is very often occurring, e.g. a very simple model of wage determination with unobserved heterogeneity is:

$$\log(wage_{it}) = \gamma \log(wage_{i,t-1}) + \mu_i + \varepsilon_{it} \tag{11.56}$$

for $t = 1, 2, \cdots, T$; in many cases people are interested to verify how persistent wages are (as measured by the size of γ) after controlling for unobserved heterogeneity ξ_i (e.g. individual productivity); for $y_{it} := \log(wage_{it})$, a standard assumption would be $E(\varepsilon_{it} \mid y_{i,t-1}, \cdots, y_{i0}, \mu_i) = 0$,which means that all of the dynamics are captured by the current lag;

c) *Program evaluation*: a standard model for estimating the effects of job training or other programs on subsequent wages is:

$$\log(wage_{it}) = \theta_t + \mathbf{z}_{it}'\boldsymbol{\delta} + \zeta_1 prog_{it} + \mu_i + \varepsilon_{it}, \tag{11.57}$$

where θ_t is again a time-varying intercept, $\mathbf{z}_{it}$ is a vector of (exogenous) observable characteristics that affect wages and may also be correlated with program participation. Evaluation databases are often collected at two time periods. At $t = 1$, no one has participated in the program, so that $prog_{i1} = 0$ for all $i = 1, 2, \cdots, N$. Then, a subgroup is chosen to participate in the program (or the individuals choose to participate), and subsequent wages are observed for the control and treatment groups in $t = 2$. The reason for including the individual effect μ_i in (11.57) is the case of omitted variables: if individuals choose whether or not to participate in the program, that choice could be correlated with individual ability. This possibility is often called the *self-selection problem.* Alternatively, administrators might assign people based on characteristics that the econometrician cannot observe. The other issue is the strict exogeneity assumption of the explanatory variables, particularly $prog_{it}$. Although the error terms ε_{it} may generally be uncorrelated with $prog_{it}$, but what can we say about the correlation between ε_{it} and, say, $prog_{i,t+1}$? Future program participation could depend on ε_{it} if people choose to participate in the future based on shocks to their wage(s) in the past, or if administrators choose people as participants at period $t+1$ who had a low ε_{it}. Another issue is that the training program could have lasting effects; if so, then we should include lags of $prog_{it}$ in model (11.57) or the program itself might last more than one time unit, in which case $prog_{it}$ can be replaced by a series of dummy variables for how long individual i at period t has been subject to the program.

A model with lagged dependent variables is called a *dynamic panel data model* and will be discussed in this section.[18]. Estimating dynamic panel data models as in (11.56), we have to take account of the correlation between the lagged dependent variable(s) and the residuals. A general formulation of such

[18]Note that almost all distributed lag models can easily be converted into models with lagged dependent variables.

a model is (11.1). For the sake of simple exposition we choose the number of lags as one ($p = 1$), so that with one lagged dependent variable as a regressor, we consider the following simple stationary $AR(1)$ panel data model:

$$y_{it} = \gamma y_{i,t-1} + \mu_i + \varepsilon_{it} \tag{11.58}$$

with $i = 1, 2, ..., N$, $t = 1, 2, ..., T$, $|\gamma| < 1$, and $\varepsilon_{it} \overset{iid}{\sim} (0, \boldsymbol{\sigma}_\varepsilon^2)$.

Autoregressive panel data models can be analyzed using fixed effects (FE) and random effects (RE) or even mixed.

FE or dummy variables estimator

In this section we have a look at the FE dummy variable estimator of the above autoregressive panel data model. Equation (11.58) in deviations from sample means (also called the *within transformation*) yields:

$$y_{it} - \overline{y}_{i.} = \gamma \left(y_{i,t-1} - \overline{y}_{i.-1} \right) + \left(\varepsilon_{it} - \overline{\varepsilon}_{i.} \right).$$

The OLS estimator for γ (also called the 'within' estimator or as before the FE estimator) is:

$$\hat{\gamma}_{FE} = \frac{\sum_{i=1}^{N} \sum_{t=1}^{T} \left(y_{i,t} - \overline{y}_{i.} \right) \left(y_{i,t-1} - \overline{y}_{i.-1} \right)}{\sum_{i=1}^{N} \sum_{t=1}^{T} \left(y_{i,t-1} - \overline{y}_{i.-1} \right)^2}, \tag{11.59}$$

with $\overline{y}_{i.} := \frac{\sum_{t=1}^{T} y_{it}}{T}$ and $\overline{y}_{i.-1} := \frac{\sum_{t=1}^{T} y_{i,t-1}}{T}$.

Investigating the properties of $\hat{\gamma}_{FE}$, we substitute (11.58) into (11.59). This yields the following expression for the FE estimator:

$$\hat{\gamma}_{FE} = \gamma + \frac{\frac{1}{NT} \sum_{i=1}^{N} \sum_{t=1}^{T} \left(\varepsilon_{it} - \overline{\varepsilon}_{i.} \right) \left(y_{i,t-1} - \overline{y}_{i.-1} \right)}{\frac{1}{NT} \sum_{i=1}^{N} \sum_{t=1}^{T} \left(y_{i,t-1} - \overline{y}_{i.-1} \right)^2}. \tag{11.60}$$

For this estimator to be consistent, the probability limit of the second term of the expression should be zero, such that $\hat{\gamma}_{FE} = \gamma$. This is not the case: $\underset{N\to\infty}{plim} (\cdot) \neq 0$. More specifically:

$$\begin{aligned} & \underset{N\to\infty}{plim} \frac{1}{NT} \sum_{i=1}^{N} \sum_{t=1}^{T} \left(\varepsilon_{it} - \overline{\varepsilon}_{i.} \right) \left(y_{i,t-1} - \overline{y}_{i,-1} \right) \\ = & -\frac{\sigma_\varepsilon^2}{T^2} \frac{(T-1) - T\gamma + \gamma^T}{(1-\gamma)^2} \neq 0. \end{aligned} \tag{11.61}$$

If both $N \to \infty$ and $T \to \infty$, $\hat{\gamma}_{FE}$ is a consistent estimator for γ, but if T is fixed and only $N \to \infty$, which is often the case in panel databases with a large number of cross sections, the estimator is inconsistent (see expression (11.61) and also Hsiao (1986)).

To solve this inconsistency problem, a different transformation should be applied, namely by taking time differences:

$$y_{it} - y_{i,t-1} = \gamma \left(y_{i,t-1} - y_{i,t-2} \right) + \left(\varepsilon_{it} - \varepsilon_{i,t-1} \right), \; i = 1, ..., N; t = 1, ..., T. \tag{11.62}$$

Since $y_{i,t-1}$ and $\varepsilon_{i,t-1}$ are correlated ($\rho\left(y_{i,t-1}, \varepsilon_{i,t-1}\right) \neq 0$, even if $T \to \infty$), OLS of equation (11.62) is not consistent. This transformation suggests an instrumental variable (IV) approach, however, say with $y_{i,t-2}$ as an instrument for $y_{i,t-1} - y_{i,t-2}$: $\rho\left(y_{i,t-2}, y_{i,t-1} - y_{i,t-2}\right) \neq 0$ but $\rho\left(y_{i,t-2}, \varepsilon_{i,t-1}\right) = 0$ unless ε_{it} has autocorrelation. The IV estimator is:

$$\hat{\gamma}_{IV} = \frac{\sum_{i=1}^{N} \sum_{t=2}^{T} y_{i,t-2}\left(y_{i,t-1} - y_{i,t-1}\right)}{\sum_{i=1}^{N} \sum_{t=2}^{T} y_{i,t-2}\left(y_{i,t-1} - y_{i,t-2}\right)}. \tag{11.63}$$

Instead of considering $y_{i,t-2}$ as an IV for $\Delta y_{i,t-1}$,we may also consider $\Delta y_{i,t-2}$,that also leads under regularity conditions to a consistent estimator, which however requires one estimator more. Remember that an IV estimator is generally set up for the simple linear regression model $\mathbf{y} = \mathbf{X}\boldsymbol{\beta} + \boldsymbol{\varepsilon}$ as $\hat{\boldsymbol{\beta}}_{IV} = \left(\mathbf{Z}'\mathbf{X}\right)^{-1}\left(\mathbf{Z}'\mathbf{Y}\right)$ with $var\left(\hat{\boldsymbol{\beta}}_{IV}\right) = \sigma_{\varepsilon}^2\left(\left(\mathbf{X}'\mathbf{Z}\right)\left(\mathbf{Z}'\mathbf{Z}\right)\left(\mathbf{Z}'\mathbf{X}\right)\right)^{-1}$, where the $\mathbf{Z}$ variables should not be (asymptotically) correlated with the residuals: $\text{plim}\frac{\mathbf{Z}'\boldsymbol{\varepsilon}}{N} = 0$.

Thus, a necessary condition for the IV estimator in (11.63) to be consistent is:

$$\text{plim}\frac{1}{N\left(T-1\right)} \sum_{i=1}^{N} \sum_{t=2}^{T} \left(\varepsilon_{it} - \varepsilon_{i,t-1}\right) y_{i,t-2} = 0 \tag{11.64}$$

for either N or T or *both* going to ∞. In (11.62), $y_{i,t-2} - y_{i,t-3}$ may be an alternative instrument for $y_{i,t-1} - y_{i,t-2}$ instead of $y_{i,t-2}$, but requires an additional lag so that the number of sample observations is reduced (one sample period is lost which might be very harmful in panel data cases with a small number of time observations).

Notice that a method of moments (MoM) estimator can unify these estimators and eliminate the disadvantages of reduced sample sizes. According to (11.64) the underlying moment condition (see Section (2.2)):

$$E\left\{\left(\varepsilon_{it} - \varepsilon_{i,t-1}\right) y_{i,t-2}\right\} \cong \text{plim}\tfrac{1}{N(T-1)} \sum_{i=1}^{N} \sum_{t=2}^{T} \left(\varepsilon_{it} - \varepsilon_{i,t-1}\right) y_{i,t-2} = 0,$$

where $y_{i,t-2}$ is substituted by $\Delta y_{i,t-2}$ if we consider this variable as an IV. Hence, for any choice of an IV we have a separate moment condition. Following Arellano and Bond (1991) the list of instruments can vary over time. For example, for $T = 4$ we get: $E\left\{(\varepsilon_{i2} - \varepsilon_{i1})y_{i0}\right\} = 0$ for $t = 2$, $E\left\{(\varepsilon_{i3} - \varepsilon_{i2})y_{i0}\right\} = 0$ and $E\left\{(\varepsilon_{i3} - \varepsilon_{i2})y_{i1}\right\} = 0$ for $t = 3$. Finally, for $t = 4$,there are three moment conditions and three valid instruments: $E\left\{(\varepsilon_{i4} - \varepsilon_{i3})y_{i0}\right\} = 0$, $E\left\{(\varepsilon_{i4} - \varepsilon_{i3})y_{i1}\right\}$ and $E\left\{(\varepsilon_{i4} - \varepsilon_{i3})y_{i2}\right\}$. All these moment conditions can be exploited in a GMM framework by defining subsequent error differences as $\Delta\boldsymbol{\varepsilon}_i := [(\varepsilon_{i2} - \varepsilon_{i1}), \ldots, (\varepsilon_{i,T} - \varepsilon_{i,T-1})]'$ and a block diagonal matrix of IVs as $\mathbf{Z}_i$ with the vectors $y_{i0}, (y_{i0}, y_{i1})$,$(y_{i0}, y_{i1}, y_{i2}) \ldots (y_{i0}, \ldots y_{i,T-2})$ on the diagonal (block) which, taking account of (11.62), leads to the set of GMM moment conditions $E[\mathbf{Z}_i' \Delta\boldsymbol{\varepsilon}_i] = E[\mathbf{Z}_i'\left(\Delta\mathbf{y}_i - \gamma\Delta\mathbf{y}_{i,-1}\right)] = 0$. Clearly, the number of moment conditions $k \geqq m$, the number of explanatory variables, so that the resulting GMM estimator of γ can directly be derived from (2.9) as

$$\hat{\gamma}_{GMM} = \left(\sum_{i=1}^{N} \Delta\mathbf{y}_{i,-1}' \mathbf{Z}_i \hat{\mathbf{W}}_n \mathbf{Z}_i' \Delta\mathbf{y}_{i,-1}\right)^{-1} \sum_{i=1}^{N} \Delta\mathbf{y}_{i,-1}' \mathbf{Z}_i \, \hat{\mathbf{W}}_n \mathbf{Z}_i' \Delta\mathbf{y}_i, \tag{11.65}$$

where the optimal estimated weighting matrix $\hat{\mathbf{W}}_n$ is given by the estimate of the inverse variance-covariance matrix of the sample moments (moments conditons) or $\hat{\mathbf{W}}$ is given by $\hat{\mathbf{W}} := [\mathbf{var}(\mathbf{Z}_i'\Delta\varepsilon_i)]^{-1} = [E(\mathbf{Z}_i'\Delta\varepsilon_i\Delta\varepsilon_i'\mathbf{Z}_i)]^{-1}$ so that its sample analog $\hat{\mathbf{W}}_n$ is given by $\hat{\mathbf{W}}_n = \left(\frac{1}{N}\sum_{i=1}^N \mathbf{Z}_i'\Delta\hat{\varepsilon}_i\Delta\hat{\varepsilon}_i'\mathbf{Z}_i\right)^{-1}$,where $\hat{\varepsilon}_i$ is the residual vector form a first step consistent estimator, for e.g. using $\mathbf{W}_n = \mathbf{I}_k$.

Dynamic panel data models with exogenous variables

If other (exogenous) regressors are included, such as in:

$$y_{it} = \mathbf{x}_{it}'\boldsymbol{\beta} + \gamma y_{i,t-1} + \mu_i + \varepsilon_{it},$$

an instrumental variables approach which is similar to the one in the previous subsection can be used, if the $\mathbf{x}_{it}$s are strictly exogenous. If $E(\mathbf{x}_{it}\varepsilon_{is}) = 0$ $\forall i, t, s$ then also $E(\mathbf{x}_{is}\Delta\varepsilon_{it}) = 0$ $\forall i, t, s$ (moment conditions for $s < t$); $\mathbf{x}_{it}$ in first differences should simply be added to the model in first differences (11.62).

An example of this approach is found in Konings and Roodhooft (1997), who examine the labour demand for 3000 firms in Belgium:

$$\begin{aligned}\log L_{it} &= \overline{\beta}_1 + \beta_2 \log W_{it} + \beta_3 \log K_{it} + \beta_4 \log Y_{it} \\ &\quad + \beta_5 \log W_{jt} + \gamma \log L_{i,t-1} + \mu_i + \varepsilon_{i,t-1}.\end{aligned}$$

Note that we might expect that $E(\Delta \log W_{it}\Delta\varepsilon_{it}) \neq 0$.

Unit roots and cointegration in panel data models

- The principal motivation behind panel data unit root tests is to increase the power of (classic) unit root tests by increasing the sample size. This is an important issue since it is not clear whether structural breaks in the case of long time series are more serious than the problem of cross-sectional heterogeneity. It is often argued that the commonly used unit root tests such as Augmented Dickey-Fuller (ADF), Phillips-Perron (PP), and Kwiatkowski *et al.* (KPSS) are not very powerful, and that using panel data a more powerful test can be obtained.

- Concepts of unit roots, spurious regression and cointegration are long run concepts (i.e. $T \to \infty$). In many cases keeping T fixed but $N \to \infty$ circumvents such problems. But the possibility of heterogeneity implies that it is conceivable that the variables belonging to different individuals have a different time-series behavior; some may even be stationary and others non-stationary (e.g. integrated of a certain order), e.g. in cases as $y_{1t} \sim I(0)$ and $y_{2t} \sim I(1)$. To test the latter case we might consider the $AR(1)$ model (11.58) with coefficients varying over individuals and which can easily be rewritten as $\Delta y_{it} = \mu_i + \pi_i y_{i,t-1} + \varepsilon_{it}$ with $\pi_i := \gamma_i - 1$. The null is then H_0 : all series have a unit root or $H_0 : \pi_i = 0,\ \forall i$. The alternative could be that all variables are stationary (with the same mean-reversion parameter), i.e. $H_1 : \pi_i = \pi < 0,\ \forall i$.

It might occur that, assuming for all cross sections that the variables y_{it} and $\mathbf{x}_{it} \sim I(1)$ are cointegrated with common cointegrating parameter (vector) β such that $y_{it} - \mathbf{x}'_{it}\boldsymbol{\beta} \sim I(0)\,,\ \forall i$.

- Breitung and Meyer (BM-1994), however, suggest a simple panel data unit root valid for fixed T and $N \to \infty$. They consider a variant of the autoregressive model (11.58) $y_{it} = \gamma y_{i,t-1} + (1-\gamma)\mu_i + \varepsilon_{it}$ with $\varepsilon_{it} \overset{iid}{\sim} N\left(0, \boldsymbol{\sigma}_{\varepsilon}^2\right)$, $i = 1, 2, ..., N$; $t = 1, 2, ..., T$. Regressing y_{it} on $y_{i,t-1}$, ignoring μ_i, yields an OLS estimator $\hat{\gamma}$ with asymptotic bias:

$$\underset{N\to\infty}{plim}\,(\hat{\gamma} - \gamma) = \frac{(1-\gamma)\,N^{-1}\sum_{i=1}^{N}\mu_i^2}{N^{-1}\sum_{i=1}^{N}\mu_i^2 + \frac{\boldsymbol{\sigma}_{\varepsilon}^2}{(1-\gamma^2)}},$$

from which it is clear that the asymptotic bias vanishes if $\gamma = 1$. Hence, the unit root hypothesis can be tested using the t-statistic for $H_0 : \gamma = 1$ (this is e.g. the ADF's null). However, since under $H_1 : \gamma < 1$, the OLS estimator $\hat{\gamma}$ is biased, this test leads to a loss of power. BM-1994 suggest estimating the equation $(y_{it} - y_{i0}) = \gamma(y_{i,t-1} - y_{i0}) + (\varepsilon_{it} - \varepsilon_{i0})$. Denoting the resulting OLS estimator by $\widetilde{\gamma}$, they show that $p\lim_{N\to\infty}\widetilde{\gamma} = \frac{1}{2}(\gamma + 1)$. Again, under the unit root hypothesis the bias disappears. Levin and Lin (1993a and 1993b) explain why the BM-1994 approach cannot be used to analyze the influence of individual effects or serial correlation on the appropriate critical values at which to evaluate the t-tests.

- Levin and Lin (LL-1993a and 1993b) develop unit root tests for the model:

$$y_{it} = \gamma y_{i,t-1} + \delta_0 + \delta_1 t + \mu_i + v_t + \varepsilon_{it}$$

with $\varepsilon_{it} \overset{iid}{\sim} \left(0, \boldsymbol{\sigma}_{\varepsilon}^2\right), i = 1, 2, ..., N; t = 1, 2, ..., T$; hence, this model incorporates a time trend as well as individual and time specific effects. Initially, they assume *iid* errors, but they showed that under serial correlation of errors the test statistics have the same limiting distributions by the inclusion of first differences of y_{it}. LL-1993 consider the following six models:

$$\begin{array}{rlcll}
\text{(i)} & y_{it} & = & \gamma y_{i,t-1} + \varepsilon_{it}, & H_0 : \gamma = 1 \\
\text{(ii)} & y_{it} & = & \gamma y_{i,t-1} + \delta_0 + \varepsilon_{it}, & H_0 : \gamma = 1, \delta_0 = 0 \\
\text{(iii)} & y_{it} & = & \gamma y_{i,t-1} + \delta_0 + \delta_1 t + \varepsilon_{it}, & H_0 : \gamma = 1, \delta_0 = 0, \delta_1 = 0 \\
\text{(iv)} & y_{it} & = & \gamma y_{i,t-1} + v_t + \varepsilon_{it}, & H_0 : \gamma = 1 \\
\text{(v)} & y_{it} & = & \gamma y_{i,t-1} + \mu_i + \varepsilon_{it}, & H_0 : \gamma = 1, \mu_i = 0 \ \forall i \\
\text{(vi)} & y_{it} & = & \gamma y_{i,t-1} + \mu_{i0} + \mu_{i1} t + \varepsilon_{it}. &
\end{array}$$

and show that for models (i) to (iv) $T\sqrt{N}\,(\hat{\gamma} - 1) \Rightarrow N(0, 2)$; t_{γ} is converging to a standard normal distribution and for model (v), if $\sqrt{N}/T \to$

0 :

$$T\sqrt{N}\left(\hat{\gamma}-1\right)+3\sqrt{N} \Rightarrow N\left(0,10.2\right)$$
$$\sqrt{1.25}t_{\gamma}+\sqrt{1.875N} \Rightarrow N\left(0,845/112\right).$$

LL (1993a) argue that in contrast to the standard distributions of unit root test statistics for a single time series, the panel test statistics have limiting normal distributions. However, the convergence rates are faster as $T \to \infty$ (superconsistency) than as $N \to \infty$. The major limitation of the LL tests is that γ is the same for all observations. Thus, denoting by γ_i the value of γ for the ith cross-section unit, then the LL test specifies the null H_0 and alternative H_1 as:

$$H_0 : \gamma_1 = \gamma_2 = \cdots = \gamma_N = \gamma = 1$$
$$H_1 : \gamma_1 = \gamma_2 = \cdots = \gamma_N = \gamma < 1.$$

- Im, Pesaran, and Shin (IPS-1997) relax the assumption $\gamma_1 = \gamma_2 = \cdots = \gamma_N$ under H_1. To demonstrate it they consider model (vi) in Levin and Lin (1993a and 1993b) and substitute γ_i for γ. This model with a linear trend for each of the N individuals is then used for separate unit root tests fo the N cross sections. Therefore, let $t_i, i = 1, 2, \cdots, N$ denote the t-statistics for testing unit roots and let $E\left(t_i\right) = \mu$ and $var(t_i) = \sigma^2$, then:

$$\sqrt{N}\frac{t-\mu}{\sigma} \Rightarrow N\left(0,1\right).$$

The authors compute μ and σ^2 by Monte Carlo methods. Important to note is that the IPS test is a way of combining the evidence on the unit root hypothesis from the N unit root tests performed on the N cross-section units. Note that implicit in the test is the assumption that T is the same for all cross-section units, so that we are considering a balanced panel.

A drawback of the IPS test is that, not only we have to apply it on balanced panel data, but also that it depends on the assumption that there is no cross-section unit correlation among the error terms (see SUR). This assumption is very often violated in practice.

11.9 Empirical cases and exercises

11.9.1 Case 1. Investment and market value of firms

Part 1. Panel estimation

Use the data in Section 13.9.7 of Greene (2003), p. 335 (the Grunfeld data Table F13.1 at
http://pages.stern.nyu.edu/~wgreene/Text/econometricanalysis.htm) to fit the

Table 11.1: Case 1 - Part 1. Fixed Effects Model.

Dependent Variable: I
Method: Pooled Least Squares
Included observations: 20
Number of cross sections used: 5
Total panel (balanced) observations: 100

Variable	Coefficient	Std. Error	t-Statistic	Prob.
F	0.10598	0.015891	6.669182	0
C	0.34666	0.024161	14.34781	0
Fixed Effects				
F1 Intercept	-76.0668			
F2 Intercept	-29.3736			
F3 Intercept	-242.171			
F4 Intercept	-57.8994			
F5 Intercept	92.53854			
R-squared	0.937454		Mean dependent var	248.957
Adjusted R-squared	0.933419		S.D. dependent var	267.8654
S.E. of regression	69.11798		Sum squared resid	444288.4
Log likelihood	-561.847		F-statistic	232.3194
Durbin-Watson stat	0.806789		Prob(F-statistic)	0

random and fixed effect models. There are five firms and 20 years of data (1935-1954) about the variables:

I_{it} = gross investment, from Moody's Industrial Manual and annual reports of firm $i = 1, 2, 3, 4, 5$ and period $t = 1, 2, \ldots, 20$;

F_{it} = market value of the firm at the end of the previous year originating from Bank and Quotation Record and Moody's Industrial Manuel of firm $i = 1, 2, 3, 4, 5$,

C_{it} = value of the stock of plant and equipment at the end of the previous year for each firm $i = 1, 2, 3, 4, 5$.

Use the F, LM and/or Hausman test statistics to determine which model (the fixed or random effects model) is preferable for these data

Fixed Effects model The estimation of the fixed effects model is presented in Table (11.1).

Both variables F and C are statistically significant at the 1 per cent level.

Pooled OLS with common intercept is presented in Table (11.2).

The F-test statistic (11.7) is given by:

$$F_{N-1,NT-N-K+1} = \frac{\left(R^2_{FE} - R^2_{Pooled}\right)/(N-1)}{(1 - R^2_{FE})/(NT - N - K + 1)},$$

checks whether all fixed effects are equal.

Table 11.2: Case 1 - Part 1. Pooled OLS.
Dependent Variable: I
Method: Pooled Least Squares
Included observations: 20
Total panel observations 100

Variable	Coefficient	Std. Error	t-Statistic	Prob.
Intercept	-48.0297	21.48017	-2.236	0.0276
F	0.105085	0.011378	9.23598	0
C	0.305366	0.043508	7.018637	0
R-squared	0.778856	Mean dependent var		248.957
Adjusted R-squared	0.774296	S.D. dependent var		267.8654
S.E. of regression	127.2583	Sum squared resid		1570884
Log likelihood	-489.568	F-statistic		170.814
Durbin-Watson stat	0.228055	Prob(F-statistic)		0

The critical values for $F_{4,93}$ are 2.46 and 3.52 at the 5 per cent and 1 per cent level, respectively. The resulting $F_{4,93}$ statistic is 58.95571, so that we can decisively reject the null hypothesis and there is a strong firm-specific effect in the panel data.

To control for heteroskedasticity in the data it is possible to estimate the FE model with the use of cross-sectional weighting. Yet in our sample only one out of five firms exhibits heteroskedasticity. Therefore only small differences were obtained when estimating the model with cross-sectional weighting and we stay the nonweighted specification in this case.

Random Effects model The estimation of the random effects is presented in Table (11.3).

The LM-test statistic (11.19) is given by:

$$g = \frac{NT}{2(T-1)}\left[\frac{\sum_{i=1}^{N}\left(\sum_{t=1}^{T} e_{it}\right)^2}{\sum_{i=1}^{N}\sum_{t=1}^{T} e_{it}^2} - 1\right]^2,$$

is calculated from the Pooled OLS residuals e_{it} and amounts to 453.8221. Here we definitely reject the hypothesis that the random effects model is a valid way to estimate this relationship. Therefore we immediately conclude that the fixed effects model is the best way to estimate. And there is no need to employ the Hausman test to choose between the FE and RE models.

Part 2. Dynamic panel estimation

Here the extension is presented in the sense that the following dynamic panel model is estimated (see Subsection 11.8):

$$I_{it} = \beta_1 + \beta_2 F_{it} + \beta_3 C_{it} + \beta_4 I_{i,t-1} + \varepsilon_{it},$$

Table 11.3: Case 1 - Part 1. RE Model.

Dependent Variable: I
Method: GLS (Variance Components)
Sample: 1935 1954
Included observations: 20
Number of cross sections used: 5
Total panel (balanced) observations: 100

Variable	Coefficient	Std. Error	t-Statistic	Prob.
Intercept	-60.2905	54.48388	-1.10658	0.2712
F	0.104886	0.014797	7.088186	0
C	0.346016	0.024254	14.26661	0
Random Effects				
F1 Intercept	-10.3894			
F2 Intercept	31.07585			
F3 Intercept	-175.667			
F4 Intercept	3.112561			
F5 Intercept	151.8678			

		GLS Transformed Regression	
R-squared	0.933998	Mean dependent var	248.957
Adjusted R-squared	0.932637	S.D. dependent var	267.8654
S.E. of regression	69.52289	Sum squared resid	468842.9
Durbin-Watson stat	0.762426		

	Unweighted Statistics including Random Effects		
R-squared	0.937375	Mean dependent var	248.957
Adjusted R-squared	0.936084	S.D. dependent var	267.8654
S.E. of regression	67.72058	Sum squared resid	444849.5
Durbin-Watson stat	0.803548		

Table 11.4: Case 1 - Part 2. Dynamic FE Regression.

Dependent Variable: I
Sample(adjusted): 1936 1954
Included observations: 19 after adjusting endpoints
Number of cross sections used: 5
Total panel (balanced) observations: 95
White Heteroskedasticity-Consistent Standard Errors and Covariance

Variable	Coefficient	Std. Error	t-Statistic	Prob.
F	0.105257	0.017075	6.164533	0
C	0.149301	0.039341	3.795021	0.0003
I(-1)	0.620848	0.11257	5.515224	0
Fixed Effects				
F1 Intercept	-290.478			
F2 Intercept	-55.5766			
F3 Intercept	-225.45			
F4 Intercept	-68.0395			
F5 Intercept	-91.844			
R-squared	0.962487	Mean dependent var		255.5987
Adjusted R-squared	0.959468	S.D. dependent var		271.8566
S.E. of regression	54.73147	Sum squared resid		260611.4
Log likelihood	-510.852	F-statistic		1116.087
Durbin-Watson stat	1.394744	Prob(F-statistic)		0

where I_{it} is the gross investment, F_{it} the market value of the firm at the end of the previous year and C_{it} is the value of the stock of plant and equipment at the end of the previous year.

Use the F, LM and/or Hausman statistics to determine which model (the fixed or random) effects model, is preferable for these data.

Dynamic Fixed Effects model The estimation results of the dynamic fixed effects model are presented in Table (11.4).

Dynamic pooled OLS with common intercept is shown in Figure (11.5).

The $F_{4,87}$ is equal 13.59672 indicating the presense of strong individual-specific effects.

Dynamic Random Effects model The dynamic random effects model is shown in Figure (11.4).

The $LM - test = 3.472896$, which leads to the conclusion that the random effects model is applicable for modelling.

Therefore we must use the Hausman test to choose between two model specifications. The Hausman test statistic is given by:

$$m = \left(\hat{\boldsymbol{\beta}}_{sFE} - \hat{\boldsymbol{\beta}}_{sRE}\right)' (\mathbf{M}_1 - \mathbf{M}_0)^{-1} \left(\hat{\boldsymbol{\beta}}_{sFE} - \hat{\boldsymbol{\beta}}_{sRE}\right) \overset{as}{\sim} \chi^2_{(K-1)},$$

Table 11.5: Case 1 - Part 2. Dynamic Pooled OLS.

Dependent Variable: I
Sample(adjusted): 1936 1954
Included observations: 19 after adjusting endpoints
Number of cross sections used: 5
Total panel (balanced) observations: 95
White Heteroskedasticity-Consistent Standard ErrorsCovariance

Variable	Coefficient	Std. Error	t-Statistic	Prob.
Intercept	-32.0968	11.78124	-2.7244	0.0077
F	0.03153	0.007848	4.017609	0.0001
C	0.074279	0.03938	1.886214	0.0625
I(-1)	0.851202	0.07064	12.04983	0
R-squared	0.939036	Mean dependent var		255.5987
Adjusted R-squared	0.937026	S.D. dependent var		271.8566
S.E. of regression	68.22145	Sum squared resid		423529.1
Log likelihood	-533.918	F-statistic		467.2255
Durbin-Watson stat	1.437072	Prob(F-statistic)		0

Table 11.6: Case 1- Part 2. Dynamic RE Model.

Dependent Variable: I
Method: GLS (Variance Components)
Sample: 1936 1954
Included observations: 19
Number of cross sections used: 5
Total panel (balanced) observations: 95

Variable	Coefficient	Std. Error	t-Statistic	Prob.
Intercept	-23.8074	9.250697	-2.57357	0.0117
F	0.023563	0.006885	3.422322	0.0009
C	0.0521	0.026598	1.958798	0.0532
I(-1)	0.912511	0.044999	20.27844	0
Random Effects				
F1 Intercept	4.046752			
F2 Intercept	-12.4666			
F3 Intercept	23.7157			
F4 Intercept	-7.91791			
F5 Intercept	-7.37791			
R-squared	0.935089	Mean dependent var		255.5987
Adjusted R-squared	0.932949	S.D. dependent var		271.8566
S.E. of regression	70.3949	Sum squared resid		450945.2
Durbin-Watson stat	1.497437			

where $\hat{\boldsymbol{\beta}}_{sFE}$ represents the OLS dummy variables (fixed effects) estimator (11.4) of $\boldsymbol{\beta}_s$ and $\hat{\boldsymbol{\beta}}_{sRE}$ the GLS error components (random effects) estimator (11.11) of $\boldsymbol{\beta}_s$. $\mathbf{M}_1$ and $\mathbf{M}_0$ represent the (asymptotic) covariance matrices of $\hat{\boldsymbol{\beta}}_{sFE}$ and $\hat{\boldsymbol{\beta}}_{sRE}$, respectively. The value of the Hausman test is equal 27.13958. From this we conclude that the fixed and random effects amodels are not compatible and it is better to use the dynamic fixed effects model.

11.9.2 Case 2. Money flows and the performance of hedge funds

In October 2003 the alumni of the students' organization VSAE in Quantitative Economics of the University of Amsterdam organized the Econometric Game for the fifth time at the Dutch Central Bank (see:http://www.vsae.nl/econometricgame/).

Hedge funds are "skill-based" asset management firms that, for different reasons, do not qualify as mutual funds. While the mutual fund industry is well regulated and fairly transparent, the hedge fund industry is not. For example, reporting of historical returns and other information is not obligatory for hedge funds. On the other hand, access to hedge funds is typically restricted to investors with "sophisticated understanding" and investing large sums of money. So-called fund-of-funds pool investments in different hedge funds and allow for smaller investments to attract a larger group of individual investors.

Hedge funds have a broad flexibility in the type of securities they hold and the type of positions they take. Most of them follow a long/short strategy, in which they buy securities for which they expect a price increase and sell (short) securities for which they expect a decrease. This way, returns are obtained irrespective of global market movements. Accordingly, the typical goal of hedge funds is formulated in terms of absolute returns (e.g. 10 or 15%), irrespective of market developments. This way, hedge funds may provide attractive investment opportunities for investors that already hold a well-diversified portfolio.

In the mutual fund industry, the typical manager receives a management fee that is a percentage of the total asset value under management (typically 1 or 2%). In the hedge fund industry, incentives for managers are different. First, in addition to a management fee, most hedge funds have incentive fees (performance fees) of about 20% of realized profits. Second, for the majority of funds, the manager also invests personal money in the fund. The incentive fees are only paid to the manager if he is able to compensate previous losses ("high watermarks").

A final important issue is that the flows of money into and out of hedge funds are restricted. For example, there exist minimum investment periods (lock up periods), and redemption notice periods. This allows hedge fund managers to invest in less liquid assets.

In this game we focus on modeling the flows into and out of hedge funds, in relation to the historical performance of the fund. A typical finding for mutual funds is that investors' money flows tend to go to funds that recently performed well, even though there is not much evidence of persistence in performance. In

Table 11.7: Case 2. Pooled OLS.

Dependent Variable: CF
Method: Pooled Least Squares
Sample(adjusted): 2 22
Included observations: 21 after adjusting endpoints
Number of cross sections used: 9
Total panel (unbalanced) observations: 154

Variable	Coefficient	Std. Error	t-Statistic	Prob.
Intercept	0.253925	0.145227	1.748476	0.0824
RET(-1)	0.00056	0.151514	0.003697	0.9971
LNAV(-1)	-0.01198	0.008233	-1.45482	0.1478
CF(-1)	0.020274	0.081262	0.249486	0.8033
RNK(-1)	-0.08718	0.082783	-1.05306	0.294
R-squared	0.023492	Mean dependent var		0.014438
Adjusted R-squared	-0.00272	S.D. dependent var		0.204067
S.E. of regression	0.204344	Sum squared resid		6.22173
Log likelihood	28.56913	F-statistic		0.896145
Durbin-Watson stat	2.03756	Prob(F-statistic)		0.467946

fact, Berk and Greene (2002) in a recent paper argue that persistence in performance disappears because investors compete for superior performance such that any persistence is competed away. The purpose of this game is to examine money flows for hedge funds.

The Data

The data contains a part of the Econometric Game 2003 data sample. In the data we observe quarterly returns (RET), size (NAV "net asset value" in USD), cash flows (CF %, $inflow - outflow$) and rankings ($0 < RNK < 1$) of nine selected hedge funds.

Pooled OLS Regression is presented in Figure (11.7). As it can be seen, the direct pooled OLS has very weak explanatory power (which was also observed in a larger sample analyzed at the Econometric Game 2003).

Arguments for the choice of the regression setup:

Cash flows to and from a particular hedge fund in a given period of time should depend mainly on the previous performance of this fund. It is difficults to imagine that the investors can obtain a really up-to-date information about the fund's activities. Therefore, in our model we assume that the current cash flow to and from the hedge fund depend on:

i) the fund's returns in the previous quarter;
ii) the total size of the fund in the previous quarter;
iii) the previous cash flows;
iv) the fund's standing among other funds in the previous quarter.

There is one more adjustment we make in our data. The fund's returns, ranking and the cash flow are given as numbers between 0 and 1. The net asset

Table 11.8: Case 1. FE Model.

Dependent Variable: CF
Method: Pooled Least Squares
Sample(adjusted): 2 22
Included observations: 21 after adjusting endpoints
Number of cross sections used: 9
Total panel (unbalanced) observations: 154

Variable	Coefficient	Std. Error	t-Statistic	Prob.
RET(-1)	0.117984	0.14253	0.827781	0.4092
LNAV(-1)	-0.20204	0.040376	-5.00384	0
CF(-1)	-0.19189	0.082548	-2.32459	0.0215
RNK(-1)	-0.00324	0.078741	-0.04115	0.9672
Fixed Effects				
F1 Intercept	3.651298			
F2 Intercept	3.662163			
F3 Intercept	4.042082			
F4 Intercept	3.037331			
F5 Intercept	2.881177			
F6 Intercept	3.555257			
F7 Intercept	3.664024			
F8 Intercept	2.673797			
F9 Intercept	3.023777			
R-squared	0.223747		Mean dependent var	0.014438
Adjusted R-squared	0.157683		S.D. dependent var	0.204067
S.E. of regression	0.187288		Sum squared resid	4.945823
Log likelihood	46.24098		F-statistic	3.386827
Durbin-Watson stat	2.102313		Prob(F-statistic)	0.000228

value is represented by much larger numbers (in order of billions). Therefore, to adjust the scale we will use the logs of the funds' net asset value (LNAV).

Results of the fixed effects model estimation are given in Table (11.8).

The F-test statistic based on (11.7), for an unbalanced panel is given by:

$$F_{N-1,\sum_{i=1}^{N} T_i - N - K + 1} = \frac{\left(R^2_{FE} - R^2_{Pooled}\right)/(N-1)}{\left(1 - R^2_{FE}\right)/(\sum_{i=1}^{N} T_i - N - K + 1)},$$

and checks whether all fixed effects are equal (the critical values for $F_{8,\infty}$ are 1.94 and 2.53 at the 5 % and 1 % level, respectively).

We obtain the $F_{(8,141)} - test = 4.546840$. Hence, we find significant individual fixed effects.

Results of the random effects model are presented in Table (11.9).

The LM test statistic equals 1.774491. The critical values of χ^2 with 1 degree of freedom are 3.84 for $\alpha = 0.05$ and 6.63 for $\alpha = 0.01$. Hence, the null

Table 11.9: Case 2. RE Model.

Dependent Variable: CF
Method: GLS (Variance Components)
Sample: 2 22
Included observations: 21
Total panel observations 154

Variable	Coefficient	Std. Error	t-Statistic	Prob.
Intercept	0.122272	0.075462	1.620309	0.1073
RET(-1)	-0.23225	0.152851	-1.51947	0.1308
LNAV(-1)	-0.00205	0.003178	-0.64397	0.5206
CF(-1)	0.453458	0.071345	6.355885	0
RNK(-1)	-0.1378	0.084719	-1.62658	0.1059
Random Effects				
F1 Intercept	-0.17781			
F2 Intercept	-0.19808			
F3 Intercept	-0.69121			
F4 Intercept	0.13068			
F5 Intercept	0.174763			
F6 Intercept	0.459052			
F7 Intercept	-0.22493			
F8 Intercept	-0.37565			
F9 Intercept	0.473509			

		GLS Transformed Regression	
R-squared	-0.40759	Mean dependent var	0.014438
Adjusted R-squared	-0.44538	S.D. dependent var	0.204067
S.E. of regression	0.245337	Sum squared resid	8.968339
Durbin-Watson stat	2.410672		

		Unweighted Statistics including Random Effects	
R-squared	-3.38382	Mean dependent var	0.014438
Adjusted R-squared	-3.5015	S.D. dependent var	0.204067
S.E. of regression	0.432963	Sum squared resid	27.93108
Durbin-Watson stat	0.774038		

of pooled regression is not rejected (and the alternative of random effects is not accepted).

The Hausman test statistic equals 210.95. Therefore, we conclude that the fixed effects model is the most appropriate for explaining the cash flows to and from the hedge funds.

11.9.3 Exercises

Consider the panel of Latin American countries and their GDP, contained in the data file ex1.wf1 (available at www.ua.ac.be/joseph.plasmans), and let the GDP of country i for year t be denoted by GDP_{it}, and model the GDP per capita in real terms of the countries ($Rgdp$) as a function of the following variables. Exchange Rate (XR), Investment share of GDP, in percentage (I), Real Government share of GDP, in percentage (G), Population (Pop), Measure of Openness of the Economy ($Open$)

i) Estimate the FE model $GDP_{it} = \alpha + \beta GDP_{i,t-1} + \varepsilon_{it}$ and interpret the output.

ii). Estimate the same model, now with the variables in logarithmic differences. Interpret. Is there serial correlation in the residuals?

iii). Estimate the model $GDP_{it} = \alpha_i + \beta_i GDP_{i,t-1} + \varepsilon_{it}$. Test whether all $FE - \beta_i$s are equal.

iv). Estimate the panel data model $Rgdp_{it} = \alpha_i + \beta_1 * Pop_{it} + \beta_2 * I_{it} + \beta_3 * G_{it} + \beta_4 * XR_{it} + \beta_5 * Open_{it} + \eta_{it}$.

v). Make plots of the XR variables. Create now these variables in logarithmic differences and make the plots again.

vi). Estimate the model with fixed effects; analyze the residuals and their variances. Is there cross-sectional heteroscedasticity? If it is so, what can be done?

vii). Compute the correlations between the residual series from different countries.

viii). Take a look at the standard errors for the estimated fixed effects and test whether all fixed effects are equal.

ix). Estimate the model with random effects. Compare the individual random effects to the individual fixed effects and comment.

x). Which is the model that best suits the data set?

xi). Introduce a time effect in the intercept and estimate and interpret corresponding FE and RE estimators.

xii). Can different reactions in government share and exchange rate behavior between individual countries be detected?

Chapter 12

Nonlinear Panel Data Models

Many nonlinear problems studied in previous sections can be analyzed in a panel data framework. For reasons of brevety and efficiency, we will stick to limited dependent variables, and more specifically to the binary choice logit and probit models. The general binary choice models for the logistic and normal distributions presented in Chapter 9 assume that the error term ε_i is *iid* (or a WN) and that is independent of the explanatory variables.

12.1 FE estimation for logit and probit

12.1.1 Logit models and conditional MLE

Assuming that the individual effects μ_i are fixed and unknown parameters for each individual and that the residuals are uncorrelated with the explanatory variables, the model to be estimated[1] can be rewritten from (11.1) for the static case and constant β over individuals and time as the discrete choice model:

$$y_{it}^* = \boldsymbol{\beta}'\mathbf{x}_{it} + \mu_i + \varepsilon_{it} \tag{12.1}$$

with y_{it}^* a linear response latent variable crossing a certain threshold and the observed y given by $y := 1$ if $y_{it}^* > 0$ and $y := 0$ if $y_{it}^* < 0$.

We can formulate the probability of $y_{it} = 1$ as (see Section 9.1.6):

$$P(\ \varepsilon_{it} > -\boldsymbol{\beta}'\mathbf{x}_{it} - \mu_i) = \int_{-\beta' x_{it} - \mu_i}^{\infty} f(\varepsilon_{it}) d\varepsilon_{it} = 1 - F(-\boldsymbol{\beta}'\mathbf{x}_{it} - \mu_i). \tag{12.2}$$

[1]For a detailed discussion on the logit and probit models for panel data, see Hsiao (1996): pp. 223-241.

Equation (12.2) implies that both μ_i and $\boldsymbol{\beta}$ are unknown parameters for the conditional probability $P(y_{it} = 1|\mathbf{x}_i; \boldsymbol{\beta})$; therefore the MLE is consistent only when $T \to \infty$. However, for a finite T, we come across what is known as the *incidental or nuisance parameter* problem: an increase in the number N of cross section units provides no additional information about μ_i (i.e., for each individual we add to the sample a new parameter has to be estimated).

In the case of the linear model studied in Chapter 11, the incidental parameter problem does not constrain the estimation of $\boldsymbol{\beta}$ because μ_i and $\boldsymbol{\beta}$ are asymptotically independent, and one can separate the estimation of the structural parameters from that of the incidental parameter μ .For nonlinear models, on the contrary, for a fixed T, the estimation of μ and $\boldsymbol{\beta}$ are not independent from each other, and the inconsistency problem of the estimation of μ_i is transmitted into the estimation of $\boldsymbol{\beta}$, therefore finding inconsistent estimates for *both* μ and $\boldsymbol{\beta}$.[2]

A convenient way to *remove* the individual effects from the logit model is the *conditional ML estimation* principle: to find estimators for $\boldsymbol{\beta}$, we need to find functions $\varkappa(\cdot)$ that are independent of the μ_i and that have the property that for the true values of $\boldsymbol{\beta}$, the function $\varkappa(y_1, \ldots, y_N|\boldsymbol{\beta})$ converges to zero in probability as $N \to \infty$. If a minimum sufficient statistic τ_i exists for the incidental parameter and if it does not depend on $\boldsymbol{\beta}$, the conditional pdf $f(y_i|\boldsymbol{\beta}, \tau_i) = f(y_i|\boldsymbol{\beta}, \mu_i)/g(\tau_i|\boldsymbol{\beta}, \mu_i)$ is independent of μ_i and we rule out the problem by analyzing the probability of the vector $\mathbf{y}_i = (y_{i1}, y_{i2}, \ldots, y_{iT})'$ conditional on $\tau_i = \sum_{t=1}^{T} y_{it}$. The conditional pdf maximization problem yields a set of functions that do not depend on μ_i and that for $N \to \infty$, the function $\varkappa(y_1, \ldots, y_N|\beta)$ converges to zero. Hence, the minimum sufficient statistic for μ_i is obtained by taking the partial derivative of the loglikelihood function with respect to μ_i.

The conditional probability of y_i is for the logit model:

$$P(y_i|\mathbf{x}_i; \boldsymbol{\beta}) = \frac{\exp\left[\mu_i \left(\sum_{t=1}^{T} y_{it}\right) + \boldsymbol{\beta}' \left(\sum_{t=1}^{T} \mathbf{x}_{it} y_{it}\right)\right]}{\prod_{t=1}^{T} \left[1 + \exp\left(\boldsymbol{\beta}'\mathbf{x}_{it} \ + \mu_i\right)\right]},$$

so that taking a product over all individuals and logarithmizing the loglikelihood function is given by

$L(\boldsymbol{\beta}) = -\sum_{i=1}^{N}\sum_{t=1}^{T} \ \log\left[1 + \exp\left(\boldsymbol{\beta}'\mathbf{x}_{it} \ + \mu_i\right)\right] + \sum_{i=1}^{N}\sum_{t=1}^{T} \ y_{it}\left(\boldsymbol{\beta}'\mathbf{x}_{it} \ + \mu_i\right)$ with necessary maximizing conditions:

$$\frac{\partial L(\boldsymbol{\beta})}{\partial \mu_i} = \sum_{i=1}^{N}\sum_{t=1}^{T} \left[-\frac{\exp\left(\boldsymbol{\beta}'\mathbf{x}_{it} \ + \mu_i\right)}{1 + \exp\left(\boldsymbol{\beta}'\mathbf{x}_{it} \ + \mu_i\right)} + y_{it}\right] \mathbf{x}_{it} = 0.$$

Consistent estimates for $\boldsymbol{\beta}$ are obtained by maximizing the conditional loglikelihood based on all possible sequences of y_{it} such that the sum of positive outcomes is equal to τ_i (such that the loglikelihood function is conditioned on

[2] Hsiao (1996) demonstrates the inconsistency of the MLE estimator of $\boldsymbol{\beta}$ in a logit model.

$\tau_i = \sum_{t=1}^{T} y_{it}$). A minimum sufficient statistic for μ_i is obtained by solving the first order condition of this conditional maximization problem.

Following Hsiao (1996) the probability of τ_i is:

$$P(\textstyle\sum_{t=1}^{T} y_{it}) = \frac{T!}{(\sum y_{it})!(T-\sum y_{it})!} \times \frac{\exp[\mu_i(\sum y_{it})]}{\prod_{t=1}^{T}[1+\exp(\boldsymbol{\beta}'\mathbf{x}_{it}+\mu_i)]} \times \\ \times \left\{\textstyle\sum_{d\in\tilde{B}_i} \exp(\boldsymbol{\beta}'\sum_{t=1}^{T}\mathbf{x}_{it}\, d_{it})\right\}$$

where d_{it} is equal to zero or to one according to y_{it}, and $\tilde{B}_i$ is the set of all possible d_{it} such that $\sum_{t=1}^{T} d_{it} = \sum_{t=1}^{T} y_{it}$. The probability of the vector $\mathbf{y}_i = (y_{i1}, y_{i2}, \ldots, y_{iT})'$ conditional on τ_i is:

$$P(y_i|\tau_i{=}\sum_{t=1}^{T} y_{it}){=}\ell(\boldsymbol{\beta}|\tau_i){=}\frac{\exp(\boldsymbol{\beta}'\sum_{t=1}^{T}\mathbf{x}_{it}\, y_{it})}{\sum_{d\in\tilde{B}_i}\exp(\boldsymbol{\beta}'\sum_{t=1}^{T}\mathbf{x}_{it}\, d_{it})} \times \frac{(\sum y_{it})!(T-\sum y_{it})!}{T!}.$$

Since the individual effects appear in a linear form, one can easily eliminate these effects by grouping the sequences with the same τ_i.

For example, for $T = 2$, the two possibilities are $\omega_i = 1$ if $(y_{i1}, y_{i2}) = (0,1)$ and $\omega_i = 0$ if $(y_{i1}, y_{i2}) = (1,0)$.

$$P(\omega_i{=}1|y_{i1}{+}y_{i2}{=}1){=}\frac{P(\omega_i{=}1|\,\boldsymbol{\beta},\mathbf{x}_{it})}{P(\omega_i{=}1|\,\boldsymbol{\beta},\mathbf{x}_{it}){+}P(\omega_i{=}0|\,\boldsymbol{\beta},\mathbf{x}_{it})}{=}\frac{\exp[\boldsymbol{\beta}'(x_{i2}-x_{i1})]}{1{+}\exp[\boldsymbol{\beta}'(x_{i2}-x_{i1})]}.$$

The conditional loglikelihood function is:

$$L = \textstyle\sum_{i\in\tilde{B}_1}\{\omega_i \ln F[\boldsymbol{\beta}'(x_{i2}-x_{i1})] + (1-\omega_i)\ln\left[1 - F[\boldsymbol{\beta}'(x_{i2}, -x_{i1})]\right]\},$$

where $\tilde{B}_1 = \{i/y_i = (0,1) \vee (1,0)\}$ (i.e., the sequences that are identical do not contribute to the likelihood) and $F(\cdot) := \exp(\cdot)/\left(1+\exp(\cdot)\right)$.

Lee (2002) analyzes the conditional logit for $T \geqq 3$.

In this way, we have removed the individual effects from the model (i.e., the individual effects *are not estimated!*).

12.1.2 FEs in probit models and a semiparametric estimator

For the case of probit models, we cannot find simple functions for the parameter that are independent of the incidental parameter (i.e. there is no minimum sufficient statistic for μ_i in this case). Therefore, for T finite, there seems to be no consistent estimators for the probit model. As an alternative, we may use the *maximum score estimator*: inference is possible if the explanatory variables vary enough over time.

A conditional version of Manski's (1975 and 1985) maximum score estimator maximizes the sample average function:

$$H_N = \frac{1}{N}\sum_{i=1}^{N}\sum_{t=2}^{T} sgn(\mathbf{w}_{it}'\mathbf{b})(y_{it} - y_{it-1}), \tag{12.3}$$

where $\mathbf{w}_{it} := x_{it} - x_{it-1}$; $sgn(\mathbf{w}_{it}'\mathbf{b}) = 1$ if $\mathbf{w}_{it}'\mathbf{b} \geqq 0$ and $sgn(\mathbf{w}_{it}'\mathbf{b}) = -1$ if $\mathbf{w}_{it}'\mathbf{b} < 0$. Since under some regularity conditions (12.3) converges to $H(\mathbf{b}) = E\left(sgn(\mathbf{w}_{it}'\mathbf{b})(y_{it} - y_{it-1})\right)$, in which $H(\mathbf{b})$ is maximized when $\mathbf{b} = \boldsymbol{\beta}$. Manski (1985) has demonstrated that under some general regularity conditions the estimator that maximizes (12.3) is strongly consistent.

12.2 RE estimation

If the μ_is are independent of x_{it} and are random samples from a univariate distribution function $G(.)$ (indexed by a finite number of parameters $\boldsymbol{\delta}$), the loglikelihood function becomes:

$$L = \sum_{i=1}^{N} \log \int_{-\infty}^{\infty} \prod_{t=1}^{T} F(\boldsymbol{\beta}'\mathbf{x}_{it} + \mu)^{y_{it}}[1 - F(\boldsymbol{\beta}'\mathbf{x}_{it} + \mu)]^{1-y_{it}} dG(\mu_i|\boldsymbol{\delta}). \tag{12.4}$$

The maximization of (12.4) under mild regularity conditions provides a consistent and efficient estimator for $\boldsymbol{\beta}$. However, if μ_i is correlated with $\mathbf{x}_{it}$, the maximization of (12.4) does not eliminate the bias. Similar to the omitted variable case in Subsection 11.2.4, an auxiliary function, where μ_i is assumed to be linearly dependent on $\mathbf{x}_{it}$ or $\mu_i = \sum_{t=1}^{T} \boldsymbol{\alpha}_t'\mathbf{x}_{it} + \xi_i$, is considered. Note that there is an important difference with the linear case, where it was not restrictive to decompose the μ_i into its linear projection on $\mathbf{x}_{it}$ and an orthogonal residual. Here, we assume that the $E(\mu_i|\,\mathbf{x}_{it};\boldsymbol{\beta},\boldsymbol{\alpha}_t)$ is linear and that ξ_i has a specific probability distribution. The loglikelihood function (12.4) becomes then:

$$L = \sum_{i=1}^{N} \log \int \prod_{t=1}^{T} F(\boldsymbol{\beta}'\mathbf{x}_{it} + \sum_{t=1}^{T} \boldsymbol{\alpha}_t'\mathbf{x}_{it} + \xi_i)^{y_{it}} [1 - F(\boldsymbol{\beta}'\mathbf{x}_{it} + \sum_{t=1}^{T} \boldsymbol{\alpha}_t'\mathbf{x}_{it} + \xi_i)]^{1-y_{it}} dG^*(\boldsymbol{\xi}) \tag{12.5}$$

with G^*a univariate distribution function for ξ. In (12.5), we take account of the dependence between the nuisance parameter μ_i and $\mathbf{x}_{it}$. The maximization of (12.4) provides a consistent and efficient estimator of $\boldsymbol{\beta}$ for the case of μ_i independent of $\mathbf{x}_{it}$, while the maximization of (12.5) provides a consistent and efficient estimator of $\boldsymbol{\beta}$ for the case of μ_i correlated with $\mathbf{x}_{it}$.

12.2.1 ML estimation

In principle, according to Hsiao (1996), to obtain the MLE one has to evaluate T dimensional integrals. The distribution for y_{it} conditional on $\mathbf{X}_i$ but marginal on μ_i has the probit form:

$$P(y_{it} = 1 | \pi_t) = F\left[(1+\sigma_{\boldsymbol{\xi}}^2)^{-\frac{1}{2}}(\boldsymbol{\beta}'\mathbf{x}_{it} + \boldsymbol{\alpha}_t'\mathbf{x}_i)\right], \tag{12.6}$$

where $\pi_t := (1+\sigma_{\boldsymbol{\xi}}^2)^{-\frac{1}{2}}(\boldsymbol{\beta}'\mathbf{x}_{it} + \boldsymbol{\alpha}_t'\mathbf{x}_i)$ and $F(.)$ is the normal distribution function. Maximizing the loglikelihood of (12.6) for all individuals at period t, each one of the cross sectional univariate probit specifications yields the ML estimates $\hat{\pi}_t$, for $t = 1, 2, \ldots, T$.

12.3 Panel count data models

In this section the models which include fixed and random individual-specific effects are considered. These models make use of the advantages offered by the panel data (e.g. in Cincera (1997) and (2003)).

12.3.1 Fixed Effects Poisson models

For count panel data the origin of the analysis lies in the Poisson regression model with exponential mean function and multiplicative individual specific term, or $\widetilde{\lambda}_{it} := \lambda_{it}\alpha_i = \exp(\mathbf{x}_{it}'\boldsymbol{\beta} + \xi_i)$.

Coefficients $\boldsymbol{\beta}$ can be estimated using the conditional maximum likelihood method. The conditional loglikelihood function is as in (10.2):

$$\begin{aligned} L_c(\boldsymbol{\beta}) &= \sum_{i=1}^{N}\left[\ln(\sum_{t=1}^{T} y_{it})! - \sum_{t=1}^{T}\ln(y_{it}!)\right. \\ &\quad \left. + \sum_{t=1}^{T} y_{it}\ln\left(\frac{\exp(\mathbf{x}_{it}'\boldsymbol{\beta})}{\sum_{s=1}^{T}\exp(\mathbf{x}_{is}'\boldsymbol{\beta})}\right)\right]. \end{aligned}$$

Differentiation of this equation with respect to $\boldsymbol{\beta}$ yields the first order conditions

$$\sum_{i=1}^{N}\sum_{t=1}^{T}\mathbf{x}_{it}\left(y_{it} - \lambda_{i,t}\frac{\bar{y}_i}{\bar{\lambda}_i}\right) = \mathbf{0},$$

where $\bar{y}_i := \frac{1}{T}\sum_t y_{it}$ and $\bar{\lambda}_i := \frac{1}{T}\sum_t \lambda_{i,t}$ and $\lambda_{i,t} := \exp(\mathbf{x}_{it}'\boldsymbol{\beta})$.

12.3.2 Random Effects Poisson models

Specifying again $\widetilde{\lambda}_{it} = \lambda_{it}\alpha_i$, with α_i a random firm-specific effect. The Poisson parameter $\widetilde{\lambda}_{it}$ is now a random variable rather than a deterministic function of $\mathbf{x}_{it}$. Because $\widetilde{\lambda}_{it}$ must be positive, we write it in the form $\widetilde{\lambda}_{it} = \lambda_{it}\alpha_i = \exp(\mathbf{x}_{it}\boldsymbol{\beta} + \xi_0 + \xi_i)$, where ξ_i is a random firm-specific effect and ξ_0 is a random common effect.

The Poisson probability is specified then as:

$$P(y_{it} | \mathbf{x}_{it}; \boldsymbol{\beta}) = \frac{\exp\left(-\lambda_{it}\exp(\xi_i)\right)(\boldsymbol{\lambda}_{it}\exp(\xi_i))^{y_{it}}}{y_{it}!}.$$

It is obvious that different distributions of α_i lead to different distributions of $y_{i1}, ..., y_{iT}$. To derive a loglikelihood function we assume that $\alpha_i = \exp(\xi_i)$ is distributed as $\Gamma(\delta, \delta)$, so that $E(\alpha_i) = 1$ and $Var(\alpha_i) = 1/\delta$. By integrating we obtain the joint pdf for the ith firm

$$\begin{aligned} f(y_{i1}, ..., y_{iT}) &= \left[\prod_{t=1}^{T} \frac{\lambda_{it}^{y_{i,t}}}{y_{it}!}\right] \left(\frac{\delta}{\sum_t \lambda_{it} + \delta}\right)^{\delta} \\ &\quad \times \left(\sum_t \lambda_{it} + \delta\right)^{-\sum_t y_{it}} \frac{\Gamma(\sum_t \lambda_{it} + \delta)}{\Gamma(\delta)}. \end{aligned}$$

This is the density of the Poisson random effect model with gamma distributed random effects. For this distribution $E(y_{it}) = \lambda_{it}$ and $Var(y_{it}) = \lambda_{it} + \lambda_{it}^2/\delta$. Maximizing the loglikelihood function, i.e. the logarithm of this pdf with $\lambda_{it} = \exp(\mathbf{x}'_{it}\boldsymbol{\beta})$, yields the following first order conditions:

$$\sum_{i=1}^{N}\sum_{t=1}^{T} \mathbf{x}_{it}\left(y_{it} - \lambda_{it}\frac{\bar{y}_i + \delta/T}{\bar{\lambda}_i + \delta/T}\right) = \mathbf{0},$$

where, again, $\bar{y}_i := \frac{1}{T}\sum_t y_{it}$ and $\bar{\lambda}_i := \frac{1}{T}\sum_t \lambda_{it}$.

To test the difference between the FE and RE model specifications, the standard Hausman test can be applied.

12.4 Cases and exercises

12.4.1 Case 1. Firms' decision to report their R&D expenditures

This case analyzes R&D investment expenditures decisions of European firms (see Hall and Oriani (2004))

The data is given in the 1995 manufacturing panel collected by Hall and Oriani (http://emlab.berkeley.edu/users/bhhall/bhdata.html). The data description file can be obtained at
http://emlab.berkeley.edu/users/bhhall/pat/panel/pan95export.doc.

Question Investigate the factors affecting the firms' decision to report their R&D expenditures. Make use of panel properties of your data; group the firms by their two-digit industry classification codes (SIC2D variable, see data description file). Compare the performance of the FE model with the performance of pooled regression.

Solution In this case we investigate the factors affecting the firms when taking decisions on R&D expenditure and group the firms based on their two-digit industry classification codes (SIC 2 D).

Each binary decision can be represented by a dichotomous variable assuming a value of one if the event occurs and zero if the event does not occur. The

Table 12.1: Pooled regression results

Nonlinear OLS Parameter Estimates (PROBIT Model)

Parameter	Estimate	Approx Std Err	t Value	Approx Pr > |t|
a	0.151038	0.0278	5.44	<.0001
bls	-0.02126	0.00702	-3.03	0.0025
bli	0.187263	0.00482	38.83	<.0001
ble	-0.17823	0.00644	-27.68	<.0001

Equation	DF Model	DF Error	SSE	R-Square	Adj R-Sq
invevent	4	68106	15979.9	0.0282	0.0282

Nonlinear OLS Parameter Estimates (LOGIT Model)

Parameter	Estimate	Approx Std Err	t Value	Approx Pr > |t|
a	0.23656	0.0452	5.24	<.0001
bls	-0.03339	0.0114	-2.92	0.0035
bli	0.306693	0.00791	38.75	<.0001
ble	-0.29403	0.0105	-28.06	<.0001

Equation	DF Model	DF Error	SSE	R-Square	Adj R-Sq
invevent	4	68106	15973.6	0.0286	0.0286

probability is assumed to depend on a vector of explanatory variables and on a vector of unknown parameters. In this case both the probit and the logit models for repeated observations will be used.

First, we need to define the variables influencing expenditure in R&D. It is usually expected that investment, sales and employment are positively correlated with R&D. Therefore, the variables considered are: *ls*- log of sales, *le*- log of employment, *li*- log of investment and the dependent variable is *r*- R&D expenditure (when $r = 0$ it means that the firms does not invest in R&D expenditure).

The resulting equation is $r = a + bls \times lns + bli \times lni + ble \times lne$. We now turn to the estimation of each one of these models to see which one is more suitable for the database.The results from the estimation are presented in Table 12.1.

The logit model has a higher adjusted R^2 than the probit model. Therefore, one can infer that the logit model is more appropriate for this database. In both models sales and employment affect positively the R&D expenditure.

Our event is defined when the R&D variable equals zero, which means that the influence of both employment and sales variables result in the firms not investing in R&D. Both variables have more effect in the logit model.

From the negativity of the intercepts one would think *a priori* that the fixed effect of firms on R&D expenditure does not exist, but this is not correct. We investigate therefore the fixed effect between groups on SIC2D variables.

Second, we estimate the fixed effects in both probit and logit models. Fixed effects represent the specific-effect of the firm on the R&D expenditure.

From the probit specification of the model presented in Tables (12.2) and (12.3), one can observe that firm-specific fixed effects do exist. Moreover improvement in the adjusted R^2 is quite dramatic for both models.

In the probit estimation, the group from a20 to a27, a29 to a34, a37 and a39 has a positive interaction, which means that specific characteristics of these firms influence their R&D expenditure.

Also, since in our model the event is equal 1 when R&D equals zero, this implies that these firms normally do not invest in R&D. Besides, all the firms that have a positive alfa, are normally from sartorial and primary sectors. Finally, only the employment variable influences positively the R&D expenditure.

When the logit model is considered, fixed effects also exist: between a20 to a27, a29 to a34 and in a37, a39 (this means that some inherent characteristic of these firms influence their R&D expenditure). The firms in group a28, a35, a36 and a38 have a negative alfa; so that these firms invest more in R&D expenditure.

12.4.2 Case 2. Count data model for patents-R&D relationship

The Hausman *et al.* (1984) model is specified as follows:

$$E(P_{i,t}|\mathbf{X}_{i,t}) = \exp(\bar{\alpha} + \mu_i + \beta_0 \ln(k_{i,t}) + \beta_1 \ln(k_{i,t-1}) + ... + \beta_5 \ln(k_{i,t-5}) + \gamma t).$$

The estimation results are given in Table 12.4.

The model shows strong dependence between the firm's current R&D and patenting. This dependence gets substantially weaker for "older" levels of R&D. There is a negative time trend in the model indicating "a decline in the 'effectiveness' or productivity of R&D" according to Hausman *et al.* (1984).

12.4.3 Exercise

In the R&D literature there exists an opinion that the relationship between R&D expenditures and other factors can be nonlinear. Investigate a possible nonlinear model explaining the R&D investments of the firms taking the panel data properties into account. Use the data given in the manufacturing panel collected by Hall and Oriani (2004) (http://emlab.berkeley.edu/users/bhhall/bhdata.html, the description file can be obtained at http://emlab.berkeley.edu/users/bhhall/pat/panel/pan95export.doc). Estimate and test such a model. Does it perform better than a comparable linear model?

Table 12.2: Fixed effects regression results (PROBIT model)

Nonlinear OLS Parameter Estimates (PROBIT Model)

Parameter	Estimate	Approx Std Err	t Value	Approx Pr > $\lvert t\rvert$	
a20	1.132213	0.0368	30.8	<.0001	
a21	0.759074	0.0826	9.19	<.0001	
a22	0.963656	0.0391	24.65	<.0001	
a23	1.637038	0.0633	25.87	<.0001	
a24	1.400936	0.0585	23.96	<.0001	
a25	0.667693	0.0422	15.81	<.0001	
a26	0.841258	0.039	21.58	<.0001	
a27	1.634082	0.0488	33.48	<.0001	
a28	-0.35378	0.0331	-10.69	<.0001	
a29	1.128833	0.0486	23.22	<.0001	
a30	0.370802	0.0358	10.34	<.0001	
a31	1.26838	0.0728	17.43	<.0001	
a32	0.864718	0.04	21.61	<.0001	
a33	0.94233	0.0375	25.14	<.0001	
a34	0.369056	0.033	11.19	<.0001	
a35	-0.34882	0.0318	-10.97	<.0001	
a36	-0.3457	0.0309	-11.2	<.0001	
a37	0.272086	0.034	8.01	<.0001	
a38	-0.81579	0.0358	-22.76	<.0001	
a39	0.325361	0.0384	8.46	<.0001	
bls	-0.07752	0.00727	-10.66	<.0001	
bli	-0.14266	0.00513	-27.8	<.0001	
ble	0.128665	0.007	18.37	<.0001	
Equation	DF Model	DF Error	SSE	R-Square	Adj R-Sq
event	23	68087	12188.4	0.2588	0.2586

Table 12.3: Fixed effects regression results (LOGIT model)

Nonlinear OLS Parameter Estimates (LOGIT Model)

Parameter	Estimate	Approx Std Err	t Value	Approx Pr > |t|	
a20	1.871327	0.0616	30.36	<.0001	
a21	1.29036	0.1343	9.61	<.0001	
a22	1.5956	0.0651	24.49	<.0001	
a23	2.812038	0.1194	23.54	<.0001	
a24	2.34704	0.1024	22.92	<.0001	
a25	1.103311	0.0695	15.88	<.0001	
a26	1.401789	0.0645	21.72	<.0001	
a27	2.77278	0.0876	31.65	<.0001	
a28	-0.59684	0.0558	-10.7	<.0001	
a29	1.887968	0.0806	23.42	<.0001	
a30	0.626963	0.0592	10.6	<.0001	
a31	2.146751	0.1299	16.53	<.0001	
a32	1.434186	0.0662	21.65	<.0001	
a33	1.562834	0.0624	25.04	<.0001	
a34	0.623051	0.0546	11.41	<.0001	
a35	-0.565	0.0534	-10.59	<.0001	
a36	-0.56107	0.0517	-10.84	<.0001	
a37	0.476937	0.0563	8.48	<.0001	
a38	-1.37911	0.0621	-22.19	<.0001	
a39	0.550495	0.0632	8.71	<.0001	
bls	-0.13057	0.0122	-10.73	<.0001	
bli	-0.23545	0.00861	-27.36	<.0001	
ble	0.209433	0.0117	17.84	<.0001	
Equation	DF Model	DF Error	SSE	R-Square	Adj R-Sq
event	23	68087	12182.8	0.2591	0.2589

Table 12.4: Estimates of the HHG(84) Poisson regression with firm effects

	Random Effects	Fixed Effects
β_0	0.36 (0.02)	0.31 (0.04)
β_1	0.03 (0.04)	0.02 (0.05)
β_2	0.06 (0.05)	0.04 (0.06)
β_3	0.08 (0.05)	0.07 (0.06)
β_4	-0.07 (0.05)	-0.07 (0.07)
β_5	0.13 (0.03)	0.07 (0.05)
t	-0.04 (0.01)	-0.03 (0.003)
$\bar{\alpha}$	2.30 (0.16)	
Log likelihood	-3827.5	-3009.4

Appendix A

Nonlinear Optimization and Estimation

A.1 General nonlinear optimization problem

Basically, we have to solve the following problem:

$$\min_{\boldsymbol{\theta}\in\Theta} \varphi(\boldsymbol{\theta}) \text{ subject to } \Theta := \{\boldsymbol{\theta} \in R^p |\, \mathbf{g}(\boldsymbol{\theta}) \leqslant \mathbf{b}, \mathbf{q}(\boldsymbol{\theta}) = \mathbf{d}\}, \tag{A.1}$$

where $\mathbf{g}$ and $\mathbf{q}$ are twice continuously differentiable functions (hence, $\mathbf{g}, \mathbf{q} \in C^2$) mapping $\mathbf{g} : \mathbb{R}^p \Longrightarrow \mathbb{R}^m$ and $\mathbf{q} : \mathbb{R}^p \Longrightarrow \mathbb{R}^k$, respectively.

The *co-constraint qualification* (CCQ) can be defined as $B(\boldsymbol{\theta}^*) := \{i \in \{1, 2, ..., m\} \mid g^i(\boldsymbol{\theta}^*) = b_i\}$, so that CCQ implies that $g^i(.)$ is *active* at $\boldsymbol{\theta}^*$ (*binding condition*).

Consider in the first instance a simplification of (A.1):

$$\min_{\boldsymbol{\theta}\in\Theta} \varphi(\boldsymbol{\theta}), \mathbb{R}^p \overset{\varphi}{\Longrightarrow} \mathbb{R}^+ \text{ subject to } \Theta := \{\boldsymbol{\theta} \in \mathbb{R}^p |\, \mathbf{g}(\boldsymbol{\theta}) \leqslant \mathbf{b}\} \text{ with } \mathbb{R}^n \overset{\mathbf{g}}{\Longrightarrow} \mathbb{R}^m, \tag{A.2}$$

where we assume that the CCQ is satisfied. Then the famous Kuhn Tucker theorem can be formulated yielding necessary conditions for a local minimum of (A.2).

Kuhn-Tucker Theorem Given functions $\varphi, g^1, g^2 ..., g^n \in C^1$ in Θ, and for $\boldsymbol{\theta}^* \in \Theta$ a local minimum in Θ, and if Θ satisfies the CCQ in $\boldsymbol{\theta}^*$, then $\exists \boldsymbol{\lambda}^* := (\lambda_1^*, \lambda_2^*, \ldots, \lambda_m^*)'$ of Lagrange multipliers such that $\boldsymbol{\lambda}^* \geqq 0$ and

$\lambda_i^* \left(b_i - g^i(\boldsymbol{\theta}^*)\right) = 0 \;\; (i = 1, 2, ..., m)$ (*complementarity slackness*) and

$\nabla\varphi(\boldsymbol{\theta}^*) = \left(\frac{\partial\varphi}{\partial\theta_1}\Big|_{\boldsymbol{\theta}=\boldsymbol{\theta}^*}, \ldots, \frac{\partial\varphi}{\partial\theta_n}\Big|_{\boldsymbol{\theta}=\boldsymbol{\theta}^*} \right)' = -\sum_{i=1}^m \lambda_i^0 \nabla g^i(\boldsymbol{\theta}^*)$.

A.2 Gradient methods of nonlinear estimation

The problem $\min_{\boldsymbol{\theta}} \varphi(\boldsymbol{\theta})$ with $\mathbb{R}^p \overset{\varphi}{\Longrightarrow} \mathbb{R}_+$ is solved by considering a sequence of parameters $\boldsymbol{\theta}^{(k+1)} := \boldsymbol{\theta}^{(k)} + \boldsymbol{\xi}^{(k)}$, where $\boldsymbol{\xi}^{(k)}$ is the step in the kth iteration and $\boldsymbol{\theta}^{(k)}, \boldsymbol{\xi}^{(k)}, \boldsymbol{\theta}^{(k+1)} \in \mathbb{R}^p$.

The problem is now to choose a direction $\boldsymbol{\delta}^{(k)} \in \mathbb{R}^p$ and a step length $\boldsymbol{\lambda}^{(k)} > 0$ such that $\varphi\left(\boldsymbol{\theta}^{(k+1)}\right) < \varphi\left(\boldsymbol{\theta}^{(k)}\right)$, with $\boldsymbol{\theta}^{(k+1)} = \boldsymbol{\theta}^{(k)} + \boldsymbol{\lambda}^{(k)}\boldsymbol{\delta}^{(k)}$. This property is known as the *downhill* property.

Which direction $\boldsymbol{\delta}^{(k)}$ to choose ? Consider the function $\varphi\left(\boldsymbol{\theta}^{(k)} + \boldsymbol{\lambda}^{(k)}\boldsymbol{\delta}^{(k)}\right)$ as a function of the step length $\boldsymbol{\lambda}^{(k)}$ alone, so that:

$$\begin{aligned} \frac{d\varphi\left(\boldsymbol{\theta}^{(k)} + \boldsymbol{\lambda}^{(k)}\boldsymbol{\delta}^{(k)}\right)}{d\boldsymbol{\lambda}^{(k)}} &= \left.\frac{\partial\varphi\left(\boldsymbol{\theta}^{(k+1)}\right)}{\partial\boldsymbol{\theta}}\right|'_{\boldsymbol{\theta}=\boldsymbol{\theta}^{(k+1)}} \frac{d\left(\boldsymbol{\theta}^{(k)} + \boldsymbol{\lambda}^{(k)}\boldsymbol{\delta}^{(k)}\right)}{d\boldsymbol{\lambda}^{(k)}} + \cdots \\ &\simeq \left(\left.\frac{\partial\varphi}{\partial\boldsymbol{\theta}}\right|_{\boldsymbol{\theta}=\boldsymbol{\theta}^{(k+1)}}\right)' \boldsymbol{\delta}^{(k)} = \mathbf{g}^{(k+1)\prime}\boldsymbol{\delta}(k) < 0, \end{aligned}$$

where $\mathbf{g}^{(k)}$ is the gradient vector of the objective function at iteration k,or $\mathbf{g}^{(k)} := \nabla\varphi\left(\boldsymbol{\theta}^{(k)}\right)$. Hence, a direction with downhill property is found when the new gradient is making an obtuse angle with the gradient. In fact, the direction being opposite to that of the new gradient is the best one.

Lemma A.1 *The objective function decreases most in the direction precisely opposite to that of the gradient.*

Proof Consider an objective function which can be approximated by a first order Taylor expansion at the old point $\boldsymbol{\theta}^{(k)}$ (thus leading to a pure gradient method):

$$\begin{aligned} \varphi\left(\boldsymbol{\theta}^{(k+1)}\right) &= \varphi\left(\boldsymbol{\theta}^{(k)}\right) + \left.\frac{\partial\varphi(\theta)}{\partial\boldsymbol{\theta}}\right|'_{\boldsymbol{\theta}=\boldsymbol{\theta}^{(k)}} \left(\boldsymbol{\theta}^{(k+1)} - \boldsymbol{\theta}^{(k)}\right) + \cdots \\ &\simeq \varphi\left(\boldsymbol{\theta}^{(k)}\right) + (\mathbf{g}^{(k)})'\left(\boldsymbol{\theta}^{(k+1)} - \boldsymbol{\theta}^{(k)}\right). \end{aligned}$$

A *step vector* is now defined as $\boldsymbol{\xi}^{(k)} := \boldsymbol{\theta}^{(k+1)} - \boldsymbol{\theta}^{(k)}$, so that the finite length step vector $\boldsymbol{\xi}^{(k)}$ should be found such that the objective function decreases most with a finite step:
$\max \varphi\left(\boldsymbol{\theta}^{(k)}\right) - \varphi\left(\boldsymbol{\theta}^{(k+1)}\right) = \max_{\boldsymbol{\xi}^{(k)}} \varphi\left(\boldsymbol{\theta}^{(k)}\right) - \varphi\left(\boldsymbol{\theta}^{(k)} + \boldsymbol{\xi}^{(k)}\right) \simeq \max_{\boldsymbol{\xi}^{(k)}} -\mathbf{g}^{(k)\prime}\boldsymbol{\xi}^{(k)}$ subject to $\boldsymbol{\xi}^{(k)\prime}\boldsymbol{\xi}^{(k)} \leq c < \infty$, where c is a finite constant.

The Lagrange function belonging to the above problem is:

$$\mathcal{L}\left(\boldsymbol{\xi}^{(k)},\lambda^{(k)}\right) := -\mathbf{g}^{(k)\prime}\boldsymbol{\xi}^{(k)} + \lambda^{(k)}\left(c - \boldsymbol{\xi}^{(k)\prime}\boldsymbol{\xi}^{(k)}\right),$$

with $\lambda^{(k)}$ the Lagrange parameter at iteration k. We have to find a maximin (or saddlepoint) solution of this Lagrange function, i.e. maximization w.r.t. $\boldsymbol{\xi}^{(k)}$ and minimization w.r.t. $\lambda^{(k)}$. This leads to the necessary maximization condition $\frac{\partial\mathcal{L}}{\partial\boldsymbol{\xi}}\Big|_{\boldsymbol{\xi}=\boldsymbol{\xi}^{(k)}} = -\mathbf{g}^{(k)} - 2\lambda^{(k)}\boldsymbol{\xi}^{(k)} = 0$, or for each pair of components of $\mathbf{g}^{(k)}$ and $\boldsymbol{\xi}^{(k)}$, we have:

$$\frac{g_i^{(k)}}{\xi_i^{(k)}} = \frac{\frac{\partial\varphi(\boldsymbol{\theta})}{\partial\theta_i}\Big|_{\boldsymbol{\theta}=\boldsymbol{\theta}^{(k)}}}{\xi_i^{(k)}} = -2\lambda^{(k)} = \frac{\frac{\partial\varphi(\boldsymbol{\theta})}{\partial\theta_i}\Big|_{\boldsymbol{\theta}=\boldsymbol{\theta}^{(k)}}}{\xi_j^{(k)}} = \frac{g_j^{(k)}}{\xi_j^{(k)}} \qquad \forall_{i,j} = 1,2,...,p.$$

The sufficient maximization condition is that the Hessian of the Lagrangian, $\frac{\partial^2\mathcal{L}}{\partial\boldsymbol{\xi}\partial\boldsymbol{\xi}'}\Big|_{\boldsymbol{\xi}=\boldsymbol{\xi}^{(k)}} = -2\lambda^{(k)}\mathbf{I}_p$, should be negative definite, which implies that the step length should be positive (or $\lambda^{(k)} > 0$).

Hence, the objective function decreases most if we choose a step vector being proportional to the negative gradient of $\varphi(\boldsymbol{\theta})$ at $\boldsymbol{\theta} = \boldsymbol{\theta}^{(k)}$, which leads to the *gradient principle* that always chooses a down-hill direction or $\boldsymbol{\delta}^{(k)} := -\alpha^{(k)}\mathbf{g}^{(k)} = -\mathbf{P}^{(k)}\mathbf{g}^{(k)}$ where $\alpha^{(k)} > 0$ and $\mathbf{P}^{(k)}$ is positive definite.■

Summarizing for gradient methods: Define that iteration step

$$\boldsymbol{\theta}^{(k+1)} := \boldsymbol{\theta}^{(k)} + \lambda^{(k)}\boldsymbol{\delta}^{(k)}, \tag{A.3}$$

with $\lambda^{(k)}\boldsymbol{\delta}^{(k)}$ being the *step vector*, $\lambda^{(k)}$ the *step length*, and $\boldsymbol{\delta}^{(k)}$ the *direction vector* with downhill property $\boldsymbol{\delta}^{(k)} = -\mathbf{P}^{(k)}\mathbf{g}^{(k)}$ where $\mathbf{P}^{(k)}$ is a positive definite direction matrix (which is called a metric) and $\mathbf{g}^{(k)}$ the gradient vector.

A.2.1 Steepest descent

This method is based on the first order Taylor expansion of the objective function φ.

It has the advantages that the function decreases most in the direction of the negative gradient and that it is a safe method if the step length is chosen sufficiently small. However, it has several disadvantages: a slow convergence, steepest descent is not invariant to parameter (re)scaling, it is not a flexible method since $\mathbf{P}^{(k)}$ cannot be adjusted in each step and, finally, it is principally advised for sphere-type problems, not for *steep valleys*.

Exercise A.1 *CES - production function*

To illustrate we use the output, labor, and capital data given in Figure A.1 (see e.g. Judge *et al.* (1990), pp. 209-210).

Figure A.1: CES production function

T	L_i	K_t	lnQ_t
1	0.228	0.802	-1.359
2	0.258	0.249	-1.695
3	0.821	0.771	0.193
4	0.767	0.511	-0.649
5	0.495	0.758	-0.165
6	0.487	0.425	-0.270
7	0.678	0.452	-0.473
8	0.748	0.817	0.031
9	0.727	0.845	-0.563
10	0.695	0.958	-0.125
11	0.458	0.084	-2.218
12	0.981	0.021	-3.633
13	0.002	0.295	-5.586
14	0.429	0.277	-0.773
15	0.231	0.546	-1.315
16	0.664	0.129	-1.678
17	0.631	0.017	-3.879
18	0.059	0.906	-2.301
19	0.811	0.223	-1.377
20	0.758	0.145	-2.270
21	0.050	0.161	-2.539
22	0.823	0.006	-5.150
23	0.483	0.836	-0.324
24	0.682	0.521	-0.253
25	0.116	0.930	-1.530
26	0.440	0.495	-0.614
27	0.456	0.185	-1.151
28	0.342	0.092	-2.089
29	0.358	0.485	-0.951
30	0.162	0.934	-1.275

Source: Judge *et al.* (1990), pp. 209-210.

The labor and capital data are random samples from a uniform distribution on the interval (0,1) and $\ln Q_t$ is computed by adding an *iid* error ε_t, generated from a normal distribution with mean zero and standard deviation 0.25, to the CES specification -0.5ln $\left(0.3L_t^{-2} + 0.7K_t^{-2}\right)$ for each $t = 1, 2, ..., 30$ (see Example 1.3). Thus $\beta_0^* = 0$, $\beta_1^* = -0.5$, $\beta_2^* = .3$, $\beta_3^* = -2$ and $\sigma^{*2} = 0.0625$.

Starting with arbitrarily chosen initial values $\beta_0 = 1; \beta_1 = -1;\ \beta_2 = 0.5$ and $\beta_3 = -1$, the steepest descent algorithm terminated after 40 iterations without reaching the global minimum of the sum of squares function. The last 20 iterations are presented in Figure A.2.

Progress is very slow although the iterations are not close to the minimum, as we will see later. To guarantee positive arguments of the logarithm, we restricted the values for β_2 to the interval $[0, 1]$.

Figure A.2: Iterations of the Steepest Descent algorithm

True Parameter Values	β_0*=0.0	β_1*=-0.5	β_2*=0.3	β_3*=-2.0	
Iterartion	β_0	β_1	β_2	β_3	$S(\beta)$
21	0.22282869	-1.06629056	0.40915651	-1.09263870	2.14819748
22	0.22019703	-1.05781151	0.40910690	-1.08388148	2.11001625
23	0.19636534	-1.06104723	0.40801302	-1.08963337	2.08631253
24	0.19527758	-1.05243868	0.40801826	-1.08077867	2.04932979
25	0.17192604	-1.05593736	0.40691240	-1.08761891	2.04780190
26	0.17225266	-1.04733044	0.40696972	-1.07878857	2.00689787
27	0.16037233	-1.04780308	0.40639450	-1.08165294	1.99692671
28	0.15922856	-1.04339570	0.40635495	-1.07755628	1.98908863
29	0.14813284	-1.04520171	0.40579539	-1.08227361	1.98896682
30	0.14818368	-1.03638455	0.40582105	-1.07379932	1.98637455
31	0.14264469	-1.04314485	0.40551648	-1.08221663	1.98369970
32	0.14310683	-1.03432048	0.40555767	-1.07378493	1.98234469
33	0.13796403	-1.04120824	0.40527184	-1.08232826	1.97973417
34	0.13878359	-1.03238385	0.40532641	-1.07394317	1.97936713
35	0.13397475	-1.03937208	0.40505660	-1.08258130	1.97669824
36	0.13453838	-1.03496179	0.40508958	-1.07841297	1.96586504
37	0.12338095	-1.03161520	0.40446625	-1.08178996	1.96256149
38	0.12362889	-1.02925600	0.40447288	-1.07989821	1.96009223
39	0.11904837	-1.02738382	0.40415996	-1.08337737	1.95960516
40	0.11939822	-1.02494985	0.40416776	-1.08160696	1.95826454

Figure A.3: Data for Example A.1

t	y_t	x_{t1}	x_{t2}	x_{t3}
1	4.284	1.000	0.286	0.645
2	4.149	1.000	0.973	0.585
3	3.877	1.000	0.384	0.310
4	0.533	1.000	0.276	0.058
5	2.211	1.000	0.973	0.455
6	2.389	1.000	0.543	0.779
7	2.145	1.000	0.957	0.259
8	3.231	1.000	0.948	0.202
9	1.998	1.000	0.543	0.028
10	1.379	1.000	0.797	0.099
11	2.106	1.000	0.936	0.142
12	1.428	1.000	0.889	0.296
13	1.011	1.000	0.006	0.175
14	2.179	1.000	0.828	0.180
15	2.858	1.000	0.399	0.842
16	1.388	1.000	0.617	0.039
17	1.651	1.000	0.939	0.103
18	1.593	1.000	0.784	0.620
19	1.046	1.000	0.072	0.158
20	2.152	1.000	0.889	0.704

Source: Judge *et al.* (1985), pp. 956-958, and Judge *et al.* (1990) pp. 956-958

A.2.2 Newton (-Raphson)

We are looking for a quicker algorithm, taking account of the curvature of the objective function. Consider therefore a second order Taylor expansion of the objective function around the old point; or for any iteration we get:

$$\varphi(\boldsymbol{\theta}^{(k+1)}) \cong \varphi(\boldsymbol{\theta}^{(k)}) + \mathbf{g}^{(k)\prime}(\boldsymbol{\theta}^{(k+1)}\text{-}\boldsymbol{\theta}^{(k)}) + \tfrac{1}{2}(\boldsymbol{\theta}^{(k+1)}\text{-}\boldsymbol{\theta}^{(k)})'\mathbf{H}^{(k)}(\boldsymbol{\theta}^{(k+1)}\text{-}\boldsymbol{\theta}^{(k)})$$

so that it follows from (A.3) that $\boldsymbol{\theta}^{(k+1)} = \boldsymbol{\theta}^{(k)} - \left(\mathbf{H}^{(k)}\right)^{-1}\mathbf{g}^{(k)}$.

Example A.1 *Newton procedure based on Example 1.2 in Section 1.2.1*

The data used for this example is given in Figure A.3. The x_{t2}s and x_{t3}s are pseudo-random numbers from a uniform distribution on the unit interval and the y_ts are computed by adding normal pseudo-random numbers to $\theta_1^* + \theta_2^* x_{t2} + \theta_2^{*2} x_{t3}$, where the true parameters θ_1^* and θ_2^* were chosen to be $\theta_1^* = \theta_2^* = 1$, i.e. a random error has actually been added to $1 + x_{t2} + x_{t3}$.

To determine the least squares estimates of $\boldsymbol{\theta}$ we have to minimize $\varphi(\boldsymbol{\theta}) = [\mathbf{y} - \mathbf{f}(\boldsymbol{\theta})]'[\mathbf{y} - \mathbf{f}(\boldsymbol{\theta})]$ using the Newton iterations for $\boldsymbol{\theta}^{(k+1)}$ and three different starting values, we obtained the results given in Figure *A.4*.

The algorithm terminates at two different points in the parameter space. This outcome is not surprising since $\varphi(\boldsymbol{\theta})$ has two different local minima 16.08 and 20.98. The example shows that an optimization algorithm does not necessarily converge to the global minimum.

Newton iterations generally lead to growth convergence, but pay attention to

Figure A.4: Iterations of the Newton algorithm

n	$\theta_{n,1}$	$\theta_{n,2}$	$H(\theta_n)$
1	3.000000	2.000000	264.3918
2	-0.084033	1.811210	20.6328
3	0.625029	1.423940	16.5105
4	0.817259	1.272776	16.0961
5	0.862590	1.237516	16.0818
6	0.864782	1.235753	16.0817
7	0.864787	1.235748	16.0817
8	0.864787	1.235748	16.0817
1	0.000000	2.000000	29.2758
2	0.334936	1.600435	17.7382
3	0.735040	1.336953	16.1955
4	0.849677	1.247743	16.0832
5	0.864541	1.235946	16.0817
6	0.864787	1.235749	16.0817
7	0.864787	1.235748	16.0817
8	0.864787	1.235748	16.0817
1	1.500000	0.500000	20.2951
2	2.256853	0.007135	20.7735
3	2.467047	-0.436460	21.0312
4	2.316982	-0.202435	20.9467
5	2.359743	-0.320579	20.9809
6	2.354457	-0.319153	20.9805
7	2.354471	-0.319186	20.9805
8	2.354471	-0.319186	20.9805

the importance of the selection of the starting values (multiple modi) and to the implied positive definitess of the Hessian in each iteration k.

Exercise A.2 *Compare the steepest descent and Newton algorithms for the following cases:*

Case 1 Assume the quadratic function $\varphi(\boldsymbol{\theta}) = 2\theta_1^2+6\theta_2^2+2\theta_1\theta_2+2\theta_1+3\theta_2+3$. Rewrite this function as a quadratic form $\varphi(\boldsymbol{\theta}) = a + \mathbf{b}'\boldsymbol{\theta} + \frac{1}{2}\boldsymbol{\theta}'\mathbf{Q}\boldsymbol{\theta}$, with $a :=$ $3; \mathbf{b} := \begin{pmatrix} 2 \\ 3 \end{pmatrix}; \mathbf{Q} := \begin{pmatrix} 4 & 2 \\ 2 & 12 \end{pmatrix}$.

Case 2 Assume the quadratic function $\varphi(\boldsymbol{\theta}) = (\theta_1 - \theta_2 + \theta_3)^2 + (-\theta_1 + \theta_2 + \theta_3)^2 + (\theta_1+\theta_2-\theta_3)^2$ with the vectors of the underlying quadratic form $a := 0$, $\mathbf{b} := \mathbf{0}$,

$$\mathbf{Q} := \begin{pmatrix} 6 & -2 & -2 \\ -2 & 6 & -2 \\ -2 & -2 & 6 \end{pmatrix}.$$

A.2.3 Quasi-Newton procedures

Since the positive definiteness of the Hessian matrix in each iteration is not guaranteed in many cases, we can propose a class of procedures to approach the inverse of this Hessian of the objective function by a direction matrix, involving a *variable metric* $\mathbf{P}^{(k)}$ and a correction matrix $\mathbf{M}^{(k)}$ in $\mathbf{P}^{(k+1)} := \mathbf{P}^{(k)} + \mathbf{M}^{(k)}$ with $\mathbf{P}^{(k)}$ the approximation for $(\mathbf{H}^{(k)})^{-1}$ in iteration k.

A reasonable choice for $\mathbf{P}^{(k+1)}$ can be derived from the first order approximation of the gradient at a *new point*, i.e. from $\boldsymbol{\theta}^{(k+1)} := \boldsymbol{\theta}^{(k)} + \lambda^{(k)}\boldsymbol{\delta}^{(k)}$ with $\boldsymbol{\delta}^{(k)} := -\mathbf{P}^{(k)}\mathbf{g}^{(k)}$ we find:

$$\mathbf{g}^{(k)} \sim \mathbf{g}^{(k+1)} + \mathbf{H}^{(k+1)}\left(\boldsymbol{\theta}^{(k)} - \boldsymbol{\theta}^{(k+1)}\right)$$

with $\left(\mathbf{H}^{(k+1)}\right)^{-1} = \mathbf{P}^{(k)} + \mathbf{M}^{(k)} =: \mathbf{P}^{(k+1)}$ or:

$$\mathbf{M}^{(k)}\left(\mathbf{g}^{(k+1)} - \mathbf{g}^{(k)}\right) \simeq \boldsymbol{\eta}^{(k)}$$

with $\boldsymbol{\eta}^{(k)} := \boldsymbol{\zeta}^{(k)} - \mathbf{P}^{(k)}\left(\mathbf{g}^{(k+1)} - \mathbf{g}^{(k)}\right)$ and $\boldsymbol{\xi}^{(k)} := \boldsymbol{\theta}^{(k+1)} - \boldsymbol{\theta}^{(k)}$, which leads to $\mathbf{M}^{(k)} \simeq \frac{\boldsymbol{\eta}^{(k)}\boldsymbol{\eta}^{(k)\prime}}{\boldsymbol{\eta}^{(k)\prime}\left(\mathbf{g}^{(k+1)} - \mathbf{g}^{(k)}\right)}$. This is the *rank one correction method.*

Alternatively, instead of these iterations on the correction matrices $\mathbf{M}^{(k)}$,we can rewrite them as:

$$\mathbf{M}^{(k)} \simeq \frac{\boldsymbol{\zeta}^{(k)}\boldsymbol{\zeta}^{(k)\prime}}{\boldsymbol{\zeta}^{(k)\prime}\left(\mathbf{g}^{(k+1)} - \mathbf{g}^{(k)}\right)} - \frac{\mathbf{P}^{(k)}\left(\mathbf{g}^{(k+1)} - \mathbf{g}^{(k)}\right)\left(\mathbf{g}^{(k+1)} - \mathbf{g}^{(k)}\right)'\mathbf{P}^{(k)}}{\left(\mathbf{g}^{(k+1)} - \mathbf{g}^{(k)}\right)'\mathbf{P}^{(k)}\left(\mathbf{g}^{(k+1)} - \mathbf{g}^{(k)}\right)}, \quad \text{(A.4)}$$

which is the *Davidon-Fletcher-Powell* (DFP) correction method.

Example A.2 *Apply the Newton and quasi-Newton algorithms on the following numerical example:*

$F(\boldsymbol{\theta}) = \frac{3}{2}\theta_1^2 + \frac{1}{2}\theta_2^2 - \theta_1\theta_2 - 2\theta_1 = (-2, 0)(\theta_1, \theta_2)' + \frac{1}{2}\boldsymbol{\theta}'\begin{pmatrix} 3 & -1 \\ -1 & 1 \end{pmatrix}\boldsymbol{\theta}$. Select as starting vector $\boldsymbol{\theta}^0 := (-2, 4)'$.

1. *With the method of Newton:* Iteration 1:

$$\mathbf{g}^{(0)} = \begin{pmatrix} 3\theta_1^{(0)} - \theta_2^{(0)} - 2 \\ \theta_2^{(0)} - \theta_1^{(0)} \end{pmatrix} = \begin{pmatrix} -12 \\ 6 \end{pmatrix}, \mathbf{H}^{(0)} = \begin{pmatrix} 3 & -1 \\ -1 & 1 \end{pmatrix} \text{ so that}$$

$$\boldsymbol{\delta}^{(o)} = -\mathbf{P}^{(0)}\mathbf{g}^{(0)} = -\begin{pmatrix} 1 & 0 \\ 0 & 1 \end{pmatrix}\begin{pmatrix} -12 \\ 6 \end{pmatrix} = \begin{pmatrix} 12 \\ -6 \end{pmatrix} \text{ or:}$$

$$\min_{\lambda^{(0)}} \varphi\left(\boldsymbol{\theta}^{(0)} + \lambda^{(0)}\boldsymbol{\delta}^{(0)}\right) = \min_{\lambda^{(0)}}\left[306\lambda^{(0)^2} - 180\lambda^{(0)} + 26\right].$$

The first order condition (FOC) of this minimization is $\frac{d\varphi}{d\lambda} = 612\lambda^{(0)} - 180 = 0$, or $\lambda^{(0)} = \frac{5}{17}$, so that $\boldsymbol{\theta}^{(1)} = (26/17, 38/17)'$ and $\boldsymbol{\xi}^{(0)} = \boldsymbol{\theta}^{(1)} - \boldsymbol{\theta}^{(0)} = (60/17, -30/17)'$.

Iteration 2: $\mathbf{g}^{(1)} = (6/17, 12/17)'$ and $\mathbf{P}^{(1)} = \left(\mathbf{H}^{(1)}\right)^{-1}$.

$$\mathbf{H}^{(0)} = \mathbf{H}^{(1)}, \text{ or } \mathbf{H}^{(1)} = \begin{pmatrix} 3 & -1 \\ -1 & 1 \end{pmatrix}.$$

Then, $\left(\mathbf{H}^{(1)}\right)^{-1} = \begin{pmatrix} 0.5 & 0.5 \\ 0.5 & 1.5 \end{pmatrix}$ and $\boldsymbol{\delta}^{(1)} = -\mathbf{P}^{(1)}\mathbf{g}^{(1)} = -\begin{pmatrix} 0.5 & 0.5 \\ 0.5 & 1.5 \end{pmatrix}\begin{pmatrix} 6/17 \\ 12/17 \end{pmatrix} = \begin{pmatrix} -0.5294 \\ -1.2353 \end{pmatrix}$

$$\min_{\lambda^{(0)}} \varphi\left(\boldsymbol{\theta}^{(1)} + \lambda^{(1)}\boldsymbol{\delta}^{(1)}\right) = \min_{\lambda^{(0)}} \left[0.5294\lambda^{(1)^2} - 1.0588\lambda^{(1)} - 0.4705\right]$$

with FOC $2 \times 0.5294\lambda^{(1)} - 1.0588 = 0$. Then, $\lambda^{(1)} = 1$ and $\boldsymbol{\theta}^{(2)} = (1,1)^{'}$, and $\boldsymbol{\xi}^{(1)} = \boldsymbol{\theta}^{(2)} - \boldsymbol{\theta}^{(1)} = (-0.5294, -1.2953)^{'}$.

Iteration 3: For $\mathbf{g}^{(2)} = (0,0)^{'}$ and $\boldsymbol{\delta}^{(2)} = 0$, the solution is $\varphi\left(\boldsymbol{\theta}^{*}\right) = -1$.

2. *With the method of Quasi-Newton (e.g. the DFP algorithm):*

Iteration 1 For $\boldsymbol{\theta}^{(0)} = (-2,4)^{'}$, $\mathbf{g}^{(0)} = (3\theta_1 - \theta_2 - 2, \theta_2 - \theta_1)^{'} = (-12,6)^{'}$; then

$$\boldsymbol{\delta}^{(o)} = -\mathbf{P}^{(0)}\mathbf{g}^{(0)} = -\begin{pmatrix} 1 & 0 \\ 0 & 1 \end{pmatrix}\begin{pmatrix} -12 \\ 6 \end{pmatrix} = \begin{pmatrix} 12 \\ -6 \end{pmatrix}$$

$$\min_{\lambda^{(0)}} \varphi\left(\boldsymbol{\theta}^{(0)} + \lambda^{(0)}\boldsymbol{\delta}^{(0)}\right) = \min_{\lambda^{(0)}} \left[306\lambda^{(0)^2} - 180\lambda^{(0)} + 26\right].$$

The FOC is $\frac{d\varphi}{d\lambda} = 612\lambda^{(0)} - 180 = 0$, then $\lambda^{(0)} = \frac{5}{17}$; $\boldsymbol{\theta}^{(1)} = (26/17, 38/17)^{'}$ and $\boldsymbol{\xi}^{(0)} = \boldsymbol{\theta}^{(1)} - \boldsymbol{\theta}^{(0)} = (60/17, -30/17)'$

Iteration 2: $\boldsymbol{\theta}^{(1)} = (26/17, 38/17)^{'}$, $\mathbf{g}^{(1)} = (6/17, 12/17)^{'}$, so that $\mathbf{z}^{(0)} = \mathbf{g}^{(1)} - \mathbf{g}^{(0)} = (210/17, -90/17)^{'}$ and

$$\begin{aligned} \mathbf{P}^{(1)} &= \mathbf{P}^{(0)} + \mathbf{M}^{(0)} \\ &= \begin{pmatrix} 1 & 0 \\ 0 & 1 \end{pmatrix} + \begin{pmatrix} 4/17 & -2/17 \\ -2/17 & 1/17 \end{pmatrix} - \begin{pmatrix} 49/58 & -21/58 \\ -21/58 & 9/58 \end{pmatrix} \\ &= \frac{1}{986}\begin{pmatrix} 385 & 241 \\ 241 & 891 \end{pmatrix}, \end{aligned}$$

so that the first direction satisfies $\boldsymbol{\delta}^{(1)} := -\mathbf{P}^{(1)}\mathbf{g}^{(1)} = (-9/29, -21/29)'$. Then,

$$\min_{\lambda^{(0)}} \varphi\left(\boldsymbol{\theta}^{(1)} + \lambda^{(1)}\boldsymbol{\delta}^{(1)}\right) = \min_{\lambda^{(0)}} \left[\tfrac{306}{1682}\lambda^{(1)^2} - \tfrac{90}{493}\lambda^{(1)} - 136\right]$$

The FOC is $\frac{d\varphi}{d\lambda} = 2\frac{306}{1682}\lambda^{(1)} - \frac{90}{493} = 0$ or $\lambda^{(1)} = \frac{145}{289}$. Furthermore, $\boldsymbol{\theta}^{(2)} = (397/289, 541/289)^{'}$ and $\boldsymbol{\xi}^{(1)} = \boldsymbol{\theta}^{(2)} - \boldsymbol{\theta}^{(1)} = (-45/289, -105/289)'$.

Remark: Control on conjugate gradients (orthogonality condition as a check on the correctness of the computations) Since subsequent directions should be orthogonal with respect to the Hessian matrix, we can always check this to verify the correctness of the computations.

At this stage, $(\boldsymbol{\delta}^{(1)})'\mathbf{H}^{(1)}\boldsymbol{\delta}^{(0)} = 0$, which is true since $\boldsymbol{\delta}^{(1)} = \begin{pmatrix} -9/29 \\ -21/29 \end{pmatrix}$, $\mathbf{H}^{(1)} = \begin{pmatrix} 3 & -1 \\ -1 & 1 \end{pmatrix}$ .and $\boldsymbol{\delta}^{(0)} = \begin{pmatrix} 12 \\ -6 \end{pmatrix}$.

Iteration 3: For $\boldsymbol{\theta}^{(2)} = (397/289, 541/289)' = (1.3737, 1.8720)'$ and $\mathbf{g}^{(2)} = (72/289, 144/289)'$we get: $\mathbf{z}^{(1)} = \mathbf{g}^{(2)} - \mathbf{g}^{(1)} = (-30/289, -60/289)'$ and

$$\begin{aligned}\mathbf{P}^{(2)} &= \mathbf{P}^{(1)} + \mathbf{M}^{(1)} \\ &= \frac{1}{986}\begin{pmatrix} 385 & 241 \\ 241 & 891 \end{pmatrix} + \frac{\boldsymbol{\xi}^{(1)}\boldsymbol{\xi}^{(1)\prime}}{\boldsymbol{\xi}^{(1)\prime}\mathbf{z}^{(1)}} - \frac{\mathbf{P}^{(1)}\mathbf{z}^{(1)}\mathbf{z}^{(1)\prime}\mathbf{P}^{(1)}}{\mathbf{z}^{(1)\prime}\mathbf{P}^{(1)}\mathbf{z}^{(1)}} \\ &= \begin{pmatrix} 0.4988 & 0.4990 \\ 0.5083 & 1.5347 \end{pmatrix},\end{aligned}$$

so that $\boldsymbol{\delta}^{(2)} = -\mathbf{P}^{(2)}\mathbf{g}^{(2)} = (-0.3729, -0.8914)'$. The (next) minimization is:

$$\min_{\lambda^{(0)}} \varphi\left(\theta^{(2)} + \delta^{(2)} * \lambda^{(2)}\right) = \min_{\lambda^{(0)}}\left[0.204\lambda^{(2)^2} - 0.5539\lambda^{(2)} - 1.4905\right]$$

The FOC is $\frac{d\varphi}{d\lambda} = 2 \times 0.204\lambda^{(2)} - 0.5539 = 0$ so that $\lambda^{(2)} = 1.3576, \boldsymbol{\theta}^{(3)} = (0.8675, 0.6618)'$ and $\boldsymbol{\xi}^{(2)} = \boldsymbol{\theta}^{(3)} - \boldsymbol{\theta}^{(2)} = (-0.5062, -1.2102)'$.

Iteration 4: For $\boldsymbol{\theta}^{(3)} = (0.8675, 0.6618)'$ and $\mathbf{g}^{(3)} = (-0.0593, -0.2057)'$, $\mathbf{z}^{(2)} = \mathbf{g}^{(3)} - \mathbf{g}^{(2)} = (-0.3084, -0.7040)'$ and $\mathbf{P}^{(3)} = \mathbf{P}^{(2)} + \mathbf{M}^{(2)} = \begin{pmatrix} 0.4919 & 0.5104 \\ 0.4687 & 1.5086 \end{pmatrix}$, so that $\boldsymbol{\delta}^{(3)} = -\mathbf{P}^{(3)}\mathbf{g}^{(3)} = (0.1342, 0.3381)'$. Then:

$$\min_{\lambda^{(0)}} \varphi\left(\boldsymbol{\theta}^{(3)} + \lambda^{(3)}\boldsymbol{\delta}^{(3)}\right) = \min_{\lambda^{(0)}}\left[0.0388\lambda^{(3)^2} - 0.0774\lambda^{(3)} - 0.9613\right]$$

with FOC $\frac{d\varphi}{d\lambda} = 2\times 0.0388\lambda^{(3)} - 0.0774 = 0$ or $\lambda^{(3)} = 0.9974, \boldsymbol{\theta}^{(4)} = (1.0014, 0.9990)'$ and $\boldsymbol{\xi}^{(3)} = \boldsymbol{\theta}^{(4)} - \boldsymbol{\theta}^{(3)} = (0.1339, 0.3372)'$.

Iteration 5: For $\boldsymbol{\theta}^{(4)} = (1.0014, 0.9990)'$ and $\mathbf{g}^{(4)} = (0.0052, 0.0024)'$we have that $\mathbf{z}^{(3)} = \mathbf{g}^{(4)} - \mathbf{g}^{(3)} = (0.0645, 0.2081)'$ and $\mathbf{P}^{(4)} = \mathbf{P}^{(3)} + \mathbf{M}^{(3)} = \begin{pmatrix} 0.4979 & 0.4903 \\ 0.4896 & 1.4697 \end{pmatrix}$, or $\boldsymbol{\delta}^{(4)} = -\mathbf{P}^{(4)}\mathbf{g}^{(4)} = (-0.0038, -0.0061)'$.

$$\min_{\lambda^{(0)}} \varphi\left(\theta^{(4)} + \delta^{(4)} * \lambda^{(4)}\right) = \min_{\lambda^{(0)}}\left[0.00002\lambda^{(4)^2} - 1\right]$$

with FOC $\frac{d\varphi}{d\lambda} = 280.00002\lambda = 0$, i.e. $\lambda = 0$. Hence we reach the solution (local minimum): $\boldsymbol{\theta}^{(*)} = (1, 1)'$ and $\varphi(\boldsymbol{\theta}^*) = -1$ after four interations.

A.3 Constrained nonlinear optimization methods

Constrained nonlinear optimization methods can be subdivided into three types of procedures:

1. Transformation of independent variables, which boils down to transformation of parameters (reparameterization).

2. Transformation of the objective function into a Lagrange function, leading to barrier function and penalty function methods.

3. Transformation of the optimization procedure itself, which gives rise to the extended or augmented Lagrangian procedure.

A.3.1 Reparameterization

1. Consider the minimization problem:

$$\min_{\boldsymbol{\theta}} \varphi(\theta_1, \theta_2, ..., \theta_p) \quad \text{with } \boldsymbol{\theta} \geqq \mathbf{0}.$$

For solving this problem we need *auxiliary parameters* $\eta_i(\boldsymbol{\theta})$, for $i = 1, 2, ..., p$. Several examples of this auxiliary parameters could be considered: $\theta_i := \eta_i^2$, $\theta_i := e^{\eta_i}$, $\theta_i := |\eta_i|$, $\theta_i := \sin^2 \eta_i = 1 - \cos^2 \eta_i$, and $\theta_i := \frac{e^{\eta_i}}{e^{\eta_i} + e^{-\eta_i}}$.

The originally constrained minimization problem is now equivalent to the unconstrained minimization problem: $\min_{\{\eta_i\}} \varphi(\eta_1, \eta_2, ..., \eta_p)$ for $i = 1, 2, ..., p$.

The main disadvantage of this approach is that in many cases non-uniqueness occurs.

2. For the minimization problem:

$$\min_{\boldsymbol{\theta}} \varphi(\theta_1, \theta_2, ..., \theta_p) \text{ for } \theta_i \geqq a_i \qquad (i = 1, 2, ..., p).$$

Redefine new parameters: $\eta_i^2 = \theta_i - a_i$, or $\eta_i = \pm\sqrt{\theta_i - a_i}$ or $e^{\eta_i} = \theta_i - a_i$, then $\eta_i = \ln(\theta_i - a_i)$ or other transformations. The above constrained problem is equivalent to the unconstrained minimization of $\varphi(\eta_1, \eta_2, ..., \eta_p)$.

3. Consider the constrained minimization problem:

$$\min_{\{\theta_i\}} \varphi(\theta_1, \theta_2, ..., \theta_p) \text{ s.t. } a_i \leqq \theta_i \leqq b_i,\ i = 1, 2, ..., p.$$

Reparameterize $\theta_i = a_i + (b_i - a_i)\sin^2 \eta_i$, $i = 1, 2, \ldots, p$ or $\eta_i = \arcsin\sqrt{\frac{\theta_i - a_i}{b_i - a_i}}$. This leads to the unconstrained minimization of $\varphi(\eta_1, \eta_2, ..., \eta_p)$:

$$\min \varphi(\theta_1, \theta_2) \quad \text{s.t. } \theta_i \geqq 0\ (i = 1, 2),\ \theta_1 + \theta_2 \leqq 1.$$

Reparameterize using polar coordinates (r, γ):
$\theta_1 = r\cos\gamma$ and $\theta_2 = r\sin\gamma$ with $0 \leqq \gamma \leqq \frac{\pi}{2}$ and $0 \leqq r \leqq R$. Hence, $\theta_1 + \theta_2 = r(\sin\gamma + \cos\gamma)$ and $R = \frac{1}{\sin\gamma + \cos\gamma}$, so that

$$\begin{array}{r} r = 0 + (R - 0)\sin^2\eta_1 = R\sin^2\eta_1 \\ \gamma = 0 + \left(\frac{\pi - 0}{2}\right)\sin^2\eta_2 = \frac{\pi}{2}\sin^2\eta_2 \end{array}; \text{ then: } \quad \begin{array}{l} \eta_1 = \arcsin\sqrt{\frac{r}{R}} \\ \eta_2 = \arcsin\sqrt{\frac{2\gamma}{\pi}}, \end{array}$$

so that the above constrained minimization problem is equivalent to the unconstrained minimization of $\varphi(\eta_1, \eta_2)$ w.r.t. η_1 and η_2.

A.3.2 Transformation of the objective function

There are two possible transformations of the objective function: first the barrier function, which is an *interior* point (minimization) problem and second, the penalty function, which is an *exterior* point (minimization) problem. The minimization problem $\min_{\boldsymbol{\beta}} \varphi(\boldsymbol{\beta})$ s.t. $\mathbf{q}(\boldsymbol{\beta}) \geqq 0$ is transformed as $\varphi(\boldsymbol{\beta}) + B(\boldsymbol{\beta})$.

1. *Interior* point method (*barrier function* method) with transformation: $B(\boldsymbol{\beta}) := \frac{c}{q(\boldsymbol{\beta})}$ for one restriction (c being a small positive number).

   In general $B(\boldsymbol{\beta}) := c\sum_{j=1}^{m} \rho(q_j(\boldsymbol{\beta}))$ with $\rho(\cdot)$ a continuously differentiable function on $\mathbb{R}_+$ with $\rho(q_j(\boldsymbol{\beta})) \to \infty$ if $q_j(\boldsymbol{\beta}) \to 0$.

2. *Exterior* point method (*penalty function* method) with transformation for one restriction $B(\boldsymbol{\beta}) := d\{\min(0, q(\boldsymbol{\beta}))\}^2$. In general $B(\boldsymbol{\beta}) := d\sum_{j=1}^{m} \eta(q_j(\boldsymbol{\beta}))$ where $\eta(\cdot)$ is a continuously differentiable function on $\mathbb{R}_+$ with:

$$\begin{array}{l} \eta(q_j(\boldsymbol{\beta})) = 0 \ \text{ if } \ q_j(\boldsymbol{\beta}) \geqq 0 \\ \eta(q_j(\boldsymbol{\beta})) > 0 \ \text{ if } \ q_j(\boldsymbol{\beta}) < 0. \end{array}$$

Example A.3 *Consider the CES production function:*

$$\ln Q_t = \beta_0 + \beta_1 \ln\left[\beta_2 L_t^{\beta_3} + (1 - \beta_2) K_t^{\beta_3}\right] + e_t$$

with $\mathbf{q}(\boldsymbol{\beta}) = \binom{\beta_2}{1-\beta_2} > 0$,then the barrier function method involves $B(\boldsymbol{\beta}) := c\left(\frac{1}{\beta_2} + \frac{1}{1-\beta_2}\right)$, or alternatively, $B(\boldsymbol{\beta}) := -c(\ln\beta_2 + \ln(1 - \beta_2))$. The penalty function method implies $B(\boldsymbol{\beta}) := d\left\{[\min(0, \beta_2)]^2 + [\min(0, 1 - \beta_2)]^2\right\}$.

Summary A.1 *For barrier and penalty function methods (in principle) an infinite row of ordered, positive and increasing numbers (c_n, d_n) should be determined to obtain a feasible minimization of the objective function; hence, many cases (c_n, d_n) have to be computed.*

A disadvantage of these methods is that the Hessian of the new objective function does not exist at the edge of the feasible region (at least not in the barrier function method).

A.3.3 Dual methods

Given the problem $\min_{\boldsymbol{\theta}} \{\varphi(\boldsymbol{\theta}) | \, q_j(\boldsymbol{\theta}) \geqq 0, \text{ for } j = 1, 2, ...\}$,the dual problem is $\max_{\boldsymbol{\mu}} \left[\gamma_{(\boldsymbol{\mu})} = \min_{\boldsymbol{\theta}} \{\mathcal{L}(\boldsymbol{\theta}, \boldsymbol{\mu})\} \right]$ with $\mathcal{L}(\boldsymbol{\theta}, \mu) := \varphi(\boldsymbol{\theta}) - \sum_{j=1}^{m} \mu_j q_j(\boldsymbol{\theta})$ the Lagrange function for which a maximin or saddle point solution should be found. The main disadvantage is that in many cases the Hessian is not positive definite in $\mathbb{R}^{pxp}$.

A.3.4 Extended (augmented) Lagrange function method

According to Rockafellar (1976), we first consider problem (A.1) with equality constraints only and $\mathbf{d} = \mathbf{0}$. The augmented (extended) Lagrangian is then:

$$\mathcal{L}_c(\boldsymbol{\theta}, \boldsymbol{\mu}) := \varphi(\boldsymbol{\theta}) - \sum_{j=1}^{k} \mu_j q_j(\boldsymbol{\theta}) + \tfrac{c}{2} \sum_{j=1}^{k} (q_j(\boldsymbol{\theta}))^2 ,$$

with $c > 0, \boldsymbol{\theta} \in \mathbb{R}^p, \boldsymbol{\mu} \in \mathbb{R}^k$, i.e. $\mathcal{L}_c(\boldsymbol{\theta}, \boldsymbol{\mu})$ is the Lagrange function of the problem:

$$\min_{\boldsymbol{\theta}} \varphi(\boldsymbol{\theta}) + \frac{1}{2} c \sum_{j=1}^{k} (q_j(\boldsymbol{\theta}))^2 \Big| \, q_j(\boldsymbol{\theta}) = 0 \;\; (j = 1, 2, ..., k) . \tag{A.5}$$

The Kuhn-Tucker conditions for (A.5) are identical to those of the relevant part of (A.1), but the interesting aspect is that the Hessian of the Lagrange function belonging to (A.5) is positive definite in the solution $\left(\boldsymbol{\theta}^{*\prime}, \boldsymbol{\mu}^{*\prime}\right)' \in \mathbb{R}^{2p}$ for sufficiently large c, say $c > \bar{c}$.

Now, we consider inequality restrictions which are introduced with the help of slack variables for the general problem :

$$\min_{\boldsymbol{\theta}} \varphi(\boldsymbol{\theta}) | \, q_i(\boldsymbol{\theta}) = 0; q_j(\boldsymbol{\theta}) \geqq 0 \, (i = 1, 2, ..., k = m_1; j = m_1 + 1, \ldots, m) . \tag{A.6}$$

Solving this problem, the conditions $q_j(\boldsymbol{\theta}) \geqq 0$ are retransformed in equality conditions as $q_j(\boldsymbol{\theta}) - z_j = 0, \;\; z_j \geqq 0 \;\; (j = m_1 + 1, ..., m)$, where z_j are *slack variables*, so that the augmented Lagrange function of (A.6) becomes:

$$\begin{aligned}\mathcal{L}_c(\boldsymbol{\theta}, \mathbf{z}, \boldsymbol{\mu}) = \varphi(\boldsymbol{\theta}) &- \sum_{i=1}^{m_1} \mu_i q_i(\boldsymbol{\theta}) - \sum_{j=m_1+1}^{m} \mu_j (q_j(\boldsymbol{\theta}) - z_j) \\ &+ \tfrac{1}{2} c \sum_{i=1}^{m_1} q_i^2(\boldsymbol{\theta}) + \tfrac{1}{2} c \sum_{j=m_1+1}^{m} (q_j(\boldsymbol{\theta}) - z_j)^2 .\end{aligned}$$

This augmented Lagrangian $\mathcal{L}_c(\boldsymbol{\theta}, \mathbf{z}, \boldsymbol{\mu})$ should be minimized as a function of $\boldsymbol{\theta}$ and $\mathbf{z}$ and maximized as a function of $\boldsymbol{\mu}$. The part of $\mathcal{L}_c(\boldsymbol{\theta}, \mathbf{z}, \boldsymbol{\mu})$ which depends on the slack variables z_j is:

$$- \sum_{j=m_1+1}^{m} \mu_j (q_j(\boldsymbol{\theta}) - z_j) + \frac{1}{2} c \sum_{j=m_1+1}^{m} (q_j(\boldsymbol{\theta}) - z_j)^2 . \tag{A.7}$$

Minimization of (A.7) w.r.t. z_j (under the condition that $z_j \geqq 0$): $\frac{\partial(\text{A.7})}{\partial z_j} = \mu_j - c\left(q_j(\boldsymbol{\theta}) - z_j\right) = 0$; then, $cz_j = cq_j(\boldsymbol{\theta}) - \mu_j$, or $z_j = q_j(\boldsymbol{\theta}) - \frac{\mu_j}{c}$ i.e. $z_j = \max\left\{0, q_j(\boldsymbol{\theta}) - \frac{\mu_j}{c}\right\}$, so that substitution of this z_j in (A.7) yields:

$$
\begin{aligned}
&\textstyle\sum_{j=m_1+1}^{m} \mu_j \left(q_j(\boldsymbol{\theta}) - \max\left\{0, q_j(\boldsymbol{\theta}) - \frac{\mu_j}{c}\right\}\right) \\
&+\frac{1}{2}c \textstyle\sum_{j=m_1+1}^{m} \left(q_j(\boldsymbol{\theta}) - \max\left\{0, q_j(\boldsymbol{\theta}) - \frac{\mu_j}{c}\right\}\right)^2 \\
= & -\frac{\sum\limits_{j=m_1+1}^{m} \mu_j^2}{c} + \frac{\sum\limits_{j=m_1+1}^{m} \mu_j^2}{2c} = -\frac{1}{2}\frac{\sum\limits_{j=m_1+1}^{m} \mu_j^2}{c} \text{ if } q_j(\boldsymbol{\theta}) > \frac{\mu_j}{c} \quad (\forall_j) \\
= & -\sum_{j=m_1+1}^{m} \mu_j q_j(\boldsymbol{\theta}) + \frac{1}{2}c \sum_{j=m_1+1}^{m} q_j^2(\boldsymbol{\theta}) \\
= & \frac{1}{2}c \sum_{j=m_1+1}^{m} \left(q_j(\boldsymbol{\theta})\right)^2 - \sum_{j=m_1+1}^{m} \mu_j q_j(\theta) \text{ if } q_j(\boldsymbol{\theta}) \leqq \frac{\mu_j}{c} \quad (\forall_j).
\end{aligned}
$$

Summarizing:

$$
= \frac{1}{2}c \left\{ \min \left[0, \sum_{j=m_1+1}^{m} \left(q_j(\boldsymbol{\theta}) - \frac{\mu_j}{c}\right)\right]\right\}^2 - \frac{\sum\limits_{j=m_1+1}^{m} \mu_j^2}{2c} \tag{A.8}
$$

Substituting (A.8) for (A.7) in $\mathcal{L}_c(\boldsymbol{\theta}, \mathbf{z}, \boldsymbol{\mu})$ yields the augmented Lagrangian for (A.6):

$$
\begin{aligned}
\mathcal{L}_c(\boldsymbol{\theta}, \boldsymbol{\mu}) &= F(\boldsymbol{\theta}) - \textstyle\sum_{i=1}^{m_1} \mu_i q_i(\boldsymbol{\theta}) + \frac{1}{2}c \sum_{i=1}^{m_1} \left(q_i(\boldsymbol{\theta})\right)^2 \\
&+\frac{1}{2}c \sum_{j=m_1+1}^{m} \left\{\min\left[0, q_j(\boldsymbol{\theta}) - \frac{\mu_j}{c}\right]\right\}^2 - \sum_{j=m_1+1}^{m} \frac{\mu_j^2}{2c}
\end{aligned}
$$

Now we are going to minimize $\mathcal{L}_c(\boldsymbol{\theta}, \boldsymbol{\mu})$ unconditionally as a function of $\boldsymbol{\theta}$ (or maximizing it as a function of $\boldsymbol{\mu}$) using the following algorithm:

1. Choose $\boldsymbol{\mu}^{(0)} \in \mathbb{R}^m$ and $c^{(0)} \in \mathbb{R}_+$
2. For iteration $k = 0, 1, 2, \ldots$:
   - step 1: determine $\boldsymbol{\theta}\left(\boldsymbol{\mu}^{(k)}, c^{(k)}\right)$ as a minimand for $\min\limits_{\boldsymbol{\theta}} \mathcal{L}_{c^{(k)}}\left(\boldsymbol{\theta}, \boldsymbol{\mu}^{(k)}\right)$;
   - step 2: determine $c^{(k+1)}$ and $\boldsymbol{\mu}^{(k+1)}$ for which $\boldsymbol{\mu}^{(k+1)}$ is an approximation to the maximand of the second order approximation of the dual function in the neighborhood of $\boldsymbol{\mu}^{(k)}$, i.e. a solution of a second order approximation to $\max\limits_{\boldsymbol{\mu}} \left[\varphi_{(\boldsymbol{\mu})} = \min\limits_{\boldsymbol{\theta}} \mathcal{L}_{c^{(k)}}\left(\boldsymbol{\theta}, \boldsymbol{\mu}^{(k)}\right)\right]$:
     $\max\limits_{\boldsymbol{\mu}} \varphi(\boldsymbol{\mu}) \simeq \max\limits_{\boldsymbol{\mu}} \varphi\left(\boldsymbol{\mu}^{(k)}\right) + (\mathbf{g}^{(k)})'\left(\boldsymbol{\mu}\text{-}\boldsymbol{\mu}^{(k)}\right) + \frac{1}{2}\left(\boldsymbol{\mu}\text{-}\boldsymbol{\mu}^{(k)}\right)' \mathbf{H}^{(k)}\left(\boldsymbol{\mu}\text{-}\boldsymbol{\mu}^{(k)}\right),$

with $\mathbf{g}^{(k)} := \left.\frac{\partial\varphi}{\partial\boldsymbol{\mu}}\right|_{\boldsymbol{\mu}=\boldsymbol{\mu}^{(k)}}$ and $\mathbf{H}^{(k)} := \left.\frac{\partial^2\varphi}{\partial\boldsymbol{\mu}\partial\boldsymbol{\mu}'}\right|_{\boldsymbol{\mu}=\boldsymbol{\mu}^{(k)}}$ so that a new point $\boldsymbol{\mu}^{(k+1)}$ is obtained;

- step 3: test on convergence, and go to step 1, step 2,...if necessary.

Try to keep $c^{(k)}$ as small as possible such that the condition for the problem to be solved in step 1 becomes not too bad. Then the row $\boldsymbol{\mu}^{(k)}$ converges to $\boldsymbol{\mu}^*$, which is the Lagrangian multiplier belonging to (A.6) $(k = 1, 2, ...)$.

Then $\boldsymbol{\theta}\left(\boldsymbol{\mu}^{(k)}, c^{(k)}\right) \Rightarrow \boldsymbol{\theta}^*$, which is the minimand of (A.6). Hence, the row $\theta^{(k)}$converges to θ^* (from solving $\min_{\boldsymbol{\theta}} \mathcal{L}_{c^{(k)}}\left(\boldsymbol{\theta}, \boldsymbol{\mu}^{(k)}\right)$), and the row

$$\boldsymbol{\mu}^{(k)} \Rightarrow \boldsymbol{\mu}^*(\max_{\boldsymbol{\mu}}\varphi\left(\boldsymbol{\mu}\right) = \max_{\boldsymbol{\mu}}\min_{\boldsymbol{\theta}}\mathcal{L}_{c^{(k)}}\left(\boldsymbol{\theta}^{(k)}, \boldsymbol{\mu}\right)).$$

This can be formulated as a quadratic programming problem where the row of solutions $\left\{\boldsymbol{\theta}^{(k)}\right\}$ converges to $\boldsymbol{\theta}^*$ (e.g. by Newton (-Raphson) if the Hessian exists) and the row of the corresponding dual variables $\left\{\boldsymbol{\mu}^{(k)}\right\}$ converges to $\boldsymbol{\mu}^*$.

This quadratic programming problem is defined as:

$$\min_{\boldsymbol{\zeta}} \mathcal{L}\left(\boldsymbol{\theta}^{(k)}+\boldsymbol{\zeta}^{(k)}, \mu^{(k+1)}\right) \cong \min_{\boldsymbol{\zeta}}\mathcal{L}\left(\theta^{(k)}, \boldsymbol{\mu}^{(k+1)}\right) + \boldsymbol{\zeta}' \nabla_{\boldsymbol{\theta}} \mathcal{L}\left(\boldsymbol{\theta}^{(k+1)}, \boldsymbol{\mu}^{(k+1)}\right) + \tfrac{1}{2}\boldsymbol{\zeta}'\mathbf{H}\left(\boldsymbol{\theta}^{(k)}, \boldsymbol{\mu}^{(k+1)}\right)\boldsymbol{\zeta}.$$

Appendix B

Mathematical Formulation of GMM

GMM is an estimation method based on moment equations that impose the nullity of the mathematical expectation of a vector function of the observations and the parameters of interest as

$$E_{\boldsymbol{\theta}^0}[\mathbf{h}(\mathbf{y}, \mathbf{X}; \boldsymbol{\theta}^0)] = \mathbf{0},$$

which is generally called the *GMM moment* (or *orthogonality*) *condition.* In this condition $\mathbf{h} \in \mathbb{R}^k$, $\boldsymbol{\theta} \in \mathbb{R}^p$ with true value $\boldsymbol{\theta}^0$ and $k \geqq p$, and $E_{\boldsymbol{\theta}^0}$ is the mathematical expectation operator with respect to the true joint probability distribution $F(\mathbf{y}, \mathbf{X}; \boldsymbol{\theta}^0)$ of $(\mathbf{y}, \mathbf{X})$. Note that the vector function $\mathbf{h}$ is generally a nonlinear function in $\boldsymbol{\theta}^0$. As also mentioned in Section 2.3 this GMM moment condition has in general no unique solution; in effect there even exist $\binom{k}{p}$ possible solutions or different sets of estimates that can be produced.

B.1 Redefining GMM

GMM selects that value of the unknown parameter $\boldsymbol{\theta}$ for which the sample mean: $\overline{\mathbf{h}}_n := \frac{1}{n}\sum_{i=1}^n \mathbf{h}(y_i, \mathbf{x}_i; \boldsymbol{\theta})$ is closest to zero.

Hansen (1982) has shown that just minimizing the simple least squares criterion $\sum_{i=1}^n \mathbf{h}'(y_i, \mathbf{x}_i; \boldsymbol{\theta})\mathbf{h}(y_i, \mathbf{x}_i; \boldsymbol{\theta})$ yields a consistent but not an asymptotically efficient estimator. Therefore, we arrive at a *precise definition of GMM.*

Definition B.1 *Let $\mathbf{W}_n$ be an $(r \times r)$ symmetric positive definite matrix, which depends on the sample observations. Then, the GMM estimator associated with $\mathbf{W}_n$ is a solution to the problem:*

$$\min_{\boldsymbol{\theta}} \left[\textstyle\sum_{i=1}^n \mathbf{h}(y_i, \mathbf{x}_i; \boldsymbol{\theta})\right] \mathbf{W}_n \left[\textstyle\sum_{i=1}^n \mathbf{h}(y_i, \mathbf{x}_i; \boldsymbol{\theta})\right].$$

Compare this result to that of the linear IV problem in Section 2.2.2, with $\mathbf{V}^{-1} = \frac{n^2}{\sigma^2}(\mathbf{Z}'\mathbf{Z})^{-1}$.Hence, $\mathbf{W}_n$ is an approximation to the inverse covariance matrix of $\sum_{i=1}^{n} \mathbf{h}(y_i, \mathbf{x}_i; \boldsymbol{\theta})$ corrected for White (1984) and/or Newey-West (1987) autocorrelation and/or heteroskedasticity consistent covariance matrix.[1]

B.2 Assumptions and properties

According to Gouriéroux and Monfort (1995) we may formulate the following hypotheses:

$H1$: The stochastic variables $(y_i, \mathbf{x}_i), i = 1, 2, \ldots, n$ are *iid*.

$H2$: The mathematical expectation $E_{\boldsymbol{\theta}^0}\{\mathbf{h}(\mathbf{y}, \mathbf{X}; \boldsymbol{\theta}^0)\}$exists and is zero when $\boldsymbol{\theta}$ is equal to the true value $\boldsymbol{\theta}^0$ of the parameter of interest (GMM moment condition).

$H3$. The sample matrix converges almost surely to a non-random matrix $\mathbf{W}_0$.

$H4$: The true parameter $\boldsymbol{\theta}^0$ is identified from the GMM moment or orthogonality condition, i.e. $E_{\boldsymbol{\theta}^0}\{\mathbf{h}(\mathbf{y}, \mathbf{X}; \boldsymbol{\theta})\}\mathbf{W}_0 E_{\boldsymbol{\theta}^0}\{\mathbf{h}(\mathbf{y}, \mathbf{X}; \boldsymbol{\theta})\} = 0 \Longrightarrow \boldsymbol{\theta} = \boldsymbol{\theta}^0$.

$H5$: The true parameter value is known to belong to a compact set.

$H6$: The sample mean $\overline{\mathbf{h}}_n := \frac{1}{n}\sum_{i=1}^{n} \mathbf{h}(y_i, \mathbf{X}_i; \boldsymbol{\theta})$ converges almost surely and uniquely in $\boldsymbol{\theta}$ to $E_{\boldsymbol{\theta}^0}\{\mathbf{h}(y, \mathbf{X}; \boldsymbol{\theta})\}$.

$H7$: The function $\mathbf{h}(y, \mathbf{X}; \boldsymbol{\theta})$ is continuous in $\boldsymbol{\theta}$.

Property 1 Under assumptions $H1 - H7$, the GMM estimator as a function of $\mathbf{W}_n$, denoted as $\hat{\boldsymbol{\theta}}(\mathbf{W}_n)$, exists (asymptotically) and strongly converges to the true value $\boldsymbol{\theta}^0$ of the unknown parameter.

If we assume now in addition to the above assumptions $H1 - H7$:

$H8$: The true parameter value belongs to the interior of Θ (the parameter set).

$H9$: The function $h(y, \mathbf{X}; \boldsymbol{\theta})$ is continuously differentiable in $\boldsymbol{\theta}$.

Then, the GMM estimator $\hat{\boldsymbol{\theta}}(\mathbf{W}_n)$ *belongs asymptotically to a neighborhood of the true parameter value* $\boldsymbol{\theta}^0$.

In particular, it satisfies the following first order conditions derived from the quadratic minimization problem in Definition B.1:

$$\left[\frac{1}{n}\sum_{i=1}^{n} \frac{\partial h(y_i, \mathbf{x}_i; \boldsymbol{\theta})}{\partial \theta}\bigg|_{\boldsymbol{\theta}=\hat{\boldsymbol{\theta}}(\mathbf{W}_n)}\right]' \mathbf{W}_n \left[\frac{1}{n}\sum_{i=1}^{n} \mathbf{h}(y_i, \mathbf{x}_i; \hat{\boldsymbol{\theta}}(\mathbf{W}_n))\right] = 0. \quad \text{(B.1)}$$

An expression similar to this has been applied when deriving the IV estimator in Section 2.2.2.

[1]Notice, however, that for cross-section studies, heteroskedasticity might be a serious problem. Then heteroskedasticity consistent standard errors, or simply White standard errors, might be provided, while for time-series models, residual autocorrelation might also be a serious problem, so that, in this case, heteroskedasticity- and autocorrelation-consistent standard errors or Newey-West standard errors might be useful. See also Example 2.6 for an empirical application.

In order to establish the *asymptotic normality of a GMM estimator*, we introduce the following three assumptions:

$H10$: The sample quantity $\frac{1}{n}\sum_{i=1}^{n}\frac{\partial \mathbf{h}(y_i,\mathbf{x}_i;\boldsymbol{\theta})}{\partial \boldsymbol{\theta}}$ converges almost surely and uniformly in $\boldsymbol{\theta}$ to $\mathbf{G}_0 := E_{\boldsymbol{\theta}^0}\left\{ \left.\frac{\partial \mathbf{h}(\mathbf{y},\mathbf{X};\boldsymbol{\theta})}{\partial \boldsymbol{\theta}}\right|_{\boldsymbol{\theta}=\boldsymbol{\theta}^0}\right\}$.

$H11$: $E_{\boldsymbol{\theta}^0}\left\|\mathbf{h}(y,\mathbf{X};\boldsymbol{\theta}^0)\right\|^2 < 0$.

$H12$: The matrix $\mathbf{G}_0\mathbf{W}_0\mathbf{G}_0$ is nonsingular, which implies that $k \geqq p$, which is always the case under GMM.

Property 2 Under assumptions $H1 - H12$, the GMM estimator $\hat{\boldsymbol{\theta}}(\mathbf{W}_n)$ is asymptotically normally distributed:

$$\sqrt{n}\,\left(\hat{\boldsymbol{\theta}}(\mathbf{W}_n) - \boldsymbol{\theta}^0\right) \overset{as}{\sim} N(0, \boldsymbol{\Sigma}(\mathbf{W}_0)), \qquad \text{with}$$

$$\boldsymbol{\Sigma}(\mathbf{W}_0) = \left(\mathbf{G}_0'\mathbf{W}_0\mathbf{G}_0\right)^{-1}\mathbf{G}_0'\mathbf{W}_0\mathbf{V}_0[\mathbf{h}\left(y,\mathbf{X};\boldsymbol{\theta}^0\right)]\mathbf{W}_0\mathbf{G}_0\left(\mathbf{G}_0'\mathbf{W}_0\mathbf{G}_0\right)^{-1}.$$

Now, we are wondering ourselves whether there exists an optimal choice for the sequence of matrices $\mathbf{W}_n$. Since, by property 2, the asymptotic covariance matrix of a GMM estimator depends only on the non random limit matrix $\mathbf{W}_0$, this issue reduces to knowing whether there exists an optimal matrix $\mathbf{W}_0$.

Property 3 *There exists a best GMM estimator*

It is obtained when the limit matrix satisfies: $\mathbf{W}_0^* = \left[\mathbf{V}_0\mathbf{h}\left(\mathbf{y},\mathbf{X};\boldsymbol{\theta}^0\right)\right]^{-1}$, so that, from property 2, the asymptotic covariance matrix of the GMM estimator simplifies to:

$$\Sigma(\mathbf{W}_0^*) = G_0'\mathbf{W}_0^*\mathbf{G}_0$$

Note that a consistent estimator of this optimal limit matrix $\mathbf{W}_0^*$ is given by:

$$\hat{\mathbf{W}}_n^* = \left(\frac{1}{n}\sum_{i=1}^{n}\mathbf{h}(y_i,\mathbf{x}_i;\hat{\boldsymbol{\theta}})\mathbf{h}'(y_i,\mathbf{x}_i;\hat{\boldsymbol{\theta}})\right)^{-1},$$

where $\hat{\boldsymbol{\theta}}$ is a GMM estimator of the unknown parameter $\boldsymbol{\theta}$, which is obtained using an initial arbitrary sequence of matrices, such as $\mathbf{W}_n = \mathbf{I}_n$, with $\mathbf{I}_n$ the identity matrix.

B.3 Nonlinear two stage least squares

The NLS estimator of the nonlinear model $y_i = f(\mathbf{x}_i;\boldsymbol{\theta}) + \varepsilon_i$ $(i = 1, 2, \ldots, n)$ is an inconsistent estimator if there is endogeneity bias. If there exist IVs $\{\mathbf{z}_i\}$,contained in an appropriate matrix $\mathbf{Z} := (\mathbf{z}_1, \mathbf{z}_2, \ldots, \mathbf{z}_r)$, a consistent estimator may result.

Since the IVs should be (highly) correlated with the explanatory variables and uncorrelated with the error term, we should require that $E_{\boldsymbol{\theta}^0} = \{\mathbf{V}_0\mathbf{z}'(\mathbf{y} - \mathbf{f}(\mathbf{X};\boldsymbol{\theta}))\} =$

$\mathbf{0}$, so that the best GMM estimator has a limit matrix: $\mathbf{W}_0^* = [\mathbf{V}_0(\mathbf{z}'\boldsymbol{\varepsilon})]^{-1}$(see Property 3 in Section B.2).

If all the error terms ε_i $(i = 1, 2, \ldots, n)$ satisfy $E_{\boldsymbol{\theta}^0}(\varepsilon_i^2 \mid \mathbf{Z}) = \sigma_0^2$, then $\mathbf{W}_0^* = \left[\sigma_0^2 E_{\boldsymbol{\theta}^0}(\mathbf{Z}'\mathbf{Z})\right]^{-1}$, so that the best GMM estimator is obtained as a solution to:

$$\min_{\boldsymbol{\theta}} \left\{ \sum_{i=1}^n \mathbf{z}_i' \left(y_i - f(\mathbf{x}_i; \boldsymbol{\theta})\right)' \frac{1}{\hat{\sigma}_n^2} \left(\frac{1}{n} \sum \mathbf{z}_i' \mathbf{z}_i \right)^{-1} \sum_{i=1}^n \mathbf{z}_i' \left(y_i - f(\mathbf{x}_i; \boldsymbol{\theta})\right) \right\},$$

which can conveniently be written with the help of matrix notation as:

$$\min_{\boldsymbol{\theta}} \left\{ (\mathbf{y} - \mathbf{f}(\mathbf{X}; \boldsymbol{\theta}))' \, \mathbf{Z}(\mathbf{Z}'\mathbf{Z})^{-1}\mathbf{Z}' \, (\mathbf{y} - \mathbf{f}(\mathbf{X}; \boldsymbol{\theta})) \right\}. \tag{B.2}$$

The nonlinear GMM estimator $\hat{\boldsymbol{\theta}}$ is now a solution to the latter minimization problem, with a consistent estimator of the asymptotic covariance matrix:

$$\hat{\sigma}_n^2 \left(\frac{1}{n} \frac{\partial \mathbf{f}'(\mathbf{X}; \boldsymbol{\theta})}{\partial \boldsymbol{\theta}} \bigg|_{\boldsymbol{\theta} = \hat{\boldsymbol{\theta}}} \mathbf{Z}(\mathbf{Z}'\mathbf{Z})^{-1}\mathbf{Z}' \frac{\partial \mathbf{f}(\mathbf{X}; \boldsymbol{\theta})}{\partial \boldsymbol{\theta}} \bigg|_{\boldsymbol{\theta} = \hat{\boldsymbol{\theta}}} \right)^{-1},$$

where the residual variance $\hat{\sigma}_\varepsilon^2 = \frac{1}{n} \sum_{i=1}^n \left(y_i - f(\mathbf{x}_i; \hat{\boldsymbol{\theta}}) \right)^2$ is involved. Hence, the GMM estimator of this model is precisely the nonlinear 2SLS estimator.

Appendix C

Stability Criteria for $AR(p)$ Models

Consider the $AR(p)$ model $y_t = \alpha_1 y_{t-1} + \alpha_2 y_{t-2} + \cdots + \alpha_p y_{t-p} + \varepsilon_t$ where $\varepsilon_t \in WN$ with variance σ_ε^2, and define a matrix $\mathbf{A}$ and vectors $\mathbf{v}_t$ and $\boldsymbol{\zeta}_t$ as follows:

$$\mathbf{A} := \begin{pmatrix} \alpha_1 & \alpha_2 & \cdots & \alpha_{p-1} & \alpha_p \\ 1 & 0 & \cdots & 0 & 0 \\ 0 & 1 & \cdots & 0 & 0 \\ \vdots & \vdots & \ddots & \vdots & \vdots \\ 0 & 0 & \cdots & 1 & 0 \end{pmatrix}, \ \mathbf{v}_t := \begin{pmatrix} y_t \\ y_{t-1} \\ y_{t-2} \\ \vdots \\ y_{t-p+1} \end{pmatrix}, \text{ and } \boldsymbol{\zeta}_t := \begin{pmatrix} \varepsilon_t \\ 0 \\ 0 \\ \vdots \\ 0 \end{pmatrix},$$

so that the $AR(p)$ model can be rewritten as a first order stochastic difference equation $\mathbf{v}_t = \mathbf{A}\mathbf{v}_{t-1} + \boldsymbol{\zeta}_t$.

Eigenvalues of the matrix $\mathbf{A}$ are those numbers λ for which $|\mathbf{A} - \lambda \mathbf{I}_p| = 0$.

For example, for $p = 2$, the eigenvalues are the solutions to:

$$\left| \begin{pmatrix} \alpha_1 & \alpha_2 \\ 1 & 0 \end{pmatrix} - \begin{pmatrix} \lambda & 0 \\ 0 & \lambda \end{pmatrix} \right| = 0, \tag{C.1}$$

or $\begin{vmatrix} \alpha_1 - \lambda & \alpha_2 \\ 1 & -\lambda \end{vmatrix} = \lambda^2 - \alpha_1 \lambda - \alpha_2 = 0$. The two eigenvalues of $\mathbf{A}$ for a second order difference equation are thus given by $\lambda_1, \lambda_2 = \frac{\alpha_1 \pm \sqrt{\alpha_1^2 + 4\alpha_2}}{2}$.

Examples

1. For the $AR(2)$ model $y_t = 0.6 y_{t-1} + 0.2 y_{t-2} + \varepsilon_t$, the eigenvalues are given by $\lambda_1, \lambda_2 = \frac{0.6 \pm \sqrt{0.6^2 + 4 \times 0.2}}{2}$, or $\lambda_1 = 0.84$ and $\lambda_2 = -0.24$.
2. On the contrary, for the $AR(2)$ model $y_t = 0.5 y_{t-1} - 0.8 y_{t-2} + \varepsilon_t$, the eigenvalues are given by $\lambda_1, \lambda_2 = \frac{0.5 \pm \sqrt{0.5^2 + 4 \times (-0.8)}}{2} = 0.25 \pm 0.86i$ with modulus $R = \sqrt{(0.25)^2 + (0.86)^2} = 0.9$, which implies damped oscillations.

For arbitrary p The determinant in a generalization of (C.1) is a pth order polynomial in λ whose p solutions characterize the p eigenvalues of $\mathbf{A}$.

Proposition C.1 *The eigenvalues of the matrix* $\mathbf{A}$ *are the values of* λ *that satisfy:*

$$\lambda^p - \alpha_1\lambda^{p-1} - \alpha_2\lambda^{p-2} - \cdots - \alpha_{p-1}\lambda - \alpha_p = 0.$$

Proof The eigenvalues of $\mathbf{A}$ are derived from $|\mathbf{A} - \lambda\mathbf{I}_p| = 0$ and since by multiplying the pth column of the matrix $[\mathbf{A} - \lambda\mathbf{I}_p]$ by a constant $\frac{1}{\lambda}$ and adding the result to the $(p-1)^{th}$ column, the determinant of the matrix does not change, we have:

$$|\mathbf{A}\text{-}\lambda\mathbf{I}_p| = \begin{vmatrix} \alpha_1\text{-}\lambda & \alpha_2 & \cdots & \alpha_{p\text{-}1} & \alpha_p \\ 0 & -\lambda & \cdots & 0 & 0 \\ 0 & 1 & \cdots & 0 & 0 \\ \vdots & \vdots & \ddots & \vdots & \vdots \\ 0 & 0 & \cdots & 1 & -\lambda \end{vmatrix} = \begin{vmatrix} \alpha_1\text{-}\lambda & \alpha_2 & \cdots & \alpha_{p-2} & \alpha_{p\text{-}1}+\frac{\alpha_p}{\lambda} & \alpha_p \\ 0 & -\lambda & \cdots & 0 & 0 & 0 \\ 0 & 1 & \cdots & 0 & 0 & 0 \\ \vdots & \vdots & \ddots & \vdots & \vdots & \vdots \\ 0 & 0 & \cdots & 1 & -\lambda & 0 \\ 0 & 0 & \cdots & 0 & 0 & -\lambda \end{vmatrix}.$$

We now multiply the $(p-1)^{th}$ column by $\frac{1}{\lambda}$ and add the result to the $(p-2)^{th}$ column, and continuing in this manner leads to the determinant of an upper triangular matrix:

$$|\mathbf{A}\text{-}\lambda\mathbf{I}_p| = \begin{vmatrix} \alpha_1\text{-}\lambda+\frac{\alpha_2}{\lambda}+\frac{\alpha_3}{\lambda^2}+\cdots+\frac{\alpha_p}{\lambda^{p-1}} & \alpha_2+\frac{\alpha_3}{\lambda}+\frac{\alpha_4}{\lambda^2}+\cdots+\frac{\alpha_p}{\lambda^{p-2}} & \cdots & \alpha_{p\text{-}1}+\frac{\alpha_p}{\lambda} & \alpha_p \\ 0 & -\lambda & \cdots & 0 & 0 \\ 0 & 0 & \cdots & 0 & 0 \\ \vdots & \vdots & \ddots & \vdots & \vdots \\ 0 & 0 & \cdots & -\lambda & 0 \\ 0 & 0 & \cdots & 0 & -\lambda \end{vmatrix}.$$

Since the determinant of an upper triangular matrix is the product of the diagonal terms, we get:

$$\begin{aligned} |\mathbf{A} - \lambda\mathbf{I}_p| &= \left(\alpha_1 - \lambda + \frac{\alpha_2}{\lambda} + \frac{\alpha_3}{\lambda^2} + \cdots + \frac{\alpha_p}{\lambda^{p-1}}\right)(-\lambda)^{p-1} \\ &= (-1)^p\left(\lambda^p - \alpha_1\lambda^{p-1} - \alpha_2\lambda^{p-2} - \cdots - \alpha_p\right) = 0. \blacksquare \end{aligned}$$

The $AR(p)$ model is considered to be stable or stationary if all the eigenvalues defined in Proposition C.1 lie inside the unit circle.

Appendix D

MLE of the RSM with Endogenous Prices

In this appendix we estimate (8.10) under the normality assumption for both independent and correlated structural error terms.

To this end we rewrite ((8.5) – (8.8)) as:

$$\begin{aligned} D_t &= X'_{d\hat{t}}\beta_1 + \alpha_1 P_t + u_{dt} && \text{(D.1)} \\ S_t &= X'_{st}\beta_2 + \alpha_2 P_t + u_{st} && \text{(D.2)} \\ Q_t &:= \min(D_t, S_t) && \text{(D.3)} \\ \Delta P_t &= \eta (D_t - S_t) + X'_{pt}\beta_3 + u_{pt} && \text{(D.4)} \end{aligned}$$

Either the error terms in (D.1), (D.2) and (D.4) are *iid* or they are correlated.

1. ***Case of uncorrelated structural error terms*** In order to estimate the unknown parameters, assume in the first instance that u_{it} are *iid* normally distributed with zero mean and variance σ_i^2, $(i = d, s, p)$.

a. The error terms in the case of excess supply (or demand regime) satisfy:

$$\begin{bmatrix} u_{dt} \\ u_{st} \\ u_{pt} - \eta u_{st} \end{bmatrix} \overset{iid}{\backsim} N\left(\begin{bmatrix} 0 \\ 0 \\ 0 \end{bmatrix}, \begin{bmatrix} \sigma_d^2 & 0 & 0 \\ 0 & \sigma_s^2 & -\eta\sigma_s^2 \\ 0 & -\eta\sigma_s^2 & \underbrace{\sigma_p^2 + \eta^2\sigma_s^2}_{\sigma_h^2} \end{bmatrix} \right).$$

Hence, representing $\{u_{st} | u_{dt}, u_{pt} - \eta u_{st}\} \sim N(\boldsymbol{\mu}, \boldsymbol{\Omega})$ we have, dropping the index t: $(u_s | (u_d, u_p - \eta u_s)) \backsim N(\boldsymbol{\mu}, \boldsymbol{\Omega})$

or $(u_s | u_d = \psi_1, u_p - \eta u_s = \tau_1) \backsim N(\mu_1 + \Omega_{12}\Omega_{22}^{-1}(y - \mu_2), \Omega_{11} - \Omega_{12}\Omega_{22}^{-1}\Omega_{21})$.

Partitioning:

$$\begin{bmatrix} \sigma_s^2 & 0 & -\eta\sigma_s^2 \\ 0 & \sigma_d^2 & 0 \\ -\eta\sigma_s^2 & 0 & \sigma_h^2 \end{bmatrix} \text{ as } \Omega_{11} := \sigma_s^2,\ \boldsymbol{\Omega}_{12} := \begin{bmatrix} 0 & -\eta\sigma_s^2 \end{bmatrix},\ \boldsymbol{\Omega}_{21} := \begin{bmatrix} 0 \\ -\eta\sigma_s^2 \end{bmatrix},$$

we get:

$$\mu_1 + \mathbf{\Omega}_{12}\mathbf{\Omega}_{22}^{-1}(y-\mu_2) = 0 + \left(0 \quad -\eta\sigma_s^2\right|\begin{pmatrix} \sigma_d^2 & 0 \\ 0 & \sigma_h^2 \end{pmatrix}^{-1}\begin{bmatrix} \psi_1 - 0 \\ \tau_1 - 0 \end{bmatrix}$$
$$= 0 + \left[0 \quad -\frac{\eta\sigma_s^2}{\sigma_h^2}\right]\begin{bmatrix} \psi_1 \\ \tau_1 \end{bmatrix} = \frac{-\eta\sigma_s^2\tau_1}{\sigma_h^2}, \text{ and:}$$

$$\Omega_{11} - \mathbf{\Omega}_{12}\mathbf{\Omega}_{22}^{-1}\mathbf{\Omega}_{21} = \sigma_s^2 - \left[0 \quad -\eta\sigma_s^2\right]\begin{bmatrix} \sigma_d^2 & 0 \\ 0 & \sigma_h^2 \end{bmatrix}^{-1}\begin{bmatrix} 0 \\ -\eta\sigma_s^2 \end{bmatrix}$$
$$= \sigma_s^2 - \left[0\text{-}\frac{\eta\sigma_h^2}{\sigma_h^2}\right]\left[0, \eta\sigma_s^2\right]' = \sigma_s^2 - \frac{\eta^2\sigma_s^4}{\sigma_h^2} = \sigma_s^2\left(1 - \frac{\eta^2\sigma_s^2}{\sigma_h^2}\right) =: \sigma_1^2.$$

Hence, the likelihood function can be defined for f the standard normal pdf and F the standard normal distribution function:

$$\ell_d = \sigma_d^{-1}f\left(\frac{\tau_1}{\sigma_d}\right)\cdot\sigma_h^{-1}f\left(\frac{\tau_1}{\sigma_h}\right)\left[1 - F\left(\frac{\psi_1-\mu_1}{\sigma_1}\right)\right]$$
$$= \frac{1}{\sigma_d}\frac{1}{\sqrt{2\pi}}e^{-\frac{\psi_1^2}{2\sigma_d^2}}\cdot\frac{1}{\sigma_h}\frac{1}{\sqrt{2\pi}}e^{\frac{-\tau_1^2}{2\sigma_h^2}}\left[1 - \int_{-\infty}^{Q_t}\frac{1}{\sqrt{2\pi}\sqrt{\sigma_1^2}}\cdot e^{-\frac{1}{2}\left(\frac{\psi_1-\mu_1}{\sigma_1}\right)^2}dS\right]$$
$$= \frac{1}{\sqrt{2\pi}\sigma_d}e^{-\frac{1}{2\sigma_d^2}\left(Q-\mathbf{x}_d'\boldsymbol{\beta}_1-\alpha_1 P\right)^2}.$$

$$\frac{1}{\sqrt{2\pi}\sigma_h}e^{-\frac{1}{2\left(\sigma_p^2+\eta^2\sigma_s^2\right)}\left(\Delta P-\mathbf{x}_p'\boldsymbol{\beta}_3-\eta\psi_1\right)^2}\left[1-\int_{-\infty}^{Q_t}\frac{1}{\sqrt{2\pi}\sigma_1}e^{\left[\frac{\left(Q-X_s'\beta_2-\alpha_2P+\frac{\eta\sigma_s^2\tau_1}{\sigma_h^2}\right)^2}{\sigma_1^2}\right]}dS_t\right]$$

$$= g_1(Q_t)\left(1 - G_2(Q_t)\right), \text{with } \sigma_1 := \sqrt{\left(\sigma_s^2\left(1 - \frac{\eta^2\sigma_s^2}{\sigma_h^2}\right)\right)}$$

b. In the case of excess demand (or supply regime), the error terms satisfy: $(u_d|\, u_s = \psi_2, u_p + \eta u_d = \tau_2) \backsim N\left(\mu_2, \sigma_2^2\right)$

Partitioning $\begin{bmatrix} \sigma_d^2 & 0 & \eta\sigma_d^2 \\ 0 & \sigma_s^2 & 0 \\ \eta\sigma_d^2 & 0 & \sigma_g^2 \end{bmatrix}$ as $\Omega_{11} := \sigma_d^2, \Omega_{12} := \begin{bmatrix} 0 & \eta\sigma_d^2 \end{bmatrix}, \Omega_{21} :=$ $\begin{bmatrix} 0 \\ \eta\sigma_d^2 \end{bmatrix}, \Omega_{22} := \begin{bmatrix} \sigma_s^2 & 0 \\ 0 & \sigma_g^2 \end{bmatrix}$

we get:

$$\mu_2 = 0 + \left(0 \ \ \eta\sigma_d^2\right)\begin{bmatrix} \sigma_s^2 & 0 \\ 0 & \sigma_g^2 \end{bmatrix}^{-1}\begin{bmatrix} \psi_2 - 0 \\ \tau_2 - 0 \end{bmatrix} = \left(0 \ \ \eta\sigma_d^2\left(\sigma_g^2\right)^{-1}\right)\begin{bmatrix} \psi_2 \\ \tau_2 \end{bmatrix} = \frac{\eta\sigma_d^2\tau_2}{\sigma_g^2}$$

and $\sigma_2^2 = \sigma_d^2 \text{-} \begin{pmatrix} 0 & n\sigma_d^2 \end{pmatrix}\begin{pmatrix} \sigma_s^2 & 0 \\ 0 & \sigma_g^2 \end{pmatrix}^{-1}\begin{pmatrix} 0 \\ \eta\sigma_d^2 \end{pmatrix} = \sigma_d^2 \text{-} \begin{pmatrix} 0 & \frac{\eta\sigma_d^2}{\sigma_g^2} \end{pmatrix}\begin{pmatrix} 0 \\ \eta\sigma_d^2 \end{pmatrix}$

$$= \sigma_d^2 - \frac{\eta^2\sigma_d^4}{\sigma_g^2} = \sigma_d^2\left(1 - \frac{\eta^2\sigma_d^2}{\sigma_g^2}\right),$$

so that the likelihood function satisfies $\ell_s = \frac{1}{\sigma_s} f\left(\frac{\psi_2}{\sigma_s}\right) \frac{1}{\sigma_g} f\left(\frac{\tau_2}{\sigma_g}\right)\left[1 - F\left(\frac{(\psi_2-\mu_2)}{\sigma_2}\right)\right] = g_2(Q_t)\left[1 - G_1(Q_t)\right]$. Finally, $\ell := \ell_d + \ell_s$.

2. *If covariances exist between all the structural form error terms we assume*:

$$\begin{pmatrix} u_{dt} \\ u_{st} \\ u_{pt} \end{pmatrix} \backsim N\left(\begin{bmatrix} 0 \\ 0 \\ 0 \end{bmatrix}, \begin{bmatrix} \sigma_d^2 & \sigma_{ds} & \sigma_{dp} \\ \sigma_{ds} & \sigma_s^2 & \sigma_{sp} \\ \sigma_{dp} & \sigma_{sp} & \sigma_p^2 \end{bmatrix}\right)$$

so that:

$$\begin{pmatrix} u_d \\ u_s \\ u_p - \eta u_{st} \end{pmatrix} \backsim N\left(\begin{bmatrix} 0 \\ 0 \\ 0 \end{bmatrix}, \begin{bmatrix} \sigma_d^2 & \sigma_{ds} & \sigma_{dp}\text{-}\eta\sigma_{ds} \\ \sigma_{ds} & \sigma_s^2 & \sigma_{sp}\text{-}\eta\sigma_s^2 \\ \sigma_{dp}\text{-}\eta\sigma_{ds} & \sigma_{sp}\text{-}\eta\sigma_s^2 & \sigma_p^2+\eta^2\sigma_s^2\text{-}2\eta\sigma_{sp} \end{bmatrix}\right).$$

Partitioning $\begin{bmatrix} \sigma_s^2 & \sigma_{ds} & \sigma_{sp} - \eta\sigma_s^2 \\ \sigma_{ds} & \sigma_d^2 & \sigma_{dp} - \eta\sigma_{ds} \\ \sigma_{sp} - \eta\sigma_s^2 & \sigma_{dp} - \eta\sigma_{ds} & \sigma_p^2 + \eta^2\sigma_s^2 - 2\eta\sigma_{sp} \end{bmatrix}$,
for σ_{h_1}:=σ_{dp}-$\eta\sigma_{ds}$, σ_{h_2}:=$\sigma_p^2 + \eta^2\sigma_s^2$-$2\eta\sigma_{sp}$ and σ_{h_3}:=σ_{sp}-$\eta\sigma_s^2$ we can define $\Omega_{11} := \sigma_s^2, \mathbf{\Omega}_{12} := \begin{bmatrix} \sigma_{ds} & \sigma_{h_3} \end{bmatrix}, \mathbf{\Omega}_{21} := \begin{bmatrix} \sigma_{ds} \\ \sigma_{h_3} \end{bmatrix}$ and $\mathbf{\Omega}_{22} := \begin{bmatrix} \sigma_d^2 & \sigma_{h_1} \\ \sigma_{h_1} & \sigma_{h_2} \end{bmatrix}$, or $(u_s|\, u_d = \psi_1, u_p - \eta u_s = \tau_1) \backsim N\left(\mu_3, \sigma_3^2\right)$ with

$$\begin{aligned}
\mu_3 &= 0 + \begin{bmatrix} \sigma_{ds} & \sigma_{h_3} \end{bmatrix} \begin{bmatrix} \sigma_d^2 & \sigma_{h_1} \\ \sigma_{h_1} & \sigma_{h_2} \end{bmatrix}^{-1} \begin{bmatrix} \psi_1 - 0 \\ \tau_1 - 0 \end{bmatrix} \\
&= \begin{bmatrix} \sigma_{ds} & \sigma_{h_3} \end{bmatrix} \frac{1}{\left(\sigma_d^2\sigma_{h_2} - \sigma_{h_1}^2\right)} \begin{bmatrix} \sigma_{h_2} & -\sigma_{h_1} \\ -\sigma_{h_1} & \sigma_d^2 \end{bmatrix} \begin{bmatrix} \psi_1 \\ \tau_1 \end{bmatrix} \\
&= \frac{1}{\left(\sigma_d^2\sigma_{h_2} - \sigma_{h_1}^2\right)} \begin{bmatrix} \sigma_{ds}\sigma_{h_2} - \sigma_{h_1}\sigma_{h_3} & -\sigma_{ds}\sigma_{h_1} + \sigma_{h_3}\sigma_d^2 \end{bmatrix} \begin{bmatrix} \psi_1 \\ \tau_1 \end{bmatrix} \\
&= \frac{1}{\left(\sigma_d^2\sigma_{h_2} - \sigma_{h_1}^2\right)} \begin{bmatrix} \sigma_{ds}\sigma_{h_2}\psi_1 - \sigma_{h_1}\sigma_{h_3}\psi_1 & -\sigma_{ds}\sigma_{h_1}\tau_1 + \sigma_{h_3}\sigma_d^2\tau_1 \end{bmatrix}
\end{aligned}$$

and

$$\begin{aligned}
\sigma_3^2 &= \Omega_{11} - \mathbf{\Omega}_{12}\mathbf{\Omega}_{22}^{-1}\mathbf{\Omega}_{21} \\
&= \sigma_s^2 - \tfrac{1}{a} \begin{bmatrix} \sigma_{ds} & \sigma_{h_3} \end{bmatrix} \begin{bmatrix} \sigma_{h_2} & -\sigma_{h_1} \\ -\sigma_{h_1} & \sigma_d^2 \end{bmatrix} \begin{bmatrix} \sigma_{ds} \\ \sigma_{h_3} \end{bmatrix} \text{ (with } a{:=}\sigma_d^2\sigma_{h_2} - \sigma_{h_1}^2\text{)} \\
&= \sigma_s^2 - \frac{1}{\sigma_d^2\sigma_{h_2} - \sigma_{h_1}^2} \begin{bmatrix} \sigma_{ds}\sigma_{h_2} - \sigma_{h_1}\sigma_{h_3} & -\sigma_{ds}\sigma_{h_1} + \sigma_{h_3}\sigma_d^2 \end{bmatrix} \begin{bmatrix} \sigma_{ds} \\ \sigma_{h_3} \end{bmatrix} \\
&= \sigma_s^2 - \frac{1}{\sigma_d^2\sigma_{h_2} - \sigma_{h_1}^2} \left[\sigma_{ds}^2\sigma_{h_2} - \sigma_{h_1}\sigma_{h_3}\sigma_{ds} - \sigma_{ds}\sigma_{h_1}\sigma_{h_3} + \sigma_{h_3}^2\sigma_d^2\right] \\
&= \sigma_s^2 - \frac{1}{\sigma_d^2\sigma_{h_2} - \sigma_{h_1}^2} \left(\sigma_{ds}^2\sigma_{h_2} - 2\sigma_{h_1}\sigma_{h_3}\sigma_{ds} + \sigma_{h_3}^2\sigma_d^2\right).
\end{aligned}$$

Further: $(u_d|\, u_p - \eta u_s = \tau_1|) \backsim N\left(\mu_4, \sigma_4^2\right)$,
and $\begin{bmatrix} u_d \\ u_p - \eta u_s \end{bmatrix} \backsim N\left(\begin{bmatrix} 0 \\ 0 \end{bmatrix}, \begin{bmatrix} \sigma_d^2 & \sigma_{h_3} \\ \sigma_{h_1} & \sigma_{h_2} \end{bmatrix}\right)$; μ_4=$0 + \sigma_{h_1}\left(\sigma_{h_2}\right)^{-1}\tau_1$=$\frac{\sigma_{h_1}\tau_1}{\sigma_{h_2}}$
and $\sigma_4^2 = \sigma_d^2 - \frac{\sigma_{h_1}^2}{\sigma_{h_2}}$, so that:

$$\ell_d = \frac{1}{\sigma_4} f\left(\frac{(\psi_1 - \mu_4)}{\sigma_4}\right) \frac{1}{\sigma_{h2}} f\left(\frac{\tau_1}{\sigma_{h_2}}\right)\left[1 - F\left(\frac{(\psi_1 - \mu_3)}{\sigma_3}\right)\right],$$

where $f\left(\frac{(\psi_1-\mu_4)}{\sigma_4}\right)$ is a (standard) normal pdf and $F\left(\frac{(\psi_1-\mu_3)}{\sigma_3}\right)$ is a (standard) normal distribution function.

Appendix E

Volatility Modeling

E.1 Detection and reduction of additive outliers

For a $GARCH(1,1)$ model we have the following five steps method:

1. We estimate the parameters $\alpha_o, \alpha_1, \beta_1$of the model:
$r_t = \eta_t h_t^{0.5}$
$h_t = \alpha_o + \alpha_1 r_{t-1}^2 + \beta_1 h_{t-1}$
with η_t standard normally or student t-distributed by ML, and herewith we estimate the series (h_t, η_t) and construct the time series $\hat{z}_t = \hat{r}_t^2 - \hat{h}_t^2$.[1]

2. We have then that $\hat{\pi}(L) = \frac{1-(\hat{\alpha}_1+\hat{\beta}_1)L}{1-\beta_1 L}$. For each $t = \tau$ we compute the regression in model (5.38) and the unconditional variance of the $\hat{z}_t$ series with the 'omit one' method. The values of the $\hat{\tau}$-statistic can then be calculated.

3. The observation with the largest $\hat{\tau}$-value is replaced by $\hat{z}_t$.

4. With this $\hat{z}_t$ we can construct $r_t^{*2} = \hat{z}_t^{*2} + \hat{h}_t$ at the period $t = \tau$, and also the AO corrected returns for $t \neq \tau$ if $r_t^* = r_t$ and for $t = \tau$ if $r_t^* = sign(r_t)(r_t^{*2})^{0.5}$, in order to avoid sign reversal.

5. This procedure can be repeated till no single $\hat{\tau}$-value is still significant. The envisaged parameters $\hat{\alpha}_o^*, \hat{\alpha}_1^*, \hat{\beta}_1^*$, with which forecasts are made are obtained now.

Franses and Ghijsels (1999) find for their empirical applications for this AO correction method that the adjustment of a number of AOs implies the introduction of conditional normality in the return series, and that, simultaneously, the $GARCH(1,1)$ property is preserved. They also find strong evidence that the forecasts of the adjusted model is improved.

[1]Here we recognise the $ARMA(1,1)$ reformulation of the $GARCH(1,1)$ model: $r_t^2 = \alpha_o + (\alpha_1 + \beta_1) r_{t-1}^2 + z_t - \beta_1 z_{t-1}$, where $z_t = r_t^2 - h_t$.

E.2 Forecasting in $GARCH(p,q)$ models

Following Baillie and Bollerslev (1992) and considering that the $GARCH(p,q)$ model implies an $ARMA(m,p)$ model for the quadratic innovations: $\varepsilon_t = \omega + \sum_{i=1}^{m}(\alpha_i + \beta_i)\varepsilon_{t-1}^2 - \sum_{j=1}^{p}\beta_j v_{t-j} + v_v$, where $v_t = \varepsilon_t^2 - \sigma_t^2 = (z_t^2 - 1)\sigma_t^2$, and $z_t \overset{iid}{\sim} N(0,1)$, $m := \max(p,q), \alpha_i = 0$ for $i > q$ and $\beta_j = 0$ for $i > p$.

Hence, the $GARCH(p,q)$ model can be expressed in vector notation as:

$\mathbf{V}_t^2 = \omega + e_1 + \Gamma\mathbf{V}_{t-1}^2 + (e_1 + e_{m+1})\mathbf{v}_t$

For the $GARCH(p,q)$ model the s-step minimum mean squared error forecast for the conditional variance has the form:

$E_t\{\varepsilon_{t+s}^2\} = E_t\{\sigma_{t+s}^2\} = \omega_s + \sum_{i=0}^{p-1}\delta_i\sigma_{t-i}^2 + \sum_{i=0}^{m-1}\rho_i\varepsilon_{t-i}^2$ where $\omega_s = \mathbf{e}_1'(\mathbf{I}+\Gamma+\ldots+\Gamma^{s-1})$, $\delta_i = -\mathbf{e}_1'\Gamma^s\mathbf{e}_{m+i+1}$ for $i = 0,1,\ldots,p-1$; $\rho_i = \mathbf{e}_1'\Gamma^s(\mathbf{e}_{i+1} + \mathbf{e}_{m+i+1})$ for $i = 0,1,\ldots,p-1$; $\rho_i = \mathbf{e}_1'\Gamma^s\mathbf{e}_{i+1}$ for $i = p, p+1,\ldots,m-1$.

Further, the forecast error for the s-step forecast of the conditional variance σ_{t+s}^2 in the $GARCH(p,q)$ model is equal to:

$\mathbf{v}_{t+s} = \sigma_{t+s}^2 - E_t\{\sigma_{t+s}^2\} = \sum_{i=1}^{s-1}\chi_{s-i}\mathbf{v}_{t+i}$.

Consider for example the simple $GARCH(1,1)$ model, then:

$$\begin{aligned} E_t\{\sigma_{t+s}^2\} &= \omega + \textstyle\sum_{i=1}^{s-1}(\alpha+\beta)^i + (\alpha+\beta)^{s-1}\alpha\varepsilon_t^2 + (\alpha+\beta)^{s-1}\beta\sigma_t^2 \\ &= \omega + \textstyle\sum_{i=1}^{s-1}(\alpha+\beta)^i + (\alpha+\beta)^{s-1} + (\alpha+\beta)^{s-1}\beta\sigma_{t+1}^2, \end{aligned}$$

which can also directly be derived from the $GARCH(1,1)$ process:
$\sigma_t^2 = \omega\beta\sigma_{t+1}^2 + \alpha\varepsilon_{t-1}^2 = \omega + (\alpha+\beta)\sigma_{t+1}^2 + \alpha(\varepsilon_{t-1}^2 - \sigma_{t-1}^2) \Leftrightarrow (1-(\alpha+\beta)L)\sigma_t^2 = \omega + \alpha(\varepsilon_{t-1}^2 - \sigma_{t-1}^2)$.Then:

$$\begin{aligned} E_t\{\sigma_{t+s}^2\} &= E_t\{\omega + (\alpha+\beta)\sigma_{t+s-1}^2 + \alpha(\varepsilon_{t+s-1}^2 - \sigma_{t+s-1}^2)\} \\ &= E_t\{\omega + (\alpha+\beta)\sigma_{t+s-1}^2\} \\ &= \omega + (\alpha+\beta)E_t\{\sigma_{t+s-1}^2\} = \omega + (\alpha+\beta)(\omega + (\alpha+\beta)E_t\{\sigma_{t+s-2}^2\} \\ &= \ldots \\ &= \omega + \textstyle\sum_{i=1}^{s-1}(\alpha+\beta)^i + (\alpha+\beta)^{s-1}\sigma_{t+1}^2. \end{aligned}$$

If the model is stationary with $\sigma_\varepsilon^2 = \omega + (1-\alpha-\beta)^{-1}$,the optimal predictor for σ_{t+s}^2can be expressed as: $E_t\{\sigma_{t+s}^2\} = \sigma_\varepsilon^2 + (\alpha+\beta)^{s-1})(\sigma_{t+1}^2 - \sigma_\varepsilon^2)$

E.3 Generalized exponential distribution

The normal and Laplace distribution are both special cases of the more general *power (or generalized) exponential distribution (GED)*. The standardized GED has the following representation:

$f_{St.GED}(\varepsilon_t^2;\kappa) = \frac{\kappa}{2^{(1/\kappa+1)}\Gamma(1/\kappa)\gamma\sigma_t}\exp\left(-\frac{1}{2}\left|\frac{\varepsilon_t}{\gamma\sigma_t}\right|^\kappa\right)$, where $\gamma := \sqrt{\frac{\Gamma(1/\kappa)}{2^{(2/\kappa)}\Gamma(3/\kappa)}}$ and $\kappa \geqq 0$.The standardized normal distribution is obtained by setting $\kappa = 2$,while the standardized Laplace distribution requires $\kappa = 1$.The fatness of the tails is negatively oriented with the value of κ. The kurtosis coefficient is $\Gamma(5/\kappa)\Gamma(1/\kappa)\Gamma(3/\kappa)^2$ and the loglikelihood function is given by:

$\ln\ell(\kappa,\theta;\varepsilon_1,\ldots,\varepsilon_T) = T\ln(\kappa) - T(\frac{1}{\kappa}+1)\ln 2 - T\ln\Gamma(\frac{1}{\kappa}) - T\ln\gamma - \frac{1}{2}\sum_{t=1}^{T}\ln\sigma_t^2 - \frac{1}{2}\sum_{t=1}^{T}\left|\frac{\varepsilon_t}{\gamma\sigma_t}\right|^\kappa$.

E.4 Generalized Student t-distribution

The generalized Student t-distribution (GSD) encompasses both the GED and the Student t-distribution due to the two shape parameters. The formula for the standardized GSD is given by:

$$f_{St.GED}(\varepsilon_t^2;\kappa) = \varepsilon_t; v, \kappa) = \frac{C}{\sigma_t}\left(1+\frac{\kappa}{v-2}\theta^{-\kappa}\left|\frac{\varepsilon_t}{\sigma_t}\right|^{\kappa}\right)^{-(v+1)/\kappa}, \text{ where}$$

$C = 0.5\kappa B(\frac{1}{\kappa},\frac{v}{\kappa})^{-3/2}B(\frac{3}{\kappa},\frac{v-2}{\kappa})^{1/2}$ and $\theta = (\kappa(v\text{-}2))^{1/\kappa}B(\frac{1}{\kappa},\frac{v}{\kappa})^{1/2}B(\frac{3}{\kappa},\frac{v-2}{\kappa})^{-1/2}$ and B is the beta function, defined as $B(x,y) := \frac{\Gamma(x)\Gamma(y)}{\Gamma(x/y)}$. The kurtosis coefficient equals $\frac{B(1/\kappa,v/\kappa)B(5/\kappa),(v-4)/\kappa}{B(3/\kappa,(v-2)/\kappa)^2}$. Setting $\kappa = 2$ results in the student t-distribution with v as the usual degrees of freedom. When $v \to \infty$ the GED is obtained. The loglikelihood function equals:

$$\ln \ell(v,\kappa,\theta;\varepsilon_1,\ldots,\varepsilon_T) = \ln C - \tfrac{1}{2}\sum_{t=1}^{T}\ln\sigma_t^2 - \tfrac{v+1}{\kappa}\sum_{t=1}^{T}\ln\left(1+\tfrac{\kappa}{v-2}\theta^{-\kappa}\left|\tfrac{\varepsilon_t}{\sigma_t}\right|^{\kappa}\right).$$

The GSD is a special case of the even more general skewed generalized Student distribution (SGSD). In a SGSD a third parameter determines the skewness; by setting it to zero the GSD is obtained.

Note: the loglikelihood function for the multivariate t-GARCH satisfies:

$$L_t = g(v) - \tfrac{1}{2}\ln|H_t| - (v+1)/2\ln(1+(v-2)^{-1}\varepsilon_t' H_t^{-1}\varepsilon_t), \text{ where}$$

$g(v) = T\ln[\Gamma(\frac{v+1}{2})] - T\ln\Gamma(\frac{v}{2})]T/2\ln(v-2)$, or the complete sample: $L{=}\sum_{t=1}^{T} L_t{=}Tg(v)\text{-}\frac{1}{2}\sum_{t=1}^{T}\ln\det H_t\text{-}\frac{(v+1)}{2}\sum_{t=1}^{T}\ln(1{+}(v\text{-}2)^{-1}\varepsilon_t' H_t^{-1}\varepsilon_t)$, where the appropiate multivariate $t - GARCH$ model can now be substituted.

E.5 Aggregation of $GARCH$ processes

Following Drost and Nijman (1993) we consider the simple $GARCH(1,1)$ process, i.e. the h-daily returns $\{y_{(h)t}\}_{t=1}^{T/h}$ follow a weak $GARCH(1,1)$ process with $\sigma_{(h)t}^2{=}\omega_{(h)}{+}\alpha_{(h)}\varepsilon_{t-1}^2{+}\beta_{(h)}\sigma_{(h)t-1}^2$ where $\omega_{(h)}{=}h\omega(\frac{1-(\alpha+\beta)^h}{1-(\alpha+\beta)})$, $\alpha_{(h)}{=}(\alpha{+}\beta)^h\text{-}\beta_{(h)}$ and $\kappa_{(h)y} = 3 + \frac{(\kappa_y-3)}{h} + \frac{6(\kappa_y-1)\{h-1-h(\alpha+\beta)+(\alpha+\beta)^2\}\{\alpha-\beta\alpha(\beta+\alpha)\}}{h^2(1-\beta-\alpha)^2(1-\beta^2-2\beta\alpha)}$ and $\left|\beta_{(h)}\right|$ is the solution to the quadratic equation $\frac{\beta_{(h)}}{1+\beta_{(h)}^2} = \frac{a(\alpha+\beta)^h-b}{a(1+(\alpha+\beta)^{2h})-2b}$, where

$$a = h(1-\beta)^2 + \frac{2h(h-1)(1-\alpha-\beta)^2(1-\beta^2-2\beta\alpha)}{(\kappa-1)(1-(\alpha+\beta)^2)} + 4\left(h\text{-}1\text{-}h(\alpha+\beta)+(\alpha+\beta)^h\right)\frac{\alpha-\beta\alpha(\alpha+\beta)}{1-(\alpha+\beta)^2},$$

and $b = \alpha - \beta\alpha(\alpha+\beta)\frac{(1-(\alpha+\beta)^{2h})}{1-(\alpha+\beta)^2}$.

Drost and Nijman (1993) suggest that for strong $GARCH$ processes the distribution of the innovations is usually given. The relation between the kurtosis and the innovations $\kappa_\varepsilon = E\{\varepsilon^4\}$ and κ_y is given by: $\kappa_y = \kappa_\varepsilon\frac{1-(\alpha+\beta)^2}{(1-(\alpha+\beta)^2)-(\kappa_\varepsilon-1)\alpha^2}$.

E.4 Generalized Student t-distribution

References

Acemoglu, D. and A. Scott (1994), "Consumer Confidence and Rational Expectations: Are agents' Beliefs Consistent with the Theory?", *Economic Journal*, vol. 104, pp. 1-17.

Acker, D. (1998), "Good News is No News", Discussion Paper, No. 98/453, University of Bristol, 26 pages.

Amemiya, T. (1984), "Tobit Models: A Survey", *Journal of Econometrics*, vol. 24, pp. 3-61.

Anderson, T.G., T. Bollerslev, and F.X. Diebold (2002), "Parametric and Nonparametric Volatility Measurement", NBER Technical Working Paper 279, 64 pages.

Anderson, T.W. (1959), "On Asymptotic Distributions of Estimates of Parameters of Stochastic Difference Equations", *Annals of Mathematical Statistics*, vol. 30, pp. 676-687.

Andrews, D.W.K. (2004), "The Block-block Bootstrap: Improved Asymptotic Refinements", *Econometrica*, vol. 72(3), pp. 673-700.

Arellano, M. and S. Bond (1991), "Some Tests of Specification for Panel Data: Monte Carlo Evidence and an Application to Emplyment Equations", *Review of Economic Studies*, vol. 58, pp. 277-294.

Baillie, R.T.and T. Bollerslev (1992), "Prediction in dynamic models with time-dependent conditional variances," *Journal of Econometrics*, vol. 52(1), pp. 91-113.

Bala, L. and G. Premaratne (2004), "Stock Market Volatility: Examining North America, Europe and Asia", Econometric Society 2004 Far Eastern Meetings Paper Number 479, Econometric Society, 41 pages.

Balasubramanyan, L. (2004). "Do Time-Varying Covariances, Volatility Comovement and Spillover Matter?", Department of Finance, Pennsylvania State University.

Baltagi, B.H. (2003), *Econometric Analysis of Panel Data*, John Wiley and Sons, New York.

Balvers, R.J., T.F. Cosimano and B. Mc. Donald (1990), "Predicting Stock Returns in an Efficient Market", *Journal of Finance*, vol. 45 (4), pp. 1109-1128.

Barro, R. and X. Sala-I-Martin (1995), *Economic Growth*, McGraw-Hill, New York, 359 pages.

Beckers, S. (1991), "Stocks, Bonds, and Inflation in the World Market: Implications for Pensions Fund Management", *Journal of Fixed Income*, vol. 1 (3), pp. 18-30.

Bera, A.K. and M.L. Higgins (1993), "ARCH Models: Properties, Estimation and Testing", *Journal of Economic Surveys*, vol. 7, pp. 305-366.

M.L. Berenson, D.Levine , T.C. Krehbiel and D. Stephan (eds.) (2003), Basic Business Statistics (International Edition), Prentice Hall, New York, 880 pages.

Bernard, H. and S. Gerlach (1998), "Does the Term Structure Predict Recessions: the International Evidence", *International Journal of Finance & Economics*, vol. 3(3), pages 195-215.

Berndt, E., B. Hall, R. Hall, and J. Hausman (1974), "Estimation and Inference in Nonlinear Structural Models", *Annals of Economic and Social Measurement*, vol. 3/4, pp. 653-665.

Black, F. (1976), "Studies of Stock Price Volatility Changes", in *Proceedings of the American Statistical Association*, Business and Economic Statistics Section, pp. 177-181.

Black, F. and M. Scholes (1973), "The Pricing of Options and Corporate Liabilities", *Journal of Political Economy*, vol. 81, pp. 637-654.

Blanchard, O. J. and D. Quah (1989), "The Dynamic Effects of Aggregate Demand and Supply Disturbances", *American Economic Review*, vol. 79 (4), pp.655-673.

Bollerslev, T. (1986), "Generalized Autoregressive Conditional Heteroscedasticity", *Journal of Econometrics*, vol. 31, pp. 307-327.

Bollerslev, T. (1990), "Modelling the coherence in short run nominal exchange rates: A multivariate generalised ARCH model", *Review of Economics and Statistics*, vol. 72, pp. 498-505.

Bollerslev, T., R.F. Engle and J. Wooldridge (1988), "A Capital Asset Pricing Model with Time Varying Covariances", *Journal of Political Economy*, vol. 96, pp. 116-131.

Box, G. E. P. and G.M. Jenkins (1970). *Time Series Analysis: Forecasting and Control*, Holden-Day, San Francisco, 592 pages.

Box, G. E. P. and G.Tiao (1975), " Intervention analysis with applications to economic and environmental problems", *Journal of the American Statistical Association*, vol. 70, pp. 70-79.

Breitung, J. and W. Meyer (1994), "Testing for unit root in panel data. Are wages on different bargaining levels cointegrated", *Applied Economics*, vol. 26, pp. 353-361.

Breusch, T. and A. Pagan (1980), "A Simple Test for Heteroscedasticity and Random Coefficient Variation", *Econometrica*, vol. 47 (5), pp. 1287-1294.

Brown, R.L. (1962), *Smoothing, forecasting and prediction*, Prentice Hall, New York, 480 pages.

Brown, R.L., J. Durbin, and J.M. Evans (1975), "Techniques for Testing the Constancy of Regression Relationships over Time", *Journal of the Royal Statistical Society*, Series B, vol. 37, pp. 149-192.

Bruneau, C. and O. De Bandt (2003), "Monetary and fiscal policy in the transition to EMU: what do $SVAR$ models tell us?", *Economic Modelling*, vol. 20, pp. 959-985.

Burnside, C. A. (1994), "Hansen-Jagannathan Bounds as Classical Tests of Asset Pricing Models", *Journal of Business and Economic Statistics*, January, 12 (1), 57-59.

Butler, J.S. and R. Moffitt (1982), "A Computationally Efficient Quadrature Procedure for the One-Factor Multinomial Probit Model", *Econometrica*, vol.50 (3), pp. 761-764.

Cameron, A C. and F.A.G Windmeijer (1996), "R-Squared Measures for Count Data Regression Models with Applications to Health-Care Utilization",

Journal of Business & Economic Statistics, American Statistical Association, vol. 14 (2), pp. 209-220.

Cameron, A.C. and P.K. Trivedi (1998), *Regression Analysis of Count Data*, Cambridge University Press, 432 pages.

Campbell, J.Y., A.W Lo and A.C. MacKinlay (1997), *The Econometrics of Financial Markets*, Princeton University Press, 611 pages.

Canova, F. and B.E. Hansen (1995), "Are Seasonal Patterns Constant over Time: A Test for Seasonal Stability", *Journal of Business and Economic Statistics*, vol. 13, pp. 237-252.

Cechetti, S. C., P. S. Lam and N. C. Mark (1994), "Testing Volatility Restrictions on Intertemporal Marginal Rates of Substitution Implied by Euler Equations and Asset Returns", *Journal of Finance*, vol. 49 (1), pp. 123-152.

Chang, I. H., G.C.Tiao and C. Chen, (1988). "Estimation of time series parameters in the presence of outliers". *Technometrics*, vol. 30 (2), pp.193-204.

Chatfield, C. (2001), *Time Series Forecasting*, Chapman and Hall, London, 352 pages.

Chen, C. and L.-M. Liu (1993), "Joint Estimation of Model Parameters and Outlier Effects in Time Series", *Journal of the American Statistical Association*, vol. 88, pp. 284-297.

Chen, N.F., R. Roll and S. Ross (1986), Economic forces and the Stock Market, *Journal of Business*, vol. 59 (3), pp. 383-403.

Choi, I., and B. Yu (1997) , "A General Framework for Testing I(1) against I(m+k)", *Journal of Economic Theory and Econometrics*, vol. 3, pp. 103-138.

Chow, G.C. (1960), "Test of Equality between Sets of Coefficients in Two Linear Regressions", *Econometrica*, vol. 28, pp. 591-605.

Christoffersen, P.F. (1998), "Evaluating interval forecasts", *International Economic Review*, vol. 39, pp. 841-862.

Cincera, M. (1997), "Patents, R&D, and Technological Spillovers at the Firm Level: Some Evidence from Econometric Count Models for Panel Data", *Journal of Applied Econometrics*, vol. 12 (3), pp. 265-280.

Cincera, M. (2003), "Financing constraints, fixed capital and R&D investment decisions of Belgian firms", in P. Butzen, and C. Fuss (eds.), *Firms' Investment and Finance Decisions: Theory and Empirical Methodology*, Edward Elgar, Cheltenham.

Clements, M.P. (2004), "Evaluating the Bank of England Density Forecasts of Inflation", *The Economic Journal*, vol. 114, pp. 844-866.

Cochrane, J.H. (1991), "Production-Based Asset Pricing and the Link Between Stock Returns and Economic Fluctuations", *Journal of Finance*, vol. 56, (1), pp. 209-237.

Cochrane, J.H. (1996), "A Cross-Sectional Test of an Investment-Based Asset Pricing Model", *Journal of Political Economy*, vol. 104, pp. 572-621.

Cochrane, J.H. (1997), "Time Series for Macroeconomics and Finance", *http://www-gsb.uchicago.edu/fac/john.cochrane/research/Papers/timeser1.pdf*, 125 pages.

Corsetti, G., L. Dedola and S. Leduc (2004),"International Risk-Sharing and the Transmission of Productivity Shocks", ECB Working Paper Series 308, 49 pages. [Available at http://www.ecb.int/].

Cox, J. C., J. E. Ingersoll and S. A. Ross (1985a), "An Intertemporal General Equilibrium Model of Asset Prices", *Econometrica*, 53 (2), pp. 363 - 384.

Cox, J. C., J. E. Ingersoll and S. A. Ross (1985b), "A Theory of the Term Structure for Interest Rates", *Econometrica*, 53 (2), pp. 385-408.

Davidson, J.E.H., D.F. Hendry, F. Srba, and S. Yeo (1978), "Econometric Modeling of the Aggregate Time-Series Relationship Between Consumer's Expenditure and Income in the United Kingdom", *Economic Journal*, vol. 88, pp. 661-692.

Davies, N., C. M. Triggs and P. Newbold (1977), "Significance Levels of the Box-Pierce Portmanteau Statistic in Finite Samples", *Biometrika*, vol. 64, pp. 517-522.

Deaton, A. (1985), "A Profitable Approach to Labor Supply and Commodity Demands over the Life-Cycle", *Econometrica*, vol. 53 (3), pp. 503-543.

Dickey, D. A. and Pantula, S.G. (1987), "Determining the Order of Differencing in AR Processes", *Journal of Business and Economic Statistics*, vol. 5, pp. 455-461.

Dickey, D.A. and W.A. Fuller (1979), "Distribution of the estimators for autoregressive time series with a unit root", *Journal of the American Statistical Association,* vol. 74, pp. 427-431.

Dickey, D.A. and W.A. Fuller (1981) "Likelihood Ratio Statistics for Autorregressive Time Series With a Unit Root", *Econometrica*, vol. 49, pp.1057-1072.

Dickey, D.A., H.P. Hasza, and W.A. Fuller (1984), "Testing for Unit Roots in Seasonal Time Series", *Journal of the American Statistical Association*, vol.79, pp. 355-367.

Diebold, F.X., T.A. Gunther and A.S. Tay (1998), "Evaluating density forecasts with applications to financial risk management", *International Economic Review,* vol. 39, pp. 863-883.

Diebold, F.X. and J.A. Lopez (1995), "Modeling Volatility Dynamics", in K.V. Hoover (ed.), *Macroeconometrics: Developments, Tensions and Prospects*, pp. 427-472, Kluwer Academic Press, Boston.

Diebold, F.X., A.S. Tay and K.F. Wallis (1999), "Evaluating density forecasts of inflation: the Survey of Professional Forecasters", in R.F. Engle and H. White (eds.), *Cointegration, Causality, and Forecasting: A Festschrift in Honour of Clive W.J. Granger*, Oxford University Press, pp. 76-90.

Drost, F.C. and T.E. Nijman, (1993), "Temporal Aggregation of GARCH Processes", *Econometrica*, vol. 61, pp. 909-927.

Durbin, J. (1970), "Testing for Serial Correlation in Least Squares Regressions When Some of the Regressors are Lagged Dependent Variables", *Econometrica*, vol. 38, pp. 410-421.

Durbin, J. and G.S. Watson (1950), "Testing for Serial Correlation in Least Squares Regressions I", *Biometrika*, vol. 37, pp. 409-428.

Durbin, J. and G.S. Watson (1951), "Testing for Serial Correlation in Least Squares Regressions II", *Biometrika*, vol. 38, pp. 159-178.

Enders, W. (1995), *Applied Econometric Time Series*, John Wiley and Sons, New York, 134 pages.

Engle, R.F. (1982), "Autoregressive Conditional Heteroscedasticity with Estimates of the Variance of UK Inflation", *Econometrica*, vol. 50 (4), pp. 987-1008.

Engle, R.F. (2000), "Dynamic Conditional Correlation - A Simple Class of Multivariate GARCH Models", University of California at San Diego, Economics Working Paper Series 2000-09, Department of Economics, UC San Diego, 28 pages.

Engle, R.F. and C.W.J. Granger (1987), "Co-Integration and Error Correction: Representation, Estimation, and Testing", *Econometrica*, vol. 55 (2), pp. 251-276.

Engle, R.F. and D. Kraft (1983), "Multiperiod forecast error variances of inflation estimated from ARCH models", in A. Zellner (ed.) *Applied time series analysis of economic data*, Bureau of the Census, Washington, DC, pp. 293-302.

Engle, R.F. and K.F. Kroner (1995), "Multivariate Simultaneous Generalized ARCH", *Econometric Theory*, vol. 11, pp. 122-150.

Epstein, L.G. and S. E. Zin (1991), "Substitution, Risk Aversion, and the Temporal Behavior of Consumption and Asset Returns: an Empirical Analysis, *Journal of Political Economy*, vol. 99 (2), pp. 263-286.

Estrella, A. (1998), "Why Do Interest Rates Predict Macro Outcomes? A Unified Theory of Inflation, Output, Interest and Policy", Research Paper 9717, Federal Reserve Bank of New York, 36 pages.

Estrella, A. and F. S. Mishkin (1998), "Predicting US recessions: Financial Variables as Leading Indicators", *Review of Economics and Statistics*, 80 (1), pp. 45-61.

Fisher, F.M. (1970), "Tests of Equality between Sets of Coefficients in Two Linear Regressions: An Expository Note", *Econometrica*, vol. 38, pp. 361-366.

Franses, P.H. and D. van Dijk (2000), *Non-linear time series models in empirical finance*, Cambridge University Press, 271 pages.

Franses, Philip Hans and Ghijsels, H. (1999), "Additive Outliers, GARCH and Forecasting Volatility", *International Journal of Forecasting*, vol. 15, pp. 1-9.

Franses, P.H. (1998), *Time Series Models for Business and Economic Forecasting*, Cambridge University Press, Cambridge, 280 pages.

Friedman, M. (1976) *Price Theory*, Aldine de Gruyter, New York, 353 pages.

Fuhrer, J. and G. Moore (1992), Monetary Policy Rules and the Indicator Properties of Asset Prices, *Journal of Monetary Economics*, vol. 29, pp. 303-306.

Fuller, W.A. (1976), *Introduction to Statistical Time Series*, Wiley, New York.

Fuller, W.A. (1996) *Introduction to Statistical Time Series*, 2nd Edition, Wiley, New York.

Garman, M. and M. Klass(1980), "On the Estimation of Security Price Volatilities from Historical Data", *Journal of Business*, vol. 53, pp. 67-78.

Gertel, H.R., M. De Santis, and Cristina, A.D. (2002), "Who chooses to become a teacher in Argentina?" Proceedings of the XXXVIII Annual Meeting of AAEP, 20 pages.

Geweke, J., R. Meese and W. Dent (1983), "Comparing Alternative Tests of Causality in Temporal Systems: Analytic Results and Experimental Evidence", *Journal of Econometrics*, vol. 21, pp. 161-194.

Glosten, L., R.Jagannathan, and D. Runkle (1993), "On the Relation between the Expected Value and the Volatility of the Nominal Excess Returns on Stocks", *Journal of Finance*, vol. 48(5), pp. 1779-1801.

Gouriéroux, C. and A. Monfort (1995), *Statistics and Econometric Models*, Cambridge University Press, 522 pages.

Gouriéroux, C. (1997), *ARCH Models and Financial Applications*, Springer, New York, 228 pages.

Gouriéroux, C. and A. Monfort (1997), *Time Series and Dynamic Models.* Cambridge University Press, 654 pages.

Gouriéroux C, A. Monfort , and A. Trognon (1984), "Pseudo maximum likelihood methods: Applications to Poisson models", *Econometrica*, vol. 52(3), pp. 701-720.

Granger, C.W.J.(1969), "Investigating Causal Relations by Econometric Models and Cross-Spectral Methods", *Econometrica*, vol. 37, pp. 424-438.

Granger, C.W.J. (1981), "Some Properties of Time Series Data and Their Use in Econometric Model Specification", *Journal of Econometrics*, vol. 16, pp. 121-130.

Granger, C.W.J. (1983), "Co-Integrated Variables and Error-Correcting Models", Discussion Paper 83-13, University of California, San Diego.

Granger, C.W.J. and P. Newbold (1974), "Spurious Regression in Econometrics", *Journal of Econometrics*, vol. 2, pp. 111-120.

Greene, W. (2000), *Econometric Analysis*, 4th Edition, Prentice Hall, New York, 1004 pages.

Greene, W. (2003), *Econometric Analysis*, 5th Edition, Prentice Hall, New York, 1026 pages.

Haavelmo, T. (1943), "The Statistical Implications of a System of Simultaneous Equations", *Econometrica,* vol. 11 (1), pp. 1 - 12.

Hagiwara, M. and M. A. Herce (1997), "Risk aversion and Stock Price Sensitivity to Dividends", *American Economic Review*, September, 87 (4), 738-745.

Hall, A. (1993), Some Aspects of Generalised Method of Moments Estimation, In Maddala, C.R. Rao and H.D. Vinod (eds.), *Handbook of Statistics*, vol. 11, pp. 393-417, Elsevier Publishers, Amsterdam.

Hall, B.H. and R. Oriani (2004), "Does the Market Value R&D Investment by European Firms? Evidence from a Panel of Manufacturing Firms in France, Germany, and Italy", NBER Working Paper 10408, 42 pages.

Hamilton, J.D. (1994), *Time Series Analysis*, Princeton University Press, 799 pages.

Hansen, L.P. (1982), "Large Sample Properties of Generalised Method of Moments Estimators", *Econometrica*, vol. 50, pp. 1029-1054.

Hansen, L.P. and K.J Singleton, (1982), "Generalized Instrumental Variables Estimation of Nonlinear Rational Expectations Models", *Econometrica*, vol. 50(5), pp. 1269-1286 and Errata (1984), *Econometrica*, vol. 52 (1), pp. 267-268.

Hansen, L.P. and K.J Singleton, (1983), "Stochastic Consumption, Risk Aversion and the Temporal Behavior of Asset Returns", *Journal of Political Economy*, vol. 91 (2), pp. 249-265.

Hausman, J.A. (1978), "Specification tests in econometrics", *Econometrica*, 46, pp. 1251–1271

Hausman, J.A., B.H. Hall, and Z. Griliches, (1984), Econometric Models for Count Data with an Application to the Patents-R&D Relationship, *Econometrica*, vol. 52, pp. 909-938.

Hayashi, F. (2000), *Econometrics*, Princeton University Press, 712 pages.

Heathfield, D. F. and S. Wibe (1987), *An Introduction to Cost and Production Functions*, MacMillan Education Ltd, London, UK, 193 pages.

Heckman, J.J. (1979), "Sample Selection Bias as a Specification Error", *Econometrica*, vol.47 (1), pp. 153-161.

Hildreth, C. and C. Houck (1968), "Some estimators for a linear model with random coefficients", Journal of the American Statistical Association, vol. 63, pp. 584.

Hsiao, C. (1974), "Statistical Inference for a Model with Both Random Cross-Sectional and Time Effects", *International Economic Review*, vol. 15, pp. 12-30.

Hsiao, C. (1975), "Some Estimation Methods for a Random Coefficient Model", *Econometrica*, vol. 43, pp. 305-325.

Hsiao, C. (1986), *Analysis of Panel Data*, Cambridge University Press.

Hsiao, C. (1996), " Logit and probit models" in Mátyás, L. and P. Sevestre (eds.), *The Econometrics of Panel Data*, Kluwer Academic Publisher, pp. 410-427.

Hull, J. and A. White (1987), "The Pricing of Options on Assets with Stochastic Volatility", *Journal of Finance*, vol. 42 (2), pp. 281-300.

Hurwicz, L. (1962), "On the Structural form of Interdependent Systems", in in E. Nagel, P. Suppes, and A. Tarski, (eds.), *Logic, Methodology and Philosophy of Science*, pp. 232-239, Stanford University Press, Stanford.

Hylleberg, S., R.F. Engle, C.W.J. Granger, and B.S. Yoo (1990), "Seasonal Integration and Cointegration", *Journal of Econometrics*, vol. 44, pp. 215-238.

Im, K.S., M.H. Pesaran, and Y. Shin, (1997), "Testing for Unit Roots in Heterogeneous Panels", Revised Version of Department of Applied Economics (DAE) Working Paper 9526, Cambridge University.

Johansen, S. (1988), "Statistical analysis of cointegrating vectors", *Journal of Economic Dynamics and Control*, vol. 12, pp. 231-254.

Johansen, S. (1991), "Estimation and Hypothesis Testing of Cointegration Vectors in Gaussian Vector Autoregressive Models", *Econometrica*, vol. 59, pp. 1551-1580.

Johnston, J. and J. Dinardo (1997), *Econometric Methods*, pp. 327-347, McGraw-Hill, New York.

Jorgenson, D.W.(1986), "Econometric Methods for Modeling Producer Behavior" in J.J. Heckman and E.E. Leamer (eds.), *Handbook of Econometrics*, vol 3, chapter 31, North-Holland, Amsterdam.

Judge, G. G., W.E. Griffits, R. Carter Hill, H. Lütkepohl and T.C. Lee (1980), *The Theory and Practice of Econometrics*, first edition, John Wiley and Sons, USA

Judge, G. G., W.E. Griffits, R. Carter Hill, H. Lütkepohl and T.C. Lee (1985) The Theory and Practice of Econometrics. Second Edition, John Wiley and Sons: USA; 1019 pages.

Kendall, M., and J. K. Ord (1990), *Time Series*, 3rd edition, Griffin, London and Oxford University Press, New York.

Klein, P. (2000), "Using the generalized Schur form to solve a multivariate linear rational expectations model", *Journal of Economic Dynamics & Control*, vol. 24, pp. 1405-1423.

Konings, J. and F. Roodhooft (1997), "How elastic is the demand for labour in Belgian enterprises? Evidence from firm level data, 1987-94", *De Economist*, vol. 145, pp. 229-241.

Koopman, S.J., B. Jungbacker, and E. Hol (2004), "Forecasting daily variability of the S&P100 stock index using historical, realized and implied volatility measurements", Tinbergen Institute Discussion Paper 04-016/4, 34 pages.

Kräger, H. and P. Kugler (1993), "Nonlinearities in foreign exchange markets: A different perspective", *Journal of International Money and Finance*, vol. 12, pp. 195-208.

Kullback, S. and H.M. Rosenblatt (1957), "On the Analysis of Multiple Regression in k Cathegories", *Biometrika*, vol. 44 (1), pp. 67-83.

Kwiatkowski , D., P.C.B. Philips, P.Schmidt and Y. Shin (1992), "Testing the Null Hypothesis of Stationarity Against the Alternative of a Unit Root: How Sure Are We That Economic Time Series Have a Unit Root?", *Journal of Econometrics*, vol. 54, pp. 159-178.

Lambert, D. (1992), "Zero-Inflated Poisson Regression, with an Application to Defects in Manufacturing", *Technometrics*, vol. 34, pp. 1-14.

Lastrapes, W.D. (2002), "Real wages and aggregate demand shocks: contradictory evidence from VARs", *Journal of Economics and Business*, vol. 54, pp. 389-413.

LeBaron, B. (1992), "Some relationships between volatility and serial correlations in stock market returns", *Journal of Business*, vol. 65, pp. 199-219.

Lee, M.-J.(2002). *Panel data econometrics. Methods of Moments and Limited Dependent Variables.* Academic Press, USA, 195 pages.

Levin, A. and C.-F. Lin (1993a), "Unit Root Tests in Panel Data: Asymptotic and Finite Sample Properties", Discussion Paper 93-23, University of California San Diego.

Levin, A. and C.-F. Lin (1993b), "Unit Root Tests in Panel Data: New Results", Discussion Paper 93-56, University of California San Diego.

Ljung, G.M. and G.E.P. Box (1978). "On a Measure of Lack of Fit in Time Series Models", *Biometrika*, vol. 67, pp. 297-303.

Long, J. S. (1997). Regression Models for Categorical and Limited Dependent Variables. Sage Publications, USA, 296 pages.

Lukach, R. and J. Plasmans (2005), "International Knowledge Flows from and into a Small Open Economy: Patent Citation Analysis", Chapter 13 in A. Spithoven and P. Teirlinck (eds.), *Beyond Borders. Internationalisation of R&D and Policy Implications for Small Open Economies,* Elsevier, pp. 331-357.

MacKinnon, J. (1991), "Critical values for cointegration tests", in R.F. Engle and C.W.J. Granger (eds.), *Long Run Economic Relationships,* Oxford University Press.

Maddala, G.S. (1971), "The Use of Variance Component Models in Pooling Cross Section and Time Series Data", *Econometrica,* vol. 39, pp. 341-358.

Maddala, G. S. (1983). Limited Dependent and Qualitative Variables in Econometrics. Cambridge: Cambridge University Press, 395 pages.

Maddala, G. S. and I.-M. Kim (1998), Unit Roots, Cointegration, and Structural Change. Cambridge: Cambridge University Press, 491 pages.

Mandelbrot, B. (1963), "The Variation of Certain Speculative Prices", Journal of Business, vol. 36, pp. 394-419.

Manski, C.F. (1975), " Maximum Score Estimation of the Stochastic Utility Model of Choice", *Journal of Econometrics,* vol. 3 pp. 205-228.

Manski, C.F. (1985), " Semiparametric Analysis of Discrete Response: Asymptotic Properties of the Maximum Score Estimator", *Journal of Econometrics,* vol. 27, pp. 313-333.

McCullagh, P. and J.A. Nelder (1989), *Generalized linear models,* 2nd edition, Chapman and Hall, New York.

Merton, R. (1980), "On Estimating the Expected Return on the Market: An Exploratory Investigation", *Journal of Financial Economics,* vol. 8, pp. 323-361.

Mills, T.C. (1993), *The econometric modelling of financial time series,* Cambridge University Press, 247 pages.

Mundlak, Y. (1978), "On the Pooling of Time Series and Cross Sectional Data", *Econometrica,* vol. 56, pp. 69-86.

Nabeya, S. and K. Tanaka (1988), "Asymptotic theory of a test for the constancy of regression coefficients against the random walk alternative", *Annals of Statistics,* vol. 16, pp. 218-235.

Nelson, C.R. and C.I. Plosser (1982), "Trends and Random Walks in Macroeconomic Time Series", *Journal of Monetary Economics,* vol. 10, pp. 139-162.

Nelson, D.B. (1990), "Stationarity and Persistence in the GARCH(1,1) Model", *Econometric Theory,* vol. 6, pp. 318-334.

Nelson, D.B. (1991), "Conditional Heteroscedasticity in Asset Returns: A New Approach",*Econometrica,* vol. 59, pp. 347-370.

Nelson, D.B. (1992), "Filtering and Forecasting with Misspecified ARCH Models I: Getting the Right Variance with the Wrong Model", *Journal of Econometrics,* vol. 52, pp. 61-90.

Nelson, D.B. and D. Foster, (1994), "Asymptotic Filtering Theory for Univariate ARCH Models", *Econometrica*, vol. 62, pp. 561-573.

Newey, W. and K. West (1987), " A Simple Positive Semi-Definite, Heteroscedasticity and Autocorrelation Consistent Covariance Matrix", *Econometrica*, vol. 55, pp. 703-708.

Nijman, T.E. and E. Santana (1996), "Marginalization and Contemporaneous Aggregation in Multivariate Garch Processes", *Journal of Econometrics*, vol. 71, pp. 71-86.

Nijman, T.E. and M.Verbeek (1990), "Estimation of Time Dependent Parameters Using Cross Sections, Panels or Both", *Journal of Econometrics*, vol. 46, pp. 333-346.

Officer, R. (1973), "The Variability of the Market Factor of the New York Stock Exchange",*Journal of Business*, vol. 46, pp. 434-453.

Parkinson, M. (1980), "The Extreme Value Method for Estimating the Variance of the Rate of Return",*Journal of Business*, vol. 53, pp. 61-65.

Peersman, G. and R. Straub (2004),"Technology Shocks and Robust Sign Restictions in a Euro Area SVAR",ECB Working Paper Series 373, 34 pages.

Pesaran, M.H. (1974), "On the General Problem of Model Selection", *Review of Economic Studies*, vol. 41, pp. 153-171.

Pesaran, M.H. and A.S. Deaton (1978), "Testing Non-Nested Nonlinear Regression Models", *Econometrica*, vol. 46 (3), pp. 677 - 694.

Phillips, P.C.B. (1987) "Time Series Regression with a Unit Root", *Econometrica,* vol 55, pp. 277-301.

Priestley, M.B. (1988), *Nonlinear and Nonstationary Time Series Analysis*, Academic Press, London.

Quandt, R. (1960), "Tests of the Hypothesis that a Linear Regression System Obeys Two Separate Regimes", *Journal of the American Statistical Association*, vol 55, pp. 324-330.

Rabemananjara, R. and J.M. Zakoian (1993), "Threshold ARCH Models and Asymmetries in Volatility", *Journal of Applied Econometrics*, vol. 8, pp. 31-49.

Rao, C.R. (1952), *Advanced Statistical Methods in Biometric Research*, John Wiley, New York.

Rao, C.R. (1965), *Linear Statistical Inference and Its Applications,* John Wiley & Sons.

Rao, M.M. (1961), "Consistency and Limit Distributions of Estimators of Parameters in Explosive Stochastic Difference Equations", *Annals of Mathematical Statistics*, vol. 32, pp. 195-218.

Rockafellar, R.T. (1976), "Augmented Lagrangians and Applications of the Proximal Point Algorithm in Convex programming", *Mathematics of Operations Research*, vol. 1, no. 2, pp. 97-116.

Sargan, J.D. (1964), "Wages and prices in the United Kingdom: A Study in Econometric Methodology" in P.E. Hart, G. Mills, and J.K. Whitaker (eds.) (1984), *Econometric Analysis for National Economic Planning*, Butterworth, London; reprinted in D.F. Hendry and K.F. Wallis (eds.), *Econometrics and Quantitative Economics*, Basil Blackwell, Oxford.

Schiff, A.F. and P.C.B. Phillips (2000), "Forecasting New Zealand's real GDP", Cowles Foundation Paper 1020, Cowles Foundation, Yale University, 24 pages.

Sims, C.A. (1972), "Money, Income, and Causality", *American Economic Review*, vol. 62, pp.540-552.

Sims, C.A. (1980), "Macroeconomics and Reality", *Econometrica*, vol. 48, pp. 1-48.

Sims, C.A. (1982), "Policy Analysis with Econometric Models", *Brookings Papers on Economic Activity* 1, pp. 107-152.

Sims, C.A. (2002), "Structural VARs", mimeo, Princeton University (see http://www.princeton.edu/~sims/#jep)

StataCorp (2003), Stata Statistical Software: Release 8, Stata Corporation, College Station TX.

Swamy, P.A.V.B. (1970), "Efficient inference in a random coefficient regression model", *Econometrica*, vol. 38, pp. 311-323.

Swamy, P. A. V. B. and S. S. Arora. (1972), "The exact finite sample properties of the estimators of coefficients in theerror components regression models", *Econometrica*, vol. 40, pp. 643–657.

Taub, A.J. (1979), "Prediction in the context of the variance-components model", *Journal of Econometrics*, vol. 10, pp. 103–108.

Tauchen, G.(1986), "Statistical Properties of Generalized Method-of-Moments Estimators of Structural Parameters Obtained from Financial Market Data: Reply", *Journal of Business & Economic Statistics,* vol. 4(4), pp. 423-25.

Teräsvirta, T. (1994), "Specification, estimation, and evaluation of smooth transition autoregressive models", *Journal of the American Statistical Association*, vol. 89, pp. 208-218.

Tobin, J. (1958), " Liquidity Preference as Behavior Towards Risk", *Review of Economic Studies*, vol. 25 (1), pp. 65-86.

Tong, H. (1978), "On a threshold model" in C.H. Chen (ed.), *Pattern Recognition and Signal Processing*, Sijthoff & Noordhoff, Amsterdam, pp. 101-141.

Tong, H. (1990), *Non-Linear Time Series: A Dynamical Systems Approach*, Oxford University Press.

Tse, Y. K. (2000), "A test for constant correlations in a multivariate GARCH model", *Journal of Econometrics*, vol. 98, pp. 107-127.

Tse, Y. K., and K.C.Tsui (2002), "A multivariate generalized autoregressive conditional heteroscedasticity model with time-varying correlations", *Journal of Business and Economic Statistics*, vol. 20, pp. 351-362.

Valckx, N. and J.E.J. Plasmans (1998), "Equilibrium asset returns and the macroeconomy", in J.Annaert (ed.) *Financiering en belegging*, vol. 21, pp. 219-238.

Verbeek, M. (2000), *A Guide to Modern Econometrics*, John Wiley & Sons, Chichester, 386 pages.

Verbeek, M. and T.E. Nijman (1992), "Testing for Selectivity Bias in Panel Data Models", *International Economic Review*, vol. 33, pp. 681-703.

Verbeek, M. and T.E. Nijman (1996), "Incomplete Panel and Selection Bias", in Mátyás, L. and P. Sevestre (eds.), *The Econometrics of Panel Data*, Kluwer

Academic Publisher, pp. 449-490.

Wallis, K.F. (2003), "Chi-squared tests of interval and density forecasts, and the Bank of England's fan charts", *International Journal of Forecasting*, vol. 19, pp. 165-175.

White, H. (1982), "Maximum likelihood estimation of misspecified models", *Econometrica*, vol. 50, pp. 1-25.

White, H. (1984), *Asymptotic Theory For Econometricians*, Academic Press, New York.

White, J.S. (1958), "The Limiting Distribution of the Serial Correlation Coefficient in the Explosive Case", *Annals of Mathematical Statistics*, vol. 29, pp. 1188-1197.

White, J.S. (1959), "The Limiting Distribution of the Serial Correlation Coefficient in the Explosive Case II", *Annals of Mathematical Statistics*, vol. 30, pp. 831-834.

Wooldridge, J. (2002), *Econometric Analysis of Cross Section and Panel Data*, MIT Press, New York.

Working, E.J. (1927), "What do statistical "demand curves" show?", *Quarterly Journal of Economics*, vol. 41(1), pp. 212-235.

Yule, G.U. (1926), "Why Do We Sometimes Get Nonsense Correlations Between Time Series? A Study in Sampling and the Nature of Time Series", *Journal of Royal Statistical Society*, vol. 89, pp. 1-64.

Index